To Measure the Sky

An Introduction to Observational Astronomy

Second Edition

The second edition of this popular text provides undergraduates with a quantitative yet accessible introduction to the physical principles underlying the collection and analysis of observational data in contemporary optical and infrared astronomy. The text clearly links recent developments in ground- and space-based telescopes, observatory and instrument design, adaptive optics, and detector technologies to the more modest telescopes and detectors students may use themselves. Beginning with reviews of the most relevant physical concepts and an introduction to elementary statistics, students are given the firm theoretical foundation they need. New topics, including an expanded treatment of spectroscopy, Gaia, the Large Synoptic Survey Telescope, and photometry at large redshifts bring the text up to date. Historical development of topics and quotations emphasize that astronomy is both a scientific and a human endeavor, while extensive end-of-chapter exercises facilitate the students' practical learning experience.

Frederick R. Chromey is Professor of Astronomy on the Matthew Vassar, Jr. Chair at Vassar College, and Director of the Vassar College Observatory. He has almost 45 years' experience in observational astronomy research in the optical, radio, and near infrared on stars, gaseous nebulae, and galaxies, and has taught astronomy to undergraduates at Brooklyn College and Vassar.

To Measure the Sky

An Introduction to Observational Astronomy

Second Edition

Frederick R. Chromey

Vassar College, New York

CAMBRIDGE
UNIVERSITY PRESS

Shaftesbury Road, Cambridge CB2 8EA, United Kingdom

One Liberty Plaza, 20th Floor, New York, NY 10006, USA

477 Williamstown Road, Port Melbourne, VIC 3207, Australia

314–321, 3rd Floor, Plot 3, Splendor Forum, Jasola District Centre, New Delhi – 110025, India

103 Penang Road, #05–06/07, Visioncrest Commercial, Singapore 238467

Cambridge University Press is part of Cambridge University Press & Assessment, a department of the University of Cambridge.

We share the University's mission to contribute to society through the pursuit of education, learning and research at the highest international levels of excellence.

www.cambridge.org
Information on this title: www.cambridge.org/9781107572560

First published 2016
6th printing 2022

A catalogue record for this publication is available from the British Library

Library of Congress Cataloging-in-Publication data
Names: Chromey, Frederick R., 1944-
Title: To measure the sky : an introduction to observational astronomy / Frederick R. Chromey, Vassar College, New York.
Other titles: Observational astronomy
Description: Second edition. | Cambridge : Cambridge University Press, [2016] | Includes bibliographical references and index.
Identifiers: LCCN 2016012710 | ISBN 9781107572560 (Hardback : alk. paper)
Subjects: LCSH: Astronomy–Observations–Textbooks. | Astronomy–Technique–Textbooks. | Astronomy–Textbooks.
Classification: LCC QB145 .C525 2016 | DDC 522–dc23 LC record available at http://lccn.loc.gov/2016012710

ISBN   978-1-107-57256-0   Paperback

To Molly

Contents

Preface

There is an old joke: a lawyer, a priest, and an observational astronomer walk into a bar. The bartender turns out to be a visiting extraterrestrial who presents the trio with a complicated-looking black box. The alien first demonstrates that when a bucketful of garbage is fed into the entrance chute of the box, a small bag of high-quality diamonds and a gallon of pure water appear at its output. Then, assuring the three that the machine is his gift to them, the bartender vanishes.

The lawyer says, "Boys, we're rich! It's the goose that lays the golden egg! We need to form a limited partnership so we can keep this thing secret and share the profits."

The priest says, "No, no, my brothers, we need to take this to the United Nations, so it can benefit all humanity."

"We can decide all that later," the observational astronomer says. "Get me a screwdriver. I need to take this thing apart and see how it works."

The first edition of this text grew out of 16 years of teaching observational astronomy to undergraduates, and this second edition benefited from six years of using that edition in my classes and from hearing from colleagues who had done the same. In both editions, my intent has been partly to satisfy – but mainly to cultivate – my students' need to look inside black boxes. The text introduces the primary tools for making astronomical observations at visible and infrared wavelengths: telescopes, detectors, cameras, and spectrometers, as well as the methods for securing and understanding the quantitative measurements they make. I hope that after this introduction, none of these tools will remain a completely black box, and that the reader will be ready to use them to pry into other boxes. The second edition has brought the discussion a bit more up to date with current technology and practices, and has added a few recent examples of discoveries that relied on careful application of fundamental practices.

The book, then, aims at an audience similar to my students: nominally second- or third-year science majors, but with a sizable minority containing advanced first-year students, non-science students, and adult amateur astronomers. About three-quarters of those in my classes are *not* bound for graduate school in astronomy or physics, and the text has that set of backgrounds in mind.

I assume my students have little or no preparation in astronomy, but do presume that each has had one year of college-level physics and an introduction to integral and differential calculus. A course in modern physics, although very helpful, is not essential. I make the same assumptions about readers of this book. Since readers' mastery of physics varies, I include reviews of the most relevant physical concepts: optics, atomic structure, and solid-state physics. I also include a brief introduction to elementary statistics. I have written qualitative chapter summaries, but the problems posed at the end of each chapter are all quantitative exercises meant to strengthen and further develop student understanding.

My approach is to be rather thorough on fundamental topics in astronomy, in the belief that individual instructors will supply enrichment in specialized areas as they see fit. The table of contents indicates that my choice of topics is blatantly selective and slanted toward the kind of observations students might make themselves at a campus observatory.

The text lends itself to either a one- or two-semester course. I personally use the book for a two-semester sequence, where, in addition to most of the text and a selection of its end-of-chapter problems, I incorporate a number of at-the-telescope projects both for individuals and for "research teams" of students. I try to vary the large team projects: these have included a photometric time series of a variable object (in different years an eclipsing exoplanetary system, a Cepheid, and a blazar), an H–R diagram, and spectroscopy of the atmosphere of a Jovian planet. I am mindful that astronomers who teach with this text will have their own special interests in particular objects or techniques, and will have their own limitations and capabilities for student access to telescopes and equipment. My very firm belief, though, is that this book will be most effective if the instructor can devise appropriate exercises that require students to put their hands on actual hardware to measure actual photons from the sky.

To use the text for a one-semester course, the instructor will have to judiciously skip many topics. Certainly, if students are well prepared in physics and mathematics, one can dispense with much of Chapters 5 and 6 (geometrical optics and telescopes), Chapter 7 (atomic and solid-state physics), and possibly all detectors (Chapter 8) except the CCD. One would still need to choose between a more thorough treatment of photometry (skipping Chapter 11, on spectrometers) and the inclusion of spectrometry with exclusion of some photometric topics (compressing the early sections of both Chapters 9 and 10).

Compared with other texts, this book has strengths and counterbalancing weaknesses. I have taken some care with the physical and mathematical treatment of basic topics, like detection, uncertainty, telescope design, astronomical seeing, and array processing, but at the cost of a more descriptive or encyclopedic survey of specialized areas of concern to observers (e.g. little treatment of the details of astrometry or of variable star observing). I believe the book is an excellent fit for courses in which students will do their own optical/infrared observing. Because I confine myself to that wavelength region, I can develop ideas more systematically, beginning with those that arise from fundamental astronomical questions like position, brightness, and spectrum. But that narrowness makes the book less suitable for a more general survey that includes radio or X-ray techniques.

The sheer number of people and institutions contributing to the production of both editions of this book makes an adequate acknowledgment of all those to whom I am indebted impossible. Inadequate thanks are better than none, and I am deeply grateful to all who helped along the way.

A book requires an audience. The audience I had uppermost in mind was filled with those students brave enough to enroll in my Astronomy 240–340 courses at Vassar College. Over the years, more than a hundred of these students have challenged and rewarded me. All made contributions that found their way into this text, but I especially thank those who asked the hardest questions (so sorry this list is incomplete): Liz Blanton, Megan Vogelaar, Claire Webb, Deep Anand, Sherri Stephan, David Hasselbacher, Trent Adams, Leslie Sherman, Kate Eberwein, Olivia Johnson, Iulia Deneva, Laura Ruocco, Ben Knowles, Aaron Warren, Jessica Warren, Gabe Lubell, Scott Fleming, Alex Burke, Colin Wilson, Charles Wisotzkey, Peter Robinson, Tom Ferguson, David Vollbach, Krista Romita, Ximena Fernandez, Max Fagin, Jenna Lemonias, Max Marcus, Rachel Wagner-Kaiser, Tim Taber, Max Fagin, Roni Teich, Zeeve Rogozinski, Alex Shvonski, Nico Mongillo, Lauren Bearden, Angelica Rivera, Megan Lewis, Sean Sellers, Alex Trunnell, Caitlin Rose, and Liz McGrath.

I owe particular thanks to Jay Pasachoff, without whose constant encouragement and timely assistance this book would probably not exist. Likewise, Tom Balonek, who introduced me to CCD astronomy, has shared ideas, data, students, and friendship over many years. I am grateful as well to my astronomical colleagues in the Keck Northeast Astronomical Consortium; all provided crucial discussions on how to thrive as an astronomer at a small college, and many, like Tom and Jay, have read or used portions of the manuscript and the completed first edition in their observational courses. The entire text, and especially the second edition, have benefited from their feedback. I thank every Keckie, but especially Frank Winkler, Eric Jensen, Lee Hawkins, Karen Kwitter, Steve Sousa, Ed Moran, Bill Herbst, Kim McLeod, and Allyson Sheffield.

The anonymous reviewers of my proposal for the second edition made many very helpful suggestions, as did Colette Salyk and Gautham Narayan. I appreciate all the readers of the first edition who have alerted me to errors, and the many excellent conversations with Zosia Krusberg about how to teach science.

Debra Elmegreen, my colleague at Vassar, collaborated with me on multiple research projects and on the notable enterprise of building a campus observatory. Much of our joint experience found its way into this volume. Vassar College, financially and communally, has been a superb environment for both my teaching and my practice of astronomy, and deserves my gratitude. My editors at Cambridge University Press have been uniformly helpful and skilled.

My family and friends have had to bear some of the burden of this writing. Clara Bargellini and Gabriel Camera opened their home to me and my laptop during extended visits, as did my sisters, Nancy and Tina Chromey. Ann Congelton supplied useful quotations and spirited discussions. I thank my children, Kate and Anthony, who gently remind me that what is best in life is not in a book.

Finally, I thank my wife, Molly Shanley, for just about everything.

Chapter 1
Light

Always the laws of light are the same, but the modes and degrees of seeing vary.

– Henry David Thoreau, *A Week on the Concord and Merrimack Rivers*, 1849

Astronomy is not for the faint of heart. Almost everything it cares for is forbiddingly remote, tantalizingly untouchable, and invisible in the daytime, when most sensible people do their work. Nevertheless, many – including you, brave reader – have enough curiosity and courage to collect the flimsy evidence that trickles in from the universe outside our atmosphere and hope it may hold a message.

In this chapter we introduce you to astronomical evidence. Some is in the form of material, like meteorites, but most is in the form of light from faraway objects. Accordingly, we begin with three familiar theories describing the behavior of light: light as a wave, light as a quantum entity called a photon, and light as a geometrical ray. The ray picture is simplest, and we use it to introduce some basic ideas about measuring the brightness of a source. Most information in astronomy, however, comes from analyzing how brightness varies with wavelength, so we next introduce the important idea of spectroscopy. We end with a discussion of the astronomical magnitude system. We begin, however, with a few thoughts on the nature of astronomy as an intellectual enterprise.

1.1 The story

... as I say, the world itself has changed. ... For this is the great secret, which was known by all educated men in our day: that by what men think, we create the world around us, daily new.

– Marion Zimmer Bradley, *The Mists of Avalon*, 1982

Astronomers are storytellers. They spin tales of the universe and of its important parts. Sometimes they envision landscapes of another place, like the roiling liquid-metal core of the planet Jupiter. Sometimes they describe another time, like the era before Earth when dense buds of gas first flowered into stars, and a darkening universe filled with the sudden blooms of galaxies. Often the stories solve mysteries or illuminate something commonplace or account for something monstrous: How is it that stars shine, age, or explode? Some of the best stories tread the same ground as myth: What threw up the mountains of the Moon? How did the skin of our Earth come to teem with life? Sometimes there are fantasies: What would happen if a comet hit the Earth? Sometimes there are prophecies: How will the universe end?

Like all stories, creation of astronomical tales demands imagination. Like all storytellers, astronomers are restricted in their creations by many conventions of language as well as by the characters and plots already in the literature. Astronomers are no less a product of their upbringing, heritage, and society than any other crafts people. Astronomers, however, think their stories are special, that they hold a larger dose of "truth" about the universe than any others. Clearly, the subject matter of astronomy – the universe and its important parts – does not belong only to astronomers. Many others speak with authority about just these things: theologians, philosophers, and poets, for example. Is there some characteristic of astronomers, besides arrogance, that sets them apart from these others? Which story about the origin of the Moon, for example, is the truer: the astronomical story about a collision 4500 million years ago between the proto-Earth and a somewhat smaller proto-planet, or the mythological story about the birth of the Sumerian/Babylonian deity Nanna-Sin (a rather formidable fellow who had a beard of lapis lazuli and rode a winged bull)?

This question of which is the "truer" story is not an idle one. Over the centuries, people have discovered (by being proved wrong) that it is very difficult to have a commonsense understanding of what the whole universe and its most important parts are like. Common sense just isn't up to the task. For that reason, as Morgan le Fay tells us in *The Mists of Avalon*, created stories about the universe themselves actually *create* the universe the listener lives in. The real universe (like most scientists, you and I behave as if there is one) is not silent, but whispers very softly to us humans. Many whispers go unheard, and the real universe is probably very different from the one you read about today in any book that claims to tell its story. People, nevertheless, must act. Most recognize that the bases for their actions are fallible stories, and they must therefore select the most trustworthy stories that they can find.

Most of you won't have to be convinced that it is better to talk about colliding planets than about Nanna-Sin if your aim is to understand the Moon or perhaps plan a visit. Still, it is useful to ask the question: what is it, if anything, that makes astronomical stories a more reliable basis for action, and in that sense more truthful or factual than any others? Only one thing, I think: *discipline*. Astronomers feel an obligation to tell their story with great care, following a rather strict, scientific, discipline.

Scientists, philosophers, and sociologists have written about what it is that makes science different from other human endeavors. There is much discussion and disagreement about the necessity of making scientific stories "broad and deep and simple," about the centrality of paradigms, the importance of predictions, the strength or relevance of motivations, and the inevitability of conformity to social norms and professional hierarchies.

But most agree on the perhaps obvious point that a scientist, in creating a story (scientists usually call them "theories" or "models") of, say, the Moon, must pay a great deal of attention to all the relevant evidence. A scientist, unlike a science-fiction writer, may only fashion a theory that never, ever, violates that evidence.

This is a book about how to collect and interpret relevant evidence in astronomy. Most of that evidence is in the form of light arriving from far, far away.

1.2 Models for the behavior of light

Some (not astronomers!) regard astronomy as applied physics. There is some justification for this, since astronomers, to help tell some astronomical story, persistently drag out theories proposed by physicists. Physics and astronomy differ partly because astronomers are interested in telling the story of an object, whereas physicists are interested in uncovering the most fundamental rules of the natural world. Astronomers tend to find physics useful but sterile; physicists tend to find astronomy messy and mired in detail. We now ponder the question: how does light behave? More specifically, what properties of light are important in making meaningful astronomical observations and predictions? Physics has the answers.

1.2.1 Electromagnetic waves

> ... we may be allowed to infer, that homogeneous light, at certain equal distances in the direction of its motion, is possessed of opposite qualities, capable of neutralizing or destroying each other, and extinguishing the light, where they happen to be united; ...
>
> – Thomas Young, *Philosophical Transactions, The Bakerian Lecture*, 1804

Electromagnetic waves are a model for the behavior of light. We know this model is incorrect (*incomplete* is perhaps a better term). Nevertheless, since the wave theory precisely describes so much of light's behavior, we need to review its claims. Christian Huygens,[1] in his 1678 book, *Traité de la Lumière*, made the first serious argument that visible light is best regarded as a *wave* phenomenon.

---

[1] Huygens (1629–95), a Dutch natural philosopher and major figure in seventeenth-century science, had an early interest in lens grinding. He discovered the rings of Saturn and its large satellite, Titan, in 1655–56, with a refracting telescope of his manufacture. At about the same time, he invented the pendulum clock and formulated a theory of elastic bodies. He developed his wave theory of light later in his career, after he moved from The Hague to the more cosmopolitan environment of Paris. Near the end of his life, he wrote a treatise on the possibility of extraterrestrial life.

A *wave* is a disturbance that propagates through space. If some property of the environment (say, the level of the water in your bathtub) is disturbed at one place (perhaps by a splash), a wave is present if that disturbance moves continuously from place to place in the environment (ripples from one end of your bathtub to the other, for example). Material particles, like bullets or ping-pong balls, also propagate from place to place. Waves and particles share many characteristic behaviors – both can *reflect* (change directions at an interface), *refract* (change speed or direction in response to a change in the transmitting medium), and can carry energy from place to place.

However, waves exhibit two characteristic behaviors not shared by particles. *Diffraction* is the ability to bend around obstacles. A water wave entering a narrow opening, for example, will travel not only in the "shadow" of the opening but will spread in all directions on the far side. *Interference* is the ability to combine with other waves in predictable ways. Two water waves can, for example, destructively interfere if they combine so that the troughs of one always coincide with the peaks of the other – the same phenomenon that permits noise-cancelling earphones.

Although Huygens knew that light exhibited the properties of diffraction and interference, he unfortunately did not discuss them in his book. Isaac Newton, his younger contemporary, opposed Huygens' wave hypothesis and argued that light was composed of tiny solid particles. Newton's reputation was such that his view prevailed until the early part of the nineteenth century, when Thomas Young and Augustin Fresnel drew attention to diffraction and interference in light. Soon the evidence for "light waves" proved irresistible.

Well-behaved waves exhibit certain measurable qualities: amplitude, wavelength, frequency, and wave speed. Physicists in the generation following Fresnel were able to measure these quantities for visible light waves. Since light was a wave, and since waves are disturbances that propagate, it was natural to ask: "What 'stuff' does a light wave disturb?" In one of the major triumphs of nineteenth-century physics, James Clerk Maxwell proposed an answer in 1873.

Maxwell (1831–79), a Scot, is a major figure in the history of physics, comparable to Newton and Einstein. His doctoral thesis demonstrated that the rings of Saturn (discovered by Huygens) must be made of many small solid particles in order to be gravitationally stable. He conceived the kinetic theory of gases in 1866 (Ludwig Boltzmann did similar work independently) and transformed thermodynamics into a science based on statistics rather than determinism. His most important achievement was the mathematical formulation of the laws of electricity and magnetism in the form of four partial differential equations. Published in 1873, *Maxwell's equations* completely accounted for separate electric and magnetic phenomena and also demonstrated the connection between the two forces. Maxwell's work is the culmination of classical physics, and its limits led to both the theory of relativity and the theory of quantum mechanics.

Maxwell proposed that light is a propagating **_electric and magnetic_** disturbance. The following example illustrates his idea.

Consider a single, motionless electron, electron A, attached to the rest of an atom by means of a spring. (The spring is just a mechanical model for the electrostatic attraction that holds the electron to the nucleus.) This pair of charges, the negative electron and the positive ion, constitute a dipole. A second electron, electron B, is also attached to the rest of its atom by a spring, but this second dipole is at some distance from A. Electron A repels B, and B's stationary position in its atom is in part determined by the location of A. The two atoms are sketched in Figure 1.1. Now to make a wave: Set electron A vibrating on its spring. Electron B must respond to this vibration, since the force it feels is changing direction. It moves in a way that will echo the motion of A. The lower part of Figure 1.1 shows the changing electric force on B as A moves through a cycle of its vibration.

The disturbance of dipole A has propagated to B in a way that suggests a wave is operating. Electron B behaves like an object floating in your bathtub that moves in response to the rising and falling level of a water wave.

In trying to imagine the actual thing that a vibrating dipole disturbs, you might envision the water in a bathtub. Now imagine some stuff that fills space around the electrons, the way a fluid would, so a disturbance caused by moving one electron can propagate from place to place. The physicist Michael Faraday[2] supplied the very useful idea of a **_field_** – an abstract _entity_ (not a material fluid at all) created by any charged particle. The field permeates space and gives other charged particles instructions about what force they should experience. In this conception, electron B consults the local field in order to decide how to move. You are probably familiar with understanding magnetic and gravitational forces as also arising from their corresponding fields. Shaking (accelerating) the electron at A distorts the field in its vicinity, and this distortion propagates to vast distances, just like the ripples from a rock dropped into a calm and infinite ocean.

The details of propagating a field disturbance turned out to be a little complicated. Hans Christian Oersted and André Marie Ampère in 1820 had shown experimentally that a changing electric field, such as the one generated by an accelerated electron, produces a magnetic field. Acting on his intuition of an underlying unity in physical forces, Faraday experimentally confirmed his guess

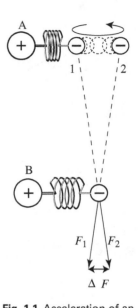

Fig. 1.1 Acceleration of an electron produces a wave. The electrons in initially undisturbed atoms are in stationary positions. Each electron is attached to the rest of the atom (the heavy, positively charged ion) by some force, which we represent as a spring. If the electron in the source atom (A) is disturbed so that it oscillates between positions (1) and (2), then the electron in the receiver (B) experiences a force that changes from F_1 to F_2 in the course of A's oscillation. The difference ΔF, sets the amplitude of the changing part of the electric force seen by B.

---

[2] Michael Faraday (1791–1867), considered by many the greatest experimentalist in history, began his career as a bookbinder with minimal formal education. His amateur interest in chemistry led to a position in the laboratory of the renowned chemist, Sir Humphrey Davy, at the Royal Institution in London. Faraday continued work as a chemist for most of his productive life, but conducted an impressive series of experiments in electromagnetism in the period 1834–55. His ideas, although largely rejected by physicists on the Continent, eventually formed the empirical basis for Maxwell's theory of electromagnetism.

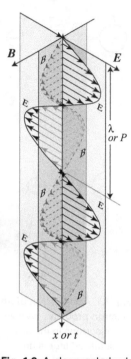

Fig. 1.2 A plane-polarized electromagnetic wave. The electric and magnetic field strengths are drawn as vectors that vary in both space and time. The illustrated waves are said to be plane-polarized because all electric vectors are confined to the x–y plane.

that a changing magnetic field must in turn generate an electric field. Maxwell had the genius to realize that his equations implied that the electric and magnetic field changes in a vibrating dipole would support one another and produce a wavelike self-propagating disturbance. Change the electric field and you thereby create a magnetic field, which then creates a different electric field, which creates a magnetic field, and so on, forever. Thus, it is proper to speak of the waves produced by an accelerated charged particle as **electromagnetic**. Figure 1.2 shows a schematic version of an electromagnetic wave. The changes in the two fields, electric and magnetic, vary at right angles to one another and the direction of propagation is at right angles to both (a **transverse wave**).

Thus, a disturbance in the electric field does indeed seem to produce a wave. Is this *electromagnetic* wave the same thing as the *light* wave we see with our eyes?

From his four equations – the laws of electric and magnetic force – Maxwell derived the speed of any electromagnetic wave, which, in a vacuum, turned out to depend only on constants

$$c = \sqrt{\varepsilon\mu} \tag{1.1}$$

Here ε and μ are well-known constants that describe the strengths of the electric and magnetic forces. (They are, respectively, the electric permittivity and magnetic permeability of the vacuum.) When he entered the experimental values for ε and μ in the above equation, Maxwell computed the electromagnetic wave speed, which turned out to be numerically identical to the speed of light, a quantity that had been experimentally measured with improving precision over the preceding century. This equality of predicted and experimentally measured speeds was a quite convincing argument that light waves and electromagnetic waves were the same thing. Maxwell had shown that three different entities, electricity, magnetism, and light, were really tightly related.

Other predictions based on Maxwell's theory further strengthened this view of the nature of light. For one thing, one can note that for any well-behaved wave the speed of the wave is the product of its frequency and wavelength:

$$c = \lambda\upsilon \tag{1.2}$$

There is only one speed that electromagnetic waves can have in a vacuum; therefore, there should be a one-dimensional classification of electromagnetic waves (the **electromagnetic spectrum**). In this spectrum, each wave is characterized only by its particular wavelength (or frequency). A single light wave of a particular wavelength is usually represented as the harmonic function

$$E(x, t) = E_0 \sin\left\{\frac{2\pi}{\lambda}(x - ct)\right\} = E_0 \sin\{\phi\} \tag{1.3}$$

where E_0 and ϕ are, respectively, the **amplitude** and the **phase** of the wave. Table 1.1 gives the modern names for various portions or **bands** of the

Table 1.1 *The electromagnetic spectrum. Region boundaries are not well defined, so there is some overlap. Subdivisions are based in part on distinct detection methods.*

Band	Wavelength range	Frequency range	Subdivisions (long λ – short λ)
Radio	> 1 mm	< 300 GHz	VLF–AM–VHF–UHF
Microwave	0.1 mm–3 cm	100 MHz–3000 GHz	Millimeter–Submillimeter
Infrared	700 nm–1 mm	3×10^{11}–4×10^{14} Hz	Far–Middle–Near
Visible	300 nm–800 nm	4×10^{14}–1×10^{15} Hz	Red–Blue
Ultraviolet	10 nm–400 nm	7×10^{14}–3×10^{16} Hz	Near–Extreme
X-rays	0.001 nm–10 nm	3×10^{16}–3×10^{20} Hz	Soft–Hard
Gamma ray	< 0.1 nm	> 3×10^{18} Hz	Soft–Hard

electromagnetic spectrum. William Herschel and Johann Wilhelm Ritter had already discovered infrared and ultraviolet "light," respectively, in 1800–01 – well before Maxwell's theory. In 1888, Heinrich Hertz demonstrated the production of radio waves based on Maxwell's principles. These experimental confirmations convinced physicists that Maxwell had discovered the secret of light. Humanity had made a tremendous leap in understanding reality. This leap to new heights, however, soon revealed that Maxwell had discovered only a part of the secret.

The wave theory of light very accurately describes the way light behaves in most macroscopic situations. In summary, the theory says:

1. Light exhibits all the properties of classical, well-behaved waves, namely: reflection at interfaces, refraction upon changes in the medium, diffraction around edges, interference with other light waves, and polarization in a particular direction (plane of vibration of the electric vector).
2. A light wave can have any positive wavelength. The range of possible wavelengths constitutes the electromagnetic spectrum. Frequency and wavelength are related by Equation (1.2).
3. In a vacuum, light waves travel in a straight line at speed c. Travel in other media is slower and subject to refraction and absorption.
4. A light wave carries energy whose magnitude depends on the squares of the amplitudes of the electric and magnetic waves.

1.2.2 Quantum mechanics and light

It is very important to know that light behaves like particles, especially for those of you who have gone to school, where you were probably told something about light behaving like waves. I'm telling you the way it does behave – like particles.

– Richard Feynman: *Q.E.D.*, 1985

Toward the end of the nineteenth century, physicists realized that electromagnetic theory could not account for certain behaviors of light. The theory that eventually replaced it, **quantum mechanics**, postulates that light possesses the properties of a particle as well as the wavelike properties described by Maxwell's theory. Quantum mechanics insists that there are situations in which we cannot think of light as a wave, but must think of it as a collection of particles, like bullets shot out of the source at the speed of light. These particles are termed **photons**. Each photon "contains" a particular amount of energy, E, that depends on the frequency it possesses when it exhibits its wavelike properties:

$$E = h\nu = \frac{hc}{\lambda} \tag{1.4}$$

Here h is Planck's constant (6.626×10^{-34} J s) and ν is the frequency of the wave. Thus a single radio photon (low frequency) contains a small amount of energy, and a single gamma-ray photon (high frequency) contains a lot. A convenient unit for the energy of a photon is the **electronvolt** (1 eV = 1.602×10^{-19} J)

The quantum theory of light gives an elegant and successful picture of the interaction between light and matter on the microscopic scale. In this view, atoms no longer have electrons bound to nuclei by springs or (what is equivalent in classical physics) electric fields. Electrons in an atom, rather, have certain permitted energy states described by a wave function – in this theory, everything, including electrons, has a wave as well as a particle nature. An electron changing from one of these permitted states to another explains the generation or absorption of light by atoms. Energy is conserved: energy lost when an atom makes the transition from a higher to a lower state is exactly matched by the energy of the photon emitted. In summary, the quantum theory says:

1. Light exhibits all the properties described in the wave theory in situations where wave properties are measured.
2. Light behaves, in other circumstances, as if it were composed of massless particles called photons, each containing an amount of energy equal to its frequency times Planck's constant.
3. The interaction between light and matter involves creation and destruction of individual photons and the corresponding changes of energy states of charged particles (usually electrons).

We will make great use of the quantum theory in later chapters, but for now, our needs are more modest.

1.2.3 A geometric approximation: light rays

> By Light Rays I understand its least Parts ... Mathematicians usually consider the Rays of Light to be Lines reaching from the luminous Body to the Body illuminated ...
>
> – Isaac Newton, *Opticks*, 1704

Since the quantum picture of light is as close as we can get to the real nature of light, you might think quantum mechanics would be the only theory worth considering. However, except in simple situations, application of the theory demands complex and lengthy computation. Fortunately, it is often possible to ignore much of what we know about light and use a very rudimentary picture which pays attention only to those few properties of light necessary to understand much of the information brought to us by photons from out there. In this geometric approximation, we treat light as if it traveled in "rays" or streams that obey the laws of reflection and refraction as described by geometrical optics. It is helpful to imagine a ray as the path taken by a single photon of a particular wavelength.

We might then imagine a stream of photons, each tracing a ray from the source to an observer's detector. Sometimes it is essential to recognize the discrete nature of the particles. We might then think of astronomical measurements as acts of *counting* and classifying the individual photons as they hit our detector like sparse raindrops tapping on a tin roof.

Sometimes, we can ignore the lumpy nature of the photon stream and just assume it behaves like a smooth fluid that carries energy from source to detector along the rays. In this case, we think of astronomical measurements as recording smoothly varying quantities – like measuring the volume of rain that falls into a bucket in one day. We might be aware that the rain arrived as discrete drops, but it is safe to ignore the fact.

We will adopt this simplified ray picture for much of the discussion that follows, adjusting our awareness of the discrete nature of the photon stream or its wave properties as circumstances warrant. For the rest of this chapter, we use the ray picture to discuss two of the basic measurements important in astronomy: *photometry,* which measures the amount of energy arriving from a source, and *spectrometry,* which measures the distribution of this energy with wavelength. Incidentally, our use of the word "wavelength" does not mean we are going to think deeply about the wave theory just yet. It will be sufficient to think of wavelength as a property of a light ray that can be measured – by noting which ray a photon follows when sent through a spectrograph, for example.

Besides photometry and spectroscopy, the other general categories of measurement are *imaging* and *astrometry*, which are concerned with the appearance and positions of objects in the sky, and *polarimetry*, which is concerned with the polarization of light from the source.

1.3 Measurements of light rays

Twinkle, twinkle, little star,
Flux says just how bright you are.
> – Anonymous, *c*. 1980

1.3.1 Luminosity and brightness

Astronomers have to construct the story of a distant object using only the tiny whisper of electromagnetic radiation it sends us. We define the (electromagnetic) luminosity, L, as the total amount of energy that leaves the surface of the source per unit time in the form of photons. Energy per unit time is called power, so we can measure L in physicists' units for power (SI units), joules per second or watts. Alternatively, it might be useful to compare the object with the Sun, and we then might measure the luminosity in solar units:

$$L = Luminosity = \text{Energy per unit time emitted by the entire source}$$

$$L_\odot = \text{Luminosity of the sun} = 3.827 \times 10^{26}\text{W}.$$

The luminosity of a source is an important clue about its nature. One way to measure luminosity is to surround the source completely with a box or (since this is physics) sphere of perfectly energy-absorbing material, then use an "energy gauge" to measure the total amount of energy intercepted by this enclosure during some time interval. Figure 1.3 illustrates the method. Luminosity is the amount of energy absorbed divided by the time interval over which the energy accumulates. The astronomer, however, cannot measure luminosity in this way. She is too distant from the source to put it inside a sphere, even in the unlikely case she has one big enough. Fortunately, there is a quantity related to luminosity, called the **apparent brightness** of the source, which is much easier to measure.

Measuring apparent brightness is a local operation. The astronomer holds up a scrap of perfectly absorbing material of known area so that its surface is perpendicular to the line of sight to the source. She measures how much energy from the source accumulates in this material in a known time interval. Apparent brightness, F, is defined as the total energy per unit time per unit area that arrives from the source:

$$F = \frac{E}{tA} \tag{1.5}$$

Fig. 1.3 Measuring luminosity by intercepting all the power from a source.

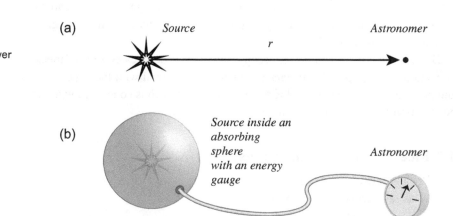

(a)   Source   Astronomer
                              r

(b)   Source inside an
                             absorbing
                             sphere   Astronomer
                             with an energy
                             gauge

This quantity, F, is usually known as the **flux** or the **flux density** in the astronomical literature. In the physics literature, the same quantity is usually called the **irradiance** (or, in studies restricted to visual light, the **illuminance**). To make matters not only complex but also confusing, what astronomers call luminosity, L, physicists call the **radiant flux**.

Whatever one calls it, F will have units of power per unit area, or Wm^{-2}. For example, the average flux from the Sun at the top of the Earth's atmosphere (the apparent brightness of the Sun) is about 1361 Wm^{-2}, a quantity known as the **solar constant.** The instantaneous value of the flux from the Sun, the **total solar irradiance,** varies by about 7% because of the Earth's elliptical orbit and by perhaps 0.2% over longer historical periods because of intrinsic solar variations.

1.3.2 The inverse square law of brightness

Refer to Figure 1.4 to derive the relationship between the flux from a source and the source's luminosity. We choose to determine the flux by measuring the power intercepted by the surface of a very large sphere of radius r centered on the source. The astronomer is on the surface of this sphere. Since this surface is everywhere perpendicular to the line of sight to the source, the apparent brightness, according to Equation (1.5) is simply the total power absorbed by the large sphere divided by its area. But surrounding the source with a sphere is exactly what we did in Figure 1.3, so the total power absorbed by the large sphere must be the luminosity, L, of the source. We assume that there is nothing located between the source and the spherical surface that absorbs light – no dark cloud or planet. The brightness, averaged over the whole sphere, then, is:

$$\langle F \rangle = \frac{L}{4\pi r^2}$$

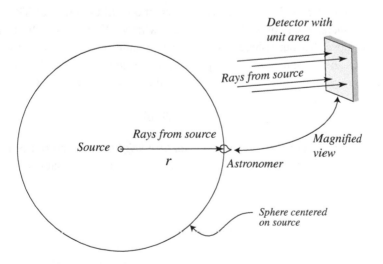

Detector with
unit area

Rays from source

Rays from source

Source

r

Astronomer

Magnified
view

Sphere centered
on source

Fig. 1.4 Measuring the apparent brightness of an isotropic source that is at distance r. The astronomer locally detects the power reaching a unit area oriented perpendicular to the direction of the source. If there is no intervening absorber, then the source luminosity is equal to the apparent brightness multiplied by the area of a sphere of radius r.

Now make the additional assumption that the radiation from the source is isotropic (the same in all directions). Then the *average* brightness is the same as the brightness measured locally, using any convenient small surface:

$$F = \frac{L}{4\pi r^2} \tag{1.6}$$

Both assumptions, isotropy and the absence of absorption, can be violated in reality. Nevertheless, in its simple form, Equation (1.6) not only represents one of the fundamental relationships in astronomy, it also reveals one of the central problems in our science.

The problem is that the left-hand side of Equation (1.6) is the flux, a quantity that can be determined by direct observation. However, the right-hand side contains two unknowns, luminosity and distance. Without further information these two cannot be disentangled. This is a frustration – you can't say, for example, how much power a quasar is producing without knowing its distance, and you can't know its distance unless you know how much power it is producing. A fundamental problem in astronomy is determining the third dimension.

1.3.3 Surface brightness

One observable quantity that does not depend on the distance of a source is its surface brightness on the sky. Consider the simple case of a uniform spherical source of radius a and luminosity L. On the surface, the amount of power leaving a unit area is called the ***radiant exitance***:

$$s = \frac{L}{4\pi a^2} \tag{1.7}$$

Note that s has the same dimensions, W m^{-2}, as F, the apparent brightness seen by a distant observer. The two are very different quantities, however. The value of s is characteristic only of the object itself, whereas F changes with distance. Now, suppose that our sphere has a detectable angular size – that our eye or telescope can distinguish it from a point source: it looks like a disk. The ***solid angle***, in ***steradians*** subtended by a disk of radius a and distance r is (for $a \ll r$):

$$\Omega \cong \frac{\pi a^2}{r^2} \, [\text{steradians}]. \tag{1.8}$$

Now we write down σ, the apparent ***surface brightness*** of the source on the sky, that is, the flux from the disk divided by the solid angle it subtends:

$$\sigma = \frac{F}{\Omega} = \frac{s}{\pi} \tag{1.9}$$

So, in this example, σ depends only on the radiant exitance of the source and so is ***independent of distance***. A more careful analysis of non-spherical, non-uniform resolved objects supports the same conclusion: σ does not change with distance. Ordinary optical telescopes can measure Ω with accuracy only if it has a value larger than a few square arc seconds (about 10^{-10} steradians), mainly because of turbulence in the Earth's atmosphere. Space telescopes and ground-based systems with adaptive optics can resolve solid angles perhaps 100 times smaller. Unfortunately, the majority of even the nearest stars have angular sizes too small (diameters of a few milliarc seconds) to resolve with present instruments, so that for them Ω (and therefore σ) cannot be measured directly. Astronomers do routinely measure σ for "extended" or "non-stellar" images of objects like planets, gaseous nebulae, and galaxies, and find these values immensely useful.

1.4 Spectra

> If a question on an astronomy exam starts with the phrase 'how do we know …'
> then the answer is probably 'spectrometry.'
>
> – Anonymous, *c.* 1950

Astronomers usually learn most about a source not from its flux, surface brightness, or even luminosity, but from its spectrum: the way in which light is distributed with wavelength. Measuring its luminosity is like reading the title of a book about the source. Measuring its spectrum is like opening the book and skimming a few chapters, chapters that might explain the source's chemical composition, pressure, density, temperature, rotation speed, or radial velocity. (You seldom get to read the whole book.) Although evidence in astronomy is usually meager, the most satisfying and eloquent evidence is spectroscopic.

1.4.1 Monochromatic flux

Consider measuring the flux from a source in the usual fashion, with a set-up like the one in Figure 1.4. Arrange our detector to register only photons that have frequencies between v and $v+dv$, where dv is an infinitesimally small frequency interval. Write the result of this measurement as $F(v, v + dv)$. Keep in mind that dv and $F(v, v + dv)$ are the *limits* of finite quantities called Δv and $F(v, v + \Delta v)$. We then define **monochromatic flux** or **monochromatic brightness** as

$$f_v = \frac{F(v, v + dv)}{dv} = \lim_{\Delta v \to 0} \frac{F(v, v + \Delta v)}{\Delta v} \tag{1.10}$$

The complete function, f_v, running over all frequencies, (or even over a limited range of frequencies) is called the **spectrum** of the object. It has units $[\mathrm{Wm^{-2}Hz^{-1}}]$. The extreme right-hand side of Equation (1.10) reminds us that f_v is the limiting value of the ratio as the quantity Δv (and correspondingly, $F(v, v + \Delta v)$) become indefinitely small. In practice, Δv must have a finite size,

Fig. 1.5 Two forms of the
ultraviolet and visible
outside-the-atmosphere
spectrum of Vega.
The two curves convey
the same information,
but have very different
shapes. Units on the
vertical axes are arbitrary.

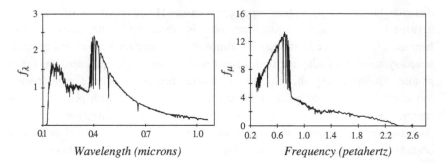

Fig. 1.5 Two forms of the ultraviolet and visible outside-the-atmosphere spectrum of Vega. The two curves convey the same information, but have very different shapes. Units on the vertical axes are arbitrary.

since $F(v, v + \Delta v)$ must be large enough to register on a detector. If Δv is large, the detailed wiggles and jumps in the spectrum will be smoothed out, and one is said to have measured a **low-resolution spectrum**. Likewise, a **high-resolution spectrum** will more faithfully show the details of the limiting function, f_v.

If we choose the wavelength as the important characteristic of light, we can define a different monochromatic brightness, f_λ. Symbolize the flux between wavelengths λ and $\lambda + d\lambda$ as $F(\lambda, \lambda + d\lambda)$ and write

$$f_\lambda = \frac{F(\lambda, \lambda + d\lambda)}{d\lambda} \tag{1.11}$$

Although the functions f_v and f_λ are each called the spectrum, they differ from one another in numerical value and overall appearance for the same object. Figure 1.5 shows schematic low-resolution spectra of the bright star, Vega, plotted over the same range of wavelengths, first as f_λ, then as f_v.

1.4.2 Flux within a band

> Less is more.
> – Robert Browning, "Andrea del Sarto," 1855, often quoted by L. Mies van der Rohe

An **ideal bolometer** is a detector that responds to all wavelengths with perfect efficiency. In a unit time, a bolometer would record every photon reaching it from a source, regardless of wavelength. We could symbolize the **bolometric flux** thereby recorded as the integral:

$$F_{\text{bol}} = \int_0^\infty f_\lambda d\lambda \tag{1.12}$$

Real bolometers operate by monitoring the temperature of a highly absorbing (i.e. black) object of low thermal mass. They are imperfect in part because it is difficult to design an object that is "black" at all wavelengths. More commonly, practical instruments for measuring brightness can only detect light within a limited range of wavelengths or frequencies. Suppose a detector registers light between wavelengths λ_1 and λ_2, and nothing outside this range. In the notation of the previous section, we might then write the flux in the 1, 2 pass-band as:

$$F(\lambda_1, \lambda_2) = \int_{\lambda_1}^{\lambda_2} f_\lambda \mathrm{d}\lambda = F(\nu_2, \nu_1) = \int_{\nu_2}^{\nu_1} f_\nu \mathrm{d}\nu \tag{1.13}$$

Usually, the situation is even more complex. Practical detectors vary in detecting efficiency over any band. If $R_A(\lambda)$ is the fraction of the incident flux of wavelength λ that is eventually detected by instrument A, then the flux actually recorded by such a system might be represented as

$$F_A = \int_0^\infty R_A(\lambda) f_\lambda \mathrm{d}\lambda \tag{1.14}$$

The function $R_A(\lambda)$ may be imposed in part by the environment rather than by the instrument. The Earth's atmosphere, for example, is (imperfectly) transparent only in the visible and near-infrared pass-band between about 0.32 and 1 micron (extending in restricted bands to 25 µm at very dry, high-altitude sites), and in the microwave–radio pass-band between about 0.5 mm and 50 m.

Astronomers routinely restrict the range of a detector's sensitivity intentionally, by using a **filter** to control the form of the function R_A. Why? First, a well-defined standard band makes it easier for different astronomers to compare measurements. Second, a filter can block troublesome wavelengths, ones where the background is high, perhaps, or where atmospheric transmission is low. Finally, comparison of two or more different bandpass fluxes for the same source is akin to measuring a very low-resolution spectrum, and thus can provide some of the information, like temperature or chemical composition, that a spectrum conveys.

Hundreds of bands have found use in astronomy. Table 1.2 lists the broad-band filters (i.e. filters where the bandwidth, $\Delta\lambda = \lambda_2 - \lambda_1$, is large) that are most commonly encountered in the visible–near-infrared window. Standardization of bands is less common in radio and high-energy observations.

1.4.3 Spectrum analysis

[With regard to stars] . . . we would never know how to study by any means their chemical composition. . . . In a word, our positive knowledge with respect to stars is necessarily limited solely to geometrical and mechanical phenomena. . . .

– Auguste Comte, *Cours de Philosophie Positive* II, 19th Lesson, 1835

. . . I made some observations which disclose an unexpected explanation of the origin of Fraunhofer's lines, and authorize conclusions therefrom respecting the material constitution of the atmosphere of the sun, and perhaps also of that of the brighter fixed stars.

- Gustav R. Kirchhoff, Letter to the Academy of Science at Berlin, 1859

Table 1.2 *Common broad bandpasses in the visible (UBVRI) and short-
(JHK), mid- (LM), long- (N), and very long- (Q) wavelength infrared.
Chapter 10 discusses standard bands in greater detail.*

Name	λ_c (μm)	Width (μm)	Rationale
U	0.365	0.068	Ultraviolet
B	0.44	0.098	Blue
V	0.55	0.089	Visual
R	0.70	0.22	Red
I	0.90	0.24	Infrared
J	1.25	0.38	SWIR
H	1.63	0.31	SWIR
K	2.2	0.48	SWIR
L	3.4	0.70	MWIR
M	5.0	1.123	MWIR
N	10.2	4.31	LWIR
Q	21.0	8	VLWIR

Astronomers are fond of juxtaposing Comte's pronouncement about the impossibility of knowing the chemistry of stars with Kirchhoff's breakthrough a
generation later. Me too. Comte deserves better, since he wrote quite thought-
fully about the philosophy of science and would certainly have been among the
first to applaud the powerful new techniques of spectrum analysis developed
later in the century. Nevertheless, the failure of his dictum about what is
knowable is a caution against pomposity for all.

Had science been quicker to investigate spectra, Comte might have been
spared posthumous deflation. In 1666, Newton observed the dispersion of
visible "white" sunlight into its component colors by glass prisms, but subsequent applications of spectroscopy were very slow to develop. It was not until
1802 that the English physicist William Wollaston noted the presence of dark
lines in the visible solar spectrum. Lines are very narrow wavelength bands
where the value of function f_v drops almost discontinuously, then rises back
to the previous "continuum." (see Figures 1.6 and 1.7). The term "line" arises
because in visual spectroscopy, one actually examines the image of a narrow slit
at each wavelength. If the intensity is unusually low at a particular wavelength,
then the image of the slit there looks like a dark line.

The Fraunhofer spectrum

Unaware of Wollaston's work, Joseph von Fraunhofer (1787–1826), used a
much superior spectroscope to produce an extensive map of the solar absorption
lines in around 1812.

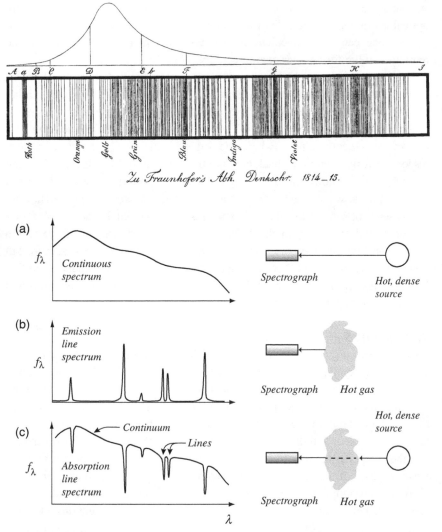

Fig. 1.6 A much-reduced reproduction of one of Fraunhofer's drawings of the solar spectrum. Frequency increases to the right, and the stronger absorption lines are labeled with his designations. Modern designations differ slightly.

Fig. 1.7 Three types of spectra and the situations that produce them. (a) A solid, liquid, or dense gas produces a continuous spectrum. (b) A rarefied hot gas like a flame or a spark produces an emission-line spectrum. (c) A continuous spectrum viewed through a rarefied gas produces an absorption line spectrum.

Of humble birth, Fraunhofer began his career as an apprentice at a glassmaking factory located in an abandoned monastery in Benediktbeuern, outside Munich. By talent and fate (he survived a serious industrial accident), he advanced quickly in the firm. The business became quite successful and famous because of a secret process for making large blanks of high-quality crown and flint glass, which had important military and civil uses. Observing the solar spectrum with the ultimate goal of improving optical instruments, Fraunhofer pursued what he believed to be his discovery of solar absorption lines with characteristic enthusiasm and thoroughness. By 1814, he had given precise positions for 350 lines and approximate positions for another 225 fainter lines (see Figure 1.6). By 1823, Fraunhofer was reporting on the spectra of bright stars and planets, although his main enthusiasm was the creation of high-quality optical instruments. Tragically, Fraunhofer died from tuberculosis at the age of 39 and one can only speculate on the development of astrophysics had he remained active for another quarter century.

Fraunhofer designated the ten most prominent of the dark lines he observed in the solar spectrum with letters (Figure 1.6, and Appendix B.2). He noted that the two dark lines he labeled with the letter D occurred at wavelengths identical to the two bright emission lines produced by a candle flame. (*Emission lines* are narrow wavelength regions where the value of function f_λ increases almost discontinuously, then drops back down to the continuum. See Figure 1.7.) Soon several observers noted these two bright lines, which occur in the yellow part of the spectrum at wavelengths of 589.0 and 589.6 nanometers, always arise from the presence of sodium in a flame. Several researchers (John Herschel, W.H. Fox Talbot, David Brewster) in the 1820s and 1830s suggested that there was a connection between the composition of an object and its flame spectrum, but none could describe it precisely. At about this same time, still others noted that heated solids, unlike the gases in flames, produce *continuous spectra* (no bright or dark lines – again, see Figure 1.7).

The Kirchhoff–Bunsen results

Spectroscopy languished for the 30 years following Fraunhofer's death in 1826. Then, in Heidelberg in 1859, physicist Gustav Kirchhoff and chemist Robert Bunsen performed a crucial experiment. They passed a beam of sunlight through a sodium flame, initially to measure the precision with which the solar absorption and flame emission lines coincided. What they observed instead was that the bright lines faded, and the dark D lines became darker still. Kirchhoff reasoned that the hot gas in the flame had both absorbing and emitting properties at the D wavelengths, but that the absorption became more apparent as more light to be absorbed was supplied, whereas the emitting properties remained constant. This suggested that absorption lines would always be seen in situations like that sketched on the right of Figure 1.7c, so long as a sufficiently bright source were observed through a gas. If the background source were too weak or altogether absent, then the situation sketched in Figure 1.7b would hold and emission lines would be prominent.

Kirchhoff, moreover, proposed an explanation of all the Fraunhofer lines: The Sun consists of a bright source – a very dense gas, as it turns out – that emits a continuous spectrum. A low-density, gaseous atmosphere overlies this dense region. As the light from the continuous source passes through the atmosphere, its chemicals each absorb their characteristic wavelengths. One could then conclude, for example, that the Fraunhofer D lines demonstrated the presence of sodium in the solar atmosphere. Identification of other chemicals in the solar atmosphere became a matter of obtaining an emission line "fingerprint" from a laboratory flame or spark spectrum, then searching for the corresponding absorption line or lines at identical wavelengths in the Fraunhofer spectrum. Kirchhoff and Bunsen quickly confirmed the presence of potassium, iron, and calcium and the absence (or very low abundance) of lithium in the solar atmosphere. The Kirchhoff–Bunsen results were not limited to the Sun. The spectra of almost all stars turned out to be absorption spectra, and it was easy to identify many of the lines present.

Quantitative chemical analysis of solar and stellar atmospheres became possible in the 1940s, after the development of astrophysics in the early twentieth century. At that time astronomers showed that most stars were composed of hydrogen and helium in a roughly 12 to 1 ratio by number, with small additions of other elements. However, the early qualitative results of Kirchhoff and Bunsen had already demonstrated to the world that stars were made of ordinary matter, and that one could hope to learn their exact composition by spectrometry. By the 1860s they had replaced the "truth" that stars were inherently unknowable with the new "truth" that stars were made of ordinary stuff.

Blackbody spectra

In 1860, Kirchhoff discussed the ratio of absorption to emission in hot objects by first considering the behavior of a perfect absorber, an object that would absorb all light falling on its surface. He called such an object a **blackbody** (*kürzer schwarze*), since it by definition would reflect nothing. Blackbodies, however, must emit (otherwise their temperatures would always increase from absorbing ambient radiation). One can construct a simple blackbody by drilling a small hole into a uniform oven. The hole is the blackbody. The black walls of the oven will always absorb light entering the hole, so that the hole is a perfect absorber. The spectrum of the light *emitted* by the hole will depend on the temperature of the oven (and, it turns out, on nothing else). The blackbody spectrum is usually a good approximation to the spectrum emitted by any solid, liquid, or dense gas. (Low-density gases produce line spectra.)

In 1878, Josef Stefan found experimentally that the surface brightness of a blackbody (total power emitted per unit area – the luminous exitance) depends only on the fourth power of its temperature, and, in 1884, Ludwig Boltzmann supplied a theoretical understanding of this relation. **The *Stefan–Boltzmann law*** is

$$s = \sigma T^4 \qquad\qquad (1.15)$$

where σ = the Stefan–Boltzmann constant = 5.6696×10^{-8} W m^{-2} K^{-4}.

Laboratory studies of blackbodies at about this time showed that although their spectra change with temperature, all have a similar shape: a smooth curve with one maximum (see Figure 1.8). This peak in the monochromatic flux curve (either f_λ or f_ν) shifts to shorter wavelengths with increasing

Fig. 1.8 Blackbody spectra
$B(\lambda, T)$ for objects at
several different surface
temperatures. (a) Both
axes are linear. The
wavelength at which the
5800 K (effective surface
temperature of the Sun)
spectrum peaks is
indicated. (b) Both axes
are logarithmic. The
dashed line traces the
peaks in the spectra.

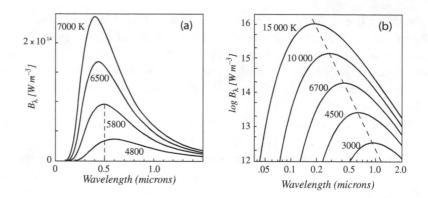

temperature, following a relation called **Wien's displacement law** (1893).
Wien's law states that for f_λ

$$T\lambda_{\text{MAX}} = 2.8979 \times 10^{-3} m \cdot K \qquad (1.16)$$

or equivalently, for f_ν:

$$\frac{T}{\nu_{\text{MAX}}} = 1.7344 \times 10^{-11} Hz^{-1}K \qquad (1.17)$$

Thus blackbodies are never black: the color of a glowing object shifts from
red to yellow to blue as it is heated.

Max Planck presented the actual functional form for the blackbody spectrum
at a Physical Society meeting in Berlin in 1900. His subsequent attempt to
supply a theoretical understanding of the empirical **"Planck function"** led him
to introduce the quantum hypothesis – that energy can only radiate in discrete
packets. Later work by Einstein and Bohr eventually showed the significance of
this hypothesis as a fundamental principle of quantum mechanics.

The Planck function gives the **specific intensity**, that is, the monochromatic
flux per unit solid angle, usually symbolized as $B(\nu.T)$ or $B_\nu(T)$. The total power
emitted by a unit surface blackbody over all angles is just $s(\nu,T) = \pi B(\nu.T)$. The
Planck function is

$$B(\nu, T) = \frac{2h\nu^3}{c^2}\left(e^{\left(\frac{h\nu}{kT}\right)} - 1\right)^{-1} \qquad (1.18)$$

$$B(\lambda, T) = \frac{2hc^2}{\lambda^5}\left(e^{\left(\frac{hc}{\lambda kT}\right)} - 1\right)^{-1} \qquad (1.19)$$

For astronomers, this means that we can observe the shape of an object's
spectrum and from it deduce the object's temperature (provided the object behaves
like a blackbody). Figure 1.8 shows the Planck function for several temperatures.
Note that even if the wavelength of the peak of the spectrum cannot be observed,
the slope of the spectrum gives a measure of the temperature.

It is useful to note that at long wavelengths or very high temperatures, a blackbody's specific intensity (and therefore angular surface brightness) is directly proportional to its temperature. This is the ***Rayleigh–Jeans approximation*** to the tail of the Planck function

$$B(\lambda, T) = 2\lambda^{-4}ckT \qquad (1.20)$$

$$B(\lambda, T) = 2c^{-2}k\nu^2 T \qquad (1.21)$$

In radio astronomy, where this approximation holds, and where one often observes extended sources, these equations suggest that observed brightness is a linear measure of the ***brightness temperature*** of a source.

1.4.4 Spectra of stars

When astronomers first examined the great variety in the absorption line spectra of different stars, they did not fully understand what they saw. Flooded with puzzling but presumably significant observations, most scientists have the (good) impulse to look for similarities and patterns: to sort the large number of observations into a small number of classes. Astronomers based their initial sorting of stellar spectra on the overall simplicity of the pattern of lines, assigning the simplest to class A, next simplest to B, and so on through the alphabet. Only after a great number of stars had been so classified from photographs[3] (in the production of the Henry Draper Catalog – see Chapters 4 and 11) did astronomers come to understand, through the new science of astrophysics, that the great variety arises mainly from temperature differences.

There is an important secondary effect due to surface gravity, as well as some subtle effects due to variations in chemical abundance. The chemical differences usually involved only the minor constituents – the elements other than hydrogen and helium.

The spectral type of a star, then, is basically an indication of its ***effective temperature*** – that is, the temperature of the blackbody that would produce the same amount of radiation per unit surface area as the star does. If the spectral type is sufficiently precise, it might also indicate the surface gravity or relative diameter or luminosity (if two stars have the same temperature and mass, the one with the larger diameter has the lower surface gravity as well as the higher luminosity). In order of decreasing effective temperature, the modern spectral classes are:

O ($T > 30\,000$ K)**, B, A, F, G, K, M, L, T, Y** ($T < 500$ K)

---

[3] Antonia Maury suggested in 1897 that the correct sequence of types should be O through M (the first seven in Table 11.1 – although Maury used a different notation scheme). Annie Cannon, in 1901, justified this order on the basis of continuity, not temperature. Cannon's system solidified the modern notation and was quickly adopted by the astronomical community. In 1921, Megh Nad Saha used atomic theory to explain the Cannon sequence as one of stellar temperature.

The *spectral type* of a star consists of three designations: (a) a letter, indicating the general temperature class, (b) a decimal subclass number between 0 and 9.9 refining the temperature estimate (0 indicating the hottest subclass), and (c) a roman numeral indicating the relative surface gravity or luminosity class. Luminosity class I (the *supergiants*) is most luminous, III (the *giants*) is intermediate, and V (the *dwarves*) is least luminous and most common. The Sun has spectral type G2 V. We discuss stellar spectra in greater detail in Chapter 11, and give more detailed data on spectral types in Table 11.1 and in Appendix K.

1.5 Magnitudes

1.5.1 Apparent magnitudes

When Hipparchus of Rhodes (*c.* 190–120 BCE), arguably the greatest astronomer in the Hellenistic school, published his catalog of 600 stars, he included an estimate of the brightness of each – our quantity F. Strictly, what Hipparchus and all visual observers estimate is F_{vis}, the flux in the visual bandpass, the band corresponding to the response of the human eye. The eye has two different response functions, corresponding to two different types of receptor cells – rods and cones. At high levels of illumination, only the cones operate (*photopic vision*), and the eye is relatively sensitive to red light. At low light levels (*scotopic vision*) only the rods operate, and sensitivity shifts to the blue. Greatest sensitivity is at about 555 nm (yellow) for cones and 505 nm (green) for rods. Except for extreme red wavelengths, scotopic vision is more sensitive than photopic and most closely corresponds to the Hipparchus system. See Appendix B3.

Hipparchus cataloged brightness by assigning each star to one of six classes, the first class (or first *magnitude*) being the brightest, the sixth class the faintest. The choice of six classes, rather than some other number – ten, for example – is a curious one, and may be tied to earlier Babylonian mysticism, which held six to be a significant number. For the next two millennia, astronomers perpetuated this system, eventually extending it to fainter stars at higher magnitudes: magnitudes 7, 8, 9, etc. could only be seen with a telescope. With the introduction of photometers in the nineteenth century, William Pogson (*c.* 1856 CE) discovered that Hipparchus' classes were in fact approximately a geometric progression in F, with each class two or three times fainter than the preceding. Pogson proposed regularizing the system so that *a magnitude difference of 5 corresponds to a brightness ratio of* **100:1**, a proposal eventually adopted by international agreement early in the twentieth century.

Astronomers who observe in the visual and near infrared persist in using this system. It has advantages: for example, all astronomical bodies have apparent magnitudes that fall in the restricted and easy-to-comprehend range of about

−26 (the Sun) to +32 (the faintest telescopic objects). The faintest magnitude detected with the unaided eye depends critically on sky brightness (and whose eye) but the limit seems to be about 8.0 under the very best conditions.

However, when Hipparchus and Pogson assigned the more positive magnitudes to the fainter objects, they were asking for trouble. Avoid the trouble and remember that *smaller* (more negative) magnitudes mean *brighter* objects. Those who work at other wavelengths are less burdened by tradition and use less confusing (but sometimes less convenient) units. Such units linearly relate to the apparent brightness, F, or to the monochromatic brightness, f_λ. In radio astronomy, for example, one often encounters the ***Jansky*** ($1 \text{ Jy} = 10^{-26} \text{ W m}^{-2} \text{ Hz}^{-1}$) as a unit for f_ν.

The relationship between apparent magnitude, m, and brightness, F, is:

$$m = -2.5\log_{10}(F) + K \tag{1.22}$$

The constant K is often chosen so that modern measurements agree, more or less, with the older catalogs, all the way back to Hipparchus. For example, the bright star, Vega, has $m \approx 0$ in modern magnitude systems. If the flux in Equation (1.22) is the total or bolometric flux (see Section 1.4) then the magnitude defined is called the ***apparent bolometric magnitude***, m_{bol}. Most practical measurements are made in a restricted bandpass, but the definition of such a bandpass magnitude remains as in Equation (1.22), even to the extent that K is often (not always) chosen so that Vega has $m \approx 0$ in any band. This standardization has many practical advantages but is potentially confusing, since the integral in Equation (1.14) will be very different in different bands for Vega (see Figure 1.5) or any other star. In most practical systems, the constant K is specified by defining the values of m for some set of ***standard stars***. Absolute calibration of such a system, so that magnitudes can be converted into energy units, requires comparison of at least one of the standard stars to a source of known brightness, like a blackbody at a known, stable temperature.

Equation (1.22) implies the *magnitude difference* between two sources is

$$\Delta m = m_1 - m_2 = -2.5\log_{10}\left(\frac{F_1}{F_2}\right) \tag{1.23}$$

This equation holds for both bolometric and bandpass magnitudes. It should be clear from Equation (1.23) that once you define the magnitudes of a set of standard stars, measuring the magnitude of an unknown is a matter of measuring a flux ratio – or magnitude difference – between the standard and the unknown. Magnitudes are almost always measured in this ***differential*** fashion without resorting to absolute energy units. Inverting Equation (1.23) gives the flux ratio as a function of magnitude difference:

$$\frac{F_1}{F_2} = 10^{-0.4(m_1 - m_2)} = 10^{-0.4\Delta m} \tag{1.24}$$

A word about notation: You can write magnitudes measured in a bandpass in two ways, by (1) using the bandpass name as a subscript to the letter "*m*", or (2) by using the name itself as the symbol. So for example, the B band apparent magnitudes of a certain star could be written as $m_B = 5.67$ or as $B = 5.67$.

1.5.2 Absolute magnitudes

The magnitude system can also be used to express the luminosity of a source. The *absolute magnitude* of a source is defined to be the apparent magnitude (either bolometric or bandpass) that the source would have if it were at the standard distance of 10 parsecs in empty space. (1 parsec $= 3.086 \times 10^{16}$ m $= 206$ 265 au – see Chapter 3). The relation between the apparent and absolute magnitudes of the same object is

$$m - M = 5 \log r - 5 \tag{1.25}$$

where M is the absolute magnitude and r is the actual distance to the source in parsecs. The quantity $(m - M)$ on the left-hand side of Equation (1.25) is called the *distance modulus* of the source. You should recognize this equation as the equivalent of the inverse square law relation between apparent brightness and luminosity. Equation (1.25) must be modified if the source is not isotropic or if there is absorption along the path between it and the observer.

To symbolize the absolute magnitude in a bandpass, use the band name as a subscript to the symbol M. The Sun, for example, has absolute magnitudes: $M_B = 5.515$, $M_v = 4.862$. The International Astronomical Union (IAU) has defined the bolometric absolute magnitude scale so that $M_{bol} = 0$ corresponds to $L = 3.055 \times 10^{28}$ W $= 79.8$ $L_\odot$. This implies that for the Sun, $M_{bol} = 4.756$

1.5.3 The bolometric correction

For a particular source in any magnitude system, the difference between the bolometric magnitude (either apparent or absolute) and the magnitude in a particular band is termed the *bolometric correction*. The bolometric correction is usually tabulated for the V band

$$BC = M_{bol} - M_v \tag{1.26}$$

and is a strong function of spectral type. Some authors vary in their definition of the BC, but the values tabulated in Appendix K are consistent with the *IAU* definition of the absolute magnitude zero point, which implies a V-band BC for the Sun of -0.11.

1.5.4 Apparent magnitudes from images

In Chapter 10 we will consider in detail how to measure the apparent magnitude of a source. Right now, it is helpful to have at least a simplified description.

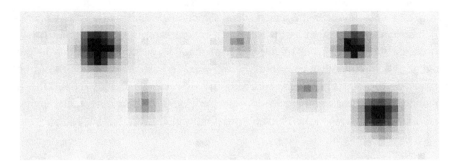

Fig. 1.9 A digital image. The width of each star image, despite appearances, is the same. Brighter stars have a higher peak and half-peak gray level, but every star image has the same width at the half-peak level.

Imagine a detector similar to the sensor in a black-and-white digital camera. Attach this to a telescope and take a picture.

Our picture, shown in Figure 1.9, is a visual light image of a few stars, and is composed of a grid of many little square elements called *pic*ture *el*ements or **pixels**. Each pixel corresponds to a detector in the sensor, and stores a number that is proportional to the energy that reaches that small detector during the exposure. Figure 1.9 displays these data by mimicking a photographic negative: each pixel location is painted a shade of gray, with the pixels that store the largest numbers painted darkest.

The image does *not* show the surfaces of the stars. The images of the stars are most intense in the center and fade out over several pixels. Several effects cause this – the finite resolving power of any lens, turbulence in the Earth's atmosphere, scattering of photons by particles in the air, or the scattering of photons from pixel to pixel within the detector itself. The apparent "diameter" of a star image in the picture has to do with the strength of the blurring effect and the choice of gray-scale mapping – not with the physical size of the star. The size of every star image is actually the same (about 4 pixels) when scaled by its peak brightness.

Suppose we manage to take a single picture with our camera that records an image of the star Vega (or some other standard star) as well as that of a star whose brightness we wish to measure. To compute the brightness of any star, we can add up the energies it deposits in each pixel. If E_{xy} is the energy recorded in pixel x, y due to light rays *from the star*, then the brightness of the star will be

$$F = \frac{1}{tA} \sum_{x,y}^{image} E_{xy} \qquad (1.27)$$

where t is the exposure time in seconds, A is the area of the telescope lens or mirror, and the sum is understood to include only pixels in the star image. Measuring F is just a matter of adding up those E_{xy} s.

Of course things are not quite so simple. One problem is that the detector cannot distinguish between light rays coming from the star and light rays coming from any other source in the same general direction. A faint star or galaxy nearly in the same line of sight as the star, or a moonlit terrestrial dust grain or air molecule floating in front of the star, can make unwelcome additions to the signal. All such sources contribute *background* light rays that reach the same pixels as the star's light. In addition, the detector itself may contribute a background that has nothing to do with the sky. Therefore, the actual signal we measure in pixel x, y, will be the sum of the signal from the star, E_{xy}, and that from the background, B_{xy}, or

$$S_{xy} = E_{xy} + B_{xy} \tag{1.28}$$

The task then is to determine B_{xy} so we can subtract it from each S_{xy}. You can do this by measuring the energy reaching some pixels near but not within a star image, and taking some appropriate average (call it B). Then assume that everywhere in the star image, B_{xy} is equal to B. Sometimes this is a good assumption, sometimes not so good. Granting the assumption, then the brightness of the star is

$$F = \frac{1}{tA} \sum_{x,y}^{image} \left[S_{xy} - B \right] \tag{1.29}$$

If we acknowledge that the estimated backgrounds for the two might be different, then the apparent magnitude difference between the unknown star and the standard star is

$$m_{star} - m_{std} = -2.5 \log_{10} \frac{F_{star}}{F_{std}} = -2.5 \log_{10} \left\{ \frac{\displaystyle\sum_{x,y}^{star} \left[S_{xy} - B_{star} \right]}{\displaystyle\sum_{x,y}^{std} \left[S_{xy} - B_{std} \right]} \right\} \tag{1.30}$$

Notice that the exposure time t, and the area, A, cancel on the right side of Equation (1.30). Also notice that since a ratio is involved, the pixels need not record units of energy – anything proportional to energy will do. Since scaled star images are the same size, the sums in Equation (1.30) should contain the same number of pixels.

The operation described in Equation (1.30) is called ***differential photometry.*** Because they appear in the same picture, the images of both the standard and the unknown star are subject to identical effects: atmospheric transparency, exposure time, telescope condition, sensor temperature, etc.

Differential photometry is only possible if a standard and unknown can be recorded on the same image. The alternative, ***all-sky photometry,*** requires two images, one of the unknown, the second of a standard star, with conditions as similar as possible in the two exposures. Ground-based all-sky photometry can be difficult in the optical-IR window, because you must look through different paths in the air to observe the standard and program stars. Any variability in the atmosphere (e.g. clouds) defeats the technique. In the radio window, clouds are less problematic and the all-sky technique is usually appropriate. In space, of course, there are no atmospheric effects, and all-sky photometry is usually appropriate at every wavelength.

1.5.5 Example problem

The bright star Betelgeuse is 150 pc from the Sun. It is a massive red supergiant that will probably end its evolution in a supernova explosion of type II. These supernovae, when brightest, reach a bolometric luminosity of 6×10^{35} W, with a spectrum similar to a blackbody of effective temperature 15 000 K. Assume Betelgeuse explodes as a supernova with these characteristics. At maximum light compute (a) its absolute bolometric magnitude, (b) its physical radius (assume a sphere) in solar units, and (c) its apparent visual magnitude, m_V, as seen from Earth. Compare the latter with the apparent brightness of the full moon (-12.74) and Venus (-4.6 at brightest). Assume the bolometric correction in the V band at maximum light is -1.3.

(a) Since we are given the bolometric luminosity, we can compute the absolute magnitude by comparing the supernova with the IAU standard for $M_{bol} = 0$. Or, since we eventually want to compare the SN with the Sun

$$M_{bol} - M_{sun} = -2.5 \log \left[\frac{L}{L_{sun}} \right] = -2.5 \log \left[\frac{6 \times 10^{35} \text{ W}}{3.8 \times 10^{26} \text{ W}} \right] = -23.0 \qquad (1.31)$$

$$M_{bol} = -23 + 4.76 = -18.24$$

(b) The effective temperature is that of a blackbody of the same size and luminosity, so that by Stefan's Law: $L = (\text{surface area})\sigma T_{eff}^4 = 4\pi R_{sn}^2 \sigma T_{eff}^4$. Again, it is helpful to compare the luminosities of the supernova and the Sun ($T_{sun} = 5800$ K):

$$\frac{L}{L_{sun}} = \frac{R_{sn}^2 T_{eff}^4}{R_{sun}^2 T_{sun}^4}$$

$$R_{sn} = \left[\frac{T_{sun}}{T_{sn}} \right]^2 \left[\frac{L}{L_{sun}} \right]^{\frac{1}{2}} R_{sun} = \left[\frac{15}{5.8} \right]^2 (1.58 \times 10^9)^{\frac{1}{2}} R_{sun}$$

$$R_{sn} = 2.7 \times 10^5 R_{sun} = 1.85 \times 10^{11} \text{km} = 1200 \text{ au}$$

(c) The absolute visual magnitude of the supernova follows from the bolometric correction: $M_v = M_{bol} - BC = -16.9$, and the apparent visual magnitude follows from the distance modulus:

$$m_v = 5\log r - 5 + M_v = 5.88 - 16.9 = -11.0$$

This is 2.76 magnitudes fainter than the full moon, (a factor of 12.7 in brightness) and 6.4 magnitudes (a factor of 360) brighter than Venus.

Summary

- The wave theory formulated by Maxwell envisions light as a self-propagating disturbance in the electric and magnetic fields. A light wave is characterized by wavelength or frequency and exhibits the properties of diffraction and interference. Concepts:

  $\lambda = c/v$ *electromagnetic spectrum*

  Radio, microwave, infrared, visible, ultraviolet, X-ray, gamma ray

  polarization *phase* *amplitude*

- Quantum mechanics improves on the wave theory and describes light as a stream of photons, massless particles that can exhibit wave properties. Concept:

  energy of a photon: $E = hv$

- A simple but useful description of light postulates a set of geometric rays that carry luminous energy. Concepts:

  luminosity *apparent brightness = flux = irradiance*

  inverse square law $F = L/(4\pi r^2)$

- The surface brightness of a resolved source is invariant with distance. Concepts:

  radiant exitance *brightness per unit solid angle*

- The spectrum of an object gives its brightness as a function of wavelength or frequency. Concepts:

  monochromatic flux *high- (or low-) resolution spectrum*

  flux within a band *standard bands*

  UBVRI(JHKLMNQ)

- The Kirchhoff–Bunsen rules specify the circumstances under which an object produces an emission line, absorption line, or continuous spectrum.

- Line spectra contain information about (among other things) chemical composition. However, although based on patterns of absorption lines, the spectral types of stars depend primarily on stellar temperatures. Concepts:

  OBAFGKMLTY *luminosity class*

- A blackbody emits a continuous spectrum whose shape depends only on its temperature as described by Planck's Law. Concepts:

  Wien's law: $T\lambda_{MAX} = 2.8979 \times 10^{-3} \text{m·K}$

  Stefan–Boltzmann law: $s = \sigma T^4$

- The astronomical magnitude system uses apparent and absolute magnitudes to quantify brightness measurements on a logarithmic scale. Concepts:

  $$\Delta m = m_1 - m_2 = -2.5 \log_{10}\left(\frac{F_1}{F_2}\right)$$

  distance modulus $m - M = 5\log r - 5$

  parsec *bolometric magnitude*

  differential photometry *all-sky photometry*

Exercises

> A bear, however hard he tries, gets tubby without exercise.
>
> – A.A. Milne, *Winnie the Pooh*, 1926

1. Propose a definition of astronomy that distinguishes it from other sciences like physics and geology.

2. (a) What wavelength photon would you need to ionize a hydrogen atom (ionization energy = 13.6 eV)? (b) Compute the temperature of the blackbody whose spectrum peaks at the wavelength you found in (a).

3. What are the units of the monochromatic brightness, f_λ?

4. What is the value of the ratio f_λ / f_ν for any source?

5. Consider an eclipsing binary star system viewed by a terrestrial observer located in the orbital plane of the two stars. The orbit is so small that the observer sees this system as a single object of apparent brightness, F. The larger star of the pair has a radius $R = 1$ and effective temperature T_1, the smaller has radius a and temperature T_2. Because of mutual eclipses, F is a periodic function of time called the light curve, $F(t)$. The orbit is circular. There are two brightness minima per orbit. (a) Regardless of the value of a, the primary (deeper) minimum always occurs when the hotter star is behind the cooler. Explain why. (b) Assume both stars are Lambertian (i.e. appear, in projection, to be uniform disks). If $F_{\max}$ is the maximum brightness of the system, show that the relative brightness at the two minima of the light curve are:

$$\frac{F_1}{F_{\max}} = \frac{1}{1+b}, \text{ when the smaller star is eclipsed by the larger}$$

$$\frac{F_2}{F_{\max}} = 1 - \frac{a^2}{1+b}, \text{ when the smaller transits in front of the larger}$$

Here, $b = a^2 \left(\frac{T_2}{T_1}\right)^4$

6. A certain radio source has a monochromatic flux density of 1 Jy at a frequency of 1 MHz. What is the corresponding flux density in photon number? (How many photons arrive per m^2 in one second with frequencies between 1 000 000 Hz and 1 000 001 Hz?)

7. The bolometric flux from a star with $m_{\text{bol}} = 0$ is about 2.65×10^{-8} W m^{-2} outside the Earth's atmosphere. Compute the value of the constant K in Equation (1.22) for bolometric magnitudes.

8. The monochromatic flux at the center of the B bandpass (440 nm) for a certain star is 375 Jy. (a) If this star has a blue magnitude of $m_B = 4.71$, what is the monochromatic flux, in Jy, at 440 nm for star X which has $m_B = 22.5$? (b) If the width of the B band is 2.5×10^{14} Hz, about how many photons from star X can be collected in 100 seconds by a telescope with a collecting area of 5 square meters?

9. A double star has two components of equal brightness, each with a magnitude of 8.34. If these stars are so close together that they appear to be one object, what is the apparent magnitude of the combined object?

10. A uniform gaseous nebula has an average surface brightness of 17.77 magnitudes per square second of arc.
    (a) If the nebula has an angular area of 144 square arcsec, what is its total apparent magnitude?
    (b) If the nebula were moved to twice its original distance, what would happen to its angular area, total apparent magnitude, and surface brightness?

11. Assume the nebula in 10(a) were moved to 100 times its original distance. If you observed this object with a telescope whose resolving power is 1.2 arc seconds (i.e. even point sources have an apparent diameter of 1.2 arc seconds), what would be the apparent angular area and apparent surface brightness of the nebula?

12. At maximum light, the brightest Type Ia supernovae are believed to have an absolute visual magnitude of -19.60. A supernova in the Pigpen Galaxy is observed to reach apparent visual magnitude of 13.25 at its brightest. (a) Compute the distance to the Pigpen Galaxy. (b) If you believe the Pigpen Galaxy contains clouds of dust that absorb 1.5 magnitudes of light in the V band, re-compute the distance to the galaxy.

13. Derive the distance modulus relation in Equation (1.25) from the inverse square law relation in Equation (1.6).

14. Show that, for small values of Δm, the difference in magnitude is approximately equal to the fractional difference in brightness, that is

$$\Delta m \approx \frac{\Delta F}{F} \tag{1.32}$$

Hint: consider the derivative of m with respect to F.

15. An astronomer is performing synthetic aperture photometry on a single unknown star and standard star (review Section 1.5) in the same field. The data frame is in the figure below. The unknown star is the fainter one. If the magnitude of the standard is 9.000, compute the magnitude of the unknown. Actual data numbers are listed for the frame in the table. Assume these are proportional to the number of photons counted in each pixel, and that the bandpass is narrow enough that all photons can be assumed to have the same energy. Remember that photometrically both star images have the same size.

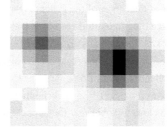

34	16	26	33	37	22	25	25	29	19	28	25
22	20	44	34	22	26	14	30	30	20	19	17
31	70	98	66	37	25	35	36	39	39	23	20
34	99	229	107	38	28	46	102	159	93	37	22
33	67	103	67	36	32	69	240	393	248	69	30
22	33	34	29	36	24	65	241	363	244	68	24
28	22	17	16	32	24	46	85	157	84	42	22
18	25	27	26	17	18	30	29	35	24	30	27
32	23	16	29	25	24	30	28	20	35	22	23
28	28	28	24	26	26	17	19	30	35	30	26

Chapter 2
Uncertainty

Errare humanum est.

– Anonymous Latin saying

Upon foundations of evidence, astronomers erect splendid narratives about the lives of stars, the prevalence of habitable planets, or the fate of the universe. Inaccurate or imprecise evidence weakens the foundation and imperils the astronomical story it supports. Incorrect ideas and theories are vital to science, which normally works by proving many, many ideas to be wrong until only one remains. Wrong data, on the other hand, are deadly.

As an astronomer you need to know how far to trust the data you have, or how much observing you need to do to achieve a particular level of trust. This chapter describes the formal distinction between accuracy and precision in measurement, and methods for estimating both. It then introduces the concepts of a population, a sample of a population, and the statistical descriptions of each. Any characteristic of a population (e.g. the masses of stars) can be described by a probability distribution (e.g. low-mass stars are more probable than high-mass stars), so we next will consider a few probability distributions important in astronomical measurements. Finally, armed with new statistical expertise, we revisit the question of estimating uncertainty.

2.1 Accuracy and precision

In common speech, we often do not distinguish between these two terms, but we will see that establishing a clear distinction is useful. An example will help.

2.1.1 An example

In the distant future, a very smart theoretical astrophysicist determines that the star Malificus might soon implode to form a black hole, and in the process destroy all life on its two inhabited planets. Careful computations show that if Malificus is fainter than magnitude $14.190 \pm .003$ by July 24, as seen from the observing station orbiting Pluto, then the implosion will not take place and its

Table 2.1 *Results of trials by four astronomers. The values for σ and s are computed from Equations (2.6) and (2.1), respectively.*

Astronomer	A	B	C	D
Trial 1	14.115	14.495	14.386	14.2
Trial 2	14.073	14.559	14.322	14.2
Trlal 3	14.137	14.566	14.187	14.2
Trial 4	14.161	14.537	14.085	14.2
Trial 5	14.109	14.503	13.970	14.2
Mean	14.119	14.532	14.190	14.2
Deviation from truth	−0.004	+0.409	+0.067	+0.077
Spread	0.088	0.071	0.418	0
s	0.033	0.032	0.174	0
σ	0.029	0.029	0.156	0
Uncertainty of the mean	0.013	0.013	0.070	(0.05)
Decision	Evacuate	Stay	Uncertain	Uncertain
Accuracy?	accurate	inaccurate	accurate	inaccurate
Precision?	precise	precise	imprecise	imprecise

planets will be spared. The Galactic government is prepared to spend the ten thousand trillion dollars necessary to construct a wormhole and evacuate the doomed populations but needs to know if the effort is really called for; it funds some astronomical research. Four astronomers and a demigod each set up experiments on the Pluto station to measure the apparent magnitude of Malificus.

The demigod performs photometry with divine perfection, obtaining a result of 14.123 010 (all the remaining digits are zeros). The truth, therefore, is that Malificus is brighter than the limit and will implode. The four astronomers, in contrast, are only human, and, fearing error, repeat their measurements – five times each. I'll refer to a single one of these five as a ***trial.*** Table 2.1 lists the results of each trial, and Figure 2.1 illustrates them.

2.1.2 Accuracy and systematic error

In our example, we are fortunate a demigod participates, so we feel perfectly confident to tell the government, sorry, Malificus is doomed, and those additional taxes are necessary. The ***accuracy*** of a measurement describes (usually numerically) how close it is to the "true" value. The demigod measures with perfect accuracy.

What is the accuracy of the human results? First, decide what we mean by "a result": Since each astronomer made five trials, we choose a single value that

Fig. 2.1 Apparent
magnitude
measurements. The arrow
points to the demigod's
result, which is the true
value. The thick gray line
marks the critical limit and
its uncertainty.

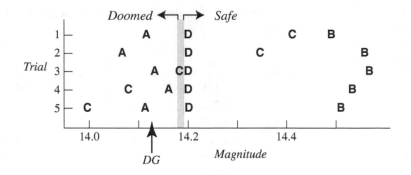

summarizes these five measurements. In this example, each astronomer chooses to compute the mean – or average – of the five, a reasonable choice. (We will see there are others.) Table 2.1 lists the mean values from each astronomer – a *statistic* that summarizes the five trials each has made.

Since we know how much each result deviates from the truth, we could express its accuracy with a sentence like: "The result of Astronomer A is 0.004 magnitude smaller than the true value." This statement is easy to make if a demigod tells us the truth, but in the real universe, how could you determine the "true" value, and hence the accuracy? In science, after all, the whole point is to discover values that are unknown at the start, and (the self-images of some astronomers notwithstanding) no demigods work at observatories.

How, then, can we judge accuracy? The alternative to divinity is variety. We can only repeat measurements using different devices, assumptions, strategies, and observers, and then check for general agreements (and disagreements) among the results. We suspect a particular set-up of inaccuracy if it disagrees with all other experiments. For example, the result of Astronomer B differs appreciably from those of his colleagues. Even in the absence of the demigod result, we would suspect that B's result is the least accurate.

If a particular set-up always produces consistent inaccuracies, if its result is always biased by about the same amount, then we say it produces a *systematic error*. Although Astronomer B's trials do not have identical outcomes, they all tend to be much too large, and are, we suspect, subject to a systematic error of around +0.4 magnitude. Systematic errors are due to some instrumental or procedural fault, or some mistake in modeling the phenomena under investigation. Astronomer B, for example, used the wrong magnitude for the standard star in his measurements. He could not improve his measurement just by repeating it – making more trials would give the same general result, and B would continue to recommend against evacuation.

In a second example of inaccuracy, suppose the astrophysicist who computed the critical value of $V = 14.190$ had made a mistake because he neglected the effect of the spin of Malificus. Then, even perfectly accurate measurements of brightness could result in a possibly disastrous decision.

2.1.3 Precision and random error

Precision differs from accuracy. The **precision** of a measurement describes how well or with what certainty a particular result is known, without regard to its truth. Precision denotes the ability to be very specific about the exact value of the measurement itself. A large number of legitimately significant digits in the numerical value, for example, indicates high precision. Because of the possibility of systematic error, of course, high precision does not mean high accuracy.

Poor precision *does* imply a great likelihood of poor accuracy. An imprecise result *could* be accurate, but the universe seldom rewards that sort of optimism. Do not expect accuracy better than your precision, and do not be shocked when, because of systematic error, it is a lot worse.

Unlike accuracy, precision is often easy to quantify without divine assistance. Just examine the degree to which multiple trials agree with one another. If the outcome of one trial differs from the outcome of the next in an unpredictable fashion, the scattering is said to arise from **stochastic, accidental,** or **random error.** (If the outcome of one trial differs from the next in a *predictable* fashion, you have uncovered a systematic effect. Hello, Astronomer C.) The term random "error" is unfortunate, since it suggests some sort of mistake or failure, whereas you should really think of it as a scattering of values due to the uncertainty inherent in the measuring process, or in the phenomenon itself. Random error limits precision and therefore limits accuracy.

To quantify random error, you could examine the **spread** in values for a collection of trials:

spread = largest trial result − smallest trial result

The spread will tend to be larger for experiments with the largest random error and lowest precision. A better description of the scatter or "dispersion" of a set of N trials, $\{x_1, x_1, \ldots x_N\}$, would depend on all N values. One useful statistic of this sort is the **estimated standard deviation**, s:

$$s = \sqrt{\frac{1}{N-1} \sum_{i=1}^{N} (x_i - \bar{x})^2} \tag{2.1}$$

We examine Equation (2.1) more carefully in later sections of this chapter. The values for s and for the spread in our example are in Table 2.1. These confirm the subjective impression from Figure 2.1 – in relative terms, the result of astronomer

> A is precise and accurate;
>
> B is precise but inaccurate;
>
> C is imprecise but accurate (to the degree expected from the precision);
>
> D is an apparently precise but actually imprecise case, discussed below.

The basic statistical techniques for coping with random error and estimating the resulting uncertainty are the subjects of this chapter. A large volume of literature deals with more advanced topics in the statistical treatment of data dominated by stochastic error – a good introduction is the book by Bevington (1969). Although most techniques apply only to stochastic error, in reality, systematic error is usually the more serious limitation to good astronomy.

Techniques for detecting and coping with systematic error are varied and indirect, and therefore difficult to discuss at an elementary level. Sometimes, one is aware of systematic error only after reconciling different methods for determining the same parameter. This is the case with Astronomer B, whose result differs from the others by more than the measured stochastic error. Sometimes, what appears to be stochastic variation turns out to be a systematic effect. This might be the case with Astronomer C, whose trial values decrease with time, suggesting perhaps some change in the instrument or environment. Although it is difficult to recognize systematic error, the fact that it is the consequence of some sort of mistake means that it is often possible to correct the mistake and improve accuracy.

Stochastic error and systematic error both contribute to the uncertainty of a particular result. That result is useless until the size of its uncertainty is known.

2.1.4 Uncertainty

> When one admits that nothing is certain one must, I think, also add that some things are more nearly certain than others.
>
> – Bertrand Russell *Am I An Atheist Or An Agnostic?* 1947

Science is about making decisions. Where do I point the telescope? What instrument should we build? What questions should I spend my life trying to answer? In our Malificus example, the decision that each astronomer recommends depends on two things: the numerical value of the result and the uncertainty the astronomer attaches to its accuracy. Astronomer A recommends evacuation because (a) her result is below the cutoff by 0.07 magnitude and because (b) the uncertainty she feels is small because her random error, as measured by s, is small compared to 0.07 and because she assumes her systematic error is also small. The assumption of a small systematic error is based mostly on A's confidence that she "knows what she is doing" and hasn't made a mistake. Later, when she is aware that both C and D agree with her result, she can be even more sanguine about this. Astronomical literature sometimes makes a distinction between ***internal error***, which is the uncertainty computed from the scatter of trials, and ***external error,*** which is the total uncertainty, including systematic effects.

Astronomer A should quote a numerical value for her ***uncertainty***. If u is the uncertainty of a result, r, then the probability that the true value is between $r + u$ and $r - u$ is 1/2. Statistical theory (see below) says that under certain broad conditions (N not too small), the ***uncertainty of the mean*** of five values is something like $s/\sqrt{N}$. Thus, the uncertainty imposed by random error (the internal error) alone for A is about 0.013. The additional uncertainty due to systematic error is harder to quantify. The astronomer should consider such things as the accuracy of the standard star magnitudes and the stability of her photometer. In the end, she might feel that her result is uncertain (external error) by 0.03 magnitudes. She concludes the chances are much greater than 50% that the limit is passed and thus, in good conscience, must recommend evacuation.

Astronomer B goes through the same analysis as A and recommends *against* evacuation with even greater (why?) conviction. Since quadrillions of dollars, disruption of space-time, and billions of lives are at stake, it would be criminal for A and B not to confront their disagreement. They must compare methods and assumptions and try to determine which (if either) of them has the accurate result.

Astronomer C shouldn't make a recommendation because his uncertainty is so large. He can't rule out the need for an evacuation, nor can he say that one is necessary. We might think C's measurements are so imprecise that they are useless, but this is not so. C's precision is sufficient to cast doubt on B's result (but not good enough to confirm A's). The astronomers thus should first concentrate on B's experimental method in their search for the source of their disagreement. C should also be suspicious of his relatively large random error compared to the others. This may represent the genuine accidental errors that limit his particular method, or it may result from a systematic effect that he could correct.

2.1.5 Digitizing effects

What about Astronomer D, who performed five trials that gave identical results? D made her measurements with a digital light meter that only reads to the nearest 0.2 magnitude, and this digitization is responsible her very uniform data.

From the above discussion, it might seem that since her scatter is zero, D's measurement is perfectly precise. This is misguided, because it ignores what D knows about her precision: rounding off every measurement produces uncertainty. D reasons that in the absence of random or systematic errors, there is a 100% chance that the true value is with ±0.1 of her measurement, and there is a 50% chance that the true value lies within 0.05 magnitudes of her measurement. Thus, D would report an uncertainty of around ±0.05 magnitude. This is a case where a known systematic error (digitization) limits precision and where stochastic error is masked by the systematic effect. Good instrument design would ensure that any digitization effect is consistent with the desired precision. Usually this means that the digitization effect is smaller than the anticipated stochastic error.

2.1.6 Significant digits

One way to indicate the uncertainty in a measurement is to retain only those digits that are warranted by the uncertainty, with the remaining insignificant digits rounded off. In general, only one digit with "considerable" uncertainty (more than ± 1) should be retained. For example, Astronomer C had measured a value of 14.194 with an uncertainty of at least $0.174/\sqrt{5} = 0.078$. He realizes that the last digit "4" has no significance whatever; the digit "1" is uncertain by almost ± 1, so the digit "9," which has considerable uncertainty, is the last that should be retained. Astronomer C should quote his result as 14.19.

Astronomer A, with the result 14.119, recognizes that her digit "1" in the hundredths place is uncertain by more than ± 1, and she rounds off her result to 14.12.

It is also very good practice to quote the actual uncertainty. Usually one or two digits in the estimate of uncertainty are all that are significant. The first three astronomers might publish (internal errors):

> A's result: $14.12 \pm .013$
> B's result: $14.53 \pm .013$
> C's result: $14.19 \pm .08$

Note that Astronomers A and C retain the same number of significant digits, even though A's result is much more certain than C's. Astronomer B, estimates his (internal) uncertainty in good faith, but is unaware of his large systematic error,

2.2 Describing populations

> As some day it may happen that a victim must be found, I've got a little list – I've got a little list
>
> – W.S. Gilbert, *The Mikado*, Act I, 1885

In the Malificus problem, our fictional astronomers used simple statistical computations to estimate both brightness and precision. We now treat more systematically the statistical analysis of observational data of all kinds and begin with the concept of a population.

Consider the problem of determining a parameter (e.g. the brightness of a star) by making several measurements under nearly identical circumstances. We

Table 2.2 *Populations and samples. Samples can be more or less representative of the population from which they are drawn.*

Population	Sample	Better sample
1000 colored marbles mixed in a container: 500 red, 499 blue, 1 purple	5 marbles drawn at random from the container	50 marbles drawn at random
The luminosities of each star in the Milky Way Galaxy (about 10^{11} values)	The luminosities of each of the nearest 100 stars (100 values)	The luminosities of 100 stars at random locations in the Galaxy (100 values)
The masses of each planet in the Galaxy	The masses of the eight planets in our Solar System	The masses of the planets orbiting the 100 stars in the above sample
The results of all possible experiments that count the number of photons arriving at your detector from the star Malificus in one second	The outcome of 1 such experiment	The outcomes of 100 such experiments

define the ***population*** under investigation as the hypothetical set of all possible measurements that could be made with an experiment substantially identical to our own. We then imagine that we make our actual measurements by drawing a finite ***sample*** (five trials, say) from this much larger population. Some populations are indefinitely large, or are so large that taking a sample is the only practical method for investigating the population. Some populations are finite in size, with some small enough to be sampled completely. Table 2.2 gives some examples of populations and samples.

2.2.1 Descriptive statistics of a finite population

> Well, that's the news from Lake Wobegon, where all the women are strong, all the men are good looking, and all the children are above average.
>
> – Garrison Keillor, *A Prairie Home Companion*, 1974–Present

Imagine a small, finite population: that is, a set of M values or members, $\{x_1, x_2, \ldots, x_M\}$. The list of salaries of the employees in a small business, like the ones in Table 2.3, is an example. We can define some statistics that summarize or describe the population as a whole.

Measures of the central value

If every value in the population is known, a familiar descriptive statistic is the ***population mean***,

Table 2.3 *Employee salaries at Astroploitcom.*

Job title (number of employees)	Salary in thousands of dollars
President (1)	2000
Vice president (1)	500
Programmer (3)	30
Astronomer (4)	15

$$\mu = \frac{1}{M} \sum_{i=1}^{M} x_i \tag{2.2}$$

Two additional statistics also measure the *central* or *representative* value of the population. The **median**, or midpoint, is the value that divides the population exactly in half: just as many members have values above as have values below the median. If $n(E)$ is the number of members of a population with a particular characteristic, E, then the median, $\mu_{1/2}$, satisfies

$$n(x_i \le \mu_{1/2}) = n(x_i \ge \mu_{1/2}) \approx \frac{M}{2} \tag{2.3}$$

Compared to the mean, the median is a bit more difficult to compute if M is large, since you have to sort the list of values. In the pre-sorted list in Table 2.3, we can see by inspection that the median salary is $30 000, quite a bit different from the mean ($300 000). The third statistic is the **mode**, which is the most common or most frequent value. In the example, the mode is clearly $15 000, the salary of the four astronomers. In a sample in which there are no identical values, you can still compute the mode by sorting the values into bins, and then searching for the bin with the most members. Symbolically, if μ_{max} is the mode, then

$$n(x_i = \mu_{max}) > n(x_i = y, y \ne \mu_{max}) \tag{2.4}$$

Which measure of the central value is the "correct" one? The mean, median, and mode all legitimately produce a central value. Which one is most relevant depends on the question being asked. In the example in Table 2.3, if you were interested in balancing the corporate accounts, then the mean would be most useful. If you are interested in organizing a workers' union, the mode might be more interesting.

Measures of dispersion

How scattered are the members of a population? Are values clustered tightly around the central value, or are many members significantly different from one another? Table 2.4 gives the speeds of stars in the direction perpendicular to the Galactic plane. Two populations differ in their chemical compositions: one set of 25 stars

Table 2.4 *Speeds perpendicular to the Galactic plane, in km s^{-1}, for 50 nearby solar-type stars.*

Group A: 25 Iron-rich stars. μ = 12.85 km s^{-1}					Group B: 25 Iron-poor stars. μ = 28.1 km s^{-1}				
0.5	7.1	9.2	14.6	18.8	0.3	7.9	16.8	35.9	48.3
1.1	7.5	10.7	15.2	19.6	0.4	10.0	18.1	38.8	55.5
5.5	7.8	12.0	16.1	24.2	2.5	10.8	23.1	42.2	61.2
5.6	7.9	14.3	17.1	26.6	4.2	14.5	26.0	42.3	67.2
6.9	8.1	14.5	18.0	32.3	6.1	15.5	32.1	46.6	76.6

(Group A) contains the nearby solar-type stars that most closely match the Sun in iron abundance. Group A is relatively iron-rich. A second group, B, contains the 25 nearby solar-type stars that have the lowest known abundances of iron in their atmospheres. Figure 2.2 summarizes the table with a histogram. Clearly, the central value of the speed is different for the two populations. Group B stars, on average, zoom through the plane at a higher speed than do members of Group A.

A second difference between these populations concerns us here. The individual values in Group B are more dispersed – spread over a wider range of values – than those in Group A. Figure 2.2 illustrates this difference decently, but we want a compact and quantitative expression for it. To compute such a statistic, we first examine the **deviation** of each member from its population's mean

$$\text{deviation from the mean} = (x_i - \mu) \tag{2.5}$$

Those values of x_i that differ most from μ will have the largest deviations. The definition of μ insures that the average of all the deviations will be zero (positive deviations will exactly balance negative deviations), so the average deviation is an uninteresting statistic. The average of all the *squares of the deviations*, in contrast, must be a positive number. This is called the **population variance**:

$$\sigma^2 = \frac{1}{M} \sum_{i=1}^{M} (x_i - \mu)^2 = \frac{1}{M} \sum_{i=1}^{M} x_i^2 - \mu^2 \tag{2.6}$$

The variance tracks the dispersion nicely – the more spread out the population members are, the larger the variance. Because the deviations enter Equation (2.6) as quadratics, the variance is especially sensitive to population members with large deviations from the mean.

The square root of the population variance is called the **standard deviation of the population**

$$\sigma = \sqrt{\frac{1}{M} \sum_{i=1}^{M} (x_i - \mu)^2} \tag{2.7}$$

The standard deviation is usually the statistic employed to measure population spread. σ has the same dimensions as the population values themselves. For

Fig. 2.2 Histogram of the data in Table 2.4. The circles and lines at the top of the plot represent summary statistics (mean and standard deviation) of the two populations.

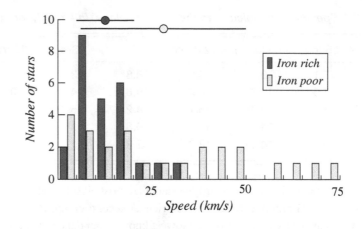

example, the variance of Group A in Table 2.4 is 57.25 km^{2}s^{-2}, and the standard deviation is 7.57 kms^{-1}, a number consistent with the distribution in Figure 2.2. Likewise, the value of $\sigma = 21.86$ looks like a reasonable metric for the spread in values for population B.

2.2.2 Estimating population statistics

Many populations, especially in astronomy, are either infinite or so large that it is impractical to tabulate all members. In this case, the strategy is to *estimate* the descriptive statistics for the population from a small, *representative* sample. For example, a sample of five trials at measuring the brightness of the star Malificus represents the population that contains *all* possible equivalent measurements of its brightness. Most scientific measurements are usually treated as samples of a much larger population of possible measurements.

In any sampling operation, we estimate the population mean from the **sample mean**, $\bar{x}$. All other things being equal, we believe a larger sample will give a better estimate. In this sense, the population mean is the limiting value of the sample mean. If the sample has N members

$$\mu = \lim_{N \to \infty} \frac{1}{N} \sum_{i=1}^{N} x_i = \lim_{N \to \infty} \bar{x} \tag{2.8}$$

Even if N is small, the sample mean is the best estimator of the population mean.

$$\mu \approx \bar{x} \tag{2.9}$$

To estimate the population variance from a sample, the best statistic is s^2, the **sample variance** computed with $(N - 1)$ weighting

$$s^2 = \frac{1}{N-1} \sum_{i=1}^{N} (x_i - \bar{x})^2 \qquad (2.10)$$

The $(N-1)^{-1}$ factor above (instead of just N^{-1}) arises because $\bar{x}$ is an *estimate* of the population mean, and is not μ itself. The difference is perhaps clearest in the case where $N = 2$. (See the exercises). In the limit of large N

$$\lim_{N \to \infty} s^2 = \lim_{N \to \infty} \frac{1}{N-1} \sum_{i=1}^{N} (x_i - \bar{x})^2 = \frac{1}{N} \sum_{i=1}^{N} (x_i - \mu)^2 = \sigma^2 \qquad (2.11)$$

Proof that (2.11) is the best estimate of σ^2 can be found in elementary references on statistics. The square root of s^2 is called the **standard deviation of the sample**. Since most astronomical measurements are samples of a population, the dispersion of the population is usually estimated as

$$s = \sqrt{\frac{1}{N-1} \sum_{i=1}^{N} (x_i - \bar{x})^2} \approx \sigma \qquad (2.12)$$

which is the expression introduced at the beginning of the chapter as Equation (2.1).

The terminology for s and σ can become imprecise. It is unfortunately common to shorten the name for s to just "the standard deviation," and to represent it with the symbol σ. You, the reader, must then discern from the context whether the statistic is an estimate from a sample (Equation (2.12)) or a description of the complete population (Equation (2.7)).

2.3 Probability distributions

> The most important questions of life are, for the most part, really only problems of probability.
>
> – Pierre-Simon Laplace, *A Philosophical Essay on Probabilities*, 1814

2.3.1 The random variable

Since scientific measurements generally sample a larger population, we consider the construction of a sample a little more carefully. Assume we have a large population, Q. For example, suppose we have a pottery urn full of small metal spheres of differing diameters, and wish to sample those diameters in a representative fashion. Imagine doing this by stirring up the contents, reaching blindly into the urn, and measuring the diameter of the first sphere you pull out. This operation is a trial, and its result is a diameter, x. We call x a ***random***

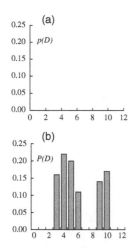

Fig. 2.3 Probability distributions of the diameters of spheres, in millimeters. (a) A continuous distribution; (b) a discrete distribution, in which only six sizes are present. Note that the dimensions on the vertical axes are different in (a) and (b).

variable – its value depends not at all on the selection method (we hope). Although the value of x is unpredictable, there clearly is a function that describes how likely it is to obtain a particular value for x in a single trial. This function, $p_Q(x)$, is called the **probability distribution** of x in Q. In the case where x can take on any value over a continuous range, we define:

$$p_Q(x)\mathrm{d}x = \text{the probability that the result of a single trial will}$$
$$\text{have a value between } x \text{ and } x + \mathrm{d}x$$

Sometimes a random variable is restricted to a discrete set of possible values. In our example, this would be the case if the urn contained only spheres whose diameters were integral multiples of 1 mm. In this case, the definition of the probability distribution function has to be a little different:

$$P_Q\left(x_j\right) = \text{the probability that the result of a single trial will have}$$
$$\text{a value } x_j, \text{ where } j = 1, 2, 3, \ldots$$

For our example, $p_Q(x)$ and $P_Q(x_i)$ might look like Figure 2.3, where (a) shows a continuous distribution in which any diameter over a continuous range is possible. Plot (b) shows a discrete distribution with only six possible sizes.

In experimental situations, we sometimes know or suspect something about the probability distribution before conducting any quantitative trials. We might, for example, look into our jar of spheres and get the impression that "there seem to be only two general sizes, large and small." Knowing something about the expected distribution before making a set of trials can be helpful in designing the experiment and in analyzing the data. Nature, in fact, favors a small number of distributions. Two particular probability distributions arise so often in astronomy that they warrant special attention.

2.3.2 The Poisson distribution

The Poisson[1] distribution describes a population encountered in certain counting experiments. These are cases in which the random variable, x, is the number of events counted in a unit time: the number of raindrops hitting a tin roof in 1 second, the number of photons hitting a light meter in 10 seconds, or the number of nuclear decays in an hour. For counting experiments where non-correlated

---

[1] Siméon Denis Poisson (1781–1840) in youth resisted his family's attempts to educate him in medicine and the law. After several failures in finding an occupation that suited him, he became aware of his uncanny aptitude for solving puzzles and embarked on a very prolific career in mathematics, becoming Laplace's favorite pupil. Poisson worked at a prodigious rate, both in mathematics and in public service in France. Given his rather undirected youth, it is ironic that he characterized his later life with his favorite phrase: "La vie, c'est le travail."

events occur at an average rate, μ, the probability of counting x events in a unit time in a single trial is

$$P_p(x,\mu) = \frac{\mu^x}{x!}e^{-\mu} \qquad (2.13)$$

Here, $P_p(x,\mu)$ is the Poisson distribution. For example, if you are listening to raindrops on the roof in a steady rain, and on average hear 3.25 per second, then $P_p(0, 3.25)$ is the probability that you will hear zero drops in the next 1- second interval. Of course, $P_p(x,\mu)$ is a discrete distribution, with x restricted to non-negative integer values (you can never hear 0.266 drops, nor could you hear -1 drops). Figure 2.4 illustrates the Poisson distribution for three different values of μ. Notice that as μ increases, so does the dispersion of the distribution. An important property of the Poisson distribution, in fact, is that its variance is exactly equal to its mean:

$$\sigma^2_{\text{Poisson}} = \mu \qquad (2.14)$$

This behavior has very important consequences for planning and analyzing experiments. For example, suppose you count the number of photons, N, that arrive at your detector in t seconds. If you count N things in a single trial, you can estimate that the average result of a single trial of length t seconds will be a count of $\mu \approx \bar{x} = N$ photons. How uncertain is this result? The uncertainty in a result can be judged by the standard deviation of the population from which the measurement is drawn. So, assuming Poisson statistics apply, the uncertainty of the measurement should be $\sigma_{\text{Poisson}} = \sqrt{\mu} \approx \sqrt{N}$. The **fractional uncertainty** is:

$$\text{Fractional uncertainty in counting } N \text{ events} = \frac{\sigma_{\text{Poisson}}}{\mu} \approx \frac{1}{\sqrt{N}} \qquad (2.15)$$

The uncertainty in a measurement is also called its noise. If the noise arises from a Poisson process, it is sometimes called **shot noise**. The reciprocal of the fractional uncertainty in a measurement is the **signal-to-noise ratio** or the **SNR**. In the Poisson case

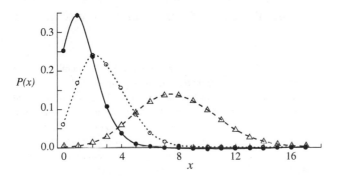

Fig. 2.4 The Poisson distribution for values of $\mu = 1.4$ (filled circles), $\mu = 2.8$ (open circles), and $\mu = 8.0$ (open triangles). Note that only the plotted symbols have meaning as probabilities. The curves merely assist the eye in distinguishing the three distributions.

$$SNR_{\text{Poisson}} = \frac{\mu}{\sigma} = \sqrt{N} \qquad (2.16)$$

Thus, to decrease the uncertainty (increase the SNR) in a measurement process dominated by photon shot noise, you should increase the number of photons you count (by increasing, for example, exposure time, number of exposures, or telescope size). To cut uncertainty in half, for example, increase the exposure time by a factor of 4.

Example: A certain space telescope/detector combination counts 10^5 photons in the V band on a 10 second exposure of a 10th magnitude star. An astronomer wants to measure the brightness of an asteroid with $m_V \approx 18.2$ to a precision of 2% (SNR = 50) with this system. Assuming that shot noise in the photons from the asteroid is the only source of noise, how long an exposure is required?

Answer: Let t = the required exposure time, and n = the average rate of photons arriving from the asteroid in photons/sec in the V band. From Equation (2.16)

$$\sqrt{nt} = 50 \Rightarrow t = \frac{2500}{n} \qquad (2.17)$$

Since we know the magnitude of the asteroid, we can compute n from the magnitude equation:

$$\frac{n}{10^5/10} = 10^{-0.4(18.2-10)} = 5.25 \times 10^{-4}$$
$$n = 5.25 \qquad (2.18)$$

So the required time is

$$t = 2500/5.25 = 476 \text{ s}$$

2.3.3 The Gaussian, or normal, distribution

The Gaussian,[2] or normal, distribution is the most important continuous distribution in the statistical analysis of data. Empirically, it seems to describe the

---

[2] Karl Friedrich Gauss (1777–1855) was a child prodigy who grew to dominate mathematics during his lifetime. He made several important contributions to geometry and number theory in his early 20s, after rediscovering many theorems because he did not have access to a good mathematics library. In January 1801 the astronomer Piazzi discovered Ceres, the first minor planet, but the object was soon lost. Gauss immediately applied a new method to only three recorded observations and computed the orbit of the lost object. The recovery of Ceres at his predicted positions led to fame and eventually a permanent position at Göttingen Observatory. At Göttingen, Gauss made important contributions to differential geometry and to many areas of physics, and was involved in the invention of the telegraph.

distribution of trials for a very large number of different experiments. Even in situations where the population itself is not described by a Gaussian (e.g. Figure 2.3), estimates of the summary statistics of the population (e.g. the mean) are described by a Gaussian.

If a population has a Gaussian distribution, then in a single trial the probability that x will have a value between x and $x + dx$ is

$$G(x, \mu, \sigma)dx = \frac{dx}{\sigma\sqrt{2\pi}} \exp\left[-\frac{1}{2}\left(\frac{x-\mu}{\sigma}\right)^2\right] \quad (2.19)$$

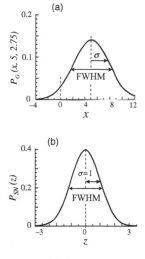

Fig. 2.5 (a) Gaussian distribution with a mean of 5 and a standard deviation of 2.75. (b) The standard normal distribution.

Figure 2.5a illustrates this distribution, a shape sometimes called a **bell curve**. In Equation (2.19), μ and σ are the mean and standard deviation of the distribution, and they are independent of one another (unlike the Poisson distribution). Sometimes we specify the **full width at half-maximum** (**FWHM**) of a Gaussian, that is, the separation in x between the two points where $G(x, \mu, \sigma) = \frac{1}{2}G(\mu, \mu, \sigma)$. The FWHM is proportional to σ

$$\text{FWHM}_{\text{Gaussian}} = 2.354\sigma \quad (2.20)$$

The dispersion of a distribution determines how a close single sample is likely to be from the population mean. One measure of dispersion, then, is the **probable error**, or P.E. By definition, a single trial has a 50% probability of lying closer to the mean than the P.E., that is $P(|x - \mu|) \le P.E.) = 1/2$. The P.E. for a Gaussian distribution is also directly proportional to σ

$$(\text{P.E.})_{\text{Gaussian}} = 0.6745\sigma = 0.2865(\text{FWHM}) \quad (2.21)$$

2.3.4 The standard normal distribution

$G(x, \mu, \sigma)$ is difficult to tabulate since its value depends not only on x, but also on the two parameters, μ and σ. This prompts us to define a new random variable:

$$z = \frac{x - \mu}{\sigma}$$
$$dz = \sigma^{-1}dx \quad (2.22)$$

Substitution into Equation (2.19) gives the **standard normal distribution**:

$$G_{SN}(z) = G(z, 0, 1) = \frac{1}{\sqrt{2\pi}}\exp\left[-\frac{z^2}{2}\right] \quad (2.23)$$

G_{SN} and its integral are tabulated in Appendix C. You can extract values for a Gaussian with a specific μ and σ from the table through Equations (2.22)

> **Example.** A population of adult freshwater crocodiles has a Gaussian distribution in mass, with a mean of 63 kg and a standard deviation of 22 kg. My mass is 67 kg. I drop randomly into a crocodile-infested lake. What is the probability that the nearest adult crocodile outweighs me?
>
> *Answer*: First, transform the mass into the variable z using (2.22). My mass corresponds to $z_o = (67-63)/22 = +0.182$. The probability that a random variable (i.e. the nearest adult) is greater than z_o is
>
> $$\int_{z_0}^{\infty} G_{SN}(z)\,\mathrm{d}z = 1 - \int_{-\infty}^{z_0} G_{SN}\,\mathrm{d}z = 1 - P_{SN}(z) \qquad (2.24)$$
>
> Where the function $P_{SN}(z)$, the cumulative probability, is tabulated in Appendix C. There we find, $P_{SN}(0.182) = .5714$ and therefore, the probability that the nearest adult crocodile will outweigh me is about $(1-0.57) = 43\%$.

2.3.5 Other distributions

Many other distributions describe populations in nature. We will not discuss these here, but only remind you of their existence. A uniform distribution, for example, describes a set of equally likely outcomes, like the number showing after the roll of a single die. Other distributions are important in elementary physics. The Maxwell–Boltzmann distribution, for example, describes the probability a molecule in a perfect gas will have energy between E and $E + \mathrm{d}E$. Several distributions (e.g. the t, χ^2, and F distributions) are useful in statistical hypothesis-testing. In astronomy, it is often the case that underlying distributions are unknown and need to be established empirically (e.g. the volume density of quasars as a function of distance).

2.4 Estimating uncertainty

We can now address the central issue of this chapter: How do you estimate the uncertainty of a particular quantitative measurement? You now recognize most measurements result from sampling the very large population of all possible measurements. You know that if a scientist samples a population by making n measurements, the sample mean is the best guess for the population mean. The question is: How good is this "best guess?" How close is the sample mean, $\bar{x}$, to the actual mean, μ, of the underlying population? What uncertainty should he attach to his measurement? And how can one estimate the uncertainly of a single measurement?

2.4.1 The central limit theorem

Return to the example of the very large population of metal spheres that have a distribution of diameters as illustrated by Figure 2.3a. This distribution is clearly not Gaussian. Nevertheless, properties of the Gaussian are relevant even for this distribution. Consider the problem of estimating the average size of a sphere. Suppose we ask Dora, our cheerful assistant, to conduct an experiment: Select five spheres *at random*, measure them, compute the average diameter, put them back. The result of such an experiment is a new random variable, $\bar{x}_5$, which is an estimate of the mean of the entire non-Gaussian population of spheres. Dora is a tireless worker. She does not stop with just five measurements but enthusiastically conducts many experiments, pulling out many spheres at random, five at a time, and tabulating many different values for $\bar{x}_5$. When we finally get her to stop measuring, Dora becomes curious about the distribution of her tabulated values. She plots the histograms shown in Figures 2.6a and 2.6b, the results for 100 and 800 (it is a very big urn) determinations of $\bar{x}_5$ respectively.

"Looks like a Gaussian," says Dora. "In fact, the more experiments I do, the more the distribution of $\bar{x}_5$ looks like a Gaussian. This is curious, because the actual distribution of diameters (the solid curve in Figure 2.6a) is not Gaussian."

Dora is correct. Suppose that $P(x)$ is the probability distribution for random variable x, where $P(x)$ is characterized by mean μ and variance σ^2, but otherwise can have any form whatsoever. In our example, $P(x)$ is the bimodal function plotted in Figure 2.3a. The **central limit theorem** states that if $\{x_1, x_2, \ldots, x_n\}$ is a sequence of n independent random variables drawn from P, then as n becomes large, the distribution of the new random variable

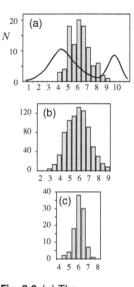

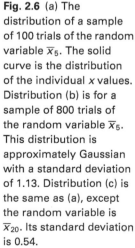

$$\bar{x}_n = \frac{1}{n} \sum_{i=1}^{n} x_i \qquad (2.25)$$

will approach a Gaussian distribution with mean μ and variance σ^2/n

To illustrate this last statement, Dora computes the values of a new random variable $\bar{x}_{20}$, which is the mean of 20 individual x s. The distribution of 100 $\bar{x}_{20}$s is shown in Figure 2.6c. As expected, the new distribution has about one-half the dispersion of the one for the 100 $\bar{x}_5$s.

Since so many measurements in science are averages of individual experiments, the central limit theorem means that the properties of the Gaussian distribution will be consequential in the analysis of experimental results. In addition, the conclusion that the variance of the average is proportional to $1/n$ relates directly to the problem of estimating uncertainty. Since s, the estimated standard deviation, is the best guess for σ, we should estimate $\sigma_\mu(n)$, the standard deviation of $\bar{x}_n$, the mean, as

Fig. 2.6 (a) The distribution of a sample of 100 trials of the random variable $\bar{x}_5$. The solid curve is the distribution of the individual x values. Distribution (b) is for a sample of 800 trials of the random variable $\bar{x}_5$. This distribution is approximately Gaussian with a standard deviation of 1.13. Distribution (c) is the same as (a), except the random variable is $\bar{x}_{20}$. Its standard deviation is 0.54.

$$\sigma_\mu(n) = \frac{s}{\sqrt{n}} = \frac{1}{\sqrt{n}} \left[\frac{1}{n-1} \sum_{i=1}^{n} (x_i - \mu)^2 \right]^{\frac{1}{2}} \tag{2.26}$$

Here, s is computed from the scatter in the n individual measurements. You will be careful to distinguish s, the standard deviation of the sample from $\sigma_\mu(n)$, the **standard deviation of the mean.** It is common to simply quote the value of $\sigma_\mu(n)$ as the uncertainty in a measurement. The interpretation of this uncertainty is clear because the central limit theorem implies that $\sigma_\mu(n)$ is the standard deviation of an approximately *Gaussian* distribution – one then knows, for example, that there is a 50% probability that $\bar{x}_n$ is within $0.6745\sigma_\mu(n)$ of the "true" value, μ.

2.4.2 Reducing uncertainty

The central limit theorem, which applies to all distributions, as well as the elementary properties of the Poisson distribution, which applies to counting experiments, both suggest that the way to reduce the uncertainty (and increase both precision and, we hope, accuracy) in any estimate of the population mean is *repetition*. Either increase the number of trials, or increase the number of things counted. If N is either the number of repetitions, *or* the number of things counted, then the basic rule is:

$$\text{relative uncertainty} \propto \frac{1}{\sqrt{N}} \tag{2.27}$$

Improving experimental precision means making N large. You can be precise if you have a large telescope (so you can collect many photons) for a long time (so you can make many measurements). But keep a number of very important cautions in mind while pondering the lesson of Equation (2.27).

- Improved precision is expensive. To decrease uncertainty by a factor of 100, for example, you have to increase the number of experiments (the amount of telescope time, or the area of its light-gathering element) by a factor of 10 000. At some point the cost becomes too high.
- Equation (2.27) only works for experiments or observations that are completely independent of one another and sample a stationary population. In real life, this need not be the case. For example, one measurement can have an influence on another by sensitizing or desensitizing a detector, or the brightness of an object can change with time. In such cases, the validity of Equation (2.27) is limited.
- Equation (2.26) only describes uncertainties introduced by scatter in the parent population. You should always treat this as the very minimum possible uncertainty. Systematic errors will make an additional contribution, and often dominate the uncertainty.

2.5 Propagation of uncertainty

Practical measurements are the combined result of several processes (the number of photons from the object, the efficiency of the detector, the brightness of the background, etc.), each having its associated uncertainty. Planning observations and evaluating measurements requires an understating of how these uncertainties combine.

2.5.1 Combining several variables

We consider first the special case where the quantity of interest is the sum or difference of more than one measured quantity. For example, in differential photometry, you are interested in a magnitude difference, $\Delta m = m_1 - m_2$. Here m_1 is the measured magnitude of an unknown object, and m_2 is the measured magnitude of a standard or comparison object in the same photometric system. The uncertainty in Δm depends on the uncertainties in both m_1 and m_2. If these uncertainties are known to be σ_1 and σ_2, then the uncertainty in Δm is given by

$$\sigma^2 = \sigma_1^2 + \sigma_2^2 \tag{2.28}$$

This could be could be stated as: "the variance of a sum (or difference) is the sum of the variances," or as "the uncertainties in a sum or difference add in quadrature." In the example given, Equation (2.28) certainly agrees with our intuition that the uncertainty of a magnitude difference will be larger than the uncertainty in either magnitude.

A second illustration of combining uncertainties concerns products or ratios of measured quantities. If, for example, one were interested in the ratio, R, between two fluxes, $F_1 \pm \sigma_1$ and $F_2 \pm \sigma_2$: $R = F_1/F_2$

$$\left(\frac{\sigma_R}{R}\right)^2 = \left(\frac{\sigma_1}{F_1}\right)^2 + \left(\frac{\sigma_2}{F_2}\right)^2 \tag{2.29}$$

Equations (2.28) and (2.29) hold only for random, uncorrelated errors. Many systematic errors cancel each other if a ratio or difference is computed. Astronomers, as you know, rely on such *differential measurements* as a way of reducing systematic errors.

2.5.2 General rule

In general, if a quantity, G, is a function of n variables, $G = G(x_1, x_2, x_3, \ldots, x_n)$, and each variable has uncertainty (or standard deviation) $\sigma_1, \sigma_2, \sigma_3, \ldots, \sigma_n$, then the variance in G is given by:

$$\sigma_G^2 \cong \sum_{i=1}^{n} \left(\frac{\partial G}{\partial x_i}\right)^2 \sigma_{x_i}^2 + \left[2 \sum_{j=2,\,k=1}^{j>k} \sigma_{jk}^2 \frac{\partial G}{\partial x_j} \frac{\partial G}{\partial x_k}\right] \qquad (2.30)$$

The term σ_{jk}^2 measures the strength of the correlation between the deviation of x_j and x_k. If these variables are independent, then this factor is zero and Equation (2.30) reduces to

$$\sigma_G^2 = \sum_{i=1}^{n} \left(\frac{\partial G}{\partial x_i}\right)^2 \sigma_{x_i}^2 \qquad (2.31)$$

You should be able to verify that Equations (2.28) and (2.29) follow from this expression.

2.5.3 Several measurements of a single variable

We can apply the result in Equation (2.31) to the following situation: Several astronomers measure the same quantity (the brightness of a star, say) and wish to combine their measurements but realize that some results are more reliable than others. Rather than just averaging the measurements, they compute a **weighted mean**:

$$\bar{y}_w = \left[\sum_{i=1}^{n} w_i\right]^{-1} \sum_{i=1}^{n} w_i y_i \qquad (2.32)$$

where $\{y_1, y_2 \ldots y_n\}$ are the measurements and $\{w_1, w_2 \ldots w_n\}$ are the weights of each, with higher weights assigned to more reliable measurements. One way to do this is to set

$$w_i = \frac{1}{\sigma_i^2} \qquad (2.33)$$

Application of Equation (2.31) then gives the uncertainty of the weighted mean:

$$\sigma_w^2 = \left[\sum_{i=1}^{n} \sigma_i^2\right]^{-1} \qquad (2.34)$$

Example: Suppose three different methods for determining the distance to the center of our Galaxy yield values 8.0 ± 0.3, 7.8 ± 0.7, and 8.25 ± 0.20 kiloparsecs. Combine these measurements to estimate of the distance, and its uncertainty.

Answer: Ignoring the quoted uncertainties (not a good idea), we could just average the three values and compute a value for s from their scatter: $\bar{y} = 7.98$, $\sigma_\mu \simeq s/\sqrt{3} = 0.19$. The proper approach would use the quoted uncertainties to assign weights 11.1, 2.0, and 25 to the three measurements, resulting in a weighted mean of

$$\bar{y}_w = \frac{[11.1(8) + 2(7.7) + 25(8.25)]}{11.1 + 2 + 25} = 8.15 \text{ kpc}$$

and a combined uncertainty of

$$\sigma_w = (11.1 + 2 + 25)^{-1/2} = 0.16 \text{ kpc}$$

Notice that the uncertainty of the combined result is less than the uncertainty of even the best of the individual results, less than the uncertainty in the unweighted mean, and that of the three measurements, the one with the very large uncertainty (7.8 ± 0.7 kpc) has little influence on the combined statistics.

Example. A detector counts photons by generating a voltage increase of $r \approx 5.0$ µV for every incident photon (e.g. an output of 60 microvolts means 12 photons were recorded during the exposure). Tests on a series of 10-second exposures indicate that, on this timescale, the value of r fluctuates. Fluctuations are normally distributed with a relative uncertainty of $\sigma_r/\mu_r = 0.05$ and a mean of $\mu_r = 5.00$ µV. An astronomer uses this device to measure the brightness of a star in a circular aperture (the aperture blocks all the light arriving at the detector except for a small circular patch on the sky), and records a signal of $V_1 = 832$ µV. He then moves the telescope to a nearby point and records a signal of $V_2 = 427$ µV. Both exposures are 10 seconds long. What is the best estimate for the number of photons arriving from the star alone, and what is its uncertainty?

Answer: The number of photons from the star follows from the difference between the two signals:

$$N_1 = \frac{V_1}{\mu_r} = \frac{832}{5} = 166 \text{ photons,}$$

$$N_2 = \frac{V_2}{\mu_r} = \frac{427}{5} = 85 \text{ photons}$$

$$N_* = \frac{\Delta V}{\mu_r} = \frac{405}{5} = 81 \text{ photons}$$

To compute the uncertainty in N_*, we note that there will be two sources of its fluctuation, the variation in r, and the variation in the number of photons that arrive

(continued)

in each aperture in 10 seconds. We can examine how these combine by writing down how we measure N_*,

$$N_* = \frac{\Delta V}{\mu_r} \tag{2.35}$$

Now apply Equation (2.31) – or Equation (2.29) – to the above expression to get σ_*^2, the variance of N_*:

$$\sigma_*^2 = \left(\frac{\sigma(\mu_r)}{\mu_r^2}\right)^2 (\Delta V)^2 + \frac{1}{r^2}\sigma^2(\Delta V) = \left(\frac{\sigma(\mu_r)}{\mu_r^2}\right)^2 (\mu_r N_*)^2 + \frac{1}{r^2}\sigma^2(V_1 - V_2) \tag{2.36}$$

The first term on the RHS represents the fact that μ_r – the value we use in converting voltage difference to photon number – might be uncertain. However, since we have measured r many times (how else would we know σ_r?), the uncertainty of its mean is very small, so we'll assume $\sigma(\mu_r) = 0$. The last term in Equation (2.36), the variance of the voltage difference, is not known directly, but we can make an estimate by another application of Equation (2.31)

$$\sigma^2(V_1 - V_2) = \sigma^2(V_1) + \sigma^2(V_2) = \sigma^2(rN_1) + \sigma^2(rN_2) \tag{2.37}$$

Above, we cannot write $\sigma^2(V_1 - V_2) = \sigma^2(r(N_1 - N_2))$ because r varies randomly between the two exposures. Again, we use Equation (2.31) to expand the RHS of (2.37)

$$\begin{aligned}\sigma^2(V_1 - V_2) &= \sigma_r^2 N_1^2 + r^2\sigma^2(N_1) + \sigma_r^2 N_2^2 + r^2\sigma^2(N_2) \\ &= \sigma_r^2(N_1^2 + N_2^2) + r^2(N_1 + N_2)\end{aligned} \tag{2.38}$$

Here we have made use of the fact that because the Ns follow the Poisson distribution, we know their variance: $\sigma^2(N) = N$. Substituting (2.38) into (2.36) and remembering that we assume $\sigma(\mu_r) = 0$:

$$\sigma_*^2 = N_1 + N_2 + \left(\frac{\sigma_r}{r}\right)^2 \left(N_1^2 + N_2^2\right) = N_* + 2N_2 + \left(\frac{\sigma_r}{r}\right)^2 \left(N_*^2 + 2N_2(N_* + N_2)\right) \tag{2.39}$$

and

$$\sigma_* = \left[1259 + (0.05)^2(692,224 + 182,329)\right]^{\frac{1}{2}} = 59 \text{ photons} \tag{2.40}$$

2.6 Additional topics

Statistics is a vast field of study. Several topics in elementary statistics are important in the analysis of data but are beyond the scope of this introduction. The **chi-square** (χ^2) statistic measures the deviation between experimental measurements and their theoretically expected values (e.g. from an assumed population distribution). Tests based on this statistic can assign a probability to the truth of the theoretical assumptions. **Least-square fitting** methods minimize the χ^2 statistic in the case of an assumed functional fit to experimental data (e.g. brightness as a function of time, color as a function of brightness . . .).

Hogg, Tanis, and Zimmerman (2013) provide a good introduction to the very broad field of mathematical statistics. Bevington (1969) and Lyons (1991), provide more compact approaches directed at students in the physical sciences.

Wall and Jenkins (2012) give a more advanced introduction to astronomy-specific statistics and to the emerging field of **astrostatistics.** Astrostatistics addresses a vast variety of tasks in extracting information from very large data sets. With astronomical surveys now collecting millions or billions of objects, we need powerful statistical methods to decide, for example, if the objects cataloged should be divided into two or three or twenty different classes, or to decide which, if any, objects have a statistically "unusual" properties.

Summary

Precision, but not **accuracy**, can be estimated from the scatter in measurements.
Standard deviation is the square root of the **variance**. For a **population**:

$$\sigma^2 = \frac{1}{M}\sum_{i=1}^{M}(x_i - \mu)^2 = \frac{1}{M}\sum_{i=1}^{M}x_i^2 - \mu^2$$

For a **sample**, the best estimate of the population variance is

$$s^2 = \frac{1}{N-1}\sum_{i=1}^{N}(x_i - \bar{x})^2$$

Probability distributions describe the expected values of a random variable drawn from a parent population. The **Poisson distribution** describes measurements made by counting uncorrelated events like the arrival of photons. For measurements following the Poisson distribution,

$$\sigma_{\text{Poisson}}^2 = \mu_{\text{Poisson}}$$

The **Gaussian distribution** describes many populations whose values have a smooth and symmetric distribution. The **central limit theorem** contends that the mean of n random samples drawn from a population of mean μ and variance σ^2 will take on

(*continued*)

Summary (*cont.*)

a Gaussian distribution in the limit of large n whose mean approaches μ and whose variance approaches σ^2/n. For any distribution, the uncertainty (standard deviation) of the mean of n measurements of the variable x approaches

$$\sigma_\mu(n) = \frac{s}{\sqrt{n}}$$

The variance of a function of several uncorrelated variables, each with its own variance, is given by

$$\sigma_G^2 = \sum_{i=1}^{n} \left(\frac{\partial G}{\partial x_i}\right)^2 \sigma_{x_i}^2 + \text{covar}$$

For measurements of unequal variance, the **weighted mean** is

$$y_c = \sigma_c^2 \sum_{i=1}^{n} \left(y_i/y_i\sigma_i^2\right)$$

and the combined variance is

$$1/\sigma_c^2 = \sum_{i=1}^{n} \left(1/\sigma_i^2\right)$$

Exercises

1. There are some situations in which it is impossible to compute the mean value for a set of data. Consider this example. The ten crew members of the starship *Nostromo* are all exposed to an alien virus at the same time. The virus causes the deaths of nine of the crew at the following times, in days, after exposure:

$$1.2, 1.8, 2.1, 2.4, 2.6, 2.9, 3.3, 4.0, 5.4$$

The tenth crew member is still alive after 9 days, but is infected with the virus. Based only on these data:

(a) Why can't you compute the "average survival time" for victims of the *Nostromo* virus?

(b) What is the "expected survival time"? (A victim has a 50–50 chance of surviving this long.) Justify your computation of this number.

2. An experimenter makes 11 measurements of a physical quantity, X, that can only take on integer values. The measurements are

$$0, 1, 2, 3, 4, 5, 6, 7, 8, 9, 10$$

(a) Estimate the mean, median, variance (treating the set as a sample of a population) and standard deviation of this set of measurements.

(b) The same experimenter makes a new set of 25 measurements of X, and finds that the values

$$0, 1, 2, 3, 4, 5, 6, 7, 8, 9, 10$$

occur

$$0, 1, 2, 3, 4, 5, 4, 3, 2, 1, \text{and } 0$$

times respectively. Again, estimate the mean, median, variance, and standard deviation of this set of measurements.

3. Describe your best guess as to the parent distributions of the samples given in questions 2(a) and 2(b).

4. The rate of impacts by meteorites capable of penetrating the roof of a standard-issue human habitation on the planet Gonforgood is 170 per century per square kilometer. (a) What distribution do you think governs the probability of these impacts? (b) Compute the probability that a 100-square-meter roof on Gonforgood will sustain no meteorite penetration in 1 year. (c) Compute the probability that it will sustain more than one penetration in 40 years.

5. A power law distribution describes many results in observational astronomy. In this distribution, $N(a)da$ is the number of objects that have a certain property with value between a and $a + da$, and $N(a) = N_0 a^{-\gamma}$. For example, the size distribution of meteoroids follows a power law (small space rocks are more common than large space) rocks, as does the apparent magnitudes of stars on an arbitrarily deep image (at every magnitude, there are more faint stars than bright stars). Explain why a population that follows a power law can have no useful mean or standard deviation. Would you feel differently if the distribution had physically meaningful cutoffs ($N = 0$ outside a certain range – e.g. no meteoroids smaller than a molecule or larger than Ceres)?

6. Define the variable $q = \log_{10}(M/M_\oplus)$, where M is the mass of a planet and $M_\oplus$ is the mass of the Earth. Assume an "Earthlike" planet falls in the range $-0.1 < q < 0.3$. A certain method for discovering exoplanets yields a distribution of q for discovered objects that is a Gaussian with mean 0.8 and standard deviation 0.6. Compute how many Earthlike exoplanets you would expect in the next 500 discovered by this method.

7. An astronomer wishes to make a photon-counting measurement of a star's brightness that has a relative precision of 5%. (a) How many photons should she count? (b) How many should she count for a relative precision of 0.5%?

8. The astronomer in Problem 7 discovers that when she points her telescope to the blank sky near the star she is interested in, she measures a background count that is 50% of what she measures when she points to the star. She reasons that the brightness of the star (the interesting quantity) is given by

$$star = measurement - background$$

Revise your earlier estimates. How many measurement photons should she count to achieve a relative precision of 5% in her determination of the star brightness? How many for 0.5%?

9. A star cluster is a collection of gravitationally bound stars. Individual stars in the cluster move about with different velocities but the average of all these should give the velocity of the cluster as a whole. An astronomer measures the radial velocities of four stars in a cluster that contains 1000 stars. They are 74, 41, 61, and 57 km s^{-1}. How any additional stars should he measure if he wishes to achieve a precision of 2 km s^{-1} for the radial velocity of the cluster as a whole?

10. The second example in Section 2.5.3 computed a SNR, $n_*/\sigma_* = 1.37$, for a pair of 10-second exposures. Repeat the computation for a pair of 100-second exposures. Explain why you change any of the values you employ.

11. An astronomer makes five one-second measurements of a star's brightness, counting 4, 81, 9, 7, and 5 photons in these trials. What is the best estimate of the average photon arrival rate and its uncertainty? Is there any reason to believe that the second measurement is not drawn from the same population as the other 4? Can you suggest a way to compute the likelihood that measurement 2 is aberrant?

12. We repeat Problem 14 from Chapter 1, where a single unknown star and standard star are observed in the same field. The data frame is in the figure below. The unknown star is the fainter one. If the magnitude of the standard is 9.000, compute the magnitude of the unknown, as in problem 1.14, but now also compute the uncertainty of your result, in magnitudes. Again data numbers represent the number of photons counted in each pixel.

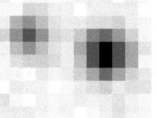

34	16	26	33	37	22	25	25	29	19	28	25
22	20	44	34	22	26	14	30	30	20	19	17
31	70	98	66	37	25	35	36	39	39	23	20
34	99	229	107	38	28	46	102	159	93	37	22
33	67	103	67	36	32	69	240	393	248	69	30
22	33	34	29	36	24	65	241	363	244	68	24
28	22	17	16	32	24	46	85	157	84	42	22
18	25	27	26	17	18	30	29	35	24	30	27
32	23	16	29	25	24	30	28	20	35	22	23
28	28	28	24	26	26	17	19	30	35	30	26

Chapter 3
Place, time, and motion

Then, just for a minute . . . he turned off the lights. . . . And then while we all still
waited I understood that the terror of my dream was not about losing just vision,
but the whole of myself, whatever that was. What you lose in blindness is the
space around you, the place where you are, and without that you might not exist.
You could be nowhere at all.

– Barbara Kingsolver, *Animal Dreams*, 1990

Where is Mars? The center of our Galaxy? The brightest X-ray source? Where,
indeed, are we? Astronomers have always needed to locate objects and events in
space. As our science evolves, it demands ever more exact locations. For
example, an astronomer discovers, with an X-ray telescope, a source that flashes
on and off with a curious rhythm. Is this source a planet, a star, or the core of a
galaxy? The exact position for the X-ray source might be the only way to
identify its otherwise unremarkable optical or radio counterpart. Astronomers
need to know where things are.

Likewise, knowing *when* something happens is often as important as *where* it
happens. The rhythms of the spinning and orbiting Earth and Moon gave astronomy an early and intimate connection to timekeeping. Because our universe has a
history, astronomers need to know what happened when and what time it is now.

The "fixed stars" are an old metaphor for the unchanging and eternal, but
positions of real celestial objects do change. Planets, stars, gas clouds, and galaxies
all trace paths decreed for them. Astronomers who measure these motions, sometimes only through the accumulated labors of many generations, can sometimes
find in their measurements the outlines of nature's decree. In the most satisfying
cases, motions reveal fundamental facts, like the age of the universe or the presence
of planets orbiting other suns. Astronomers need to know how things move.

3.1 Astronomical coordinate systems

Any problem of geometry can easily be reduced to such terms that acknowledge
of the lengths of certain straight lines is sufficient for its construction.

– René Descartes, *La Geometrie*, Book I, 1637

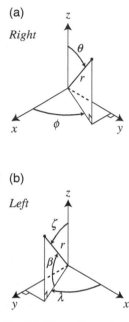

(a)

Right

(b)

Left

Fig. 3.1 Three-dimensional coordinate systems. (a) The traditional system is right-handed. (b) This system is left-handed; its axes are a mirror image of those in (a). In either system one can choose to measure the second angle from the fundamental plane (e.g. angle β) instead of from the z-axis (angles θ or ζ).

Descartes' brilliant application of coordinate systems to solve geometric problems has direct relevance to astrometry, the business of locating astronomical objects. Astrometry has venerably ancient origins,[1] and it retains a central importance in astronomy.

3.1.1 Three-dimensional coordinates

I assume you are familiar with the standard (x, y, z) Cartesian coordinate system and the related spherical coordinate system (r, ϕ, θ), illustrated in Figure 3.1. Think for a moment how you might set up such a coordinate system in practice. Many methods could lead to the same result, but consider a process that consists of four decisions:

1. Locate the origin. In astronomy, this often corresponds to identifying some distinctive real or idealized object: the centers of the Earth, Sun, or Galaxy, for example.
2. Locate the x–y plane. We will call this the "fundamental plane." The fundamental plane, again, often has physical significance: the plane defined by the Earth's equator – or the one that contains Earth's orbit – or the symmetry plane of the Galaxy, for example. The z-axis passes through the origin perpendicular to the fundamental plane.
3. Decide on the direction of the positive x-axis. We will call this the "reference direction." Sometimes the reference direction has a physical significance – the direction from the Sun to the center of the Galaxy, for example. The y-axis then lies in the fundamental plane, perpendicular to the x-axis.
4. Finally, decide on a convention for the signs of the y- and z-axes. These choices produce either a left- or right-handed system – see below.

The traditional choice for measuring the *angles* is to measure the first coordinate, ϕ (or λ), within the fundamental plane so that ϕ increases from the $+x$-axis toward the $+y$-axis. See Figure 3.1. The second angle, θ (or ζ), is measured in a plane perpendicular to the fundamental plane increasing from the positive z-axis toward the x–y plane. In this scheme, ϕ ranges, in radians, from 0 to 2π and θ ranges from 0 to π. A common alternative is to measure the second angle (β in the figure) from the x–y plane, so it ranges between $-\pi/2$ and $+\pi/2$.

The freedom to choose the signs of the y- and z-axes in step 4 of this procedure implies that there are two (and only two) kinds of coordinate systems. One, illustrated in Figure 3.1a, is **right-handed**: if you wrap the fingers of your right hand around the z-axis so the tips point in the $+\phi$ direction (that is, from the $+x$-axis toward the $+y$-axis), then your thumb will point in the $+z$-direction. In a **left-handed** system, like the (r, λ, ζ) system illustrated in Figure 3.1b, you use

---

[1] Surviving Babylonian records go back to about 650 BCE but contain copies of observations made at least 1000 years earlier as well as strong hints that the written tradition had Sumerian roots in the late third millennium. Ruins of megalithic structures with clear astronomical alignments date from as early as 8000 BCE (Warren Field, Scotland).

your left hand to find the $+z$-direction. The left-handed system is the mirror image of the right-handed system.

3.1.2 Coordinates on a spherical surface

> It is one of the things proper to geography to assume that the earth as a whole is spherical in shape, as the universe also is ...
>
> — Strabo, *Geography*, II, 2,1, *c*. 18 CE

If all points of interest are on the surface of a sphere, the r-coordinate is superfluous, and we can specify locations with just two angular coordinates like (ϕ, θ) or (λ, β). Many astronomical coordinate systems fit into this category, so it is useful to review some of the characteristics of geometry and trigonometry on a spherical surface.

1. A ***great circle*** is formed by the intersection of the sphere and a plane that contains the center of the sphere. The shortest distance between two points on the surface of a sphere is an arc of the great circle connecting the points.
2. A ***small circle*** is formed by the intersection of the sphere and a plane that does not contain the center of the sphere.
3. The ***spherical angle*** between two great circles is the angle between the planes, or the angle between the straight lines tangent to the two great circle arcs at either of their points of intersection.
4. A ***spherical triangle*** on the surface of a sphere is one whose sides are all segments of great circles. Since the sides of a spherical triangle are arcs, the sides can be measured in angular measure (i.e. radians or degrees) rather than linear measure. See Figure 3.2.
5. The ***law of cosines*** for spherical triangles in Figure 3.2 is:

$$\cos a = \cos b \cos c + \sin b \sin c \cos A \qquad (3.1)$$

or

$$\cos A = \cos B \cos C + \sin B \sin C \cos a \qquad (3.2)$$

6. The ***law of sines*** is

$$\frac{\sin a}{\sin A} = \frac{\sin b}{\sin B} = \frac{\sin c}{\sin C} \qquad (3.3)$$

3.1.3 Terrestrial latitude and longitude

> "I must be getting somewhere near the center of the Earth ... yes ... but then I wonder what Latitude and Longitude I've got to?" (Alice had not the slightest idea what Latitude was, nor Longitude either, but she thought they were nice grand words to say.)
>
> — Lewis Carroll, *Alice's Adventures in Wonderland*, 1865

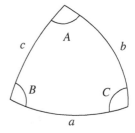

Fig. 3.2 A spherical triangle. You must imagine this figure is drawn on the surface of a sphere. *A*, *B*, and *C* are spherical angles; *a*, *b*, and *c* are arcs of great circles.

Fig. 3.3 The latitude–longitude system. The center of coordinates is at O. The fundamental direction, line OX, is defined by the intersection of the prime meridian (great circle NGX) and the equator. Latitude, β, and longitude, λ, for some point, P, are measured as shown. Latitude is positive north of the equator, negative south. Astronomical longitude for Solar System bodies is positive in the direction opposite the planet's spin. (i.e. to the west on Earth). On Earth, coordinates traditionally carry no algebraic sign, but are designated as north or south latitude, and west or east longitude. The coordinate, β, is the geocentric latitude.

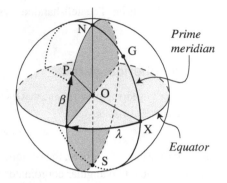

Ancient geographers introduced the seine-like latitude–longitude system for specifying locations on Earth well before the time Hipparchus of Rhodes (*c.* 190–120 BCE) wrote on geography. Figure 3.3 illustrates the basic features of the system.

In our scheme, the first steps in setting up a coordinate system are to choose an origin and fundamental plane. We can understand why Hipparchus, who believed in a geocentric cosmology, would choose the center of the Earth as the origin. Likewise, choice of the equatorial plane of the Earth as the fundamental plane makes a lot of practical sense. Although the location of the equator may not be obvious to a casual observer like Alice, it is easily determined from simple astronomical observations. Indeed, in his three-volume book on geography, Eratosthenes of Alexandria (*c.* 275 – *c.* 194 BCE) is said to have computed the location of the equator relative to the parts of the world known to him. At the time, there was considerable dispute as to the habitability of the (possibly too hot) regions near the equator, but Eratosthenes clearly had little doubt about their location.

Great circles perpendicular to the equator must pass through both poles, and such circles are termed **meridians**. The place where one of these – the **prime meridian** – intersects the equator could constitute a reference direction (*x*-axis). On Earth, there is no obvious prime meridian; for a long time, geographers simply chose a prime meridian that passed through some locally prominent or worthy place. Thus, the latitude of any point on Earth was unique, but its longitude was not, since it depended on which meridian one chose as prime. This was inconvenient. Eventually, in 1884, the "international" community (in the form of representatives of 25 industrialized countries meeting in Washington, DC, at the First International Meridian Conference) settled the zero point of longitude at the meridian of the Royal Observatory in Greenwich, located just outside London, England.

You should note that the latitude coordinate, β, just discussed, is called the **geocentric latitude**, to distinguish it from ϕ, the **geodetic latitude**. Geodetic latitude is defined in reference to an ellipsoid-of-revolution that approximates the actual shape of the Earth. It is the angle between the equatorial plane and a line perpendicular to the surface of the reference ellipsoid at the point in question.

Figure 3.4 shows the north pole, N, equator, E, and center, O, of the Earth. The geocentric and geodetic latitudes of point P are β and ϕ, respectively. Geodetic latitude is easier to determine and is the one employed in specifying positions on the Earth. The global positioning satellites (GPS) system, for example, returns geodetic latitude, longitude, and height above a reference ellipsoid. To complicate things a bit more, the most easily determined latitude is the **geographic latitude**, the angle between the local vertical and the equator. Massive objects like mountains affect the geographic but not the geodetic latitude and the two can differ by as much as an arc minute. Further complications on the sub-arc-second scale arise from short- and long-term motion of the geodetic pole itself relative to the Earth's crust due to tides, earthquakes, internal motions, and continental drift.

Planetary scientists establish latitude–longitude systems on other planets, with latitude usually easily defined by the object's rotation, while definition of longitude depends on identifying some feature to mark a prime meridian.

Which of the two poles of a spinning object is the "north" pole? In the Solar System, the IAU convention is that the "positive" pole is determined by a right-hand rule applied to the direction of spin: wrap the fingers of your right hand around the object's equator so that they point in the direction of its spin. Your thumb then points to the positive pole. However, a special rule (for historical reasons) applies to the eight planets and their larger satellites. The **ecliptic** – the plane containing the Earth's orbit – defines a fundamental plane, and a planet's "geographic north" pole is the one that lies to the (terrestrial) north side of this plane. Thus, on Venus, for example, the positive rotational pole coincides with the geographic south pole.

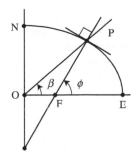

Fig. 3.4 Geocentric (β) and geodetic (ϕ) latitudes. Line PF is perpendicular to the surface of the reference spheroid, and approximately in the direction of the local vertical (local gravitational force).

3.1.4 The altitude–azimuth system

Imagine an observer, a shepherd with a tranquil flock, say, who has some leisure time on the job. Our shepherd is lying in an open field, contemplating the night-time sky. After a little consideration, our observer comes to imagine the sky as a hemisphere – an inverted bowl whose edges rest on the horizon. The observer sees astronomical objects, whatever their real distances, to be stuck onto or projected onto the inside of this bowl.

In Figure 3.5, we carry the shepherd's fiction of a hemispherical sky a bit further, and imagine that the bowl above is matched by a similar (but invisible) hemisphere below the horizon. The shepherd will naturally take himself to define the origin of a spherical coordinate system, and will find it hard to resist using the (apparently) flat Earth as the fundamental plane. This is another

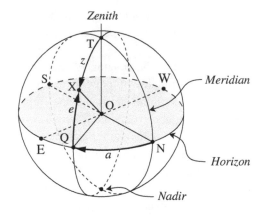

Fig. 3.5 The altitude–azimuth system. The horizon defines the fundamental plane (gray) and the north point on the horizon, N, defines the fundamental direction. Point P has coordinates *a* (azimuth), which is measured along the horizon circle from north to east, and *e* (altitude), measured upwards from the horizon. Objects with negative altitudes are below the horizon.

situation in which the *r*-coordinate becomes superfluous. At night, he'll know the direction of a star but not its distance from the origin. Astronomers face the same issue: the directions of objects are easily and accurately determined, but their distances are not. This inspires coordinate systems that ignore the *r*-coordinate and only specify the two direction angles.

Astronomers and a few shepherds use the spherical coordinate scheme illustrated in Figure 3.5. Here, the origin of the system is at O, the location of the observer. The fundamental plane is tangent to the tiny spherical Earth at point O. This fundamental plane intersects the sphere of the sky at the ***celestial horizon*** – the great circle passing through the points NES in the figure. ***Vertical circles*** are great circles on the spherical sky (the ***celestial sphere***) that are perpendicular to the fundamental plane. All vertical circles pass through the overhead point, which is called the ***zenith*** (point T in the figure), as well as the diametrically opposed point, called the ***nadir***. The vertical circle that runs in the north–south direction (circle NTS in the figure) is called the ***observer's meridian***.

The fundamental direction in the altitude–azimuth coordinate system runs directly north from the observer to the horizon (to point N in the figure). In this system, a point on the sky, X, has two coordinates:

- The **altitude**, or **elevation**, is the angular distance of X above the horizon (∠QOX or *e* in the figure). Objects below the horizon have negative altitudes.
- The **azimuth** is the angular distance from the reference direction (the north point on the horizon) to the intersection of the horizon and the vertical circle passing through the object (∠NOQ or *a* in the figure).

Instead of the altitude, astronomers sometimes use its complement, *z*, the ***zenith distance*** (∠TOX in the figure).

The (*a*, *e*) coordinates of an object clearly locate it in an observer's sky. You can readily imagine an instrument to measure these coordinates: a telescope or other sighting device mounted to rotate on vertical and horizontal circles that are marked with precise gradations.

One of the most elementary astronomical observations, noticed even by the most unobservant shepherd, is that celestial objects don't stay in the same place in the horizon coordinate system. Stars, planets, the Sun, and Moon all execute a ***diurnal motion***: they rise in the east, cross the observer's meridian, and set in the west. This, of course, is a reflection of the spin of our planet on its axis. Careful measurement will show that stars (but not the Sun and planets, which move relative to the "fixed" stars) will take about 23 hours, 56 minutes, and 4.1 seconds between successive meridian crossings. This period of time is known as one ***sidereal day***. Very careful observations would show that the sidereal day is actually getting longer, relative to a stable atomic clock, by about 0.0015 second per century. The spin rate of the Earth is slowing down.

3.1.5 The equatorial system: definition of coordinates

Because the altitude and azimuth of celestial objects change rapidly, we create another reference system, one in which the coordinates of stars do not change. In this ***equatorial coordinate system***, we further refine the fiction that all celestial objects are on a sphere centered at the center of the Earth, a planet that is insignificantly small compared to this ***celestial sphere***. From a geocentric point of view, we can account for the diurnal motion of celestial objects by presuming that the entire celestial sphere spins east to west on an axis coincident with the Earth's actual spin axis. Relative to one another, objects on the sphere never change their positions (not quite true – see below). The star patterns that make up the figures of the constellations stay put, while terrestrials observe the entire sky – the global pattern of constellations – to spin around its north–south axis once each sidereal day. Objects stuck on the celestial sphere thus appear to move east to west across the terrestrial sky, traveling in small circles centered on the nearest celestial pole.

The fictional celestial sphere is an example of a scientific model. Although the model is not the same as the reality, it has features that help one discuss, predict, and understand real behavior. (You might want to think about the meaning of the word "understand" in a situation where model and reality differ so extensively.) The celestial-sphere model allows us to specify the positions of the stars in a coordinate system, the equatorial system, which is independent of time, at least on short scales. Because positions in the equatorial coordinate system are also easy to measure from Earth, it is the system astronomers use most widely to locate objects on the sky.

The equatorial system chooses the center of the Earth as origin and the plane containing its equator as the fundamental plane. This aligns the z-axis with the Earth's spin axis, and fixes the locations of the two ***celestial poles*** at the intersections of the z-axis and the celestial sphere. The great circle defined by the intersection of the fundamental plane and the celestial sphere is called the

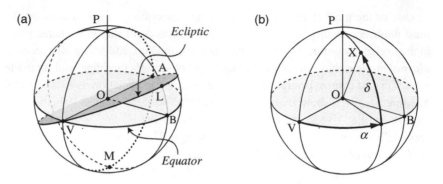

Fig. 3.6 The equatorial coordinate system. In both celestial spheres pictured, the equator is the great circle passing through points V and B, and the ecliptic is the great circle passing through points V and L. (a) shows the locations of the north (P) and south (M) celestial poles, the vernal (V) and autumnal (A) equinoxes, the summer solstice (L), and the hour circles for 0 hr (arc PVM) and 6 hr (arc PLBM) of right ascension. (b) shows the right ascension (∠VOQ, or α) and declination (∠QOP, or δ) of the point X.

celestial equator. The latitude-like coordinate measured with respect to the celestial equator is called the ***declination*** (abbreviated as Dec or δ), whose value is taken to be zero at the equator, and positive in the northern celestial hemisphere. See Figure 3.6. Circles of constant declination are called ***declination circles***.

We choose the fundamental direction in the equatorial system by observing the apparent motion of the Sun relative to the background of "fixed" stars. Because of the Earth's orbital motion, the Sun traces out a great circle on the celestial sphere in the course of a year. This circle is called the ***ecliptic*** (it is where eclipses happen) and intersects the celestial equator at an angle, ε (∠LOB in Figure 3.6a) called the ***obliquity of the ecliptic***, equal to about 23.5 degrees. The point where the Sun crosses the equator traveling from south to north is called the ***vernal equinox*** and this point marks the reference direction of the equatorial system. The coordinate angle measured in the equatorial plane is called the ***right ascension*** (abbreviated as RA or α). As shown in Figure 3.6b, the equatorial system is right-handed, with RA increasing from west to east.

For reasons that will be apparent shortly, RA is usually measured in hours: minutes:seconds rather than in degrees (24 hours of RA constitute 360 degrees of arc at the equator, so one hour of RA is 15 degrees of arc long at the equator). To deal with the confusion that arises from both the units of RA and the units of Dec having the names "minutes" and "seconds," one can speak of "minutes (or seconds) *of time*" for RA and "minutes of arc" for Dec. A line of constant RA forms half of a great circle and is called an ***hour circle***.

3.1.6 The relation between the equatorial and the horizon systems

Figure 3.8 shows the celestial sphere with some of the features of the horizon and equatorial systems superimposed. The figure assumes an observer, O, located at about 60 degrees north latitude on Earth. Note the altitude of the north celestial pole (∠NOP in Figure 3.8a). You should be able to construct (see Figure 3.7) a simple geometric argument to convince yourself that:

> *The altitude angle of the north celestial pole*
> *equals the observer's geodetic latitude.*

Observer O, motionless in the horizon system, will watch the celestial sphere turn, and see stars move along circles of constant declination. Figure 3.8a shows the declination circle of a star (*nr*) that just touches the northern horizon. Stars north of this circle (like star *cp*) never set and are termed **circumpolar**. Figure 3.8a also shows the declination circle that just touches the southern horizon circle, and otherwise lies entirely below it. Unless she changes her latitude, O can never see any of the stars south of this declination circle.

Stars that are neither circumpolar nor permanently below the horizon will rise in the east, cross, or **transit**, the observer's celestial meridian, and set in the west. When a star transits the meridian it has reached its greatest altitude above the horizon, and is said to have reached its **culmination**. Notice in Figure 3.8 that circumpolar stars cross the meridian twice each sidereal day – once when highest in the sky, and again when lowest. To avoid confusion, the observer's celestial meridian is divided into two pieces at the pole. The smaller bit visible between the pole and the horizon is called the lower meridian, and the remaining piece (arc PTMS) is called the **upper meridian**.

Figure 3.8b shows a star, X, which has crossed the upper meridian some time ago and is moving to set in the west. Its hour circle, arc PXR, is shown in the figure.

You can specify how far an object is from the meridian by giving its **hour angle**. The hour circle of an object and the upper celestial meridian intersect at the pole. The hour angle, HA, is the angle between them (∠ MPR). Application of the law of sines to a spherical right triangle shows that the hour angle could also be measured along the equator, as the arc that runs from the intersection of the meridian and equator to the intersection of the star's hour circle and the equator (arc RM). Hour angle, like right ascension, is usually measured in time units. Recalling the definition of RA we can state an alternative definition of the hour angle:

$$HA \ of \ the \ object = RA \ on \ meridian - RA \ of \ the \ object$$

The hour angle of a star tells how long ago (in the case of positive HA) or how long until (negative HA) the star crossed, or will cross, the upper meridian. The best time to observe an object is usually when it is highest in the sky, that is, when the HA is zero and the object is at culmination.

To compute the hour angle from the formula above, you realize that the RA of the object is always known – you can look it up in a catalog or read it from a star chart. How do you know the right ascension of objects on the meridian? You read that from a **sidereal clock**.

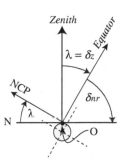

Fig. 3.7 Directions in the plane of the meridian. The observer is at positive latitude λ on the Earth, whose spin axis defines the direction of the north celestial pole (NCP) and the celestial equator. Definition of latitude ensures that the elevation angle of the NCP equals the latitude. Likewise, the declination of the zenith δ_z and the declination of the southernmost visible star δ_{nr} must be λ and $\lambda - 90$, respectively.

Fig. 3.8 The horizon and equatorial systems. Both spheres show the horizon, equator and observer's meridian, the north celestial pole at P, and the zenith at T. (a) illustrates the diurnal paths of two circumpolar stars and of a star that never rises. (b) shows the hour circle (PXR) of a star at X, as well as its declination, δ, its hour angle, HA = arc RM = ∠MPS, its altitude, e, its zenith distance, z, and its azimuth, arc NSQ.

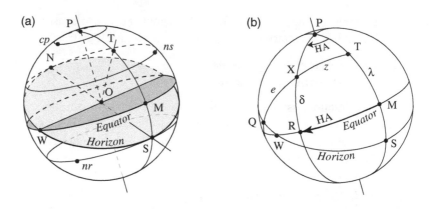

A clockmaker creates a clock that ticks off exactly 24 uniform "sidereal" hours between successive upper meridian transits by the vernal equinox (a period of about 23.93 "normal" hours, remember). If one adjusts this clock so that it reads zero hours at precisely the moment the vernal equinox transits, then it gives the correct **sidereal time**.

$$\text{Sidereal day} = \text{Time between upper meridian transits} \\ \text{by the vernal equinox}$$

A sidereal clock mimics the sky, where the hour circle of the vernal equinox can represent the single hand of a 24-hour clock, and the observer's meridian can represent the "zero hour" mark on the clockface. There is a nice correspondence between the reading of any sidereal clock and the right ascension coordinate, namely

$$\text{sidereal time} = \text{right ascension of an object on the upper meridian}$$

Clearly, we can restate the definition of hour angle as:

$$\text{HA of object} = \text{sidereal time now} - \text{sidereal time object culminates}$$

If either the sidereal time or an object's hour angle is known, one can derive the coordinate transformations between equatorial (α, δ) and the horizon (e, a) coordinates for that object. Formulas are given in Appendix D.

3.1.7 Measuring equatorial coordinates

Astronomers use the equatorial system because RA and Dec are easily determined with great precision from Earth-based observatories. You should have a general idea of how this is done. Consider a specialized instrument, called a **transit telescope** (or **meridian circle**): the transit telescope is constrained to point only at objects on an observer's celestial meridian – it rotates on an axis aligned precisely east–west. The telescope is rigidly attached to a graduated circle centered on this axis. The circle lies in the plane of the meridian and rotates with the telescope. A fixed index, established using a plumb line perhaps, always points to the zenith.

By observing where this index falls on the circle, the observer can thus determine the altitude angle at which the telescope is pointing. The observer is also equipped with a sidereal clock, synchronized to upper transits of the vernal equinox.

To use the transit telescope to determine declinations, first locate the celestial pole. Pick out a circumpolar star. Read the graduated circle when you observe the star cross the upper and then again when it crosses the lower meridian. The average of the two readings gives the location of $\pm 90°$ declination (the north or south celestial pole) on your circle. After this calibration you can then read the declination of any other transiting star directly from the circle.

To find the *difference* between the *RAs* of any two objects, note the sidereal times when each transits, and subtract them. To set the zero point for the RA coordinate, require the right ascension of the Sun to be zero when you observe its declination to be zero in the spring.

Astrometry is the branch of astronomy concerned with measuring celestial positions and changes in position. Chapter 11 of Birney et al. (2006) gives a more thorough introduction to the subject than we will here, and Van Altena (2013) gives a more advanced discussion. The *Gaia* website gives a good introduction to astrometry from space.

Observations with a transit telescope can measure arbitrarily large angles between sources, and the limits to the accuracy of *large-angle, or fundamental astrometry* are different from, and usually much more severe than, the limits in small-angle astrometry. In *small-angle astrometry*, one measures positions of sources relative to a one another in the same detector field. For example, measuring the positions of a minor planet relative to the background stars in two successive images of the same field is a task in small-angle astrometry.

The angular size and regularity of the stellar images formed by the transit telescope limit the precision of large-angle astrometry. The astronomer or her computer (modern transit observations are automated) must decide when and where the center of the image transits, a task made difficult if the image is faint, diffuse, irregular, or changing shape on a short timescale. In the optical or near infrared, atmospheric turbulence usually limits ground-based position measurements to an accuracy of about 0.05 arcsec, or 50 milliarcsec (mas).

Positional accuracy at radio wavelengths can be much greater. The technique of *very long baseline interferometry* (**VLBI**) utilizes the wave properties of light to measure positions. Figure 3.9 sketches the basic principle, a Michelson interferometer, which depends on detecting light from one source with two different telescopes. In the diagram, a radio wavefront from direction Y arrives at Telescope B, but must travel an additional time

$$\tau = \frac{b \sin \theta}{c} = \frac{n\lambda}{c} \tag{3.4}$$

before arriving at Telescope A. Measuring the position of the source, θ (or $\sin\theta$) is thus the equivalent of measuring the time delay between the arrivals of the same wavefront at the two telescopes. This can be done, for example, by

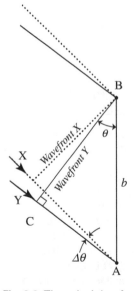

Fig. 3.9 The principle of the Michelson interferometer. Measuring the direction of signal Y depends on sensing the transit time for path CA. Signal X, from a slightly different direction, will be indistinguishable from Y if $\Delta\theta$ is less than $\sigma(\theta) = c\sigma(\tau)/b$.

introducing a delay in the connection to Telescope A sufficient to achieve constructive interference

$$\tau - \tau_{delay} = \frac{n\lambda}{c} \tag{3.5}$$

From our discussion of the propagation of error in Chapter 2, the uncertainty in position from Equation (3.4) is

$$\sigma(\theta) = \frac{c}{b\cos\theta}\sigma(\tau) \simeq \frac{c}{b}\sigma(\tau) \tag{3.6}$$

where we assume the uncertainty in b is small and that the source is near the zenith. Maximizing the baseline therefore minimizes positional uncertainty. We have ignored very many important points in the above discussion. One is that the position of the source requires us to determine two coordinates, so this method will benefit from the addition of telescopes outside the plane of Figure 3.9

A half-dozen arrays of radio telescopes are presently operating in VLBI mode, with baselines ranging from 30 to 10 000 km. Such systems can determine fundamental (large angle) coordinates for point-like radio sources (e.g. QSOs and other active galaxies) with uncertainties less than 1 mas. Relative positions can be determined to about 10 micro-arc seconds (0.01 mas) for bright sources. Although the astrometric contributions of radio VLBI, have been substantial, especially for distant galaxies, most normal stars are too faint in the radio to be detected, and their positions must be determined by optical methods.

Optical long baseline interferometers are possible, and indeed, several are in operation. Most are used for small-angle astrometry and small-angle, high-resolution imaging.

There are other sources of error in large-angle ground-based astrometry. Refraction by the atmosphere (see Figure 3.10 and Appendix D) moves the apparent positions of radio and (especially) optical sources toward the zenith. Variability of the atmosphere can produce inaccuracies in the correction made for refraction, and transmission time through the atmosphere can introduce differential phase delays for components of a VLBI array. Flexure of telescope and detector parts due to thermal expansion or variations in gravitational loading can cause serious systematic errors in transit telescopes.

Images in space are neither blurred nor refracted by the atmosphere, and telescopes there experience minimal gravitational stress and thermal cycling. Space-based large-angle astrometry has thus been responsible for huge improvements in data volume and precision. Large-angle space astrometry uses principles slightly different from ground-based programs. The European Space Agency's *Gaia* satellite,[2] which started taking data in 2014, is orbiting about the L2 point about 1.5 million km from Earth. The satellite mounts two identical

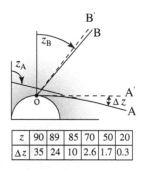

z	90	89	85	70	50	20
Δz	35	24	10	2.6	1.7	0.3

Fig. 3.10 Atmospheric refraction. The observer is on the surface of the Earth at point O. The atmosphere curves the path of a light ray from object A so that O receives light from direction A'. Likewise, the image of object B appears at B' – a smaller shift in position because both the path length and the angle of incidence are smaller. Refraction reduces the apparent zenith distance, z, of all objects. The table under the figure gives the decrease in zenith distance, Δz, in minutes of arc, as a function of z in degrees.

---

[2] Gaia originally stood for Global Astrometric Interferometer for Astrophysics. Early in the design phase, the optical interferometer concept was replaced by the more compact and sensitive two-telescope imaging technique.

telescopes perpendicular to its spin axis. These point in two directions separated by a fixed angle of 106.5°. Additional mirrors combine the beams from both telescopes onto a very large (800 Megapixel) detector. *Gaia* spins once every 6 hours and tracks each object in the combined image as it transits the detector pixels. Special segments of the detector measure apparent brightness, color, and (for brighter objects) spectra. Over its 5-year mission *Gaia* is expected to observe about 10^9 objects (nominally, every object brighter than $V = 20$) about 70 times each. The end result will be a network of objects (not just stars) separated by precisely determined large and small angles. The data rate is large – about 10 Terabytes from the spacecraft each year, and analysis will be complex. The final astrometric precision expected depends on brightness, color, and position in the sky, and (as of this writing, one year into the mission) should range from 5–16 µas for star-like objects brighter than $V = 14$ to a few hundred µas at $V = 20$. Such precision challenges the limits of hardware and software engineering: *Gaia* compensates for spin rate changes due to micrometeoroid impacts, calibrates for shifts in image position for objects of different color due to wavelength-dependent diffraction, and passively controls the 106.5-degree angle between the two telescopes. Accurate tracking of the spacecraft ($V = 21$ from Earth) is essential.

Gaia, with a total mission cost of about $\$10^9$ represents a major advance in astronomy because, in addition to positions, it will measure the distances and motions of many objects with unprecedented accuracy. More about this later in this chapter. For now it will be sufficient to compare *Gaia* with the previous astrometric gold standard, the artificial satellite *HIPPARCOS* (1989–93).[3] *HIPPARCOS* measured 120 000 stellar positions with a similar technique and achieved precisions of 1000 µas for most objects in its catalog.

Catalogs produced with large-angle astrometric methods like transit telescope observations and the *Gaia* and *HIPPARCOS* missions are usually called **fundamental catalogs**. The most important are listed in Appendix D. You should realize that the *relative* positions of some objects, especially nearby stars, do change very slowly due to their actual motion in space relative to the Sun. This **proper motion**, although small (a large proper motion would be a few arcsec per century), will cause a change in coordinates over time, and any fundamental catalog must specify both of the coordinates and the **epoch** (or date) for which they are valid. Most catalogs also specify a measurement of the object's proper motion. See Section 3.4.2 below.

3.1.8 Precession and nutation

Very distant objects (e.g. galaxies far, far away) should have essentially zero proper motion, so their relative positions on the celestial sphere never change.

---

[3] High Precision PARallax Collecting Satellite. The acronym intentionally echoes the name of the accomplished Greek astronomer, Hipparchus of Rhodes.

However, their equatorial coordinates *do* change by as much as 50 arcsec per year due to precession, nutation, and other effects.

Conservation of angular momentum might lead one to expect that the Earth's axis of rotation would maintain a stationary orientation with respect to the more distant objects. However, the Earth has a non-spherical mass distribution, and so responds to gravitational torques from the Moon (primarily) and Sun. In addition to this lunisolar effect, the other planets produce much smaller torques. All these torques set the location of the celestial pole, equator, and equinox into continuous motion relative to the fixed objects.

This is, at best, an inconvenience. What must be done when specifying any measurement of RA and Dec coordinates is to say *when* those coordinates were correct. That is, you say **which equator and equinox** were used. Usually, because the motion of the equator and equinox are well known, one computes and lists the coordinates that the celestial location *would* have at a certain date. If you know the proper motion of your object, include that in your computation. Currently, the celestial equator and origin of right ascension for 12:00 UT (see Section 3.3) on January 1, 2000 (usually denoted J2000) are the ones you are likely to use.

Astronomers separate this motion into two components: a long-term general trend called **precession** and a short-term oscillatory motion called **nutation**. Figure 3.11 illustrates precession: the north *ecliptic* pole remains fixed with respect to the distant background galaxies, while the north *celestial* pole (NCP) moves in a small circle centered on the ecliptic pole. The precessional circle has a radius equal to the average obliquity (around 23 degrees), with the NCP completing one circuit in about 26 000 years, moving at a very nearly – but not precisely – constant speed. The celestial equator, of course, moves along with the pole. The vernal equinox, which is the fundamental direction for both the equatorial and ecliptic coordinate systems, moves westward along the

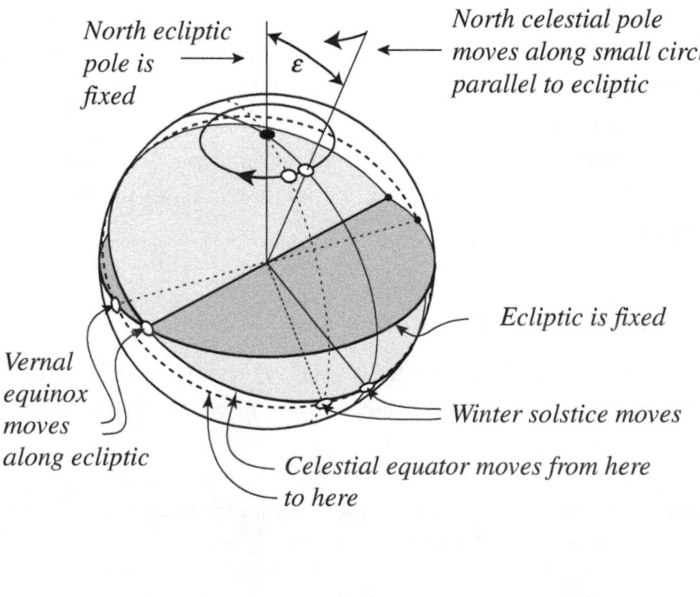

Fig. 3.11 Precession of the equinoxes. The location of the ecliptic and the ecliptic poles is fixed on the celestial sphere. The celestial equator moves so that the north celestial pole describes a small circle around the north ecliptic pole of radius equal to the mean obliquity. The figure indicates motion after about 1000 years.

North ecliptic pole is fixed

North celestial pole moves along small circle parallel to ecliptic

ε

Ecliptic is fixed

Vernal equinox moves along ecliptic

Winter solstice moves

Celestial equator moves from here to here

ecliptic at the rate (in the year 2000) of 5029.097 arcsec (about 1.4 degrees) per century. Precession will in general cause both the right ascension and declination of every star to change over time and will also cause the ecliptic longitude (but not the ecliptic latitude) to change as well.

The most influential ancient astronomer, Hipparchus of Rhodes (recorded observations 141–127 BCE) spectacularly combined the rich tradition of Babylonian astronomy, which was concerned with mathematical computation of future planetary positions from extensive historic records, with Greek astronomy, which focused on geometrical physical models that described celestial phenomena. He constructed the first quantitative geocentric models for the motion of the Sun and Moon, developed the trigonometry necessary for his theory, injected the Babylonian sexagesimal numbering system (360° in a circle) into western use, and compiled the first systematic star catalog. Hipparchus discovered lunisolar precessional motion as a steady regression of the equinoxes when he compared contemporary observations with the Babylonian records. Unfortunately, almost all his original writings are lost and we know his work mainly through the admiring Ptolemy, who lived three centuries later.

Since the time of Hipparchus, the vernal equinox has moved about 30° along the ecliptic. In fact, we still refer to the vernal equinox as the "first point of Aries," as did Hipparchus, even though it has moved out of the constellation Aries and through almost the entire length of the constellation Pisces since his time. Precession also means that the star Polaris is only temporarily located near the north celestial pole. About 4500 years ago, at about the time the Egyptians constructed the Great Pyramid, the "North Star" was Thuban, the brightest star in Draco. In 12 000 years, the star Vega will be near the pole, and Polaris will have a declination of 43°.

Unlike lunisolar precession, planetary precession actually changes the angle between the equator and ecliptic. The result is an oscillation in the obliquity so that it ranges from 22° to 24°, with a period of about 41 000 years. At present, the obliquity is decreasing from an accepted J2000 value of 23° 26′ 21.4″ at a rate of about 47 arcsec per century.

Nutation, the short period changes in the location of the NCP, is usually separated into two components. The first, nutation in longitude, is an oscillation of the equinox ahead of and behind the precessional position, with an amplitude of about 9.21 arcsec and a principal period of 18.6 years. The second, nutation in obliquity, is a change in the value of the angle between the equator and ecliptic. This also is a smaller oscillation, with an amplitude of about 6.86 arcsec and an identical principal period.

3.1.9 Barycentric coordinates

Coordinates measured with a transit telescope from the surface of the moving Earth or from a satellite in orbit are measured in a non-inertial reference frame, since the

spin and orbital motions of the Earth or satellite accelerate the telescope. These
apparent equatorial coordinates exhibit variations introduced by this non-inertial
frame, and their exact values will depend on the time of observation and the
location of the telescope. Catalogs therefore give positions in an equatorial system
whose origin is at the barycenter (center of mass) of the Solar System. The
barycentric coordinates are computed for the mean equinox of the catalog date
(a fictitious equinox which moves with precessional motion, but not nutational) by
correcting the apparent coordinates for several effects, including precession, nutation, proper motion, and for highest precision, relativity. We will discuss two
others. The first, due to the changing vantage point of the telescope as the Earth
executes its orbit, is called **heliocentric stellar parallax**. The small variation in a
nearby object's apparent coordinates due to parallax is a very important quantity
because it depends on the object's distance. We discuss parallax in Section 3.2.2.

The second effect is called the **aberration of starlight**. It produces a shift in
every object's apparent coordinates because of the finite velocity of light. The
magnitude of the shift depends only on the angle between the object's direction
and the direction of the velocity of the observer. Figure 3.12 shows a telescope in
the barycentric coordinate system, drawn so that the velocity of the telescope, at
rest on the moving Earth, is in the $+x$-direction. A photon from a distant object
enters the telescope at point A, travels at the speed of light, c, and exits at point
B. In the barycentric frame, the photon's path makes an angle θ with the x-axis.
However, if the photon is to enter and exit the moving telescope successfully, the
telescope must make an angle $\theta' = \theta - \Delta\theta$ with the x-axis in the frame fixed on
the Earth. A little geometry shows that, if V is the speed of the Earth,

$$\Delta\theta = \frac{V}{c} \sin \theta \qquad (3.7)$$

Thus aberration moves the apparent position of the source (the one measured
by a telescope on the moving Earth) toward the direction of the telescope's

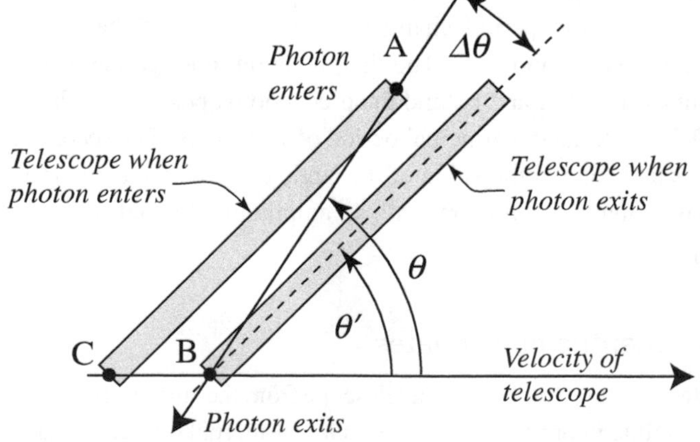

Fig. 3.12 The aberration of starlight. A telescope points toward a source. The diagram shows the telescope moving to the right in the barycentric frame. The apparent direction of the source, θ', depends on the direction and magnitude of the telescope velocity.

velocity. The magnitude of this effect is greatest when $\theta = 90°$, where it amounts to about 20.5 arcsec.

3.1.10 The ICRS

The International Astronomical Union (IAU) in 1991 recommended creation of a special coordinate system whose origin is at the barycenter of the Solar System, with a fundamental plane *approximately* coincident with the Earth's equatorial plane in epoch J2000.0. The *x*-axis of this **International Celestial Reference System (ICRS)** is taken to be in the *approximate* direction of the vernal equinox on that date. However, unlike previous barycentric systems, the axes of the ICRS are defined and fixed in space by the assigned positions of distant galaxies, not by the apparent motion of the Sun or the spin of the Earth. To emphasize this distinction one now speaks of the ***origin of right ascension***, rather than the equinox.

In practice, the ICRS2 (the most recent realization) is based on VLBI radio astronomical determinations of the positions of over 3400 compact extragalactic sources (mostly quasars) of which 295 are used to define the axes of the system. Unlike nearby stars, these distant objects have undetectable angular motions relative to one another and their relative positions do not depend on our imperfect knowledge of the Earth's rotation, precession, and nutation. Thus, the ICRS is a very good approximation of an inertial, non-rotating coordinate system. The International Earth Rotation Service in Paris coordinates the ongoing ICRS observing program. Directions of the ICRS2 axes are now specified with a precision of about 10 μas relative to the defining sources, and positions of the other sources are known to better than 100 μas. The ICRS positions of most optical sources are known primarily through HIPPARCOS and Hubble Space Telescope (HST) observations near the optical counterparts of either the defining radio sources or a larger number of other radio sources. Approximately 120 000 stars measured by HIPPARCOS thus have ICRS coordinates known with uncertainties typical of that satellite's measurements, around 1 mas. Through the *HIPPARCOS* measurements, ICRS positions can be linked to other Earth- and space-based position measurements. (See Appendix D.) Positional uncertainties for many more optical sources, of course, will be considerably reduced as the *Gaia* mission proceeds.

3.1.11 The ecliptic coordinate system

The ecliptic, the apparent path of the Sun on the celestial sphere, can also be defined as the intersection of the Earth's orbital plane with the celestial sphere. The orbital angular momentum of the Earth is much greater than its spin angular momentum, and the nature of the torques acting on each system suggests that the orbital plane is far more likely to remain invariant in space than is the equatorial plane. Moreover, the ecliptic plane is virtually coincident with the plane of symmetry of the Solar System as well as lying nearly perpendicular to the Solar System's total angular momentum vector. As such, it can be an important reference plane for observations and dynamical studies of Solar System objects.

Astronomers define a geocentric coordinate system in which the ecliptic is the fundamental plane and the vernal equinox is the fundamental direction. Measure ecliptic longitude, λ, from west to east in the fundamental plane. Measure the ecliptic latitude, β, positive northward from the ecliptic. Since the vernal equinox is also the fundamental direction of the equatorial system, the north ecliptic pole is located at $RA = 18$ hours and $Dec = 90° − \varepsilon$, where ε is the obliquity of the ecliptic.

The ecliptic is so nearly an invariant plane in an inertial system that, unlike the equatorial coordinates, the ecliptic latitudes of distant stars or galaxies will *not* change with time because of precession and nutation. Ecliptic longitudes on the other hand, are tied to the location of the equinox, which is in turn defined by the spin of the Earth, so longitudes will have a precessional change of about $50''$ per year.

3.1.12 The Galactic coordinate system

The central plane of the disk-shaped Milky Way Galaxy is another reference plane of physical significance. The IAU has specified a great circle (the **Galactic plane**) that approximates the center-line of the Milky Way on the celestial sphere to constitute the fundamental plane of the Galactic coordinate system. The origin of the system is at the Sun, and its fundamental direction toward the center of the Galaxy. Galactic latitude (b or b^{II}) is then measured positive north (the Galactic hemisphere that contains the north celestial pole) of the plane, and Galactic longitude (l or l^{II}) is measured from Galactic center so as to constitute a right-handed system.

Since this reference frame is essentially barycentric, coordinates are unaffected by terrestrial effects like precession. However, it is inconvenient to measure l and b directly, and they less intuitively relate to the horizon system. Galactic coordinates of any object are in practice derived from its equatorial coordinates. The important parameters are that the north Galactic pole ($b = +90°$) is defined to be at

$$\alpha = 12{:}51{:}26, \delta = +27° \ 7.7' \,(J2000),$$

and the Galactic center ($l = b = 0$) at

$$\alpha = 17{:}45{:}37, \delta = −28°56.17'$$

It is unlikely that the definition of the l,b system will be adjusted, even though we now have good evidence that the actual dynamical center of the

Galaxy is a massive black hole associated with the radio source Sagittarius A*, located about 4 arc minutes from the above position.

3.1.13 Transformation of coordinates

Transformation of coordinates involves a combination of rotations and (sometimes) translations. Note that for very precise work (the transformation of geocentric to ICRS coordinates, for example), some general-relativistic modeling may be needed. Some of the more common transformations are addressed in the various national almanacs, and for systems related just by rotation (equatorial and Galactic, for example), you can work transformations out by using spherical trigonometry. Some important transformations are given in Appendix D, and calculators for most can be found on the internet.

3.2 The third dimension

Determining the distance of almost any object in astronomy is notoriously difficult, and uncertainties in the coordinate r are usually enormous compared to uncertainties in direction. For example, the position of Alpha Centauri, the nearest star after the Sun, is uncertain in the ICRS by about 0.4 mas (three parts in 10^9 of a full circle), yet its distance, one of the best known, is uncertain by about one part in 2500. A more extreme example would be one of the quasars that define the ICRS, with a typical positional uncertainty of 0.02 mas (six parts in 10^{10}). Estimates of the distances to these objects depend on our understanding of the expansion and acceleration of the universe, and are probably uncertain by at least 10%. This section deals with the first two rungs in what has been called the "cosmic distance ladder," the sequence of methods and calibrations that ultimately allow us to measure distances (perhaps "estimate distances" would be a better phrase) of the most remote objects.

3.2.1 The astronomical unit

We begin in our own Solar System. Kepler's third law gives the scale of planetary orbits:

$$a = P^{2/3} \tag{3.8}$$

where a is the average distance between the planet and the Sun measured in **astronomical units** (AU, or, preferably, au) and P is the orbital period in years. This law sets the *relative* sizes of planetary orbits. One au is defined to be the mean distance between the Earth and Sun, but the mean distance in meters, and the absolute scale of the Solar System, must be measured empirically.

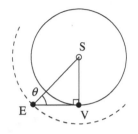

Fig. 3.13 Radar ranging to Venus. The astronomical unit is the length of the line ES, which scales with EV, the Earth-to-Venus distance.

Figure 3.13 illustrates one method for calibrating the au. The figure shows the Earth and the planet Venus when they are in a position such that apparent angular separation between Venus and the Sun, as seen from Earth, (the *elongation* of Venus) is at a maximum. At this moment, a radio (radar) pulse is sent from the Earth toward Venus, and a reflected pulse returns after elapsed time Δt. The Earth-to-Venus distance is just $c\Delta t/2$. Thus, from the right triangle in the figure, the length of the line ES is one au or:

$$1\,\text{au} = \frac{c\Delta t}{2\cos\theta}$$

Some corrections need to be made because the orbit of neither planet is a perfect circle, but the geometry is known rather precisely. Spacecraft in orbit around Venus and other planets (Mars, Jupiter, and Saturn) also provide the opportunity to measure light-travel times, and similar geometric analyses yield improved absolute orbit sizes. The size of the Earth's orbit is now known to about one part in 10^{10}. Astronomers are content that this is good enough, and have *defined* the au to be a unit of measure conveniently approximate to the Earth–Sun distance:

$$1\,\text{au} \equiv 1.495978707 \times 10^{11}\,\text{m}$$

3.2.2 Stellar parallax

> It is hard to achieve this precision [needed to measure parallax], both on account of the imperfection of astronomical instruments, which are subject to much variation, and because of the shortcomings of those who handle them with less care than is required.
>
> – Galileo Galilei, *Dialogue Concerning the Two Chief World Systems*, 1632

Once the size and shape of the Earth's orbit has been established, we can determine the distances to nearby stars through observations of **heliocentric stellar parallax**. Figure 3.14 depicts the orbit of the Earth around the Sun. For simplicity, we again assume a circular orbit. The plane of the orbit is the ecliptic plane, and we set up a Sun-centered coordinate system with the ecliptic as the fundamental plane, the z-axis pointing toward the ecliptic pole, and the y-axis chosen so that a nearby star, S, is in the y–z plane. The distance from the Sun to S is r. As the Earth travels in its orbit, the apparent position of the nearby star shifts in relation to very distant objects. Compared to the background objects, the nearby star appears to move around the perimeter of the **parallactic ellipse**, reflecting the Earth's orbital motion.

Figure 3.15 shows the plane that contains the x-axis and the star. The parallax angle, p, is half the total angular shift in the star's position (the semi-major axis

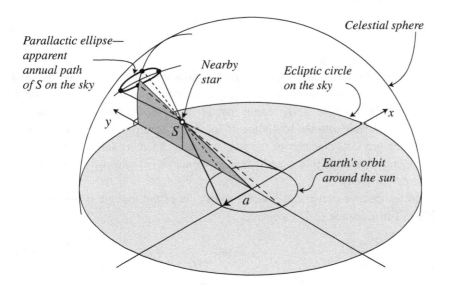

Parallactic ellipse— apparent annual path of S on the sky

Nearby star

Celestial sphere

Ecliptic circle on the sky

y

S

x

Earth's orbit around the sun

a

Fig. 3.14 The parallactic ellipse. The apparent position of the nearby star, *S*, as seen from Earth, traces out an elliptical path on the very distant celestial sphere as a result of the Earth's orbital motion.

of the parallactic ellipse in angular units). From the right triangle formed by the Sun–star–Earth:

$$\tan p = \frac{a}{r} \qquad (3.9)$$

where *a* is 1 au. Since *p* is in every case going to be very small, we make the small-angle approximation: for $p \ll 1$:

$$\tan p \simeq \sin p \simeq p \qquad (3.10)$$

So that for any right triangle where *p* is small:

$$p = \frac{a}{r} \qquad (3.11)$$

In this equation, it is understood that *a* and *r* are *measured in the same units* (aus, for example) and *p* is measured in radians. Radian measure is inconvenient for small angles, so, noting that there are about 206 265 arcsec per radian, we can rewrite the small-angle formula as

$$p[\text{arcsec}] = \left(\frac{180 \cdot 60 \cdot 60 \text{ arc sec}}{\pi \text{ radian}} \right) \frac{a}{r} \cong 206\,265 \frac{a}{r} \, [a, r \text{ in same units}] \qquad (3.12)$$

Finally, to avoid very large numbers for *r*, it is both convenient and traditional to define a new unit, the ***parsec***, with the length:

$$1 \text{ parsec} = 3.085678 \times 10^{16} \text{m} = 206265 \text{ au} = 3.2616 \text{ light years}$$

The parsec (pc) is so named because it is the distance of an object whose *par*allax is one *sec*ond of arc. With the new unit, the parallax equation becomes:

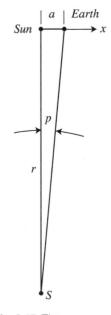

Sun | *a* | *Earth*

x

p

r

S

Fig. 3.15 The parallax angle.

$$p[\text{arcsec}] \frac{a[\text{au}]}{r[\text{pc}]} \qquad (3.13)$$

This equation represents a fundamental relationship between the small angle and the sides of the ***astronomical triangle*** (any right triangle with one very short side).

Example. Suppose a supergiant star is 20 pc away, and we measure its angular diameter with the technique of speckle interferometry as 0.023 arcsec. Then the physical diameter of the star, which is the short side of the relevant astronomical triangle (the quantity a in Equation (3.13) must be 20×0.023 pc arcsec = 0.46 au).

In the case of stellar parallax, the short side of the triangle is always 1 au. If $a = 1$ in Equation (3.13) we have:

$$p[\text{arsec}] \frac{1}{r[\text{pc}]} \qquad (3.14)$$

Note that the parallactic ellipse will have a semi-major axis equal to p, and a semi-minor axis equal to $p \sin \lambda$, where λ is the ecliptic latitude of the star. An ellipse fit to multiple observations of the position of a nearby star will therefore estimate its parallax.

As Galileo pointed out, there are uncertainties in the measurement of the parallax angle. Images of stars formed by Earth-based telescopes are typically blurred by the atmosphere and are seldom smaller than a half arc second in diameter, and are often much larger. In the early days of telescopic astronomy, a great visual observer, James Bradley (1693–1762), like many astronomers before him, undertook the task of measuring stellar parallax. Bradley could measure stellar positions with a precision of about 0.5 arcsec (500 milliarc seconds or mas). This precision was sufficient to discover the phenomena of nutation and aberration, but not to detect a stellar parallax.

A few generations later, Friedrich Wilhelm Bessel (1784–1846), a young clerk in an importer's office in Bremen, began to study navigation in order to move ahead in the business world. Instead of mercantile success, Bessel discovered his love of astronomical calculation and embarked on a quest for astrometric precision. He revised his career plans and in 1804 secured a post as assistant to the astronomer Johann Hieronymus Schröter, and began an analysis of Bradley's observations. Bessel deduced the systematic errors in Bradley's instruments (about 4 arcsec in declination, and 1 second of time in RA – much worse than Bradley's random errors, and due mostly to atmospheric refraction). By 1810, Bessel was director of the observatory at Königsberg. He demonstrated that major improvements in positional accuracy should be possible, and after correcting the positions of the 3222 stars in Bradley's catalog, measured his own positions for 62 000 other stars. In 1837, Vasilii Yakovelich Struve published his measurement (using a 9-inch Fraunhofer telescope, then the

world's largest refractor) of the parallax of Vega, but stated his result (0.125″) as
an extremely tentative one that required further observation. Bessel had sporad-
ically been trying to measure a heliocentric parallax since 1818, but inspired by
Struve's "almost" success, immediately began to monitor the double star 61
Cygni along with two "background" stars. Bessel had a supremely suitable
telescope, a Fraunhofer heliometer. A heliometer is a refractor with a rotatable
split objective that creates moveable double images. Object separations can be
read off by noting the offset between the lens halves that was required to
superimpose images. In 1838, after over a year of intense work (98 nights of
observing, at least 16 measurements per night), Bessel announced his value for
the parallax, 314 ± 20 mas, close to the modern value of 286 mas. Bessel's
reputation was such that his result was immediately hailed as the first successful
measurement of heliocentric parallax. Bessel's labor was typical of his ambition[4]
and meticulous attention to error reduction. The 61 Cygni parallax project was
Herculean, and parallaxes for any but the very nearest stars emerged only after
the introduction of photography.[5]

Beginning in the late 1880s, photography steadily transformed astronomy. It
provided an objective record and allowed the accumulation of many photons in a
long exposure. With photography, human eyesight no longer limited human
ability to detect faint objects. Photography also vastly augmented the power of
small-angle astrometry.

Astronomical photographs (negatives) were usually recorded on emulsion-
covered glass plates at the telescope, then developed in a darkroom. Away from
the telescope, astronomers could measure the positions of objects on the plate, at
first with microscopes, then with ***measuring machines*** that became increasingly
automated, precise, and expensive. Computer-controlled ***microdensitometers***,
(essentially ultra-precise digital scanners) which measured the darkness of the
image at each position on a plate became important astronomical resources.
Direct digital recording of images with electronic arrays, beginning in the 1970s,
gradually reduced the importance of photography and microdensitometers.
Photography still has a place in some specialized areas of astronomy, and
measuring machines still continue to produce important data from archival
photographs.

---

[4] Bessel pioneered mathematical analysis using the functions that now bear his name. He spent
30 years measuring the "Prussian degree" – the length, in meters, of a degree of arc of geodetic
latitude. This was part of an international effort to determine the shape of the Earth from
astronomical measurements. Bessel inspired his student, F.W. Argelander, to organize a project
to determine the transit-circle positions for all stars brighter than ninth magnitude in the northern
hemisphere – about a third of a million objects.

[5] In addition to Struve's measurement of the parallax of Vega, Thomas Henderson published the
parallax of Alpha Centauri in 1839. Bessel's measurement was the most accurate. As of 1898,
parallax had only been detected for 50 stars.

Uncertainty in p from conventional ground-based small-angle astrometry can be routinely reduced to around 5 mas with repeated measurements (50 observations of a single star are not unusual), suitable reference stars, and a stable system. A very few ground-based optical parallax measurements can approach 0.5 mas precision. For most stars though, this means that only parallaxes larger than 50 mas will have uncertainties smaller than 10%, so only those stars nearer than $1/0.05 = 20$ pc can be considered to have distances precisely known by the ground-based parallax method. There are approximately 1000 stars detected closer than 20 pc, a rather small number compared to the 10^{11} or so stars in the Milky Way. Appendix D lists the nearest stars and brown dwarves based upon current parallaxes.

About 8000 stars have been cataloged with parallaxes in the optical with uncertainties of hundreds of mas. Several hundred bright radio sources have VLBI parallaxes with much smaller (0.02 mas) uncertainties. However, space-based methods have produced the greatest volume of precision measurements. The 118 000 *HIPPARCOS* parallaxes had a median uncertainty of 0.97 mas for stars brighter than magnitude 9.0. The Hubble Space Telescope has made a much smaller number of measurements of similar or better accuracy. The *Gaia* mission anticipates precisions of 5–16 µas for $V = 7$, 10–25 µas for $V = 15$, and around 500 µas at $V = 20$. A 10 mas precision means that astronomers will be able to map the nearest half of our Galaxy with better than a 10% uncertainty.

3.3 Time

> Alice sighed wearily. "I think you might do something better with the time," she said, "than wasting it in asking riddles that have no answers."
>
> "If you knew Time as well as I do," said the Hatter, "you wouldn't talk about wasting IT. It's HIM.... I dare say you never even spoke to Time!"
>
> "Perhaps not," Alice cautiously replied; "but I know I have to beat time when I learn music."
>
> – Lewis Carroll, *Alice's Adventures in Wonderland*, 1865

Time is a physical quantity of which we have never enough, save for when we have too much and it gets on our hands. Ambition to understand its nature has consumed the time of many. It is unclear how much of it has thereby been wasted in asking riddles with no answers. Perhaps time will tell.

3.3.1 Atomic time

Measuring time is a lot easier than understanding it. The way to measure time is to "beat" it, like Alice. In grammar school, I learned to count seconds by pronouncing syllables: "Mississippi one, Mississippi two, Mississippi three...." A second of time is thus, roughly, the duration required to enunciate five syllables.

A similar definition, this one set by international agreement, invokes a more objective counting operation:

> **1 second** (*Système International*, or SI second) = the duration of 9,192,631,770 periods of the radiation corresponding to the transition between the two hyperfine levels of the ground state of the cesium-133 atom.

A device that counts the crests of a light wave and keeps a continuous total of the elapsed SI seconds is an **atomic clock.** An atomic clock located at rest on the surface of the earth keeps **TAI** or **international atomic time** (TAI = *Temps Atomique International*). Experimental atomic clocks (based on atomic transitions in strontium and ytterbium) have a precision of about one part in 10^{18} and will eventually lead to a different (but consistent) definition of the second. TAI is the basis for dynamical computations involving time as a physical parameter and for recording observations made on the surface of the Earth. Things get a little complicated if you compare an atomic clock on the surface of the Earth with one located elsewhere (like the barycenter of the Solar System). Such clocks will not run at the same rate, but, when compared will differ according to their relative velocities, accelerations, and local gravitational fields as predicted by relativity theory. Precise timekeeping accounts for relativity effects, but the starting timescale in these computations is TAI.

The **astronomical day** is defined as 86 400 SI seconds. There are, however, other kinds of days.

3.3.1 Solar time

Early timekeepers found it most pragmatic to count days, months, and years, and to subdivide these units. For those on Earth, the day is the most intrusive and practical of these units. Much of early timekeeping was a matter of counting, grouping, and subdividing days. Since the rotation of the Earth establishes the length of the day, counting days is equivalent to counting rotations.

Figure 3.16 illustrates an imaginary scheme for counting and subdividing days. The view is of the Solar System, looking down from above the Earth's north pole, which is point P. The plane of the page is the Earth's equatorial plane, and the large circle represents the equator itself. The small circle represents the position of the Sun projected onto the equatorial plane. In the figure, we assume the Sun is motionless, and we attach hour markers just outside the equator as if we were painting the face on an Earth-sized 24-hour clock. These markers are motionless as well, and are labeled so that they increase counterclockwise. The marker in the direction of the Sun is labeled 12, the one opposite the Sun, 0. Our choice of 24 hours around the circle, as well as the subdivision into 60 minutes per hour and 60 seconds per minute, originated with the ancient Babylonian sexagesimal (base 60) number system.

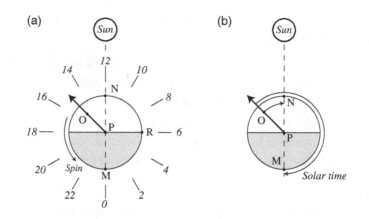

Point O in the figure is the location of a terrestrial observer projected onto the equatorial plane. This observer's meridian projects as a straight line passing through O and P. The figure extends the projected meridian as an arrow, like the hand of a clock, which will sweep around the face with the painted numbers as the Earth rotates relative to the Sun. Since we are using the Sun as the reference marker, this turning of the Earth is actually a combination of spin and orbital motion. The meridian points to the number we will call the *local apparent solar time*. Each cycle of the meridian past the zero mark (midnight) starts a new day for the observer. Every longitude has a different meridian and thus a different solar time. The local solar time, for example, is 12 hours for an observer at point N in the figure, and 6 hours for an observer at R.

Consideration of Figure 3.16b should convince you of the following definition:

> *Local Apparent Solar Time* = The hour angle of the sun as it appears
> on the sky (∠OPN), plus 12 hours.

Simple observations (for example, with a sundial) will yield the local apparent solar time, but this method of timekeeping has a serious deficiency. Compared to TAI, local apparent solar time is non-uniform, mainly because of the Earth's orbital motion. Because of the obliquity of the ecliptic, motion in the orbit has a greater east–west component at the solstices (longer days by about 20 s) than at the equinoxes. In addition, because Earth's orbit is elliptical, Earth's orbital speed varies; it is greatest when it is closest to the Sun (at perihelion, around January 4, days are about 8 s longer) and slowest when furthest away (aphelion). As a result, apparent solar days throughout the year have different lengths compared to the defined astronomical day of 86 400 SI seconds.

This non-uniformity is troublesome for precise timekeeping. To remove it, one strategy is to average out the variations by introducing the idea of the *mean Sun*: a fictitious body that moves along the celestial equator at uniform angular speed, completing one circuit in one tropical year (i.e. equinox to equinox). If we redefine the "Sun" in Figure 3.16 as the mean Sun, we can define a more uniform timescale:

Local Mean Solar Time = The hour angle of the fictitious mean sun, plus 12 hours.

The difference between the apparent and the mean solar times is called the ***equation of time:***

   Equation of time = Local Apparent Solar Time − Local Mean Solar Time

The equation of time takes on values in the range ± 15 minutes in the course of a year. See Appendix D for more information.

To circumvent the difficulty arising from the fact that every longitude on Earth will have a different mean solar time, one often records or predicts the time of an event using the reading from a mean solar clock located at the zero of longitude. This is called the universal time (UT1):

   Universal time (UT1) = mean solar time at $0°$ *longitude*

The UT1 clock, of course, is actually located in your laboratory – it is simply set to agree with the mean solar time at the origin of longitude. Thus, if the Moon were to explode, everyone on Earth would agree about the UT1 of the mishap, but only people at the same longitude would agree about the mean solar time at which it occurs.

Although a big improvement on apparent solar time, UT1 is not completely uniform. For one thing, the precession rate (needed to compute the mean Sun) is imperfectly known and changes over long timescales. The major difficulty, however, is that the spin of the Earth is not quite uniform. The largest variations are due to tidal effects that have monthly and half-monthly periods, as well as seasonal (yearly) variations probably due to thermal and meteorological effects. A smaller, random variation, with a timescale of decades, is probably due to poorly understood core–mantle interactions. Finally, over the very long term, tidal friction causes a steady slowing of the spin of the Earth. As result of this long-term trend, the mean solar day is getting longer (as measured in SI seconds) at the rate of about 0.0015 seconds per century. Thus, on the timescale of centuries, one second on the UT1 clock, (defined as $1/86\,400$ of one Earth rotation) is increasing in duration compared to the SI second, and is fluctuating in length by small amounts on shorter timescales.

In order to coordinate the Earth's rotation with TAI, the US Naval Observatory, working for the International Earth Rotation Service (IERS, in Paris), maintains the ***coordinated universal time*** (UTC) clock. Coordinated universal time approximates UT1, but uses SI seconds as its basic unit. To keep pace with UT1 to within a second, the UTC clock introduces an integral number of "leap" seconds as needed. Because of the random variations of the Earth's spin, it is not possible to know in advance when it will be necessary to add (or remove) a leap second. Between 1972 and the end of 2014, the TAI clock counted a total of 25 leap seconds that were not counted by the UTC clock. Coordinated universal time is the basis for most legal time systems. The leap seconds in the UTC clock do present real difficulties for some applications in telecommunications and navigation systems.

Unlike UTC, *local* mean solar time at least has the practical advantage of approximate coordination with local daylight: at 12 noon on the local mean solar clock, you can assume the Sun is near the meridian. However, every longitude will have a different meridian and a different local solar time. Even nearby points will use different clocks. To deal in a practical fashion with the change in mean solar time with longitude, most legal clocks keep zone time:

$$\text{zone time} = \text{UTC} + \text{longitude correction for the zone}$$

This strategy ensures that the legal time is the same everywhere inside the zone. Zones are usually about $15°$ wide in longitude, so the longitude correction is usually an integral number of hours. (Remember the Earth spins at a rate of $15°$ per hour.) For example, Eastern Standard Time (longitude $75°$) = UTC $-$ 5 hours, Pacific Standard Time (longitude $120°$) = UTC $-$ 8 hours.

Time services provide signals for setting clocks to the current UTC value. In the US, The National Institute of Standards and Technology broadcasts a radio signal (stations *WWV, WWVB, and WWVH*) at 2.5, 5, 10, 15, and 20 MHz that contain time announcements and related information. Computer networks can synchronize to UTC using standard protocols. A convenient one-time check on UTC is at the US Naval Observatory website, which is also a good source for details about various times scales. (There are *many* more.)

Sidereal time is also defined by the rotation of the Earth and its precessional variations, and therefore does not flow uniformly, but follows the variations manifest in UT1. In fact, UT1 is measured in practice by VLBI radio observations of the rotation of the Earth relative to distant quasars, since the mean Sun is fictitious:

Sidereal Time = The hour angle of the mean vernal equinox of date.

Having defined the day, astronomers find it useful to maintain a continuous count of them:

Julian date = number of elapsed *UT1* or *UTC* days since
4713 BCE January 1.5 (12 hrs UT on January 1).

It is also common to use a Julian date, rather than a UT date, to specify the date. The date of the equator and equinox in a catalog of equatorial coordinates might be specified as

J2000.0 = ``Julian epoch 2000.0'' = 2000 Jan 1.5 UT = JD 2451545.0

Appendix D summarizes some other time units.

3.4 Motion

3.4.1 Space motion

Consider an object that moves relative to the Solar System barycenter. Figure 3.17 shows the motion, that is, the displacement of such an object over a suitably long time. The plane of the figure contains both the origin of coordinates and the displacement vector. Part (a) of the figure shows the actual displacement, while part (b) shows the displacement divided by the time interval, that is, the velocity. Both displacement and velocity vectors can be decomposed into radial and tangential components. The total velocity, usually called the *space velocity*, is the vector sum of the *tangential velocity* and *the radial velocity*:

$$\vec{V} = \vec{v}_T + \vec{v}_R \tag{3.15}$$

$$V = \sqrt{v_T^2 + v_R^2} \tag{3.16}$$

Measuring the two components requires two very different observing strategies. Astronomers can measure radial velocity directly with a spectrograph, and can measure tangential velocities indirectly by observing changes in position.

3.4.2 Proper motion

Measure the position of a certain star in ICRS coordinates tonight. Wait 10 years (or 100 years) and measure its ICRS coordinates a second time, that is, measure the coordinates at a different *epoch.* If this star were truly motionless with respect to the center of the Solar System and the distant galaxies that define the axes of the ICRS, then the coordinates you measure 10 years from now will be the same as those you measure tonight. Remember, in both measurements, you will have removed effects of precession, so the star's positions are with respect to the *equator and origin of right ascension* of the J2000 ICRS.

On the other hand, most stars *do* move with respect to the ICRS axes. Especially if the star is nearby, its coordinates may very well change after only 10 years. The *rate of change* in coordinates is called the *proper motion* of the object. As the name suggests, proper motion reflects only motion with respect to the barycenter of the Solar System, and does not include those coordinate changes like aberration, precession, nutation, or heliocentric parallax that result from terrestrial motions.

Think about the objects that will *not* exhibit a proper motion over your observing interval. Certainly, these will include very distant objects like quasars, since they define the coordinate system. Also, any object that has no tangential velocity will have no proper motion. Its radial velocity is irrelevant, and cannot produce a proper motion. From Figure 3.17a you can see that the size of an object's proper motion, μ, is just

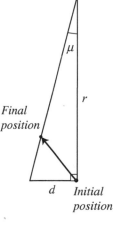

(a)

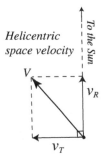

(b)

Fig. 3.17 Displacement in space and space velocity: (a) illustrates the relation between proper motion, μ, and the displacement in a unit time; (b) shows the two components of the space velocity.

$$\mu = \frac{d}{rt} = \frac{v_T}{r} \qquad (3.17)$$

Even if an object has a very large tangential velocity, you will not detect its proper motion if it is so far away that it does not change its angular position between epochs by an amount detectable by your instruments. Conversely, statistical implications of Equation (3.17) are so important they are expressed in an astronomical "proverb": *swiftness means nearness*. That is, given a group of objects with some distribution of tangential velocities, the objects with the largest values for μ (swiftness) will tend, statistically, to have the smallest values for r (nearness). Putting the quantities in Equation (3.17) in their usual units (km s^{-1} for velocity, parsecs for distance, seconds of arc per year for μ), it becomes

$$\mu = \frac{v_T}{4.74r} \qquad (3.18)$$

This means, of course, that you can compute the tangential velocity if you observe both the proper motion and the parallax (p):

$$v_T = 4.74\frac{\mu}{p} \qquad (3.19)$$

In a procedure known as *statistical parallax*, you can also use Equation (3.18) to *guess* the distance of a group of stars of known proper motion by assuming you can guess their tangential velocities. (e.g. assume on average, the magnitudes of their radial and tangential velocities are equal).

Note that μ in Equation (3.17) is the magnitude of a vector quantity: in the time between epochs, t, the object will in general change its right ascension by amount $\Delta\alpha$ and declination by amount $\Delta\delta$. Traditionally, these are quoted separately:

$$\mu = \sqrt{\mu_\alpha^2 + \mu_\delta^2} = \left\{ \left(\frac{\Delta\alpha}{t}\right)^2 + \left(\frac{\Delta\delta}{t}\right)^2 \right\}^{\frac{1}{2}} \qquad (3.20)$$

Clearly, there is a big advantage in maximizing the time between epochs. *Fundamental proper motions* are computed by comparing positions in fundamental catalogs for two different epochs, being careful to express the coordinates using the same barycentric equator and equinox. For example, the Tycho-1 Catalog computed proper motions by comparing ICRS positions determined by the *HIPPARCHOS* satellite (epoch 1991.25) with those in the Astrographic Catalog (epoch 1905). The method depends on astronomers in widely separated generations doing the hard work of assembling at least two fundamental catalogs. (Tycho-2, an improved version of Tycho-1, used 156 different catalogs.)

You can also measure proper motions using small-angle astrometry. Compare a photograph of a star field taken in 1994 with one taken with the same instrument in 1934. Align the photographs so that most of the images coincide, especially the faint background stars and galaxies. Any object that has shifted its position with respect to these "background objects" is exhibiting *relative proper motion*. The possibility that there might be some net proper motion in the background objects limits the accuracy of this sort of measurement, as does the likelihood of changes in the instrument over a 60-year span. Nevertheless, relative proper motions are more easily determined than fundamental motions, and are therefore very valuable because they are available for many more stars. You can, of course, use observations from different instruments (an old photograph and a recent CCD frame for example) to measure relative proper motions, but the analysis becomes a bit more complex and prone to systematic error.

3.4.3 Radial velocity

On May 25, 1842, Christian Doppler (1803–53) delivered a lecture to the Royal Bohemian Scientific Society in Prague. Doppler considered the situation in which an observer and a wave source are in motion relative to one another. He made the analogy between the behavior of both water and sound waves on the one hand, and of light waves on the other. Doppler correctly suggested that, in all three cases, the observer would measure a frequency or wavelength change that depended on the radial velocity of the source. The formula that expresses his argument is exact for the case of light waves from sources with small velocities:

$$z = \frac{\lambda - \lambda_o}{\lambda_o} = \frac{\Delta\lambda}{\lambda_o} \cong \frac{v_R}{c} \tag{3.21}$$

Here λ_0 is the wavelength observed when the source is motionless, λ is the wavelength observed when the source has radial velocity v_R, and c is the speed of light. z is called the **redshift parameter**, or just the redshift. If the source moves away from the observer, both v_R and z are positive, and a spectral feature in the visual (yellow-green) will be shifted to longer wavelengths (i.e. toward the red). The spectrum is then said to be **redshifted** (even if the observed feature were a microwave line that was shifted to longer wavelengths and thus *away* from the red). Likewise, if the source moves toward the observer, v_R and z are negative, and the spectrum is said to be **blueshifted.**

In his lecture, Doppler speculated that the differing radial velocities of stars were largely responsible for their different colors. To reach this conclusion, he assumed that many stars move at a considerable fraction of the speed of light relative to the Sun. This is wrong. But even though he was incorrect about the colors of the stars, the **Doppler effect**, as expressed in Equation (3.21), was soon verified experimentally, and is the basis for all astronomical direct measurements of radial velocity. It is interesting to note that first Armand Fizeau, in Paris in 1848, and then Ernst Mach, in Vienna in 1860, each independently worked out the theory of the Doppler effect without knowledge of the 1842 lecture.

Fizeau and Mach made it clear to astronomers how to *measure* a radial velocity. The idea is to observe a known absorption or emission *line* in the spectrum of a moving astronomical source, and compare its wavelength with some zero-velocity reference. The first references were simply the wavelength scales in visual spectrographs. Angelo Secci, in Paris, and William Huggins, in London, both attempted visual measurements for the brighter stars during the period 1868–76, with disappointing results. Probable errors for visual measurements were on the order of 30 km s^{-1}, a value similar to the actual velocities of most of the bright stars. James Keeler, at Lick Observatory in California, eventually was able to make precision visual measurements (errors of about 2–4 km s^{-1}), at about the same time (1888–91) that astronomers at Potsdam and Harvard first began photographing spectra.

Spectrographs (with photographic recording) soon proved vastly superior to *spectroscopes*. Observers began recording *comparison spectra*, usually from electrically activated iron arcs or hydrogen gas discharges, to provide a recorded wavelength scale. Figure 3.18 shows a photographic spectrum and comparison. A measuring engine, a microscope whose stages are moved by screws equipped with micrometer read-outs, soon became essential for determining positions of the lines in the source spectrum relative to the lines in the comparison. In current practice, astronomers record spectra and comparisons digitally and compute shifts and velocities directly from the data.

Precise radial velocities

What limits the precision of a radial velocity measurement? We consider spectrometry in detail in Chapter 11. For now, just note that, since the important measurement is the physical location of spectral lines on the detector, an astronomer certainly would want to use a detector/spectrometer capable of showing as much detail as possible. The *resolving power* of a spectrograph is the ratio:

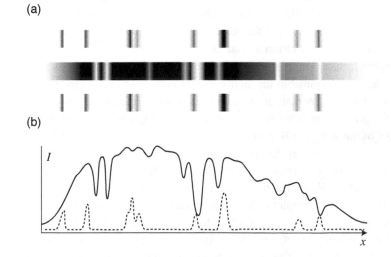

Fig. 3.18 (a) A conventional photographic spectrum. A stellar spectrum, with absorption lines, lies between two emission-line comparisons. (b) A digital spectrum of the same region, with a spectrum of the comparison plotted on the same scale. Because rest wavelengths are usually known for all lines, the observed wavelength for a stellar line is a direct function of its observed *x*-coordinate.

(a)

(b)

$$R = \frac{\lambda}{\delta\lambda} \tag{3.22}$$

where $\delta\lambda$ is wavelength resolution (i.e. two narrow spectral lines that are closer than $\delta\lambda$ in wavelength will appear as a single line in the spectrogram). Limits to resolving power will be set by the design of the spectrograph, but also by the brightness of the object being investigated, and the size and efficiency of the telescope feeding the spectrograph. As is usual in astronomy, the most precise measurements can be made on the brightest objects.

Early spectroscopists soon discovered other limits to precision. They found that errors arose if a spectrograph had poor mechanical or thermal stability, or if the path taken by light from the source was not equivalent to the path taken by light from the comparison. New spectrograph designs improved resolving power, efficiency, stability, and the reliability of wavelength calibration. At the present time, random errors of less than 100 m/s in absolute stellar radial velocities are possible with the best optical spectrographs. At radio wavelengths, even greater precision is routine.

Greater precision is possible in differential measurements. Here the astronomer is concerned only with *changes* in the velocity of the object, not the actual value. Very precise optical work, for example, has been done in connection with searches for planets orbiting solar-type stars. The presence of a planet will cause the radial velocity of its star to vary as they both orbit the barycenter of the system. Precisions at a number of observatories now approach 3 m/s or better for differential measurements of brighter stars.

Large redshifts

When the radial velocity of the source is a considerable fraction of the speed of light, special relativity replaces Equation (3.21) with the correct version:

$$z = \frac{\Delta\lambda}{\lambda_o} = \sqrt{\frac{1+\beta}{1-\beta}} - 1 \tag{3.23}$$

where

$$\beta \equiv \frac{v_R}{c} = \frac{(1+z)^2 - 1}{(1+z)^2 + 1} \tag{3.24}$$

In an early result from the spectroscopy of non-stellar objects, V. M. Slipher, in 1914, noticed that the vast majority of the spiral nebulae (galaxies) had redshifted spectra. By 1931, Milton Humason and Edwin Hubble had recorded galaxy radial velocities up to 20 000 km/s, and were able to demonstrate that the redshift of a galaxy was directly proportional to its distance. Most astronomers interpret **Hubble's law,**

$$v = H_0 d \qquad\qquad (3.25)$$

as indicating that our universe is expanding (the distances between galaxies are increasing). In Equation (3.25), it is customary to measure v in km/s and d in Megaparsecs, so H_o, which is called the **Hubble constant**, has units of km s^{-1}Mpc^{-1}. In these units, recent measurements of the Hubble constant fall in the range 67–78. Actually, the redshifts are not interpreted as due to the Doppler effect, but as the result of the expansion of space itself.

The object with the largest spectroscopic redshift (as of late 2015) is a galaxy, EGSY8p7, which has $z = 8.68$. You can expect additional detections in this range. Doppler's 1842 assumption that major components of the universe have significant shifts in their spectra was quite correct after all.

Summary

- Coordinate systems can be characterized by a particular origin, reference plane, reference direction, and sign convention.
- Astronomical coordinates are treated as coordinates on the surface of a sphere. The laws of **spherical trigonometry** apply. Concepts:

    great circle   law of sines   law of cosines

- The geocentric terrestrial **latitude and longitude** system uses the equatorial plane and **prime meridian** as references. Concepts:

    geocentric latitude   geodetic latitude   geographic latitude
    Greenwich   polar motion

- The altitude–azimuth system has its origin at the observer and uses the horizontal plane and geographic north as references. Concepts:

    vertical circle   zenith   nadir
    zenith distance   meridian   diurnal motion
    sidereal day

- The equatorial system of right ascension and declination locates objects on the celestial sphere. The Earth's equatorial plane and the vernal equinox are the references. This system rotates with respect to the altitude–azimuth system. Concepts:

    celestial pole   ecliptic   obliquity
    altitude of pole = observer's latitude   upper meridian   circumpolar star
    transit   sidereal time   hour circle
    hour angle

- Astrometry establishes the positions of celestial objects. Positions are best transformed into the International Celestial Reference Frame *(ICRS)* which is independent of motions of the Earth. Concepts:

transit telescope	*meridian circle*	*interferometer*
HIPPARCOS	*Hipparchus*	*atmospheric refraction*
fundamental catalog	*Gaia*	*VLBI*
precession	*nutation*	*apparent coordinates*
aberration of starlight	*epoch*	*ecliptic coordinates*
Galactic coordinates	*J2000*	

- Heliocentric stellar parallax is an effect that permits measurement of distances to nearby stars. Concepts:

astronomical unit (au)	*astronomical triangle*	
parallax angle	*parsec (pc)*	*Bessel*

  $$p[\text{arcsec}] = \frac{a[\text{au}]}{r[\text{pc}]}$$

- Physicists define time in terms of the behavior of light, but practical time measurements have been historically tied to the rotation of the Earth. Concepts:

atomic clock	*local apparent solar time*
TAI second	*local mean solar time*
universal time	*coordinated universal time*
zone time	*Julian date*

- The tangential component of an object's velocity in the ICRS system gives rise to a change in angular position whose rate of change is called the proper motion.

  $$v_T[\text{km/s}] = 4.74 \frac{\mu[\text{arcsec/yr}]}{p[\text{arcsec}]}$$

- The radial component of an object's velocity can be measured by a shift in its spectrum due to the Doppler effect. Similar shifts are caused by the expansion of the universe. Concepts:

  redshift parameter: $z = \Delta\lambda/\lambda \approx v_R/c$

  spectroscopic resolving power (R)

  Hubble's law: $v_R = H_0 d$ *relativistic Doppler effect*

Exercises

> Each problem that I solved became a rule which served afterwards to solve other
> problems.
>
> – René Descartes, *Discours de la Méthode* . . ., 1637

1. Two objects differ in RA by an amount $\Delta\alpha$, and have declinations δ_1 and δ_2. Show that their angular separation, θ, is given by

$$\cos\theta = \sin\delta_1 \sin\delta_2 + \cos\delta_1 \cos\delta_2 \cos\Delta\alpha$$

2. Which city is closer to New York (74°W, 41°N): Los Angeles (118°W, 34°N) or Mexico City (99°W, 19°N)? By how much? (The radius of the Earth is 6300 km.)

3. A. Kustner conducted one of the first systematic radial velocity studies. In 1905, he found that the velocity of stars in the ecliptic plane varied with an amplitude of 29.617 ± 0.057 km/s in the course of a sidereal year. Assume that the Earth's orbit is circular and use this information to derive the length (and uncertainty) of the au in kilometers.

4. Position angles are measured from north through east on the sky. For example, the figure at right shows a double star system in which component B is located in position angle θ with respect to component A. The two have an angular separation of r arc seconds. If component A has equatorial coordinates (α, δ), and B has coordinates $(\alpha + \Delta\alpha, \delta + \Delta\delta)$, derive expressions for $\Delta\alpha$ and $\Delta\delta$.

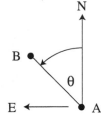

5. The field of view of the Vassar 32-inch CCD camera is a square 1000 seconds of arc on each side. Fill in the width of the field in the RA coordinate (i.e. in H:M:S units) when the telescope is pointed at declinations listed in the table:

Declination (degrees)	Width of field (minutes:seconds of RA)
0	1:06.7
20	
40	
60	
70	
80	
85	

6. The winter solstice (December 22) is the date of the longest night of the year in the northern hemisphere. However, the date of the earliest sunset in the northern hemisphere occurs much earlier in the month (at about 16:35 zone time on December 8th for longitude 0° and latitude 40°N). Examine the curve for the equation of time and suggest why this might be the case. Explain how this observation would depend upon one's exact longitude within a time zone.

7. On the date of the winter solstice, what is the approximate local sidereal time at midnight? At sunset? (Assume 40°N latitude and use a celestial sphere.)

8. A certain supernova remnant in our Galaxy is an expanding spherical shell of glowing gas. The angular diameter of the remnant, as seen from Earth, is 22.0 arc seconds. The parallax of the remnant is known to be 4.17 mas from space telescope measurements. Compute its distance in parsecs and radius in astronomical units.

9. An astronomer obtains a spectrum of the central part of the above remnant, which shows emission lines. Close examination of the line due to hydrogen near wavelength 656 nm reveals that it is actually double. The components, presumably from the front and back of the shell, are separated by 0.160 nanometers. (a) with what velocity is the nebula expanding? (b) Assuming this has remained constant, estimate the age of the remnant. (c) The astronomer compares images of the remnant taken 60 years apart, and finds that the nebula has grown in diameter from 18.4 to 22.0 arcsec. Use these data to make a new computation for the distance of the remnant independent of the radial velocity.

10. In 1840, the estimated value of the au, 1.535×10^8 km, was based upon Encke's 1824 analysis of the observations of the transits of Venus in 1761 and 1769. Encke's result should have been accorded a relative uncertainty of around 5%. If Bessel's (1838) parallax for 61 Cygni was 0.32 ± 0.04 arcsec, compute the distance and the total relative uncertainty in the distance to this star, in kilometers, from the data available in 1840. If the presently accepted value for the parallax is 287.1 ± 0.5 mas, compute the modern estimate of the distance, again in kilometers, and its uncertainty.

11. The angular diameter of the Sun is 32 arc minutes when it is at the zenith. Using the table below (you will need to interpolate), plot a curve showing the apparent shape of the Sun as it sets. You should plot the ellipticity of the apparent solar disk as a function of the elevation angle of the lower limb, for elevations between 0 and 10 degrees. (If a and b are the semi-major and semi-minor axes of an ellipse, its ellipticity ε, is $(a-b)/a$. The ellipticity varies between 0 and 1.) Is your result consistent with your visual impression of the setting Sun?

Apparent zenith distance (degrees)	75	80	83	85	86	87	88	89	89.5	90
Atmospheric refraction (arcsec)	215	320	445	590	700	860	1103	1480	1760	2123

12. The Foggy Bottom Observatory has discovered an unusual object near the ecliptic, an object some students suggest is a very nearby sub-luminous star, and others think is a trans-Neptunian asteroid. The object was near opposition on the date of discovery.

   Below are sketches of four CCD images of this object, taken 0, 3, 9, and 12 months after discovery. Sketches are oriented so that ecliptic longitude is in the horizontal direction. The small squares in the grid surrounding each frame measure 250 mas × 250 mas. Note that the alignment of the grid and stars varies from frame to frame.

   a. Why is there no frame 6 months after discovery?

   b. Compute the proper motion, parallax, and distance to this object.

    c. Is it a star or an asteroid? Explain your reasoning.

    d. Compute its tangential velocity.

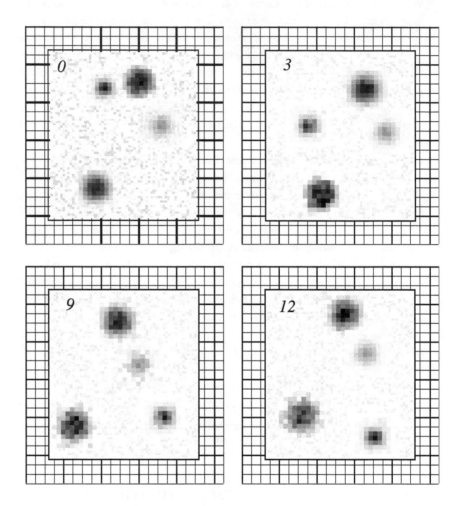

13. Your spectrograph has a resolving power of 9000. You observe a star whose spectrum has 25 lines with well-determined rest wavelengths. What is the best precision, in km s^{-1}, that can you expect if you measure the radial velocity of this star on one spectrum?

14. An astronomer measures the radial velocities of 20 stars in a nearby star cluster, and finds they have an average radial velocity of 51 km s^{-1} with a standard deviation of the sample of 16 km s^{-1}. He also determines that the proper motions of the same stars average to 14.5 arcsec per century with a standard deviation of the sample of 6.0 arcsec per century. Compute the distance of the cluster and the magnitude of its space velocity. Compute the uncertainty of your answers.

Chapter 4
Names, catalogs, and databases

> ... the descriptions which we have applied to the individual stars as parts of the constellation are not in every case the same as those of our predecessors (just as their descriptions differ from their predecessors') ... However, one has a ready means of identifying those stars which are described differently; this can be done simply by comparing the recorded positions.
>
> Claudius Ptolemy,[1] The Almagest, Book VII, H37, c. 150 CE

The names of astronomical objects bear traces of the history of astronomy – a history that starts with the mythological interpretation of the sky lingering in constellation names, and that continues to an era when comets are named after spacecraft and quasars after radio telescopes. As discoveries accumulate, so too do the names. As the number of objects of interest has risen to the hundreds of millions, tracking their identities and aliases has inflated to a daunting enterprise, made tractable only by the use of worldwide computer networks and meta-database software. In this chapter we introduce strategies for identifying a particular celestial object, but more importantly, introduce the methods for discovering what is known about it.

Very early in the history of astronomy, as Ptolemy tells us, astronomers realized the obvious. The identities of most objects in the sky, like the identities of mountains or cities, could be safely tied to their locations. However, a difficult problem arose in our Solar System (the subject of most of *The Almagest*), where objects move around the sky quickly. For 1400 years, Ptolemy was the supreme astronomical authority. He provided his students with ingenious, laborious, and ultimately inaccurate methods for predicting the positions of Solar System

---

[1] Claudius Ptolemy (*c.* 100–165 CE) spent most of his life in Alexandria, Egypt, and worked on many branches of applied mathematics. His major treatise on astronomy, written in Greek, was one of the few classical works on the subject to survive intact. This book, μεγαλη συνταξιζ (*Megale Syntaxis* – the "great composition") became *Al Magisti* ("the Greatest") in the Arabic translation. When the Arabic version reached Spain in the twelfth century, translators rendered this title as "*Almagest*" in Latin. The *Almagest* remained the unchallenged authority on astronomy until Copernicus published *De Revolutionibus* in 1543.

objects. You will find that even though you will have to deal with many more objects, modern computer technology means your experience will be less arduous than theirs. It is, of course, up to you whether or not your experience is more worthwhile.

4.1 Star names

> You can know the name of a bird in all the languages of the world, but when you're finished, you'll know absolutely nothing whatever about the bird.... I learned very early the difference between knowing the name of something and knowing something.
>
> – Richard Feynman, "What is Science?," 1966

We are about to spend several pages discussing the names of celestial objects. Bear in mind Feynman's point: a name is only a link to what others actually know about an object. It is satisfying to know all the names of the red star at Orion's shoulder, but more significant to know it is larger than the orbit of Earth, varies in brightness and size, has spots on its surface, and has nearly run out of nuclear fuel.

4.1.1 Proper names

Abd al Rahman Abu al Husain (903–86 CE), known as Al-Sufi (The Mystic), named the stars. In his *Book of the Fixed Stars* he combined traditional Arabic nomenclature with Ptolemy's catalog of 1022 stars in 48 constellations. Al-Sufi (probably a Persian) wrote in Baghdad, but his book reached Europe and became a primary source for our present names of the brightest stars. Names like Rigel, Denebola, and Altair are medieval Latin transliterations or corruptions of Al-Sufi's Arabic names, which he in turn often composed from the Greek designations collected or suggested by Ptolemy. For example, in his list for the constellation Orion, Ptolemy describes star number 2 as "the bright reddish star on the right shoulder." Al-Sufi gives the Arabic name, *Ibt al Jauzah* (Armpit of the Central One), which medieval Latin transliterates to something like *Bed Elgueze*, and hence the modern *Betelgeuse*. Independent of Al-Sufi, some modern names derive with little change from the classical Latin (Arcturus, Spica), Greek (Sirius, Procyon), or Arabic (Vega) designations. See Appendix E2 for a list of the most commonly recognized star names, and Allen (1899) for an extended discussion. In addition to the bright stars, some fainter stars have acquired enough fame to deserve "proper" names, like *Barnard's Star* (largest proper motion) and *Proxima Centauri* (nearest star to the Sun).

Professional astronomers generally avoid using such traditional names, except for the most universally recognized.

4.1.2 Bayer designations

In 1603 CE, Johann Bayer produced *Uranometria*, a star atlas based in part upon the superb positional data for 777 stars from Tycho's[2] catalog of 1598. Bayer used star names and constellation designations derived in large part from Al-Sufi. Many of the fainter stars had no traditional names, or had very obscure names, so Bayer invented a naming convention. In each constellation he assigned a Greek letter to every star. Letters went in order of brightness, with alpha allotted to the brightest star, beta to second brightest, and so on. A star's two-part name then consisted of a Greek letter followed by the Latin genitive form of its constellation name. So, for example:

Betelgeuse, the brightest star in Orion = alpha Orionis = α Ori

Alcaid, the seventh brightest star in Ursa Major = eta Ursa Majoris = η UMa

The International Astronomical Union in 1930 defined the boundaries of the 88 modern constellations and established a standard three-letter abbreviation for each (see Appendix E1 for a list). The pictorial figures these constellations are imagined to represent have various origins. Many are the 48 transmitted by Ptolemy from classical sources, and many of these are probably of very ancient origin indeed (e.g. Leo as a lion, Taurus as a bull – see Schaefer, 2006), but several are relatively new, especially in the south. Eighteenth-century astronomers proposed many of these to represent the tools of science and art (e.g. Telescopium, Microscopium, Antila, Pictor) – figures that lack deep mythic resonance.

4.1.3 Flamsteed designations

But beyond the stars of sixth magnitude you will behold through the telescope a host of other stars, which escape the unassisted sight, so numerous as to be almost beyond belief . . .

– Galileo Galilei, *The Sidereal Messenger*, 1610

---

[2] Tycho Brahe (1546–1601) may well have been the greatest observational astronomer who ever lived, and introduced several revolutionary practices that today characterize astronomy. (His name should be pronounced "Tee-ko Bra-hee." The first name is the Latinized version of the Danish "Tyge," and is usually mispronounced "Tye-ko.") Tycho used technology to push the precision of measurement to its fundamental limits. (In Tycho's case the limits were set by the resolving power of the human eye, about one arc minute. The best positional measures prior to Tycho have precisions of 10–15 arc minutes.) Over twenty years, he built and directed the great observatory complex, Uraniborg, on the island of Hveen in the Baltic. At Uraniborg, Tycho tested the reliability of new instruments by examining reproducibility of measurements and agreement with other instruments. Tycho's brand of astronomy was an expensive enterprise, and he marshalled the financial support of his own noble family as well as the state (in the person of King of Denmark, and briefly, Emperor Rudolph II of Bohemia). He practiced astronomy in an international arena, and relied on a network of scientists and technologists throughout northern Europe for advice and collaboration. See Thoren (1990) for a modern biography of Tycho.

Tycho's Catalog recorded the position and brightness of each star investigated by the greatest naked-eye observer in history. His positions are accurate to about 60 seconds of arc, a phenomenal achievement. In 1609, however, a mere six years after the publication of *Uranometria*, Galileo turned a primitive telescope to the sky. This momentous act completely transformed astronomy and coincidentally unleashed a nightmare for stellar nomenclature. Telescopes unveiled many more stars than were visible to Tycho, a number "almost beyond belief" and certainly beyond naming. Obviously, Bayer's Greek letter scheme could not gracefully be extended to include these telescopic stars. Fortunately, the telescope held out the promise of great positional accuracy, and this capability has been the key to subsequent naming practices.

The first Astronomer Royal, John Flamsteed, introduced an important convention in his *British Catalog*, published in 1722. This, the first reliable *telescopic* catalog, gave equatorial positions for 2935 stars (12 are duplicates), with names constructed within each constellation by assigning numbers in order of increasing RA, e.g.

$$\text{Betelgeuse} = 85\ \text{Ori}, \text{Alcaid} = 85\ \text{UMa}$$

4.1.4 Double stars, exoplanets, and variables

One of the earliest telescopic discoveries was that the star Mizar (ζ UMa) is in fact double. ***Double and multiple stars*** turned out to be quite common and, in the case of stars gravitationally bound to one another, astrophysically important. To distinguish the various components of a multiple star system, the IAU convention is to append a capital roman letter to the star name, *in order of brightness* (e.g. the nearest star system has the components α Cen A, α Cen B, α Cen C). In hierarchical systems append a lowercase letter for sub-components, and numbers for sub-sub components (e.g. Castor, α Gem, is a triple double, with components Aa, Ab, Ba, Bb, Ca, Cb).

Exoplanet names are normally formed by appending, *in the order of discovery,* a lowercase roman letter to the star name, starting with *b*. Applying this rule in the case of planets in a multiple star system does not distinguish planets from stars (α Cen Bb designates a planet, but mimics a star name), so the convention may evolve. The IAU has recently coordinated an assignment of proper names to certain exoplanets: e.g. 55 Cancri d = Lippershey.

One well-established modern naming convention concerns stars that change brightness. Except for the brightest (like Betelgeuse and Alcaid) ***variables*** are named *in order of discovery* by concatenating capital roman letters with the constellation name, beginning with the letters R through Z, then continuing with RR, RS, through RZ. The sequence then continues with SS to SZ, TT to TZ, and so on until ZZ. Then comes AA to AZ, BB to BZ, etc. until QZ. This provides for a total of 334 stars per constellation (the letter J is not used). If more names are required, the lettering continues with V 335, then V 336 etc. For example, all the following are variable star names:

Polaris (bright), S And, RR Lyr, V341 Cyg

Supernovae are rare stars that flare to spectacular luminosities in a single explosive episode, and then fade out forever. Almost all the supernovae discovered occur in other galaxies. They have a unique naming convention: Each is assigned a roman letter in the order in which it is discovered in a particular year. Thus, SN 1987 A was the first supernova discovered in 1987. After reaching the letter Z, the sequence continues with aa, ab . . . az, ba, bb etc. (Note the change to lower case.)

4.1.5 Durchmusterung numbers

> . . . if one seriously wants to aim at discovering all the principal planets that belong to the Solar System, the complete cataloging of stars must be carried out.
>
> – Freidrich William Bessel, Letter to the Royal Academy of Sciences in Berlin, 1824

Bessel (1784–1846) initiated the visionary project of a complete catalog of all stars brighter than ninth magnitude. Friedrich Wilhelm August Argelander (1799–1875), his student, oversaw the production and publication, in 1859–62, of the *Bonner Durchmusterung*, or BD, a catalog and atlas. The BD gives the positions and visually estimated magnitudes for 324 198 stars brighter than magnitude 9.5 and north of Dec $= -2°$. In the next generation, three other catalogs extended the stellar mapping on the remaining parts of the celestial sphere: the *Bonner Durchmusterung Extended* (in 1886, also abbreviated BD, or sometimes BDE) the *Cordoba Durchmusterung* (in 1892–1914, the CD or CoD) and the *Cape Photographic Durchmusterung* (in 1896, the CPD). The naming scheme for the *Durchmusterungs* is as follows: within each 1-degree-wide band of declination, stars are numbered consecutively in order of right ascension around the sky. The three-part name consists of the catalog abbreviation, the declination band, and the right ascension ordinal. For example:

$$\text{Betelgeuse} = \text{BD} + 70° \ 1055 = \text{BD} + 70 \ 1055, \text{Canopus} = \text{CPD} - 52 \ 914, \text{a Cen}$$
$$= \text{CD} \ 60 \ 5293$$

The BD, CD, and CPD provide designations for over a million of the brightest stars are still in widespread use in astronomical literature.

4.1.6 The nomenclature problem

There are thousands of catalogs and lists of celestial objects. Some attempt to be complete in some way (e.g. every source above a certain brightness) and many more have specialized concerns (e.g. parallaxes, colors, characteristics of variable stars, X-ray sources, radio sources, radial velocities, spectra, etc.). All catalogs result in possible names for the objects they include.

The International Astronomical Union has issued guidelines for generating a new name for any object (star, galaxy, nebula . . .) outside the Solar System. The guidelines are lengthy, but approach the problem by regularizing and restricting some of the more widespread naming conventions. They propose that new object names consist of two parts: a unique three (or more)-character acronym (like TYC, UCAC2, or GSC) followed by a sequence number. The sequence number can be the order in a list (as in the HD), or, a combination of field number and sequence (as in the BD or GSC). Preferably, the sequence number should be some specification of coordinate position in "IAU Style," that is, RA and Dec without punctuation as in JHHMMS±DDMMSS. Even with the help of internet databases, (see Section 4.5) nomenclature can get obscure and confusing. Except for bright objects, it is good astronomical practice to recall Ptolemy's advice, and give both name and position when identifying an object.

The IAU also instigated the creation of the *Reference Dictionary of Nomenclature of Celestial Objects*. The dictionary is an important resource for identifying object references as well as for avoiding ambiguity or duplication in new designations. It currently (2015) lists over 22 000 acronyms and you can access it through the *VizieR* site (vizier.inasan.ru/viz-bin/Dic).

4.1.7 Other stellar catalogs

The previous sections omitted discussion of many important star catalogs, and it would be impossible in a book like this to even list them by title. Fortunately, the power of computer networks has proved a good match to the explosive growth in the number, length, and variety of catalogs. A number of national and international data centers now hold most important catalogs online, and each site provides methods for interrogating its holdings via the internet (see Section 4.6). Table 4.1 is a small sample of the catalogs of stars (and therefore star names) that await the curious. All catalogs are most easily accessed through the SIMBAD/VizieR website (see Section 4.6).

4.2 Non-stellar objects outside the Solar System

Wonderful are certain luminous Spots or Patches, which discover themselves only by the Telescope, . . . in reality are nothing else but light coming from an extraordinary great Space in the Aether; through which a lucid Medium is diffused, that shines with its own proper Lustre. . . . These are Six in Number. . . . There are undoubtedly more of these which have not yet come to our Knowledge. . . .

– Edmund Halley, *Philosophical Transactions of the Royal Society*, vol. 4, 1721

Table 4.1 *A few examples of stellar designations from optical catalogs.*

Example name and reference	Comment
HR 2061 – *Yale Bright Star Catalog*, 5th edition, D. Hoffleit and W. H. Warren, Jr. (1991)	Widely used compilation of basic astronomical data for the 9096 stars brighter than magnitude 6.5.
HD 39801 – *The Henry Draper Catalog* Annie J. Cannon and Edward C. Pickering, *Annals of Harvard College Observatory* 91–99 (1918–1924)	Important first classification of spectra of 225 000 stars.
GSC2.3 N915–000041 – *The Hubble Guide Star Catalog II*, Lasker and Lattanzi (2008)	Ground-based photographic photometry of 9×10^8 stellar and non-stellar objects brighter than J_{PHG} = 21. Guide stars for HST and JWST.
HIP 27989, TYC 129–1873–1, UCAU2 34235809, USNO B1.0 978 1727,	Major astrometric catalogs. See Appendix D.
Giclas 123–109 *Lowell Proper Motion Survey*, H. L. Giclas et al., Lowell Observatory Bulletin (1971–1978)	12 000 stars with large proper motions.
CCDM J01016–1014AB *Un catalogue des composantes d'etoiles doubles et multiples (C.C.D.M.)*, J. Dommanget, Bull. Inf. Centre Donnees Stellaires, 24, 83–90 (1983)	Components for 34 031 double or multiple systems.
GCRV 10221. *General Catalog of Stellar Radial Velocities*, R. E. Wilson, Carnegie Institution of Washington DC, Publ. 601 (1953)	Radial velocities of 15 000 stars.

4.2.1 Bright objects

Simon Marius (1570–1624) probably used a telescope to examine astronomical objects before Galileo, and almost certainly discovered the first of Halley's "luminous spots" – the spiral galaxy in Andromeda – late in 1612. Marius described the object as "like a candle flame seen through the horn window of a lantern." One hundred years later, Halley called attention to the six extended, cloud-like objects – **nebulae** – so far discovered. Astronomers would puzzle over these fuzzy objects for two more centuries before they fully realized the great physical variety lurking behind a superficially similar appearance in the telescope. Not all lanterns hold candles.

Charles Messier (1730–1817) published the first important catalog (of 40 nebulae) in 1774. The final (1783) version lists 103 nebulae. These include relatively nearby clouds of incandescent gas, illuminated dust, star clusters in our own Galaxy, and many galaxies outside our own. (Modern authors sometimes add numbers 104–110 to Messier's list.) Messier was a comet-hunter, and had

Table 4.2 *Messier objects.*

Messier	NGC	Type	Individual name
1	1952	Supernova remnant	Crab Nebula
13	6205	Globular star cluster	Hercules Cluster
16	6611	Emission nebula	Eagle Nebula
20	6514	Emission nebula	Trifid Nebula
27	6853	Planetary nebula	Dumbbell Nebula
31	224	Spiral galaxy	Andromeda Galaxy
32	221	Elliptical galaxy	Satellite of Andromeda
33	598	Spiral galaxy	Local group galaxy
42	1976	Emission nebula	Great Nebula in Orion
44	2632	Open star cluster	Praesepe, The Beehive
45		Open star cluster	Pleiades, The Seven Sisters
51	5194	Spiral galaxy	Whirlpool Galaxy
57	6720	Planetary nebula	Ring Nebula
101	5457	Spiral galaxy	Pinwheel Galaxy

made his lists partly to avoid the embarrassment of mistaken "discoveries." His list, and not his comet discoveries, perpetuates his name, or at least his initial. The **Messier objects** include the most prominent non-stellar objects in the northern sky, many of which are commonly referred to by their Messier numbers. The great nebula in Andromeda that Marius noted is probably called "M 31" more often than anything else. Table 4.2 lists some famous Messier objects.

In 1758, Messier observed a comet and coincidently discovered M 1. At about the same time, a young musician in the Hanoverian army decided to emigrate to England rather than see further action in the Seven Years' War. In England, William Herschel (1738–1822) made his living as an organist and popular music teacher in the city of Bath. In 1772 he returned briefly to Germany to fetch his sister Caroline (1750–1848). Shortly following their return to Bath, William developed an interest in telescope-making, and after each day's work, would labor obsessively to grind or polish his latest mirror. The story goes that Caroline would place food in William's mouth and read to him while he worked on his optics. With their telescopes, Caroline discovered several comets, and William, on March 13, 1781, happened to discover the planet Uranus. This feat brought instant fame and eventual financial support from King George III. William retired from the music business in 1782 to become a full-time astronomer, assisted, as ever, by Caroline. Caroline herself was granted an annual salary in 1787, and thus became the first professional woman astronomer (that we know of). Holmes (2008) places the story of the Herschels in the context of science in the romantic age.

Familiar with Messier's lists and equipped with a superior telescope, Herschel set about discovering nebulae, and by the time of his death, he and Caroline had compiled a list of around 2000 objects. Caroline continued this work in collaboration with William's son, John Herschel (1792–1871), primarily using William's 18-inch telescope. In 1864, John presented a catalog of 5079 objects, which was further expanded by John L. E. Dreyer, a Danish–Irish astronomer, and published in 1887 as the *New General Catalog* of 7840 nebulae, listed in order of RA. Many nebulae are today known and loved by their numbers in this catalog: NGC 6822 is a faint nearby irregular galaxy, NGC 7027 is a planetary nebula, NGC 6960 is a supernova remnant (the Veil), NGC 2264 is a very young star cluster. In 1895 and 1908, Dreyer published two additional lists (6900 nebulae, total) as supplements to the NGC. Together, these lists constitute the *Index Catalog*. Only a few of these fainter objects are famous: IC 434 is the Horsehead Nebula; IC 1613 is an irregular galaxy in the local group.

4.2.2 Faint non-stellar objects

Twentieth-century astronomers soon recognized that the NGC and IC contain non-stellar objects of vastly differing physical natures, and subsequent catalogs of nebulae tended to be more specialized. Table 4.3 gives a very incomplete sampling of some of these.

4.3 Objects at non-optical wavelengths

Optical observers had a head start of several thousand years in the task of cataloging celestial objects. The opening of the electromagnetic spectrum to observation, first with radio detections from the surface of the Earth in the 1940s, then with observations at all wavelengths from space beginning in the 1960s, added a huge number of catalogs to our libraries as well as complexity and variety to astronomical names in use. Astronomers making the very first detections of objects in new bands tended to mimic the early Bayer-like naming conventions in optical astronomy: for example, the radio sources Centaurus A, Sagittarius B, etc., and X-ray sources Cygnus X-1 and Cygnus X-2.

However, as the numbers of non-visible sources accumulated, they tended to follow the IAU recommendations on nomenclature. Thus, all except the brightest sources have useful (but perhaps unromantic) names like the examples listed in Table 4.4.

4.4 Atlases, finding charts, and sky surveys

It might seem that Ptolemy had the solution to the nomenclature problem: an object's name might be obscure or confusing, but it cannot escape its position.

Table 4.3 *Catalogs of non-stellar objects. Code: HHhh = hours, decimal hours of RA, MMSS = minutes, seconds of time or arc, DDdd degrees and decimal degrees of arc, LLll, BBbb = degrees and decimal degrees of Galactic longitude and latitude; FF = field number.*

Type of object	Sample designation	Reference
287 million objects, mostly, galaxies and stars	SDSS JHHMMSS. ss+DDMMSS.s	The Sloan Digital Sky Survey, ongoing (images, photometry, and spectra). www.sdss.org/
137 Globular star clusters	GCl 101	Catalogue of Star Clusters and Associations, G. Alter, J. Ruprecht, and V. Vanysek, Akad. Kiado, Budapest, Hungary (1970) (plus supplements)
1112 Open star clusters	OCl 925	Catalogue of Star Clusters and Associations, G. Alter, J. Ruprecht, and V. Vanysek, Akad. Kiado, Budapest, Hungary (1970) (plus supplements)
1125 Emission nebulae	LBN 1090 or LBN LLL.±BB.bb	Catalogue of Bright Nebulae, B.T. Lynds, Astrophys. J., Suppl. Ser., 12, 163 (1965)
1036 Planetary nebulae	PK LLL+BB	Catalogue of Galactic Planetary Nebulae, L. Perek and L. Kohoutek, A. Pub. Czech. Acad. Sci., 1–276 (1967)
1802 Dark nebulae	LDN 1234	Catalogue of Dark Nebulae, B.T. Lynds, Astrophys. J., Suppl. Ser., 7, 1–52 (1962)
12,921 Galaxies	UGC 12345	Uppsala General Catalogue of Galaxies, P. Nilson, Nova Acta Regiae Soc. Sci. Upsaliensis, Ser. V (1973). Data for 12,921 galaxies north of delta = −23
5200 Clusters of galaxies	ACO 1234 or ACO S 1234	A Catalog of Rich Clusters of Galaxies, G.O. Abell, H.G. Corwin, Jr., and R.P. Olowin, Astrophys. J., Suppl. Ser., **70**, 1–138 (1989)

However, establishing precise coordinates is not a trivial task, nor is it trivial to point a telescope with the same precision. Often, an astronomer may wonder which of the two faint stars near the center of the field of view is the one she wants to spend the next hour observing. *Atlases* give a pictorial representation of the sky, and can be of tremendous help in identification. In practice, astronomers will routinely make use of *finding charts*, images of a small area of the sky near the object of interest. In cases where confusion is possible, astronomers will publish finding charts along with their results as aides to the identification of the objects of interest. Again, internet sites usually provide the most convenient access to atlases and the means of producing finding charts. Note that the usual practice for astronomical images duplicates the orientation on the sky, i.e. north up and *west to the right*.

Table 4.4 *A few examples of source designations at non-optical wavelengths.*

Modern designation	Source	Other designations
2MASS J05551028+0724255	Two Micron All Sky Survey (near-IR JHK photometry)	Betelgeuse
IRAS 05314+2200	Infrared Astronomical Satellite (12, 25, 60, and 100 micron photometry) 1950 position	Crab Nebula
CSC J123105.1+121611	Chandra X-ray satellite catalog	Source at the specified J2000 coordinates (M 87)
4C 02.32	Entry in the 4th Cambridge Catalog of Radio Sources	3C 273, first quasar discovered = QSO J1229+0203
FIRST J022107.4−020230	Catalog of Faint Images of the Radio Sky at Twenty centimeters	Very faint anonymous galaxy
GeV J0534+2159	Compton Gamma-Ray Observatory, EGRET instrument, J2000 position	Crab Nebula: M1, Taurus A, Tau X-1, and many other designations

The **Palomar Sky Survey** (1949–58) was, and is, an important two-color photographic atlas (POSS), originally distributed as photographic prints and glass plates. Follow-up surveys (SERC, POSS II) in the photographic era and subsequent digitization have made deep ($V < 22$) images of the entire sky available to anyone with internet access. More recently, the ongoing (since 2000) **Sloan Digital Sky Survey** is producing deeper images with more comprehensive information. Initially concerned mainly with extragalactic objects and their redshifts, four phases of the survey have supported research in many areas of astronomy. SDSS has surveyed 35% of the entire sky (mainly avoiding the Galactic plane) with an imaging five-color CCD camera, and has followed up with spectra of objects of interest. Efficient instruments and automated data reduction have so far identified about 470 million objects, and produced about 5 million spectra (mostly of galaxies brighter than $V = 17.5$, but also about 500 000 quasars and 850 000 stars). In the next decade, the **Large Synoptic Survey Telescope** (LSST) project (Chapter 6) will improve on the SDSS photometry by about 4 magnitudes in detection limit and by about a factor of 10 in precision.

Several other optical modern surveys exist (Pan-STARRS, Palomar-quest). There are also important surveys at other wavelengths, of course. Since there are strong similarities between optical and infrared techniques, we mention here only the **Two Micron All Sky Survey** (2MASS) in the J, H, and K_S infrared bands, and the **Wide-field Infrared Survey** (WISE) in the 3.4, 4.6, 12, and 22-micron bands as useful sources of images.

4.5 Solar System objects

Unlike stars and nebulae, Solar System objects move around the sky quickly. There are many examples of minor planets, comets, and small moons that were discovered and subsequently lost because their orbits were not known with sufficient accuracy.[3] There are even more examples of "new" objects that turn out to have been previously observed. This potential for confusion has led astronomers to a system of provisional designations for newly discovered small bodies, so that the provisional name can be superseded by a permanent name once the orbit is accurately known, or the object is identified as one previously cataloged.

For example, the **_Minor Planet Center_** of the IAU (Table 4.5) manages the identification process for minor planets. If an observer reports the positions of a possible discovery from at least two different nights, the Center assigns a provisional designation based on the date of the report, and attempts to identify the candidate as an object already provisionally or permanently named. If the candidate is observed on four successive oppositions without identification with a previously designated object, MPC assigns a permanent designation consisting of a catalog number and a name suggested by the discoverer.

Similar identification schemes are used for comets, small moons, and rings. Appendix E3 gives some of the rules for provisional and permanent designations.

The IAU Working Group on Planetary System Nomenclature coordinates the naming of regions and features like craters, basins, fissures, and mountains on the surfaces of large and small bodies. The group attempts to enforce an international approach in theme selection, to prevent duplication of the same name for features on different objects, to maintain a thematic structure in the names, and to avoid political or religious references.

If you know the name of an object outside the Solar System, it is relatively easy to use the SIMBAD resource to find all its aliases and produce a finding chart for it (you can also easily generate a bibliography of most of the astronomical works that mention it). To generate a finding chart for a Solar System object, however, an additional step is necessary – you need to compute the position the object will occupy at the time you wish to observe it. A table of object positions as a function of time is called an **_ephemeris_**, and the ephemerides for bright

---

[3] To determine the orbit with accuracy, one usually needs to observe the object's motion over a significant arc of the complete orbit. For a newly discovered asteroid in the belt, for example, this means following the object for about 3 or 4 weeks (about 2% of a complete orbit) in order to establish an ephemeris (prediction of future positions) for the next opposition. Only after four observed oppositions (almost a complete sidereal orbit), however, can the orbit be regarded as precisely known from ground-based observations. Opposition occurs when the RA of the Sun and the object differ by 12 hours. For periodic comets, the requirement is usually two perihelion passages, which may be separated by as long as 200 years.

Table 4.5 *Important internet data sites and services.*

Title and current URL	Description
CDS – cdsweb.u-strasbg.fr; cdsportal.u-strasbg.fr;	International center for astronomical data at Strasbourg. The portal links to important databases and bibliographic services including SIMBAD, VizieR (> 13 000 catalogs), *Aladin* (sky atlas), X-match (catalog and table crossreference), and others.
SIMBAD – simbad.u-strasbg.fr or simbad.harvard.edu	CDS astronomical object database objects. Searches by name, position, or criteria (e.g. All A0 stars with parallax > 50 mas).
NED – https://ned.ipac.caltech.edu	Extragalactic database: objects, literature, positions, data, and tools.
ADS – adsabs.harvard.edu	Searchable database of abstracts and full texts of astronomical research publications.
VAO– www.usvao.org	US virtual Observatory.
Sky View – skyview.gsfc.nasa.gov	Images from 36 surveys at 100 different wavelengths bands, with professional, amateur, and educational virtual observatory resources.
IRSA – irsa.ipac.caltech.edu	Infrared data archive from multiple sources (Spitzer, WISE, Herschel, 2MASS, and others).
MPC – www.cfa.harvard.edu/iau/ mpc.html	Minor Planet Center.
NSSDC – nssdc.gsfc.nasa.gov	NASA Space sciences data center for spacecraft investigations of Solar System objects.
Horizons – ssd.jpl.nasa.gov/? horizons	Solar System data and ephemeris generator.
Extrasolar Planets Encyclopedia – Exoplanet.eu/	Ephemerides and data on confirmed exoplanets.

objects are traditionally listed in the annual edition of the *Astronomical Almanac*. However, for these, and especially for fainter objects, online resources are the best option (an alternative is to compute the position yourself). The ***Horizons ephemeris generator*** at the Jet Propulsion Laboratory is a good example, as is the generator at the Minor Planet Center (see Table 4.5 for links).

4.6 Websites and other computer resources

The twenty-first century data storm. Each of the photographic plates of the POSS is the equivalent of a 1 Gigabyte (1000 Mb) digital image, POSS exposure times meant a raw data rate on the order of 15 Gb per clear night. However, it required about 40 years for these data to be reduced to catalog form (GSC2). Over the past two generations, accelerating developments – better telescopes, access to all parts of the electromagnetic spectrum, better detectors, and faster computers – have transformed the stately flow of data into a swirl of information. Fifty years after the POSS observations, the SDSS (*c.* 2000), with a 0.12 Gigapixel digital camera produced a comparable raw data rate (and much higher data quality), and processed images into catalog information with only an 18-month lag. Space-based and ground-based missions generate data at ever higher rates. The ground-based Pan-STARRS 1 (2010–14), for example, used a 1.4 Gigapixel camera and short exposures to generate up to 1 Terabyte/night. LSST is expected to produce 10–15 Terabytes/night.

Storms can be exhilarating, confusing, and challenging. Astronomers want to learn all we can from increasingly precise, detailed information on billions of objects. We would like to be able to point to any object and immediately learn everything known about it. We would like to map 10 billion objects in three dimensions, monitor the ones that are changing, discover which 10 000 of the 10 billion are unusual, and look for the subtle effects of elusive entities like exoplanets, dark matter, and dark energy. Computers enabled the data storm, but now provide the means of getting what we want from it.

The relatively new field of **astrostatistics** responds by developing algorithms to mine astronomical information from very large data sets. The International **Virtual Observatory** alliance makes a related response by constructing tools and services that allow researchers to interrogate data maintained at different data centers or archives, and provide astronomical information for a position, source or category: images, catalogs, and references to literature. Table 4.5 lists some of the more important data centers.

Any search of the astronomical professional research literature is best undertaken with a specialized search engine (Google and its competitors can miss a great deal). The **Astronomical Data System**, **ADS**, is probably the preferred site.

Summary

- Many systems exist for naming stars and non-stellar objects. Often, names reference a catalog entry that provides information about the object, and many names themselves contain some positional information.
- Bayer, Flamsteed, and variable star designations use constellation locations in generating a star's name as do the designations of the very brightest X-ray and radio sources.
- Other schemes simply use a sequential number, some combination of zone and sequence, or actual equatorial coordinates to produce a name. There are now IAU-sanctioned conventions in assigning new names to objects Some historically important or currently useful catalogs:

Henry Draper (HD)   *Bonner Durchmunsterung (BD, CD, CPD)*
Tycho Catalog (HIP, TYC)   *Hubble Guide Star Catalog (GSC)*
US Naval Observatory Catalogs *Messier (M) catalog*
  (AC 2000, UCAC, USNO-B)
New General Catalog (NGC)   *Sloan Digital Sky Survey (SDSS)*

- Observations at non-optical wavelengths commonly generate new catalogs and new names.
- Atlases and finding charts greatly aid the identification process, as do accurate positions of sufficient precision.
- Motion of even bright Solar System objects complicates their location and identification. Astronomers use a system of provisional names for newly discovered objects until accurate orbits can be computed, and generate ephemerides to locate known objects at any time.
- Web-accessible resources are essential for coping with problems of identification, nomenclature, and links to bibliographic information for a particular object. Very important sites are:
  ○ CDS – with subsets SIMBAD, VizieR, and *Aladin*, for the identification of objects outside the Solar System, links to basic catalog information, and a sky atlas
  ○ ADS – for searches of the astronomical research literature
  ○ JPL Horizons – for Solar System ephemerides

Exercises

1. Take the short SIMBAD tutorial at the CDS website (look under the "help" tab). (a) Use the SIMBAD *criteria query* feature to find *how many* stars brighter than $V = 6.0$ have an assigned spectral type earlier (hotter) than B1. (b) Use the criteria query and the *output options* tab to produce a list of all stars in order of distance, that have a parallax greater than 250 mas, and a proper motion greater than 3.5 arcsec/yr. Include the ICRS coordinates, radial velocity, parallax value, proper motion value, V magnitude, and spectral type in the list you produce.
2. As you know, α Cen is the nearest star. In the southern sky it forms a striking pair with β Cen. How far away is β Cen? Cite the source of this information (reference to the actual measurement).
3. Use the SIMBAD site to investigate the star cluster NGC 7790. Produce a finding chart for the cluster with the *Aladin* or Simplay application. Then:
   (a) Use VizieR to produce a list of all known and suspected variable stars within 8 arc minutes of the cluster center. Identify the three stars (one is a close double) that are Cepheid variables.

(b) Find the name and apparent magnitude at maximum brightness of the non-double Cepheid.

(c) Use SIMBAD/VizieR to look up the catalog entry in the Combined General Catalog of Variable Stars for this star and record its period and epoch of maximum light.

(d) Use a spreadsheet to predict the universal dates and times of maximum light for this star over the next month. On the same sheet, produce a calendar for the next two weeks that shows the phase (fraction of advance over a full period) the star will have at 9 pm local time on each day.

(e) Find the reference to a recent paper (use ADS) that discusses the distance to NGC 7790.

4. Use the NED/IRSA site's *Images>finder chart* tool to compare images of the galaxy M 82 from the Palomar Sky Survey, the SDSS, the 2MASS, and WISE. Comment on any differences in structure apparent at different wavelengths. (You will find it interesting to manipulate the color stretch.) (b) Use the *Simplay* or *Aladin* viewer in SIMBAD to identify the very bright star about 2 arcmin to the southwest of the center of this galaxy. Find the B-V color and spectral type of this star. (c) Use *Aladin* to examine the XXM image of M 82. Again comment on differences in structure.

5. Use the JPL Horizons site to find the position of the dwarf planets Eris and Makemake at 0 hours UT on the first day of next month. One of these should be visible at night. Generate a finding chart that would enable you to identify the dwarf planet on that date and compute the approximate local standard time at which it crosses the meridian.

Chapter 5
Optics for astronomy

> But concerning vision alone is a separate science formed among philosophers, namely, optics. . . . It is possible that some other science may be more useful, but no other science has so much sweetness and beauty of utility. Therefore it is the flower of the whole of philosophy and through it, and not without it, can the other sciences be known.
>
> – Roger Bacon, *Opus Maius*, Part V, 1266–68

Certainly Bacon's judgment that optics is the gateway to other sciences is particularly true of astronomy, since virtually all astronomical information arrives in the form of light. We devote the next two chapters to how astronomers utilize the sweetness and beauty of optical science. This chapter introduces the fundamentals.

We first examine the simple laws of reflection and refraction as basic consequences of Fermat's principle, then review the behavior of optical materials and the operation of fundamental optical elements: films, mirrors, lenses, fibers, and prisms.

Telescopes, of course, are a central concern, and we introduce the simple concept of a telescope as camera. We will see that the clarity of the image produced by a telescopic camera depends on many things: the diameter of the light-gathering element, the turbulence and refraction of the air, and, if the telescope uses lenses, the phenomenon of chromatic aberration. Concern with image quality, finally, will lead us to an extended discussion of monochromatic aberrations and the difference between the first-order and higher-order ray theories of light.

5.1 Principles of geometrical optics

This section reviews some results from geometrical optics, and assumes you have an acquaintance with this subject from an introductory physics course. Geometrical optics adopts a ray theory of light, ignoring many of its wave and all of its particle properties.

5.1.1 Rays and wavefronts in dielectric media

The speed of light in a vacuum is a constant, c, identical for all observers. Experiment shows that the ***phase velocity*** of light waves (the speed at which the changes in the electric field propagate) in transparent dielectrics like air, water, or glass is always less than c. The ***index of refraction*** measures the degree to which a particular material reduces the speed of the light wave. If v is the actual speed of light in a medium (the phase velocity), then the index of refraction of the material is

$$n(\lambda) = \frac{c}{v(\lambda)} \qquad (5.1)$$

In general, n depends on the chemical and physical (e.g. temperature) properties of the medium, as well as on wavelength. Table 5.1 lists the refractive indices of a few important dielectrics.

Table 5.1 also lists the ***chromatic dispersion***, $dn/d\lambda$, of each material. The dispersion is an important quantity – the "fire" of a diamond, for example, is due to its high chromatic dispersion – and is often itself a strong function of wavelength. Glassmakers traditionally express dispersion as the Abbe number, the quantity $v_D = (n(\lambda588) - 1)/(n(\lambda486) - n(\lambda656))$. For optical glasses, the Abbe number ranges from about 20 to 85.

Figure 5.1a shows a bundle or pencil of light rays that originate at a point source in a homogeneous dielectric medium. In a homogeneous medium, the rays are straight lines. Along each ray, you can measure s, the distance that light, moving at the phase velocity, will travel in a given time, t:

Table 5.1 *Indices of refraction. Both the index and dispersion are given at the wavelength of the Fraunhofer D line. Data after Pedrotti et al. (2006) and the Schott Glass (2009) website: www.us.schott.com/optics_devices/english/download/kataloge.html.*

Material	n ($\lambda = 588$ nm)	$dn/d\lambda$ (μm^{-1})	v_D
Air (STP)	1.00029	2×10^{-5}	85
Water	1.33	0.017	114
Calcium fluoride	1.435	0.027	95
Fused quartz	1.458	0.040	68
Fluoride glass (N-FK5)	1.487	0.040	70
Borosilicate crown glass (BK7)	1.517	0.047	64
Flint glass (F2)	1.620	0.100	36
Dense flint glass (SF4)	1.756	0.161	28
Diamond	2.42	0.250	33

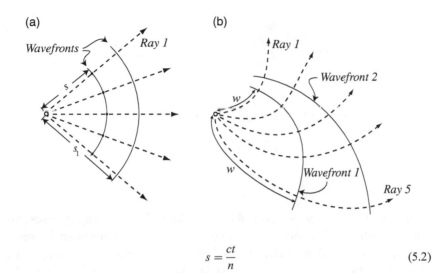

Fig. 5.1 (a) Dotted lines represent rays from the source and the smaller solid curve is a wavefront that has traveled a distance *s* in time *t*. The larger solid curve is a wavefront at distance s_1. (b) Rays and wavefronts in a medium where the index of refraction is not homogeneous. Each point on a wavefront has the same optical path length from the source.

$$s = \frac{ct}{n} \tag{5.2}$$

The figure locates some points at identical *s* values on each ray. These points delineate a surface called a ***geometrical wavefront*** and, in Figure 5.1a each wavefront is a segment of a sphere. Wavefronts are always orthogonal to rays, and provide an alternative visualization of the transmission of light: you can either picture light as energy sliding along rays, or you can imagine light propagating as a series of wavefronts that expand away from the source at the phase velocity.

Figure 5.1b shows a more complicated situation, where the medium is inhomogeneous (here *n* increases toward the upper right). Light rays are no longer straight, and wavefronts no longer spherical. For example, because it travels through a higher index, light moving along ray 1 moves a shorter physical distance than light moving for the same time along ray 5. Wavefront 1 locates photons that have left the source together. We say that wavefront 1 locates the ends of rays of equal ***optical path length***. If d*s* is an infinitesimal element of length along a path, the light-travel time along a ray is just

$$t = \int \frac{ds}{v} = \frac{1}{c} \int n ds = \frac{w}{c} \tag{5.3}$$

where the quantity

$$w = \int n ds \tag{5.4}$$

is the optical path length. Everywhere on a wavefront, then, the optical path length and travel time to the source are constants. The wavefront concept is a very useful *geometrical* concept and does not depend on the actual behavior of light as a wave phenomenon. Nevertheless, in some situations with a coherent source (where all waves are emitted in phase) the geometrical wavefronts also correspond to surfaces of constant phase (the ***phase fronts***).

Fig. 5.2 Reflection and refraction at a plane boundary. The dashed line is the normal to the interface, and the arrows show the reflected and refracted rays for a ray incident from the lower left.

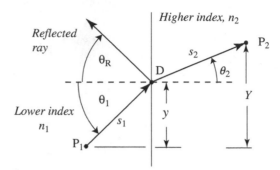

In Figure 5.2, a plane perpendicular to the plane of the diagram separates two different materials. The index of refraction is larger in the material on the right. In the plane of the diagram (the **plane of incidence**) a light ray travels upwards and to the right, striking the normal to the interface at the **angle of incidence**, θ_1. At the interface, the ray splits into two components – a reflected ray, which stays in the original material, and a refracted ray, which crosses the interface. These two rays respectively make angles θ_R and θ_2 with the normal. By convention, we measure positive angles counterclockwise from the normal.

In 1652, Pierre de Fermat[1] formulated a curious statement that can be used to determine the exact paths of the reflected and refracted rays. **Fermat's principle** asserts that the path of a ray between two points will always be the one that constitutes an extremum (i.e. a local minimum or, occasionally, maximum) in the total travel time, or, equivalently, in the total optical path length. Fermat's principle, in the form of a simple geometric argument (see Problem 1), implies the familiar law of reflection. That is,

$$\theta_1 = -\theta_R \qquad (5.5)$$

With regard to refraction, an example may convince you that Fermat's principle should lead to a change in direction upon crossing the interface. In Figure 5.3, Tarzan, who is lounging on the sand at point A, observes that Jane is about to be devoured by crocodiles in the water at point B. Tarzan knows his running speed on smooth sand is much higher than his swimming speed in crocodile-infested water. He reasons that the straight-line path ACB will actually

---

[1] Fermat (1601–65) lived quietly and published little, although he was well respected as a superb mathematician and corresponded with the leading scientists and mathematicians of his day, including Huygens and Pascal. He was moved to publish his principle in optics by a dispute with Descartes. (Fermat was correct, but Descartes had great influence and managed to damage Fermat's reputation and delay acceptance of his principle.) Many of Fermat's most important mathematical results – many without proof – were discovered in his private papers and in marginal notes in texts in his library only after his death. These included his famous "last theorem," which withstood proof until 1994.

take longer to traverse than path ADB, since ADB involves considerably less swimming and fewer vexatious reptiles. The "ray" Tarzan actually traverses is thus "refracted" at the sand–water interface. The angle of refraction (Tarzan's choice of point D) will depend on his relative speeds in sand and water.

Returning to Figure 5.2, we can apply Fermat's principle to deduce the path of the refracted ray by requiring the optical path between the fixed points P_1 and P_2 to be an extremum – a minimum in this case. Treat the distance y as the variable that locates point D. Then Fermat's principle demands

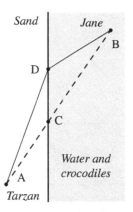

Fig. 5.3 Fermat's principle applied by Tarzan.

$$\frac{dw}{dy} = \frac{d}{dy}(s_1 n_1 + s_2 n_2) = 0 \qquad (5.6)$$

Substitution for the distances s_1 and s_2 leads (see Problem 2) to **Snell's law of refraction**:

$$n_1 \sin(\theta_1) = n_2 \sin(\theta_2) \qquad (5.7)$$

The sense of Snell's[2] law is that rays traveling from a lower index medium to a higher index medium (the situation in Figure 5.3) will bend toward the perpendicular to the interface. Note that Equation (5.7) reduces to the law of reflection if we take $n_1 = -n_2$.

An equivalent conception of refraction describes the turning of a wavefront, as in Figure 5.4a. This view suggests that when one part of a wavefront is slowed down by the medium on the right, it turns so that it can keep pace with its faster-moving portion in the left-hand region.

Snell's law applies equally well if the incident ray travels from right to left in Figure 5.3. In this case, moving from a higher to a lower index medium, the refracted ray bends away from the perpendicular to the interface. In fact, there must be a certain angle of incidence, called the **critical angle**, which produces a refracted ray that bends so far from the perpendicular that it never leaves the higher index medium. From Equation (5.7) you can see that the critical angle is given by

$$\theta_C = \sin^{-1}\left(\frac{n_1}{n_2}\right) \qquad (5.8)$$

What actually happens is called **total internal reflection** – for angles of incidence greater than critical, there is no refracted ray, and all light that reaches the interface is reflected back into the higher index medium. Snell's law is a general result that applies to interfaces of any shape, and (with reflection as a special case) can be used as the foundation of almost all of geometrical optics.

[2] Ibn Sahl (Abu Sa'd al-'Ala' ibn Sahl, *c.* 940–1000 CE) apparently published the law of refraction (in the form of a diagram with the correct ratios of the sides of right triangles) considerably before its rediscovery by the Dutch mathematician Willebrord Snellius (1580–1626) and its subsequent popularization by Descartes.

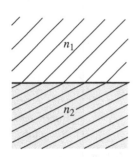

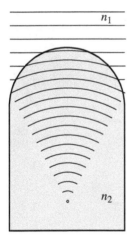

Fig. 5.4 (a) Plane wavefronts traversing a plane interface between a low index medium on the top and a higher index medium on the bottom. (b) Plane wavefronts traversing a curved interface producing a focusing effect – wavefronts become spherical after refraction.

5.1.3 Reflection and transmission coefficients

The laws governing the relative *intensities* of the incident beam that are reflected and refracted fall outside the realm of geometrical optics, and are deduced by rather messy applications of the theory of electromagnetic waves. ***Fresnel's formulas***, for the reflection and transmission coefficients, give the amplitudes of the reflected and refracted waves as a function of angle of incidence, polarization, and indices of refraction. You should be aware of a few results:

(a) Polarization makes a difference. Waves polarized with the electric field vectors perpendicular to the plane of incidence (the transverse electric, or TE, case) in general are reflected differently from waves polarized with the magnetic field perpendicular to the plane of incidence (the transverse magnetic, or TM, case).

(b) Light moving from a lower index to a higher index medium experiences a 180-degree phase shift. Light moving from higher to lower index does not undergo a phase shift.

(c) The ***reflectance***, R, is the fraction of the power of the incident wave that is reflected. At normal incidence ($\theta_1 = 0$) for all cases (TE, TM, external, or internal):

$$R = \left(\frac{n_1 - n_2}{n_1 + n_2}\right)^2 \tag{5.9}$$

(d) For both the TE and TM polarizations, the reflectance becomes large at large angles of incidence. In the external case, $R \to 1.0$ as $\theta_1 \to 90°$, and light rays that strike a surface at ***grazing incidence*** (θ_1 close to 90°) will be mostly reflected. For the internal case, $R = 1.0$ for all angles greater than the critical angle.

(e) For all values of θ_1 other than those described above, R is smaller for the TM polarization than for the TE polarization. Thus, initially unpolarized light will become partially polarized after reflection from a dielectric surface. At one particular angle (Brewster's angle, $\theta_p = \tan^{-1}(n_1/n_2)$), in fact, $R = 0$ for the TM polarization, and only one polarization is reflected.

5.1.4 Reflecting materials

An ideal mirror should have a reflectivity of 1.0 for all wavelengths of interest. The substrate should be easy to shape to an accuracy of a fraction of the shortest of those wavelengths, and once shaped, the substrate and its coating, if any, should be mechanically and chemically stable. Mirrors in astronomical telescopes are often both large and mobile, and may even need to be placed into space; so low mass is a virtue. Since telescopes are normally in environments where the temperature can change rapidly, they should have high thermal conductivity and a low coefficient of thermal expansion.

No materials match this ideal, but some are better than others. For the reflecting telescope's first two centuries, telescope makers fashioned mirrors out of ***speculum metal***, an alloy, primarily of copper and tin, that is difficult to prepare. Although speculum mirrors produced some historic discoveries (e.g. Herschel and Uranus), speculum is dense, only 45% reflective at best, and tarnishes easily. Astronomers quickly switched to silvered-glass mirrors (90% reflective) once that technology became available in the 1880s. Compared to speculum, glass is much more easily worked, has a lower density, and better mechanical and thermal stability.

Most modern optical telescope mirrors generally utilize substrates made with special glasses (e.g. Pyrex) or ceramics (Cervit or Zerodur) that have very low coefficients of thermal expansion. Often large mirrors are ribbed or honeycombed on the back face to minimize mass while retaining rigidity. Choice of surface coating depends on the application. A coating of metallic aluminum, over-coated with a protective layer of silicon monoxide, is the usual choice in the near ultraviolet and optical because of durability and low cost. Silver, which is poor in the ultraviolet, is superior to aluminum longward of 450 nm, and gold is a superb infrared reflector longward of 650 nm. Solid metal mirrors still have limited application in situations where their high thermal conductivity is especially useful. Beryllium, although toxic, is the lowest density workable metal, with excellent rigidity. The Spitzer Space Telescope has a bare polished beryllium mirror, and the James Webb Space Telescope mirror will use gold-coated beryllium segments. Extremely large ground-based telescopes now in the planning stages will probably utilize low-density materials like beryllium and silicon carbide, which, although expensive, are superior to glass for mirror substrates. Composite materials incorporating carbon fibers and nanotubes hold some promise as substrate components, but are still under development.

Very short wavelengths (extreme ultraviolet (EUV) and shorter) present two difficulties for mirrors: First, rather than being reflected, energetic photons tend to be absorbed, scattered, or transmitted by most materials. Second, for optimum images, curved mirrors need to be shaped with an accuracy of at least $\lambda/4$, which, for a 1 nm X-ray, amounts to one atomic diameter. X-Ray and EUV focusing telescopes have usually been designed with metal mirrors operating in "grazing-incidence" mode.

5.1.5 Transmitting materials

Transmitting materials form lenses, windows, correctors, prisms, filters, fibers, and many other more specialized elements. Of primary concern are index of refraction, dispersion, and absorption. Other properties of relevance to astronomical applications include: homogeneity, thermal expansion, frequency of bubbles and inclusions, and dependence of refractive index on temperature. Environmental stability is especially important for instruments (e.g. in spacecraft) that cannot be easily adjusted.

Fig. 5.5 Refractive index as a function of wavelength for some optical glasses in the Schott Catalog. Curves end where the glass becomes nearly opaque, except we show the curve for borosilicate crown glass (N-BK7) as a dotted line in the region where it is highly absorbing. SF4 is one of the most dispersive of the flints. N-FK5 is a fluoride glass. N-LAK10 is an unusual crown with low dispersion and high index. SiO₂ is fused quartz, and CaF₂ is calcium fluoride.

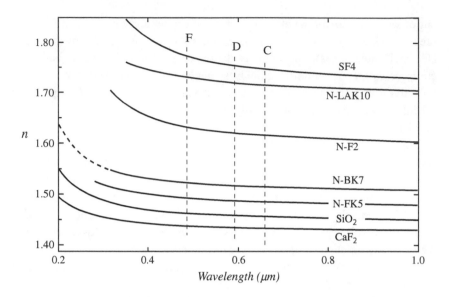

At visible wavelengths, a variety of optical glasses exhibit indices of refraction ranging from 1.5 to 1.9, and dispersions (at 588 nm, the Fraunhofer D line) in the range -0.03 to -0.18 μm^{-1}. Such a variety is important in designing systems free from chromatic aberration (see below). Generally, glasses with a high index will tend to have a dispersion with a large absolute value and are termed "flints," while those with dispersions above -0.06 μm^{-1} (closer to zero) are called "crowns." Figure 5.5 shows the run of index with wavelength for several optical glasses. Note that as the index and dispersion rise at short wavelengths, glasses become highly absorbing.

In the ultraviolet (from about 150 to 400 nm) ordinary glasses become opaque. Fused quartz (SiO_2) is the exception. It transmits well over all but the shortest wavelengths in this range, has a low thermal expansion coefficient, and can be shaped to high accuracy. All other ultraviolet-transmitting materials are not glassy, but crystalline, and more difficult to shape and more likely to chip and scratch. The most useful of these is probably calcium fluoride CaF_2, which transmits from 160 nm to 7 μm. Other fluoride crystals (BaF_2, LiF, MgF_2) have similar properties. Fused quartz and the fluorides do not transmit well below 180 nm, and some birefringent crystals (most notably sapphire, Al_2O_3) find limited use as windows in the very far ultraviolet. Below 150 nm, optics must be reflecting, and for wavelengths below about 15 nm, only grazing-incidence reflections are practical.

In the infrared, ordinary optical glasses transmit to about 2.2 μm and some special glasses transmit to 2.7 μm. Infrared-grade fused quartz transmits to about 3 μm. A large selection of crystalline materials, many identical to those useful in

Table 5.2 *Some materials transparent in the infrared.*

Material	Wavelength range (µm)	n	Comments
Sapphire	0.14–5	1.7	Slightly birefringent
LiF	0.18–6	1.4	Slowly degrades with humidity
BaF$_2$	0.2–11	1.4	Slowly degrades with humidity
ZnS	0.5–12	2.4	Strong
ZnSe	0.6–18	2.4	Soft
NaCl	0.25–16	1.5	Water soluble
Csl	0.4–45	1.7	Water soluble

the ultraviolet, transmit to much longer wavelengths, but most are soft, or fragile, or sensitive to humidity, so can only be used in protected environments. Germanium (transmits 1.8 to 12 µm) has a high index (4.0) and low dispersion, so is especially useful for making lenses. A few infrared- and ultraviolet-transmitting materials are listed in Table 5.2.

5.1.6 Thin film coatings

Coating the surface of an optical element with a thin film can exploit the wave properties of light to either increase or decrease its reflectance. A thin film ¼ wavelength thick applied to a glass surface, for example, will introduce two reflected beams, one from the front surface, and the second from the film–glass interface. The second beam will emerge one-half wavelength out of phase from the first, and the two reflected beams will destructively interfere. If their amplitudes are equal, then the interference will be total, and the reflectance reduced to zero for that wavelength. If the index of refraction of the glass is n_s, the condition for equal amplitudes is that the index of the film be

$$n_F = \sqrt{n_s} \tag{5.10}$$

For glass with an index of 1.5, this implies a coating with index 1.22. Amorphous MgF$_2$ ($n = 1.38$) or sometimes cryolite (Na$_3$AlF$_6$, $n = 1.34$) is the common practical choice for single-layer coatings. An anti-reflection coating works best only near the design wavelength, but multiple coatings of varied indices can greatly expand the width of that band. A similar treatment can *enhance* the reflectivity of a surface – multiple layers of alternating high and low index materials can improve the reflectivity of a mirror over a broad range of wavelengths.

5.1.7 Cleaning

Accumulation of dust or dirt on a surface can significantly reduce transmission or reflection and increase scattering. However, cleaning is always an opportunity for surface damage, so requires great care. Gentle washing with detergent, followed by thorough rinsing, is effective, economical, but not always practical. Pressurized streams of gas or carbon dioxide snow can remove looser particles, but organic material (fingerprints!) is generally very sticky. Solvents like acetone or alcohols are helpful. Special cleaning gels applied to the surface solidify and remove dirt particles when peeled off, a process that is quite effective, although expensive.

5.1.8 Reflection at a spherical surface

Important results in geometrical optics describe reflection and refraction at a spherical interface. We begin with reflection, since it is somewhat simpler, and since most modern astronomical telescopes use one or more curved mirrors as their primary light-gathering and image-forming elements. In Figure 5.6, parallel rays of light are traveling from left to right and reflect from a concave spherical surface of radius R whose center is at C. This figure illustrates conventions we will apply in geometrical optics. First, the horizontal dotted line that is coincident with the axis of symmetry of the system is called the **optical axis**. We set up a right-handed Cartesian coordinate system, its z-axis is coincident with the optical axis, its origin is at the **vertex** of the mirror, its x-axis goes into the page, and we assume light enters the system traveling in the direction of increasing z. Radii are positive if displacement from vertex to center (e.g. V to C in Figure 5.5) is in the $+z$-direction.

Now consider the special case of the **paraxial approximation** (also called the **Gaussian** case) – the assumption that all incident rays are nearly parallel to the optical axis, and that all angles of reflection are small. This latter assumption

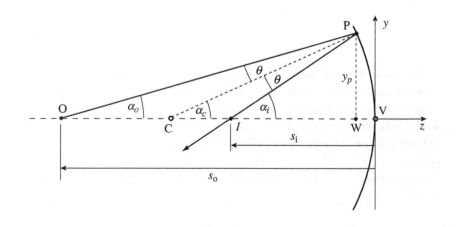

Fig. 5.6 Reflection from a spherical surface.

means that the diameter of the mirror (e.g. in the y-direction) is small compared to its radius of curvature. Some very useful relationships then apply. In Figure 5.6, consider the ray that originates at the **object** at point O on the optical axis, and is reflected to reach the **image** point I. In the paraxial approximation, all angles labeled α and θ must be very small, and the distance WV is also very small.

In this case, triangles with side y_p in common yield

$$\begin{aligned}
\alpha_o &\approx \tan(\alpha_o) \approx \frac{y_p}{s_o} \\
\alpha_c &\approx \tan(\alpha_c) \approx \frac{y_p}{R} \\
\alpha_i &\approx \tan(\alpha_i) \approx \frac{y_p}{s_i}
\end{aligned} \tag{5.11}$$

If we also consider triangles OPC and CPI, we have

$$\begin{aligned}
\theta &= \alpha_c - \alpha_o = \alpha_i - \alpha_c \\
2\alpha_c &= \alpha_o + \alpha_i
\end{aligned} \tag{5.12}$$

and substituting into the last equation from the first three approximations, we have the paraxial equation for mirrors:

$$\frac{2}{R} = \frac{1}{s_o} + \frac{1}{s_i} \tag{5.13}$$

The distance $R/2$ is termed the **focal length** of the mirror, and is often symbolized by f, so the above equation is usually written as:

$$\frac{1}{f} = \frac{1}{s_o} + \frac{1}{s_i} = -P \tag{5.14}$$

The sign convention (z value increase to the right) means that R, f, s_o, and s_i are all negative numbers. The quantity P on the right-hand side of this equation is called the **power** of the surface. (The units of P are m^{-1}, or **diopters**.) Note that if the **object distance** s_o approaches infinity, the **image distance** s_i, approaches f.

Figure 5.7 illustrates this, where every ray parallel to the axis passes through point F, and where the distance FV is the focal length, f, of the mirror. This would be the situation if a very distant source, like a star, were located on the axis at $z = -\infty$; all incident rays from that source are parallel, and we can say that the mirror **gathers** and concentrates them at F. Clearly, the **light-gathering power** (the amount of light brought to focus) of a mirror will be directly proportional to its surface area, which is in turn proportional to the square of its **aperture** (diameter). Bigger mirrors or dishes (radio astronomers use the word "dish" instead of "mirror") gather more light. Since most astronomical sources are faint, bigger is, in this one regard, better.

A convex mirror, illustrated in Figure 5.8, will disperse, rather than gather, a bundle of parallel rays. For convex mirrors, the paraxial approximation still results in the same expression, Equation (5.14) that applied to concave mirrors.

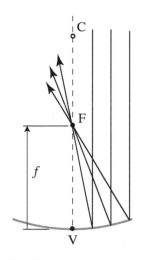

Fig. 5.7 Rays parallel to the optical axis all pass through point F, which is one focal length from the vertex.

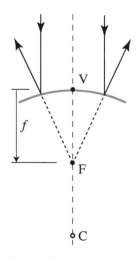

Fig. 5.8 The focal point of a convex spherical mirror.

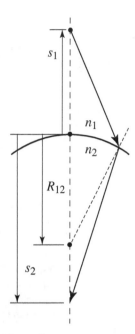

Fig. 5.9 Refraction at a spherical interface. *R* and s_2 are positive; s_1 is negative.

The sign convention means that the power, *P*, of a convex mirror is negative, its focal length, *f*, is positive. The image, at *F*, is said to be *virtual*.

5.1.9 Refraction at a spherical surface

Figure 5.9 illustrates a ray refracted at a spherical interface between media of differing indices of refraction. You can derive the paraxial equation for refraction in a way analogous to the derivation for spherical mirrors. Begin with Snell's law and assume all angles are small. The result, which we simply state without proof, is

$$\frac{n_2}{s_2} - \frac{n_1}{s_1} = \frac{(n_2 - n_1)}{R_{12}} \tag{5.15}$$

As in the case for mirrors, take the focal length, f_2, to be the value of s_2 when s_1 approaches infinity:

$$f_2 = \frac{n_2 R_{12}}{(n_2 - n_1)} \tag{5.16}$$

So, for a refractive surface, the paraxial equation for image and object distances is

$$\frac{n_2}{s_2} - \frac{n_1}{s_1} = \frac{n_2}{f_2} = -\frac{n_1}{f_1} = P_{12} = \frac{(n_2 - n_1)}{R_{12}} \tag{5.17}$$

Again, the quantity P_{12} is called the **power** of the surface. The power measures how strongly an interface will converge (or diverge, if *P* is negative) a bundle of parallel rays. For both mirrors and refracting surfaces, a plane ($R \to \infty$) surface has zero power.

5.2 Lenses, mirrors, and simple optical configurations

5.2.1 Thick lenses

Figure 5.10 shows the cross-section of a **thick lens** in air ($n_1 = n_3 = 1, n_2 = n$). If you apply Equation (5.17) to each surface in succession, the power of the combined surfaces in the Gaussian case is given by **Gullstrand's equation**

$$P = \frac{1}{f} = P_{12} + P_{23} - \frac{d}{n} P_{12} P_{23} \tag{5.18}$$

Where $P_{12} = (n - 1)/R_{12}$ and $P_{23} = (1 - n)/R_{23}$. The focal length on each side is measured from the corresponding **principal plane** for that side (*H* and *H'* in the figure). Object and image distances are given by

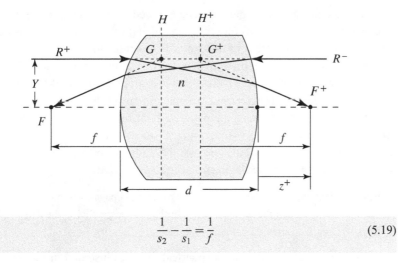

$$\frac{1}{s_2} - \frac{1}{s_1} = \frac{1}{f} \qquad (5.19)$$

Fig. 5.10 Refraction by a thick lens in air. To locate the principal plane for rays moving in the +z-direction, trace a ray (R^+) incident parallel to the axis at height Y. This ray emerges and crosses the axis at F^+. Extend this emergent ray backward (dashed line) until it reaches height Y at point G^+. The principal plane H^+ is perpendicular to the axis and contains G^+. A similar construction using ray R^- locates the other principal plane at H. The back focal distance, z^+, is measured from the right vertex to point F^+.

Here each quantity is measured from the corresponding principal plane. Figure 5.10 shows the construction that locates the principal planes. Often, it is important to compute the back focal distance, z^+:

$$z^+ = f\left(1 - \frac{d}{nf_2}\right) \qquad (5.20)$$

5.2.2 Thin lenses

If you further simplify to the limiting case of a **thin lens**, the assumption that d is negligibly small reduces Equation (5.18) to

$$P = P_{12} + P_{23} = (n-1)\left(\frac{1}{R_{12}} - \frac{1}{R_{23}}\right) = \frac{1}{f} = \frac{1}{s_2} - \frac{1}{s_1} \qquad (5.21)$$

Note that the focal length (and power) of a thick or thin lens in air (unlike a single refracting surface) is the same for rays moving in the +z-direction as for rays in the −z-direction. Likewise, note except for the sign on P and on the $1/s_1$, term, Equation (5.21) for a thin lens is identical to Equation (5.14) for a mirror. In the discussion of optical layouts that follows, then, we will often replace a mirror with a thin lens of the same aperture and power, provided we make the appropriate reflection in ray direction and sign.

5.2.3 Graphical ray tracing

The best way to evaluate an optical design is to trace the paths of many rays from an object through all the optical elements, applying Snell's law and/or the

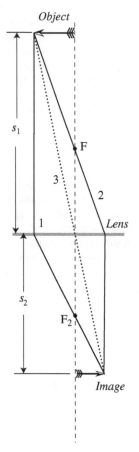

Fig. 5.11 Graphical ray tracing for a thin lens.

law of reflection at every interface encountered. ***Ray-tracing computer programs*** use exact formulations (not the paraxial approximation) to do this and are essential tools for the design of astronomical instruments.

However, the paraxial approximation itself permits a simple method suitable for rough estimates. Figure 5.11 illustrates this graphical method for ray tracing through a thin lens. Here we have an object (an arrow) located to the top of the diagram and we want to trace the paths of light rays it emits. Equation (5.19) predicts the trajectories of rays that originate on-axis at the tail of the arrow. For rays that originate off-axis, rules predict the paths of three particular rays, one more than needed to locate the image of the arrowhead:

1. Rays incident parallel to the axis emerge through the image focal point, F_2.
2. Rays incident through the object focal point, F, emerge parallel to the axis.
3. Rays through the vertex do not change direction.

A similar set of rules applies for a spherical mirror, illustrated in Figure 5.12:

1. Rays incident parallel to the optical axis are reflected through the focal point, F.
2. Rays incident through the focal point are reflected parallel to the axis.
3. Rays incident on the vertex are reflected back at an equal and opposite angle.
4. Rays incident through the center of curvature, C, are reflected back on themselves.

5.2.4 Multiple lenses

Most practical optical systems have multiple elements, and many of these are designed by ray-tracing computer programs. However, it is sometimes useful to estimate the properties of such systems through manual tracing techniques, or through algebraic formulas. For example, in the thin-lens limit, the formula for the combined power and back focal distance of two aligned lenses separated by distance d is given by Equations (5.18) and (5.20), with the index set to 1, for instance:

$$P = \frac{1}{f} = P_1 + P_2 - dP_1P_2 \qquad (5.22)$$

As with the thick lens, the focal length is measured from the appropriate principal plane.

5.2.5 The thick plane-parallel plate

A thick plane-parallel plate of index n_2 has a power of zero, so does not cause rays to change direction (see Figure 5.13). However, for a converging or diverging beam, the plate will displace the location of the focus by an amount

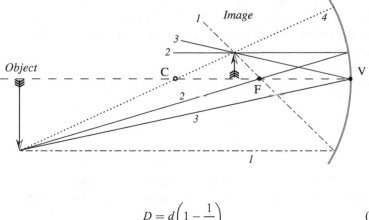

$$D = d\left(1 - \frac{1}{n_2}\right) \tag{5.23}$$

Parallel plates frequently appear in astronomical instruments as filters, windows, and specialized elements.

5.2.6 Refraction by an atmosphere

We can approximate a planet's atmosphere as a series of plane-parallel plates, and the surface as an infinite plane. In Figure 5.14a, for example, we imagine an atmosphere of just two layers that have indices, $n_2 > n_1$. A ray incident at angle α refracts at each of the two interfaces, and ultimately makes a new angle with the surface, $\alpha + \Delta\alpha$: thus, for an observer on the surface refraction shifts the apparent position of a source toward the zenith. In Figure 5.14b, we imagine that the atmosphere consists of a very large number of thin layers, so in the limit, the effect of refraction is to curve the path of the incident ray. In this limit, the plane-parallel model gives, in radians,

$$-\Delta\alpha = R_0 \tan\alpha = \frac{(n^2 - 1)}{2n^2} \tan\alpha \approx (n - 1)\tan\alpha \tag{5.24}$$

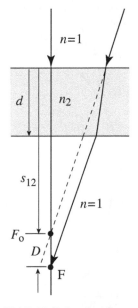

Fig. 5.13 Refraction by a thick plane-parallel plate in air.

where n is the index of refraction at the surface. The quantity $(n - 1) \times 10^6$ is called the *refractivity*. As discussed in Chapter 3, very precise correction for refraction is site-dependent, and in any case, Equation (5.24) fails near the horizon, as does the plane-parallel approximation. Figure 3.10 gave some values for $\Delta\alpha$ as a function of α.

Since the index is a function of wavelength (see Figure 5.14c and Table 5.3), rays of different colors are refracted at slightly different angles, and for Earth, images observed through air are actually very low-resolution spectra – with the blue image closer to the zenith than the red. Atmospheric dispersion (as opposed to total refraction) is quite small in the near infrared, so this effect is unimportant there. However, chromatic image distortion in the optical becomes very

(a)

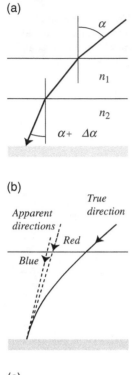

Table 5.3 *Atmospheric refraction as a function of wavelength. Data are for dry air at 0°C and standard pressure. Actual refractivity depends on humidity and scales with P/T. Data here are from Cox (1999). More detailed treatment is given by Young (2006).*

λ (nm)	$(n-1) \times 10^6$	R_0 (arcsec)	λ (μm)	$(n-1) \times 10^6$	R_0 (arcsec)
200	342	70.4	0.7	291	60.0
300	308	63.3	1	289	59.6
350	302	62.2	4	288	59.3
400	298	61.5	10	288	59.3
500	294	60.6	10 cm	355	73.2

pronounced at large zenith distances. In applications where high-resolution images are important, a telescope can be equipped with an atmospheric dispersion compensator (ADC), which is often based on counter-rotating Amici prisms – see Figure 5.16.

5.2.7 Optical fibers

The optical fiber is an important application of total internal reflection. As illustrated in Figure 5.15, a long cylinder of high-index material can serve as a guide for rays that enter one face of the cylinder and strike its side at angles greater that the critical angle. Such rays will travel the length of the cylinder, reflecting from the walls, to emerge from the other face.

Typical fibers are manufactured with a core of high-index glass or plastic enclosed in a cladding of a lower index material. Internal reflections occur at the core–cladding boundary. Although the cladding serves to protect the core from scratches and dirt (either would generate light loss), many fibers are also coated with an additional layer called the buffer, which provides further protection from mechanical and chemical hazards. The core diameters of fibers used in astronomy are typically in the 50–200 micron range, are made of special materials, are quite flexible, and have multiple applications. A single fiber, for example, can conduct light from the focus of a (moving) telescope to a large or delicate stationary instrument like a spectrograph, or many fibers can simultaneously feed light from multiple images (one fiber per image in a star cluster, for example) to the input of a multi-channel detector.

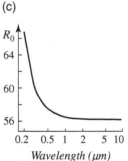

Fig. 5.14 The plane-parallel model for atmospheric refraction: (a) Refraction in a model atmosphere that has only two different uniform layers; (b) shows the limit of this model as the number of layers becomes large, with n increasing toward the surface. In (c) R_0 is given in seconds of arc as a function of wavelength (the refractivity of air is R_0 in radians times 10^6).

5.2.8 Prisms

Prisms are geometric solids with two parallel polygonal faces joined by a number of faces shaped like parallelograms. Figure 5.16 illustrates a few prisms

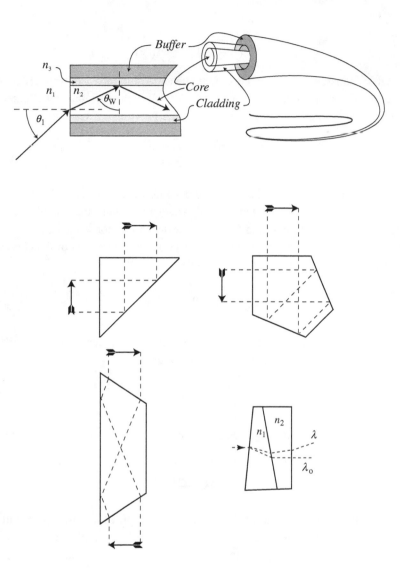

Fig. 5.15 Structure of an optical fiber. Total internal reflection occurs if θ_W is greater that the critical angle.

Fig. 5.16 Some prisms, from top to bottom: a right-angle prism bends the beam 90° and inverts the image, a pentaprism bends the beam but leaves the image unchanged, a dove (or dovetail) prism inverts the beam without changing its direction and an Amici prism disperses an incident beam, passing one wavelength in the original direction, but others at different angles.

in cross-section that are useful because their internal reflections invert an image or change the path of a beam.

Prisms are also useful because of their dispersing properties. Figure 5.17 shows the path of a ray through a triangular prism with apex angle A. After two refractions, a ray entering the left face at angle of incidence, α, emerges from the second face at angle θ from the original direction (the angular deviation). Application of Snell's law at both surfaces of this prism gives

$$\sin\left(\theta + A - \alpha\right) = \left(n^2 - \sin^2\alpha\right)^{1/2}\sin A - \sin\alpha\cos A \qquad (5.25)$$

You can show that the angular deviation is a minimum when α and the final angle of refraction are equal, in which case

(a)

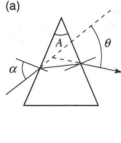

(b)

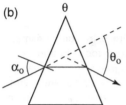

Fig. 5.17 Dispersion by an isosceles triangular prism: (a) the angular deviation, θ, depends on the wavelength because $n = n(\lambda)$. Longer wavelengths have smaller deviations. (b) The minimum angular deviation occurs when the path through the prism is parallel to its base.

$$\sin\left(\frac{\theta_0 + A}{2}\right) = n\sin\left(\frac{A}{2}\right) \tag{5.26}$$

These equations make it clear that, because the index of refraction is a function of wavelength, then so are θ and θ_0. We define the **angular dispersion** as $\partial\theta/\partial\lambda$, and note that since only n and θ are functions of wavelength

$$\frac{\partial\theta}{\partial\lambda} = \frac{\partial n}{\partial\lambda}\frac{\partial\theta}{\partial n} \tag{5.27}$$

The first factor on the right-hand side depends on the dispersion of the prism material, which is a strong function of wavelength (note the variation in the *slope* of any curve in Figure 5.5), while the second factor is only a very weak function of wavelength (see Problem 7). A good approximation of the curves in Figure 5.5 is given by the Cauchy formula:

$$n(\lambda) = K_0 + \frac{K_2}{\lambda^2} + \frac{K_4}{\lambda^4} + \dots \tag{5.28}$$

where K_0, K_2, and K_4 are constants that depend on the material. Ignoring all but the first two terms and substituting for $\partial n/\partial\lambda$ gives

$$\frac{\partial\theta}{\partial\lambda} \cong \frac{g}{\lambda^3} \tag{5.29}$$

where g is a constant that depends primarily on the prism geometry. In the case of minimum deviation,

$$g = -4K_2\frac{\sin(A/2)}{\cos(\alpha)} \tag{5.30}$$

Thus, the absolute value of the angular dispersion of a glass prism will be much higher in the blue than in the red, which can be a disadvantage for some astronomical spectroscopic applications.

5.3 Simple telescopes

> The next care to be taken, in respect of the Senses, is a supplying of their infirmities with Instruments, and, as it were, the adding of artificial Organs to the natural; this in one of them has been of late years accomplisht with prodigious benefit to all sorts of useful knowledge, by the invention of Optical Glasses. By the means of Telescopes, there is nothing so far distant but may be represented to our view.... By this means the Heavens are open'd, and a vast number of new Stars, and new Motions, and new Productions appear in them, to which all the ancient Astronomers were utterly Strangers.
>
> – Robert Hooke, *Micrographia*, 1665

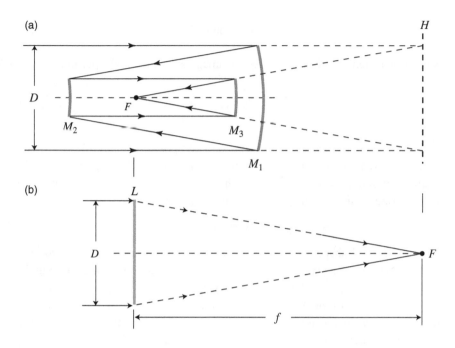

Fig. 5.18 (a) A three-mirror reflecting telescope, with primary mirror M_1 (with diameter D) and secondary and tertiary mirrors M_2 and M_3. Rays from infinity arrive at the margins of M_1 which defines the entrance pupil, and emerge at the focus, F. The dashed extensions of the rays locate the principle plane at H. A detector at F will see only rays arriving from within diameter D at H. Flipping the diagram (b) gives the equivalent telescope with a single lens, L, that has the same aperture, D, and reproduces the geometry of the converging beam with effective focal length, f.

5.3.1 Telescopes as single-element cameras

Most astronomical telescopes are used as *cameras* – they form images of objects both on and off the optical axis in a *focal plane or surface*. Although a telescope's optics can be complex, we can roughly represent its optical properties with a single "equivalent thin lens" – one that matches the aperture, principal plane, focal length, and image-forming properties of the instrument. For example, part (a) of Figure 5.18 shows a three-mirror reflecting telescope in the Paul–Baker configuration, an arrangement for wide-angle telescopes (The James Webb Space Telescope and the Large Synoptic Survey Telescope use a similar design). Ray traces locate the principle plane at H, and part (b) of the figure shows the equivalent single-thin-lens diagram for this telescope.

5.3.2 Image scale and image size

The *image scale*, s, describes the mapping of the sky by any camera. The image scale is the angular distance on the sky that corresponds to a unit linear distance in the focal plane of the camera. Figure 5.19 shows the equivalent thin-lens diagram of a camera of focal length f, which represents *any* telescope. We draw the paths followed by two rays, one from a star on the optical axis, the other from a star separated from the first by a small angle θ on the sky. Rays pass through the vertex of the lens without deviation, so assuming the paraxial approximation, $\theta \approx \tan \theta$, it should be clear from the diagram that

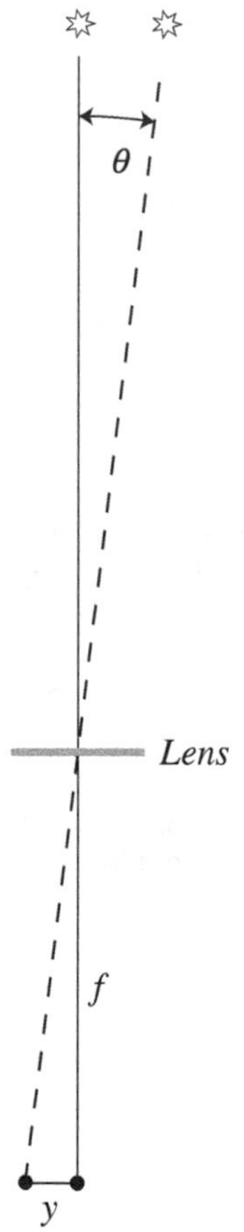

Fig. 5.19 Definition of image scale. Stars separated by angle θ on the sky form images that are separated by distance y in the focal plane.

$$s = \frac{\theta}{y} = \frac{1}{f} \; [\theta \text{ in radians}] \qquad (5.31)$$

Since it is usually convenient to express image scales in arcsec per mm, then

$$s = \frac{206\,265}{f} [\text{arcsec per unit length}] \qquad (5.32)$$

Typical modern focal-plane detectors are composed of many identical light-sensitive pixels. If the center of each (usually square) pixel is separated from its nearest neighbors by distance, d, then the **pixel scale** of a telescope (that is, the angular size on the sky imaged by one pixel) is just.

$$s_\text{p} = sd \qquad (5.33)$$

The size and shape of the detector often determine the **field of view** of a camera (the angular diameter or dimensions of a single image). A rectangular detector, for example, with physical length, l, and width, w, will have an angular field of view of sl by sw. More rarely, the detector may be oversized and the field of view set by obstructions in the telescope's optical system or by the limit of acceptable image quality at large distances from the optical axis.

5.3.3 Focal ratio and image brightness

Focal ratio is defined as dimensionless quantity

$$\Re = \frac{f}{D} = \frac{\{\text{focal length}\}}{\{\text{diameter of entrance aperture}\}} \qquad (5.34)$$

For example, the 50 cm telescope at Vassar College Observatory has a focal length of 374 cm, so $\Re = 7.5$. This is usually expressed as "$f/7.5$."

You can show that the brightness (energy per unit area in the focal plane) of the image of an *extended* source like the Moon or a nebula is proportional to $\Re^{-2}$, so that images in an $f/5$ system, for example, will be four times as bright as images in an $f/10$ system.

5.3.4 Telescopes with oculars

For their first three centuries, telescopes augmented direct human vision, and an astronomer's eyesight, persistence, skill, and sometimes even physical bravery were important factors in telescopic work. To use a telescope visually, an ocular, or eyepiece, is needed to examine the image plane of the objective (the light-gathering element). Figure 5.20 shows the arrangement where the telescope, ocular, and human eye lens are represented as thin lenses. The two stars in the diagram each produce an **afocal beam** (all rays parallel) with the two beams

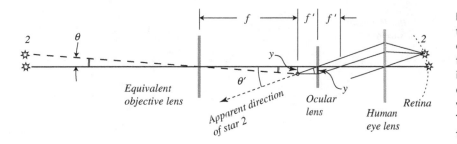

Fig. 5.20 A Keplerian telescope. A positive ocular lens located one focal length from the image plane of the objective forms two virtual images at infinity. Theses afocal beams are focused by the eye. Galileo's original telescope used a negative lens for the ocular, a design with serious disadvantages.

separated by angle θ on the sky. Viewed through the telescope plus ocular, they appear to be separated by the angle θ'. The **angular magnification** is the ratio

$$M = \left| \frac{\theta'}{\theta} \right| \tag{5.35}$$

From the diagram, making the paraxial approximation, this is just

$$M \approx \left| \frac{\tan \theta'}{\tan \theta} \right| = \frac{y/f'}{y/f} = \frac{f}{f'} \tag{5.36}$$

So the magnification is the ratio of the focal lengths of the objective and eyepiece. Oculars are subject to chromatic and other aberrations (see below), and benefit from careful design and matching with telescope optics. Most contain three to eight lenses. Rutten and van Venrooij (1988) discuss oculars at length as do many commercial and amateur web sites.

5.4 Image quality: telescopic resolution

> That telescope was christened the Star-splitter,
> Because it didn't do a thing but split
> A star in two or three the way you split
> A globule of quicksilver in your hand
> — Robert Frost, "The Star-Splitter," 1923

5.4.1. The diffraction limit

The wave properties of light set a fundamental limit on the quality of a telescopic image. Figure 5.21 illustrates the formation of an image by a telescope outside the atmosphere. Wavefronts from a point source arrive at the telescope as perfectly planar and parallel surfaces, provided the source is very distant. The **entrance aperture** is the light-gathering element of the telescope, and its diameter, D, is usually that of the mirror or lens that the incoming wave first encounters. Despite the fact that the source is a point, its image – created by a perfect telescope

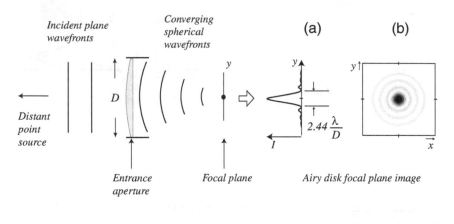

Fig. 5.21 Camera images in the absence of an atmosphere. Plane wavefronts diffract upon encountering the circular aperture, and focus as an Airy disk: a bright central spot surrounded by rings of decreasing brightness. (a) The intensity of the resulting image vs. distance on the *y*-axis. The central peak has a full width of twice the Airy radius: $2\alpha = 2.44\lambda/D$. (b) A negative of the two-dimensional diffraction pattern.

operating in empty space – will have a finite size because of diffraction of the wave. This size, the ***diffraction limit*** of the telescope, depends on both the wavelength of light and on D. The diffraction of a plane wavefront by a circular aperture is a messy problem in wave theory, and its solution, first worked out in detail by the English astronomer George Airy in 1831, says that the image is a bulls-eye-like pattern, with the majority (84%) of the light focused into a spot or "disk."[3] Concentric bright rings, whose brightness decreases with distance from the center, surround the very bright central spot, the ***Airy disk***. The angular radius of the dark ring that borders the Airy disk is

$$\alpha_A = \frac{1.22\lambda}{D} \text{ [radians]} = \frac{0.252\lambda}{D} \text{ [arcsec m } \mu\text{m}^{-1}] \tag{5.37}$$

The full width at half-maximum (FWHM) of the disk is $0.9\alpha_A$. If two point sources lie close together, their blended Airy patterns may not be distinguishable from that of a single source. If we can say for sure that a particular pattern is due to two sources, not one, the sources are ***resolved***. The ***Rayleigh criterion*** for the resolution of two sources requires that the centers of their Airy disks be no closer than α_A, the angular radius of either central disk (both radii are the same). At this limiting resolution, the maximum intensity of one pattern coincides with the first dark ring of the other; see Figure 5.22. At a wavelength of 0.5 nm, according to Equation (5.37), a 1-m telescope should have a resolution of 0.126 arcsec. Details smaller than this size will be lost.

[3] The shape of the Airy pattern for a clear, circular aperture is proportional to the function $(2J_1(x)/x)^2$, where J_1 is the first Bessel function of order 1, and $x = \pi D\lambda^{-1} \sin\alpha$. A circular central obstruction (e.g. secondary mirror) will alter the pattern by moving intensity from the central disk to the bright rings. Other obstructions or shape changes will complicate diffraction patterns in images (e.g. secondary supports will produce "diffraction spikes").

5.4.2 Rayleigh refused: atmospheric seeing and optical aberrations

Rayleigh's criterion is a good predictor of the performance of optically perfect space telescopes. On the surface of the Earth, however, turbulence causes dynamic density variations in the Earth's atmosphere. These variations distort wavefronts and limit the resolving power of all but the smallest telescopes. This loss of resolution is termed **seeing**. Seeing (measured as the angular FWHM of the image of a point source) may be as small as several tenths of a second of arc on the very best nights at the very best sites on Earth, but it can reach several seconds of arc at other sites. We postpone a discussion of seeing until the next chapter.

Most telescopes suffer from **aberrations**: optical imperfections that degrade the images they form so much that Rayleigh's criterion is far too optimistic a standard for the expected resolution. We discuss aberrations in the next section.

5.5 Aberrations

5.5.1 Chromatic aberration

> ... it is not the spherical Figures of Glasses, but the different Refrangibility of the Rays which hinders the perfection of Telescopes.... Improvement of Telescopes of a given length by Refractions is desparate.
>
> – Isaac Newton, *Opticks*, rev. edn, 1718

Since $n(\lambda)$ for optical glasses decreases with wavelength in the visible, the focal length of a convex lens will be longer for red wavelengths than for blue. Different colors in an image will focus at different spots. This inability to obtain perfect focus is called **chromatic aberration**, and limits the resolving power of a telescope with lenses; see Figure 5.23.

(a)

(b)

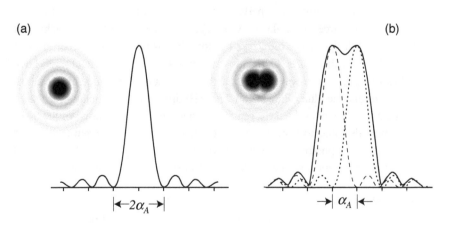

Fig. 5.22 (a) The Airy pattern as a negative image and plotted as intensity vs. radius. (b) The negative image of two identical monochromatic point sources separated by an angle equal to Rayleigh's limit. The plot shows intensity along the line joining the two images. The unblended images are plotted as the dotted and dashed curves, their sum as the solid curve.

Fig. 5.23 Chromatic aberration: (a) shows a thin lens, which, because of dispersion, focuses blue rays nearer the vertex than red rays. The image at blue focus will be a blue spot surrounded by a rainbow-hued blur, with red outermost. The image at the red focus will be a red spot surrounded by a large rainbow blur with blue outermost. A cemented achromat (b) consists of a positive and negative lens of differing powers and dispersions. Focal length differences and surface curvatures are greatly exaggerated.

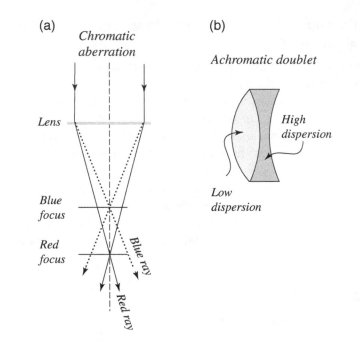

To correct chromatic aberration in a lens, the usual strategy is to cement together two lenses made of different glasses, a positive power (convex) lens with low chromatic dispersion, and a concave lens with lower absolute power, but higher chromatic dispersion. See Figure 5.23b. If powers are chosen to be inversely proportional to dispersions, then the combination will have a finite focal length but reduced dispersion over a significant part of the spectrum.

An *achromat* is a lens whose focal length is the same at two different wavelengths. The first useful achromats, *doublets* of crown (convex lens) and flint (concave lens) glass, began to appear in the 1760s in France and England, an appearance certainly delayed by Newton's declaration of its impossibility. Achromats were initially of only small aperture owing to the difficulty in producing large flint glass blanks free from flaws. With Fraunhofer's perfection of achromat production and design, large refractors became the instruments of choice at most observatories after the 1820s. Opticians usually designed these refractors to have equal focal lengths at the wavelengths of the red Fraunhofer C-line at 656.3 nm (Hydrogen α) and the blue F-line at 486.2 nm (Hydrogen β), producing nearly zero chromatic aberration over the most sensitive range (green and yellow) of the eye.

Magnificently suited to human vision, these telescopes were a poor match for the new technology of photography when it was introduced in the 1880s. They had large residual chromatic aberration in the violet and ultraviolet, where the emulsions had their sensitivity. Astronomers soon constructed refracting *astrographs* optimized to short wavelengths – for example, the large number of "standard astrographs" of 33 cm aperture built for the Carte du Ceil project begun in 1887.

Modern glasses now provide the achromat designer with a much wider choice of indices and dispersions than were available to Fraunhofer. A properly designed doublet, called an **apochromat**, can bring three different wavelengths to a common focus, but only through use of expensive materials (e.g. fluoride crowns and lanthanum flints). A properly designed triplet, called a **superapochromat**, can bring four different wavelengths to a common focus.

5.5.2 Classification of monochromatic wavefront aberrations

Chromatic aberration is present when rays of differing wavelength fail to reach a common focus. Unfortunately, some aberrations are present in monochromatic light and affect the images in both reflecting and refracting systems.

Consider a "perfect" telescope. It should transform an incident plane wavefront into a converging spherical wavefront whose center is at the focus predicted by the paraxial approximation – the **Gaussian focus**. For point sources off-axis, the perfect telescope should produce spherical wavefronts converging to a point on the **Gaussian image plane**. Moreover, straight lines in the sky should produce straight lines in the image plane. This mapping of the points and lines on the sky to points and lines in the image plane is called a **collinear transformation**. If an optical system fails to produce this collinear transformation it is said to exhibit **aberrations**.

We will discuss aberrations as differences between the actual wavefront and the perfect wavefront, and do so by tracing rays. Figure 5.24 defines a coordinate system specifying a "test" ray anywhere on a wavefront. The idea is to describe the aberrations by measuring the wavefront errors at a large number of such test locations. Figure 5.24a shows two reference rays and a test ray from the same off-axis source. After passing through an optical system represented by a single curved mirror, all three are intended to focus at F. The first ray, the **chief ray**, $V'VF$, is the one passing from the source through the center of the entrance aperture (here it is the mirror vertex). The plane of the left-hand diagram, called the **meridional**, or **tangential**, **plane**, contains the chief ray and the optical axis.

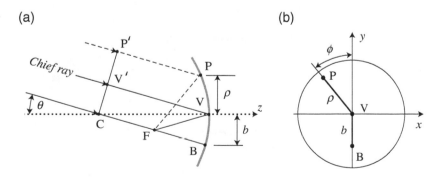

(a)   (b)

Fig. 5.24 Specification of the location of an arbitrary test ray from an off-axis source. Diagram (a) shows rays through the center of curvature (CFB) and vertex (V'V, the chief ray), which, along with the optical axis, define the plane of the diagram, known as the meridional plane. Point P is outside the plane of the diagram. Diagram (b) locates points P, V, and B in the plane of the aperture when looking down the optical axis.

The plane perpendicular to the meridional plane that contains the chief ray is called the *sagittal plane*. Tangential rays and sagittal rays are confined to their respective planes.

The second ray, CFBFC, passes through the center of curvature of the mirror and reflects back over its original path. This meridional ray will specify the off-axis direction, θ, of the source.

The third ray, P'PF, is the arbitrary "test" ray, which may be outside the plane of the diagram. In a perfect optical system, the optical path lengths V'VF, P'PF, and CBF will all be equal. Figure 5.24b views the ray geometry in the plane of the entrance aperture, looking down the optical axis, and locates points P, V, and B. In this diagram, the parameter $b \propto \sin \theta$ is the normalized field radius ($b = 1$ at the edge of the aperture). It measures how far the source is from the axis. The circular coordinates φ and ρ locate the intersection of the test ray and the aperture (ρ is also normalized to 1 at the edge). If point P is on the y-axis of this diagram, the test ray is tangential; if P is on the x-axis, the ray is sagittal.

We are concerned with any possible difference between the optical path lengths of this test ray (i.e. ray P'PF) and the chief ray through the vertex. This means tracing rays. If you trace rays with Snell's law using the approximation

$$\sin \theta \approx \tan \theta \approx \theta - \frac{\theta^3}{3!} \tag{5.38}$$

you obtain the results of *third-order aberration theory* – a much more accurate computation of where rays travel than the one given by paraxial theory, which assumes $\sin \theta = \tan \theta = \theta$.

In the third-order treatment, William Rowan Hamilton in 1833 showed that the optical path difference between the test ray and the chief ray (Figure 5.25) takes the form:

$$\Delta w (\rho, \phi, b) = C_1 \rho^4 + C_2 \rho^3 b \cos \phi + C_3 \rho^2 b^2 \cos^2 \phi + C_4 \rho^2 b^2 + C_5 \rho b^3 \cos \phi \tag{5.39}$$

where Δw is in radians (i.e. a phase error). The C_i coefficients are dimensionless and depend on the shapes of the optical surfaces and indices of refraction. Each

Fig. 5.25 The difference in optical path length between a ray through point P and one through the vertex. The perfect wavefront is a sphere centered on the Gaussian image point at F, and a perfect ray is perpendicular to this front.

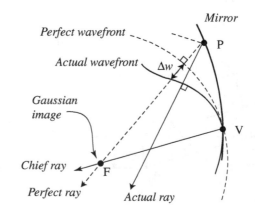

Table 5.4 *Third-order monochromatic aberrations.*

Aberration	Functional dependence
Spherical aberration (SA)	ρ^4
Coma	$\rho^3 b \cos \phi$
Astigmatism	$\rho^2 b^2 \cos^2 \phi$
Curvature of field	$\rho^2 b^2$
Distortion	$\rho b^3 \cos \phi$

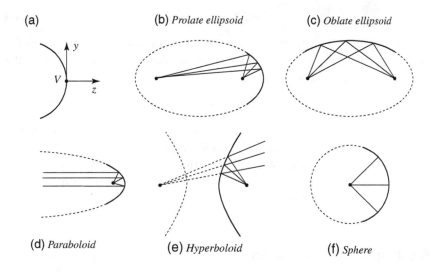

(a)

(b) *Prolate ellipsoid*

(c) *Oblate ellipsoid*

(d) *Paraboloid*

(e) *Hyperboloid*

(f) *Sphere*

Fig. 5.26 (a) Coordinate system for describing a mirror or lens surface shaped like a rotated conic section. (b)–(f) Light rays originating at one focus of a reflecting conic of revolution will reconverge at the other focus. The degenerate cases are the sphere (f), where the two foci coincide, and the parabola (d), where one focus is at infinity. In the oblate ellipsoid, conjugate foci align diametrically on a ring around the axis of rotation. In the hyperboloid, either the object or the image is virtual.

of the terms in Equation (5.39) has a different functional dependence, so we distinguish five monochromatic third-order aberrations, also known as the ***Seidel aberrations***. Table 5.4 lists the aberrations in order of importance for large telescopes, where the exponent on ρ is usually crucial. This is also in the order in which they are usually corrected. Astigmatism, for example, is only corrected after both coma and SA have been eliminated.

5.5.3 Shapes of optical surfaces

Although a spherical surface is the easiest for a lens- or mirror-maker to produce and test, other shapes are frequently required. Those most commonly used can be generated by rotating a conic section around its axis of symmetry. Usually, this axis coincides with the symmetry axis of the telescope, as in Figure 5.26a, where we show the cross-section of such a surface (e.g. a mirror) in the y–z plane of our usual coordinate system. The cross-section satisfies the equation for a conic in the y–z plane:

Table 5.5 *Conic section eccentricities and conic constants.*

Shape	Eccentricity	Conic constant
Sphere	0	0
Oblate ellipsoid	$0 < e < 1$	$-1 < K < 0$
Prolate ellipsoid	$e^2 < 0$	$K > 0$
Paraboloid	$e = 1$	$K = -1$
Hyperboloid	$e > 1$	$K < -1$

$$y^2 = 2Rz - \left(1 - e^2\right)z^2 \tag{5.40}$$

Here e is the eccentricity of the conic, and R is the radius of curvature at the vertex. In three dimensions, if this curve is rotated around the z-axis, the resulting surface satisfies the equation

$$r^2 = x^2 + y^2 = 2Rz - (1 + K)z^2 \tag{5.41}$$

where the conic constant, K, is just the value of $-e^2$ for the two-dimensional curve. Table 5.5 gives the values of e and K for specific conics. As demonstrated by Descartes in 1630, each conic will perfectly focus a single object point onto a single conjugate point. These points, as you might guess, are the two geometric foci of each conic; see Figure 5.26.

5.5.4 Spherical aberration

Except for spherical aberration, all the Seidel aberrations in Table 5.4 vanish for sources on axis ($b = 0$). For visual astronomy, where one typically examines only on-axis images, SA is the only monochromatic aberration that is troublesome.

Figure 5.27 shows this aberration in a spherical mirror. Two rays from an on-axis source; one, near the axis, reflects to the Gaussian focus at F. The second, in violation of the paraxial approximation, strikes the mirror at an angle of 30 degrees, and is brought to focus at G, a distance of 0.845($R/2$) from the vertex – considerably closer to the mirror than the Gaussian focus (see Problem 13).

For a mirror that is a conic of revolution, the focal length of a ray parallel to the axis is exactly given by the series

$$f(r, R) = \frac{R}{2} - (1 + K)\left\{\frac{r^2}{4R} + \frac{(3 + K)r^4}{16R^3} + \dots\right\} \tag{5.42}$$

An equation similar to (5.42) exists for a lens, but is more complex because it accounts for two surfaces and the index of refraction. The first term in Equation (5.42) gives the Gaussian focus. The r^2 term is the third-order aberration, and the

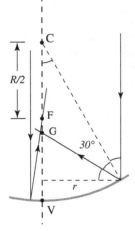

Fig. 5.27 Reflection of a ray that is not paraxial from a spherical mirror. A ray parallel to the axis strikes the mirror at a 30° angle of incidence. It crosses the optical axis at G and does not pass through the paraxial focus at F.

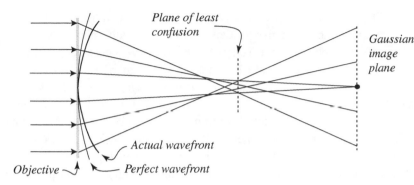

r^4 term is the fifth-order aberration. If SA is present in a mirror or lens, then the image will be blurred, and the best image is usually not at the Gaussian focus; see Figure 5.28.

It is possible to minimize, but not eliminate, SA in a spherical-surface single lens by minimizing the angles of incidence on every surface – a plano-convex lens with the curved side facing the sky is a common example. Likewise, any lens or mirror with a large enough focal ratio will approach the paraxial case closely enough that the blur due to SA can be reduced to the size of the seeing disk. Since a large focal ratio also minimizes chromatic aberration, very early (1608–1730) refracting telescope designs tended to have modest apertures and large focal lengths. The problem with these designs was reduced image brightness and, for large apertures, unwieldy telescope length.[4]

Another solution to SA recognizes that a negative power lens can remove the SA of a positive lens of a different index, and an achromatic doublet can be designed to minimize both spherical and chromatic aberration. When larger flint glass disks became available the 1780s, the stage was set for the appearance of excellent refracting telescopes free of both spherical and chromatic aberrations. These grew in aperture, beginning with 3-cm aperture "spy glasses" used during the Napoleonic wars and advancing, with Fraunhofer's[5] superb instruments, to

[4] Typical of these is the still-extant telescope that Giuseppe Campani produced in his optical shop in Rome, which has an aperture of 13 cm (5 inches) and a focal length of 10 m (34 feet). King Louis XIV purchased this instrument in 1672 for Jean Dominique Cassini and the just-commissioned Paris Observatory, where Cassini used it to discover Rhea, the satellite of Saturn. Focal lengths of seventeenth-century telescopes approached the unusable – Cassini successfully employed a 41-m (136-feet) telescope at Paris: the objective was supported by an old water tower and manipulated by assistants. "Aerial" telescopes – with no tube connecting objective and eyepiece – were not uncommon. Huygens used one 123 feet long, and Helvius operated one 150 feet in length.

[5] Friedrich Georg Wilhelm Struve used the most celebrated of these, the 24-cm telescope at Dorpat (Tartu) Observatory, to measure the angular separations of 3000 double stars, as well as the parallax of Vega. Bessel used a 16-cm Fraunhofer telescope for the definitive measures of the parallax of 61 Cygni (1838). Argelander used a 9-cm Fraunhofer telescope for all his observations for the *Bonner Durchmusterung* (1862).

the 10–24 cm range in the period 1812–25. The era of the refractor culminated with a 0.9-m (36-inch) objective for Lick Observatory, California, in 1888 and a 1.02-m (40-inch) objective for Yerkes Observatory, Wisconsin, in 1897. The legendary optical shop of Alvin Clark and Sons in Cambridge, Massachusetts, produced both objectives. A 1.25-m refractor was on display at the Paris exhibition of 1900, but never produced useful results, and a modern 1.0 m refractor was commissioned in 2002 for the Swedish Solar Telescope on La Palma. A diameter near 1 m is the upper limit for terrestrial refracting telescopes. Gravity will deform a lens larger than this to an unacceptable degree.

For a conic section mirror and a point source at infinity, SA will produce a blurred image with angular diameter (in radians)

$$ASA = \frac{(K+1)}{128 \, \mathfrak{R}^3} \tag{5.43}$$

where $\mathfrak{R}$ is the focal ratio, f/D.

For mirrors, the removal of SA is simple. Choose $K = -1$ in Equations (5.42) or (5.43): SA is absent if the mirror shape is a paraboloid. All other aberrations vanish for on-axis images (where $b = 0$). Thus, reflecting telescopes became competitive with refractors in the 1730s and briefly, popularized by William Herschel's spectacular discoveries, speculum-metal reflectors were very productive before achromatic refractors proved superior in the early nineteenth century. Finally, metal-film-on-glass parabolic reflectors gradually replaced the refractors as the superior large telescope design in the first half of the twentieth century.

5.5.5 Coma

Prior to the end of the nineteenth century, visual observers were concerned only with a telescope's on-axis performance. The advent of photography changed those concerns forever, and telescope design has since needed to satisfy more stringent optical criteria. Of the four off-axis aberrations, only the first two, coma and astigmatism, actually degrade the image resolution, while the other two only alter the image position.

Coma is the wavefront aberration that depends on the factor $\rho^3 b \cos \phi$. Unlike SA, coma increases with object distance from the axis. The $\cos \varphi$ term means that even rays from the same radial zone of the objective will fail to come to a common focus. Figure 5.29 shows the ray paths and images typical of a system with coma.

If you imagine the rays from a source passing through a particular zone of the objective (i.e. a circle for which ρ = constant), then, in the presence of coma, those rays form a ring-shaped image offset from the Gaussian focus, as in the right of Figure 5.29. Comatic images of stars have a heavy concentration of light near the Gaussian focus, and become broader and fainter for rays in the outer

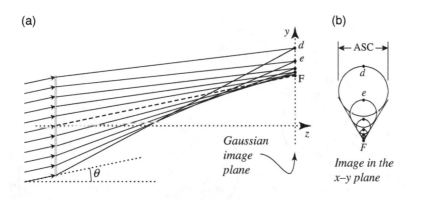

(a)

(b)

Gaussian
image
plane

Image in the
x–y plane

Fig. 5.29 (a) is in the tangential plane for a system with coma. (b) shows the image plane where each circle represents a bundle of rays from the same radial zone of the aperture. Dots on these circles mark the focus of each of the illustrated rays. Sagittal rays from the same zone strike the tops of the corresponding circles. The angular width of the image is ASC, and point F is the Gaussian focus.

circles – this comet-like appearance is the origin of the name for the aberration. For mirrors, the "point" of the comet faces the *z*-axis.

For a single paraboloidal reflecting surface, the radius of the largest ring of the comatic image in Figure 5.29b is

$$\text{ASC} = \frac{\theta}{16\,\mathfrak{R}^2} \tag{5.44}$$

where $\mathfrak{R} = f/D$, the focal ratio. The above value for ASC is roughly the diameter of the blur that contains 50% of the image power. The total length of the comatic image is 3ASC.

An optical system with neither SA nor coma is called *aplanatic*. No single-element practical aplanatic telescope is possible, either in a refractor or reflector. As with SA, large focal ratios reduce coma, but impose penalties in image brightness and telescope length. Otherwise, minimizing coma in refracting systems requires a system of lenses and, fortunately, achromatic doublet or triplet designs that minimize SA also reduce coma. Aplanatic reflecting telescope designs require two mirrors. Alternatively, a correcting lens system (usually non-conic surfaces with zero power) can also minimize coma in a single-mirror telescope. Such correcting optics may also aim to correct additional aberrations, but must take care to avoid introducing chromatic aberration.

5.5.6 Astigmatism

Astigmatism is the Seidel aberration whose wavefront distortion depends on the term $b^2\rho^2 \cos^2\phi$, so it increases more rapidly than coma for off-axis images. The cosine term means that wavefront delay is zero for rays in the sagittal plane (i.e. $\varphi = 90°$, $\varphi = 270°$), but an extremum for rays in the meridional plane. In the absence of other aberrations, astigmatic rays pass through two line segments (see Figure 5.30). One line, called the sagittal focus, is in the meridional plane and extends on either side of the Gaussian image point. The other line segment,

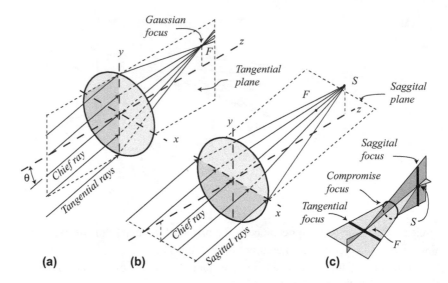

Fig. 5.30 Astigmatism for an off-axis source at infinity in direction θ. (a) Rays in the meridional plane suffer no aberration and come to a focus at the paraxial focus, *F*. (b) Rays in the sagittal plane come to a secondary focus at *S*. (c) All rays pass through two line segments that are perpendicular to the chief ray. The first, the tangential focus, is in the sagittal plane centered at *F*. The second, the sagittal focus, is in the tangential plane centered at *S*.

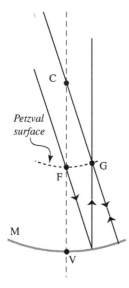

Fig. 5.31 Curvature of field. The spherical mirror M has a radius of curvature, $R = \overline{CV}$, with center of curvature at C and on-axis focus at F. If all other aberrations are absent, rays from an off-axis image will focus on a spherical surface of radius *R*/2, also centered at C.

called the meridional focus, is in the sagittal plane, centered on the focus point of the tangential rays. The best compromise focus position lies between the two focal lines, where a star image has a circular shape. The diameter of this "best" blurred image (in radians) is

$$AAS = \frac{\theta^2}{2\mathfrak{R}} \tag{5.45}$$

All uncorrected refractors and all *practical* two-mirror reflectors suffer from astigmatism. In some cases, the astigmatism is small enough to be ignored. If it can't be ignored, a corrector lens or plate located near the focal plane usually removes the astigmatism. A three-mirror reflector can also eliminate astigmatism. A telescope design free from astigmatism, coma, and SA is called an ***anastigmatic aplanat***. For small telescopes, a popular anastigmatic design is the Schmidt–Cassegrain, a two-mirror telescope with a corrector plate located at the aperture (see Chapter 6).

5.5.7 Field curvature

In the absence of other aberrations, third-order theory predicts that off-axis images (e.g. point G in Figure 5.31) will not be blurred, but will fall on a spherical surface known as the Petzval surface, not on the plane predicted in paraxial theory. The radius of the Petzval sphere depends on the curvatures and refractive indices of the mirrors and lenses in the optical system. For the simple case in Figure 5.31, with a spherical mirror of radius of curvature *R*, the Petzval surface has radius $R/2$ and has its center at the center of curvature of the mirror.

Detectors tend to be flat. A large flat detector placed tangent to the Petzval surface will necessarily record most of its images out of focus. For a small detector, this defocus might not exceed the seeing disk or diffraction limit and therefore not present a problem. A large enough detector, however, will produce blurred images because of field curvature. One solution is to bend the detector to match the Petzval surface. This has been done for many years with glass photographic plates, which can be forced, usually without breaking, into a considerable curve by a mechanical plate-holder. Large solid-state detectors like charge-coupled devices (CCDs) are mechanically quite fragile, so bending them is not an option. In many telescopes, a **corrector plate** or lens, which may also help remove other residual aberrations, usually serves to flatten the field.

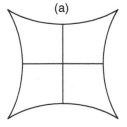

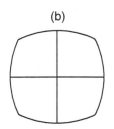

Fig. 5.32 (a) Pincushion distortion of an object that has a square outline centered on the optical axis. (b) Barrel distortion of the same object.

5.5.8 Distortion

Distortion relocates images in the focal plane so that the collinearity requirement is violated – straight lines on the sky become curved lines in the focal plane. There is no image blur. Figure 5.32 illustrates two kinds of distortion, "barrel" and "pincushion," either of which will increase more rapidly with distance from the axis than do the other Seidel aberrations. Since distortion does not change image quality, an observer can remove it from an image if he has the calibrations needed.

5.5.9 Other aberrations and ray tracing in practice

The Seidel aberrations describe imperfections introduced by particular elements in a centered optical system. More than one aberration may be present, and the image-forming behavior in an actual system may not only exhibit some combination of the Seidel aberrations but may also suffer addition degradation due to higher-order aberrations, element misalignment, spacing, and tilt. Computer ray-tracing programs are essential in modeling planned and existing systems, and in computing sensitivities to variables. Helpful outputs include plots of ray paths, maps of wavefront errors, and **spot diagrams.** The spot diagram plots the points where rays from the source intersect the image surface. Rays are chosen to sample the entrance aperture either randomly or in a uniform pattern, so the density of spots gives an indication of the brightness distribution of the final image. Figure 5.33 shows a few examples.

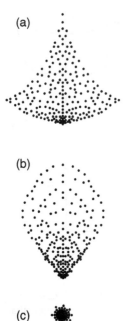

(a)

(b)

(c)

Fig. 5.33 Spot diagrams from a ray-tracing program. Three simulated star images: (a) poor alignment in a two-mirror telescope, (b) combined coma and astigmatism, (c) on-axis paraboloid.

Summary

- Geometrical optics models light as either a bundle of rays or as a sequence of geometrical wavefronts. Concepts:

  index of refraction: $v(\lambda) = c/n(\lambda)$ *chromatic dispersion*
  optical path length *Fermat's principle*
  Snell's law: $n_1 \sin\theta_1 = n_2 \sin\theta_2$ *total internal reflection*
  reflection coefficient *grazing incidence*
  optical axis and vertex *paraxial approximation*
  focal length, f, and power, P *aperture, D*
  object distance, image distance *principal plane*

- A variety of special transmitting and mirror-making materials are available for astronomical applications. Special thin film coatings can enhance reflection or transmission.

- In the paraxial approximation:

  power of a spherical surface is ($n_1 = -n_2 = 1$ for a mirror):

$$P_{12} = \frac{n_2 - n_1}{n_2 R_{12}} = \frac{n_2}{s_2} - \frac{n_1}{s_1} = \frac{n_2}{f_2} = -\frac{n_1}{f_1}$$

  thick lens, two mirrors, or two thin lenses:

$$P = \frac{1}{f} = P_{12} + P_{23} - \frac{d}{n} P_{12} P_{23}$$

  thin lens:

$$P = P_{12} = P_{23} = \frac{1}{f} = \frac{1}{s_2} - \frac{1}{s_1}$$

- Simple rules permit graphical ray tracing in the paraxial approximation. Ray-tracing computer applications employ exact rules and are important tools for optical design.

- Prisms are used for both the redirection of light and for angular dispersion. The angular dispersion of a triangular prism is proportional to λ^{-3}.

- Astronomical telescopes act as cameras, gather light, and enhance image detail. Concepts:

  image scale: $s = 206265/f$ *focal ratio: $\mathcal{R} = f/D$*
  diffraction limit *Airy disk*
  Rayleigh criterion: $\alpha_A = 1.22\lambda/D$ *seeing*
  differential atmospheric refraction *angular magnification*

  A system that fails to perform a collinear transformation between the object and its image is said to exhibit aberrations. Concepts:

chromatic aberration	*achromat*	*apochromat*
collinear transformation	*wavefront distortion*	*corrector plate*
Gaussian image	*third-order theory*	*chief ray*
meridional plane	*sagittal plane*	*Seidel aberrations*

- Most optical surfaces used in astronomy are conics of revolution, characterized by a particular conic constant.
- Removing the Seidel aberrations produced by a spherical surface is a crucial concern in telescope design:
  - *Spherical aberration* is minimized in an achromatic doublet or absent in a paraboloid mirror.
  - *Coma* is also minimized in an achromatic doublet or triplet, and in certain aplanatic two-mirror systems.
  - *Astigmatism* requires additional optical elements for elimination.
  - *Curvature of field* requires additional optical elements for elimination.
  - *Distortion* is usually not corrected in telescopes.

Exercises

> In general, any form of exercise, if pursued continuously, will help train us in perseverance. Long distance running is particularly good training in perseverance.
>
> – Mao Ze Dong, "A Study of Physical Education," 1917

1. Prove that the law of reflection follows from Fermat's principle by using the geometry illustrated in part (a) of the figure below. Let x be the unknown location of the point where the ray from P_1 to P_2 reflects from a horizontal surface. (Hint: write an expression for s, the total optical path length, set $ds/dx = 0$, and replace terms in the result with trigonometric functions of the angles of incidence and reflection.)

2. Prove that Snell's law of refraction follows from Fermat's principle by using the geometry illustrated in part (b) of the figure below. Let x be the unknown location of the point where the ray from P_1 to P_2 refracts at a horizontal surface. See the hint for Problem 1.

3. A telephoto lens (see figure in the margin) consists of a positive "objective" lens, L1, followed by a negative "telephoto element," L2. When employed in a camera, the equivalent focal length, f, of the telephoto lens can be appreciably greater than the physical length of the camera, z_c. Compute z_c and f for the case in which $f_1 = 1.0$ m, $f_2 = -0.6$ m, and $d = 0.5$ m. (Refer to Equations (5.18), (5.20), and (5.22)). Illustrate your result with a graphical ray trace that shows the location of the principal plane. If the diameter of the objective is 0.5 m, what is the focal ratio of the system? Compute the minimum diameter of the telephoto element.

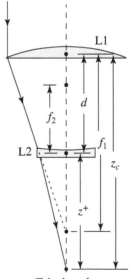

Telephoto lens

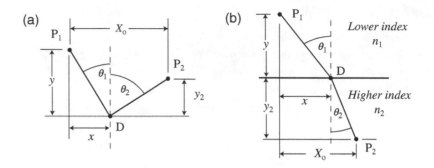

4. If the arrangement in the previous problem is such that $d = f_1 - |f_2|$ the system is afocal, and constitutes a *Galilean telescope*. The negative element is the ocular. (Compare with Figure 5.20.) Sketch the rays in a Galilean telescope to show that except for a minus sign, Equation (5.36) applies. Where is the principal plane of an afocal system like this?

5. A telescope has a focal ratio of $f/7.5$. You wish to use it with a spectrometer that requires an $f/10$ beam at its input. Compute the focal length of a 50 mm diameter lens that, when inserted in the beam 150 mm in front of the unmodified focal plane, produces the required beam.

6. Consider Figure 5.16. Show that rotating a dove prism around its long axis by angle θ will rotate a transmitted beam by angle 2θ. (Hint: redraw the figure with the prism rotated by 180°.)

7. Compute the angular dispersion in red and ultraviolet light for a prism with apex angle 30 degrees and index given by $n(\lambda) \approx 1.52 + (0.00436\,\mu m^2)\lambda^{-2}$. To do this, assume the prism is operated at minimum angular deviation for 600 nm light and compute the angular dispersion at that wavelength. Then repeat the calculation at minimum angular deviation for 350 nm light. Show that the ratio of the two dispersions is close to the value predicted by Equation (5.29).

8. Compute the frame width (in arcmin) and pixel scale (in arcsec per pixel) for CCD observations with the following telescope and instrument: aperture 1.0 m, focal ratio $f/6.0$. CCD is 1024×1024 pixels; each pixel is 18 μm square.

9. You are designing a camera for a 24-m telescope with an $f/3$ focal ratio. It is expected that an adaptive optics system will deliver a resolution of double the diffraction limit at 2 microns. What size pixels should your camera have if a star image is 2 pixels across?

10. Show that in order for a ray to be transmitted by an optical fiber, its angle of incidence on the end of the fiber must satisfy the condition:

$$\sin\theta_1 \le N.A. \equiv \sqrt{n_2^2 - n_3^2}$$

where n_2 and n_3 are the indices of the core and cladding, respectively. (Hint: refer to Figure 5.15 and require that θ_W be greater than or equal to the critical angle.) The quantity $N.A.$ is known as the **numerical aperture** of the fiber.

11. On a particular night, the planet Mars has an angular diameter of 15 arcsec and an apparent brightness of 1.0×10^{-7} Wm^{-2}. Two astronomers observe the planet, using identical CCD cameras whose pixels are 25 μm apart. Albert uses a telescope of 0.3-m aperture whose focal ratio is $f/8$. Bertha uses a telescope of 30-m aperture whose focal ratio is $f/4$. How much energy accumulates in a single pixel of Albert's CCD image of Mars in a 100 s exposure? How much in a single pixel of Bertha's image of Mars?

12. Compute the diffraction limit of the Hubble Space Telescope (2.4-m diameter) in the ultraviolet (300 nm) and near infrared (2.0 μm). Compare with the (a) the diffraction limit of the human eye at 0.5 μm, (b) the diffraction limit at 2.0 μm of a space telescope that has an 8-m diameter, and a 30-m ground-based telescope at 2.0 μm.

13. Show that for a spherical mirror, the focal length defined by a ray parallel to the optical axis is *exactly*

$$ f = R - \frac{R}{2}\left[1 - \frac{\rho^2}{R^2}\right]^{-\frac{1}{2}} $$

where R is the radius of curvature and ρ is the distance between the ray and the optical axis. Use this to verify the result quoted in the text for Figure 5.27.

Chapter 6
Astronomical telescopes

The adventure is not to see new things, but to see things with new eyes.

– Marcel Proust (1871–1922), *Remembrance of Things Past*, 1927

While I disagree with Proust about the thrill of seeing utterly new things (I'm sorry, that *is* an adventure), if I wonder about seeing things "with new eyes," telescopes immediately come to mind. No instrument has so revolutionized a science, nor so long and thoroughly dominated its practice, as has the telescope astronomy. No instrument so simple (amateurs still make their own) has produced such a sustained transformation in humanity's understanding of the universe.

In this chapter, we examine the basic features of the modern ground-based optical telescope designs. Schroeder (1987) provides a more advanced treatment. We will consider some pronounced advantages, disadvantages, and special requirements that space-based telescopes have compared to their ground-based cousins. Because it is such an important technology for the future of the ground-based telescope, we will take some trouble to understand the principles of adaptive optics, and its potential for removing some of the natural but nasty (for astronomy) consequences of living on a planet with an atmosphere. Finally, we will look at the kinds of large – and extremely large–ground-based telescopes now under construction.

We begin, however, not with the telescope, but with the apparatus that supports and points it.

6.1 Telescope mounts and drives

My brother began his series of sweeps when the instrument was yet in a very unfinished state, and my feelings were not very comfortable when every moment I was alarmed by a crack or fall, knowing him to be elevated fifteen feet or more on a temporary cross-beam instead of a safe gallery. The ladders had not even their braces at the bottom; and one night, in a very high wind, he had hardly touched the ground when the whole apparatus came down.

– Caroline Herschel, Letter, 1784

6.1.1 Altazimuth and equatorial mounts

For ground-based telescopes, the mount that supports the telescope has two important functions. First, it moves the telescope to *point* at a specific position on the celestial sphere. Second, the mount *tracks* the object pointed at – that is, it moves the telescope to follow accurately the apparent position of the object. Object position relative to the horizon changes rapidly due to diurnal motion (i.e. the spin of the Earth) and less rapidly due to changing atmospheric refraction and even proper motion (e.g. Solar System objects). In some circumstances, tracking may need to compensate for effects like telescope flexure or image rotation. Some specialized telescopes do not track: transit telescopes (see Chapter 3) point only along the meridian. Some telescopes do not even vary their pointing: a zenith tube points straight up at all times. In some specialized telescopes, pointing and tracking are accomplished by moving the detector in the image plane while the massive telescope remains stationary.

Most telescope mounts, however, are mobile on two axes, and move the entire telescope. Figure 6.1 shows the two most common forms of mount. The configuration in Figure 6.1a is called an *altazimuth* mount. This mount points to a particular position on the sky by rotating the vertical axis to the desired azimuth, and the horizontal axis to the desired elevation. To track, the altazimuth mount must move both axes at the proper rates, rates that change as the tracked object moves across the sky.

The *equatorial mount*, illustrated in Figure 6.1b, has one axis (the polar axis) pointed directly at the celestial pole, and the second axis (the declination axis) at right angles to the first. This mount points to a particular position on the sky by rotating the polar axis to acquire the desired hour angle, and rotating the declination axis to the desired declination.

The advantage of the equatorial mount is thus the simplicity of both pointing and tracking. Many manually operated equatorials, for example, are equipped with *setting circles* – graduated circles attached to each axis to indicate the declination and hour angle at which the telescope points. Since the hour angle is

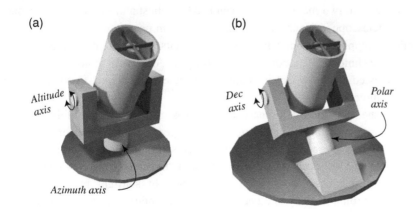

Fig. 6.1 (a) Altazimuth and (b) equatorial telescope mounts. Note the larger "footprint" of the equatorial.

just the sidereal time minus the RA of the object, astronomical coordinates translate directly into telescope position. Tracking is a simple matter of rotating the polar axis at the steady rate of one turn each sidereal day and for this reason the tracking mechanism of an equatorial is sometimes called a *clock drive*. For very precise tracking, effects like atmospheric refraction, mechanical errors, and flexure require corrections to the sidereal rate as well as small movements in declination. Many of these can be computer-controlled, but an astronomer (or an instrument) often generates such small corrections by monitoring the position of a *guide star* near the field of interest and moving the telescope as needed.

The altazimuth mount is more compact and presents fewer mechanical challenges than the equatorial, and is therefore potentially less expensive. Virtually all the ground-based telescopes with apertures above 5 m use altazimuth mounts.

The altazimuth has disadvantages. Neither of its axes corresponds to the axis of diurnal motion, so that pointing requires a complicated transformation of coordinates, and tracking requires time-varying rotation rates on both axes. Moreover, the coordinate transformation from hour angle–declination to altitude–azimuth becomes singular at the zenith. An altazimuth mount is therefore unable to track objects precisely within a few degrees of the zenith. In addition, as an altazimuth tracks, the image of the sky in the telescope focal plane rotates with respect to the telescope tube and its axes, so any image-recording detectors must be mounted behind a counter-rotating prism, mirror system (called a k-mirror) or on a mechanical stage to avoid trailing. Again, the angular speed of the "de-rotator" is not constant. Computer and autoguider control of the drive axes and rotating detector stage copes with these disadvantages, but with additional cost and complexity.

6.1.2 Telescope mounts in space

Telescopes in space must also point and track, but since gravity does not glue them to a spinning planet, at least some aspects of these tasks are less problematic. In general, two methods have been used to adjust and stabilize the orientation of a telescope in space: small rockets and spinning reaction wheels. Rockets require a supply of propellant – directed bursts of a compressed gas are a common technique. Reaction wheels require an on-board motor – when the rotation speed of the wheel is changed, the telescope begins to rotate in the opposite direction.

Since the resolution of a space telescope is generally much higher than a ground-based instrument, there are more stringent requirements for precision tracking. Space telescopes often rely on guide stars for this precision. To point the Hubble Space Telescope, for example, at a "fixed star" requires continuous telescope movement because of the aberration of starlight induced by the telescope's orbital velocity and because of torques induced by atmospheric drag and thermal effects. Some space telescopes, like Gaia, are not designed to produce steady images at all, but are mounted on spinning platforms and data accumulate as objects drift across the field of the detector. Whether a space telescope is stabilized or is freely spinning, the control system must know its orientation relative to an inertial frame – this reference is provided either by a set of gyroscopes or by star sensors.

6.2 Reflecting telescope optics

All large-aperture and most small-aperture modern optical telescopes are reflectors. The great number of possible designs usually narrows to the few practical choices discussed in this section. Wilson (1996, 1999) gives a thorough treatment of classical reflecting telescope optics in the historical context.

6.2.1 Prime focus and Newtonian focus

One very simple telescope design mounts a detector at the focus of a paraboloid ($K = -1$) mirror. This *prime focus* configuration has the advantage of both simplicity and minimum light loss (there is only one reflection), but also has some limitations. First, any observer or apparatus at the prime focus will obstruct the mirror. For this reason, the prime focus configuration is generally only found in telescopes where the diameter of the aperture exceeds the diameter of the observing apparatus by a significant amount. In these cases (i.e. telescopes larger than around 3.5 m in diameter), a prime focus *cage* is fixed on the optical axis of the telescope to carry an astronomer or, in the modern era, her remote-controlled instruments. Besides reducing the light-gathering power of the telescope and introducing a greater opportunity for scattering light, this central obstruction has only a small effect on properly focused images. Out-of-focus images will have a characteristic "doughnut" shape. The support structure that extends between the central obstruction and the side of the tube produces the artifact of radial diffraction spikes on bright stars.

   A second problem with a single-mirror telescope is wide-field image quality. Recall from Section 5.5 that for a paraboloid, coma and astigmatism blur images by the angular amounts (in radians):

$$ASC_{PAR} = \frac{\theta}{16\mathfrak{R}^2}; \quad AAS_{PAR} = \frac{\theta^2}{2\mathfrak{R}} \tag{6.1}$$

where $\mathfrak{R}$ is the focal ratio and θ, the angular distance of the image from the optical axis. Because coma can be severe even close to the optical axis for fast mirrors, prime focus cameras are usually equipped with compound aspheric refractory corrector lenses to reduce aberrations over an acceptably wide field, but at the cost of some light loss.

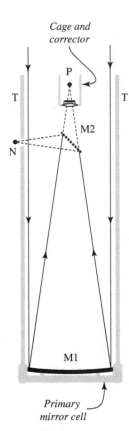

Cage and corrector

P

T

T

M2

N

M1

Primary mirror cell

Fig. 6.2 Prime focus and Newtonian configurations. Both utilize a paraboloid primary mirror, M1, in a mirror cell and tube, T. The prime focus instruments at P must be supported in a cage inside the tube. The Newtonian has a diagonal flat mirror, M2, that directs the focus outside the tube to N.

For optical telescopes where the prime focus configuration is impractical, the **Newtonian** design, which uses a flat diagonal mirror to redirect the converging beam to the side of the telescope tube (see Figure 6.2) provides a more convenient access to the focus. The diagonal mirror introduces both a central obstruction and an additional reflection, but the Newtonian design is so simple that many homemade and inexpensive telescopes use this layout. Since many amateur observers only use images near the optical axis, the aberrations need not be regarded as serious flaws. Although the wide-field performance of a Newtonian can be improved with corrector lenses near the focus, professional astronomers generally prefer a design that utilizes two curved mirrors for moderate-sized telescopes. Especially at large focal lengths, Newtonians require long tubes, powerful drives, large buildings, and an astronomer with no fear of heights – all serious disadvantages.

6.2.2 Cassegrain and Gregorian reflectors

Figure 6.3 shows two alternative two-mirror designs. Like the Newtonian, both the **Cassegrain** and the **Gregorian** utilize a paraboloid as the primary mirror. In Cassegrain's configuration, the secondary is a convex hyperboloid located on the optical axis, with one (virtual) focus coincident with the focus of the primary (point F in the figure). This means that rays converging to F will be redirected to the second focus of the hyperboloid at point F′, which is usually located behind the primary. A hole in the center of the primary allows the rays to reach this secondary focus. The Gregorian design is similar to the Cassegrain, except the secondary is a concave ellipsoid.

In a design with two curved mirrors, the application of third-order aberration theory is far less trivial than for a single "powered" surface. The multiple-lens formula (Equation (5.22)) gives the combined power of two mirrors of power P_1 and P_2:

$$P = \frac{1}{f} = P_1 + P_2 - dP_1P_2 \qquad (6.2)$$

To discuss two-mirror designs quantitatively, we define three dimensionless parameters that depend on the final focal length, the mirror spacing, and the desired back focal distance, z_F (the distance from the vertex of the primary to the final focus) – see Figure 6.4:

$$m = \frac{f}{f_1} = \frac{P_1}{P} = -\frac{s_2'}{s_2}$$

$$k = 1 - \frac{d}{f_1} = \frac{y_2}{y_1} \qquad (6.3)$$

$$\beta = \frac{z_F}{f_1}$$

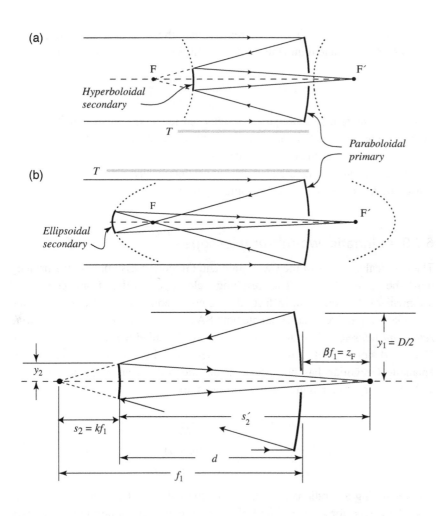

(a)

F

Hyperboloidal secondary

T

F′

Paraboloidal primary

(b)

T

F

F′

Ellipsoidal secondary

Fig. 6.3 (a) Cassegrain and (b) Gregorian mirror configurations. Primary focus at F, combined focus at F′. For the same primary, the tube length of the Gregorian is longer.

y_2

$\beta f_1 = z_F$

$y_1 = D/2$

$s_2 = kf_1$

s_2'

d

f_1

Fig. 6.4 Parameters for a two-mirror telescope. The sign convention measures s and z_F positive to the right from the mirror vertex. The value of d is always positive.

Note that the power of the secondary is determined by substitution into Equation (6.2):

$$\frac{f_2}{f_1} = -\frac{km}{m-1} \qquad (6.4)$$

The sign convention is that β is positive if the focus is behind the primary, and that f, m, and k are positive for a Cassegrain and negative for a Gregorian. The three parameters (m, β, and k) are constrained by the requirement that the foci of the conics must coincide, which means that:

$$k = \frac{(1+\beta)}{(1+m)} \qquad (6.5)$$

Thus, a designer can freely choose only two parameters. The shapes of the mirrors will drastically affect the aberrations. The primary is a paraboloid by definition (so $K_1 = -1$). A requirement that the spherical aberration (SA) remains zero then determines the conic constant, K_2, of the secondary. Specifically:

$$K_{1C} = -1, \qquad K_{2C} = -\left(\frac{m+1}{m-1}\right)^2 \tag{6.6}$$

Both Cassegrain and Gregorian designs can locate the focus conveniently – behind the primary. A Gregorian will be longer than a Cassegrain, but both will still be much shorter than a Newtonian of the same effective focal length. For telescopes of even moderate size, this advantage easily outweighs the trouble caused by the increased optical complexity of two curved surfaces.

6.2.3 Aplanatic two-mirror telescopes

The classical versions of the Cassegrain and Gregorian assume that the primary must be a paraboloid. The resulting telescopes suffer from coma and astigmatism. But one can, in fact, choose conic constants for the primary and secondary, K_1 and K_2, that eliminate both SA *and* coma, producing an *aplanatic reflecting telescope*. The aplanatic Cassegrain is called a *Ritchey–Chrétien* or *R–C*, and consists of a hyperbolic primary and hyperbolic secondary. The aplanatic Gregorian has no special name and utilizes an ellipsoidal primary and secondary. The required conic constants are

$$\begin{aligned}
K_1 &= K_{1C} - \frac{2(1+\beta)}{m^2(m-\beta)} \\
K_2 &= K_{2C} - \frac{2m(m+1)}{(m-\beta)(m-1)^3}
\end{aligned} \tag{6.7}$$

The remaining aberrations are smaller for the aplanatic Gregorian than for the R–C. However, for a given f, both the central obstruction due to the secondary and the overall length of the tube are greater for the Gregorian, and the resulting increase in expense and decrease in optical efficiency have usually been decisive for the R–C. Most modern telescopes also tend to favor the R–C over the primary focus. In addition to tube length, convenience, and weight considerations, refractive correctors for the prime focus must have many elements to remove coma and other aberrations, and thus tend to lose more light than does the single reflection from the R–C secondary. For wide-field applications, R–C telescopes frequently carry mild refracting optics to correct astigmatism and/or curvature of field.

6.2.4 Nasmyth and coudé foci

Astronomers can mount heavy equipment at the Cassegrain focus, on the strong part of the telescope tube that also supports the primary mirror cell. There are

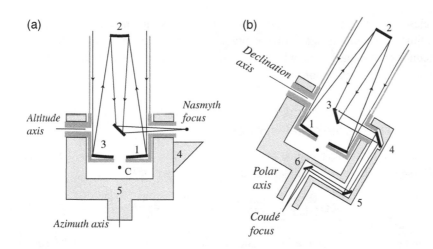

Fig. 6.5 (a) The Nasmyth configuration. Light from the secondary mirror (2) instead of passing to a Cassegrain focus (C), is redirected by a tertiary flat (3) along the hollow altitude axis and reaches a focus above the Nasmyth platform (4). The platform rotates around the azimuth axis (5) as the telescope points and tracks. (b) A coudé configuration. Light from the secondary is redirected by flat 3 to a series of flats (4–5–6) that bring the beam to the polar axis. Similar arrangements can direct the beam along the azimuth axis of an altazimuth.

limits, though. The **Nasmyth focus**, illustrated in Figure 6.5a, locates even heavier instruments directly on the altazimuth mount. (James Nasmyth (1808–90) was a British machine tool inventor and amateur astronomer.) In this arrangement, the telescope secondary produces a slightly longer final focal length than in a typical Cassegrain, and a flat mirror intercepts the beam from the secondary and directs it horizontally along the altitude axis. As the telescope tracks, this focus remains fixed relative to the mount. Equipment at the Nasmyth focus thus exerts no stress on the telescope tube and a force on the mount that will not change over time.

If an instrument is very massive or delicate, or especially if the mount is an equatorial, the **coudé focus** (French for "bent like an elbow") provides even more stability than the Nasmyth. Figure 6.5b gives an example of this arrangement, implemented in an equatorial mount. A flat mirror redirects light from the secondary along the declination axis, and then a series of flats (sometimes the beam is made temporarily afocal) conducts the beam to emerge along the polar axis, where it reaches focus at a point that does not move with respect to the Earth.

Both the Nasmyth and coudé have disadvantages. The additional reflections produce some light loss. Very frequently, aberrations are an issue because a general-purpose telescope (prime, Cassegrain, Nasmyth, and coudé all available) will be equipped with removable/interchangeable secondaries. Suppose you design a telescope to be aplanatic in the R–C configuration with the focus behind the primary, and with conic constants K_1 and K_2 given by Equations (6.7). To switch from the R–C to the Nasmyth or coudé, you swap in a new secondary to get a longer focal length, and therefore use different values for m and β. However, the existing K_1 no longer satisfies Equations (6.7), and the resulting combination, called a **hybrid two-mirror telescope**, cannot be aplanatic. In the hybrid, you can choose a value for K_2 so that SA is zero, but coma will still be present. Usually, therefore, Nasmyth and coudé instruments in such telescopes tend to work on near-axis images only. Some modern telescopes can adjust the conic constant of the primary through active optics, and thereby reduce or eliminate coma when the secondary is changed, and refractive correctors are always an option.

6.2.5 Three-mirror telescopes

The wide-field performance of the R–C telescope is superior to the classical Cassegrain, but is ultimately limited by astigmatism of the size:

$$AAS_{R-C} = \frac{\theta^2}{2\Re} \left[\frac{m(2m-1) - \beta}{2m(1+\beta)} \right] \rightarrow \frac{\theta^2}{2\Re} \left[m + \frac{1}{2} \right] \tag{6.8}$$

The approximation on the right-hand-side of the above is for $\beta \rightarrow 0$. To achieve good images over fields wider than a half a degree of arc or so, a designer can add a third mirror (instead of just a refractive corrector plate). A good example is the **LSST** (Large Synoptic Survey Telescope – currently under construction with projected completion by 2020) – see Figure 6.6. The LSST is highly compact (f/1.19 primary), with a 3.5-degree-wide field corrected for and all the Seidel aberrations except distortion. It consists of three mirrors plus a three-element refractive corrector to flatten the field without chromatic aberration. The primary and tertiary are fabricated from the same 8.4-m blank. At 3.4 m, the secondary will be the largest precision convex mirror ever

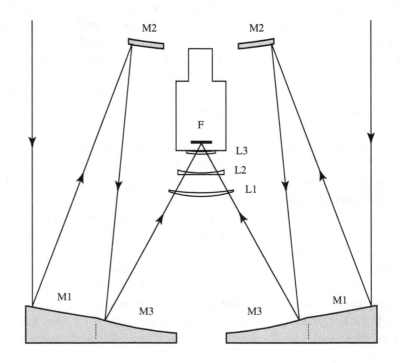

Fig. 6.6. The optical layout of the LSST. M1 is a near-paraboloid, M2 and M3 are slightly ellipsoidal. L1 and L2 are aspheric.

produced, and the design pays for the wide field with considerable central obstruction. The LSST will have the light-gathering power of a 6.7-m telescope, but will collect data from a 50 times wider area on the sky than a conventional 6.7-m R–C. The final shapes of all surfaces, some of which depart slightly from perfect conics, were optimized after extensive work with computer ray-tracing programs.

A rough measure of a telescopic camera's ability to gather information is the total rate at which it can record photons from all sources on its detector. For a camera, this is proportional to the effective area of the primary times the solid angle on the sky it records, a quantity called the étendue:

$$U = A\Omega = \frac{\pi}{4}D_{\text{eff}}^2\Omega \qquad (6.9)$$

The effective aperture and the large focal-plane array on the LSST combine to produce an étendue of $319\ \mathrm{m^2 deg^2}$. For comparison, the wide-field Dark Energy Camera currently operating at the prime focus of the Blanco 4-m telescope has an étendue of about $37\ \mathrm{m^2 deg^2}$.

6.2.6 Schmidt telescopes

> I shall now show how completely sharp images can be obtained with a spherical mirror. . . .. If the correcting plate is now brought to the center of curvature of the mirror, . . . the spherical aberration is abolished, even over the whole field.
>
> – Bernhard Schmidt, Ein lichtstares komafreies Spiegelsystem,
> *Mitteilungen der Hamburger Sternwarte*, vol. 7, no. 15, 1932

An arguably superior alternative to a three-mirror design like the LSST is limited in aperture by its refractive element – the Schmidt[1] telescope produces good images over a very large field – six to eight degrees. These telescopes became the standard instruments for many important photographic surveys during the mid- and late twentieth century. Schmidt exploited the symmetry of a spherical mirror to avoid off-axis aberrations. His design consists of three elements: a spherical primary mirror, an aperture stop located at the center of curvature of the primary, and a refracting corrector plate designed to remove spherical aberration.

Figure 6.7 shows the layout. The aperture stop insures there can be no distinction between on-axis and off-axis sources: wavefronts from different

---

[1] Bernard Schmidt (1879–1939), an Estonian, lost his right arm in a boyhood experiment with gunpowder. At a time when almost all optical work was done by hand, he nevertheless became internationally recognized as a master lens- and mirror-maker. He constructed the first Schmidt camera (36-cm aperture, f/1.7, with a 16° field) at Hamburg Observatory in 1930 and described the design in the 1932 paper quoted. Schmidt never divulged his method for the very difficult task of grinding the surface of the corrector.

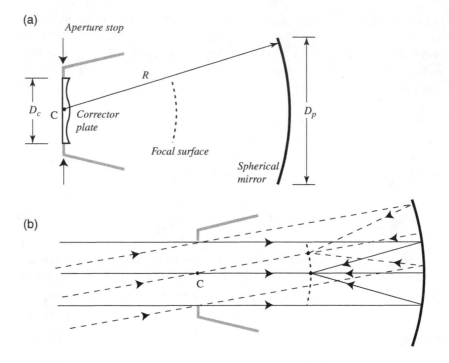

directions will illuminate slightly different parts of the mirror, but since it is a spherical mirror, all will experience an identical change in wavefront shape upon reflection. Because of the aperture stop, the chief ray from every source always passes through point C, the center of curvature. This means that points B and V in Figure 5.24 always coincide, and therefore there can be no third-order coma or astigmatism. Banishing these off-axis aberrations means that the Schmidt offers the possibility of a fast focal ratio and large field of view. The stop does not affect the curvature of field, so the Petzval surface will be the one expected for a spherical mirror.

The corrector plate, located in the plane of the aperture stop, is designed to remove SA. If you review Figure 5.27 and Equation (5.42) you can see that SA in a spherical mirror means the marginal rays (the ones near the edge of the aperture, ρ large) converge more strongly than the axial rays (the ones near the center). A Schmidt corrector, then, should be a refracting element whose power is larger (more positive) for the axial rays and smaller for the marginal. An entire family of shapes can do the job. Two possible shapes are sketched in Figure 6.8. The shape labeled (b), which is thickest at center and thinnest at 86.6 % radius, is the one usually chosen, since it minimizes the chromatic aberration introduced by the corrector plate. It is possible to further minimize chromatic aberration by using a two-element achromatic corrector. Unlike the spherical mirror, the corrector plate *does* have an optical axis and introduces some off-axis aberrations, which are of concern in systems with very fast focal ratios ($< f/2$).

The refracting corrector plate limits apertures to modest values, and the focal surface is inaccessible to a human observer, so the instrument is often called a Schmidt camera.

Gravitational sag of the corrector limits apertures to 1.5 m or less, and the tube of the telescope is very long (see Problem 6). Other than that, the location and curvature of the focal surface is the main inconvenience of the design. Until recently, the usual observing method was photographic, using a large glass photographic plate flexed to match the focal surface at prime focus inside the telescope tube.[2] Some Schmidts have been modified with Newtonian or Nasmyth focuses where it is more convenient to mount modern solid-state arrays, and the largest utilize the prime focus with modern detectors. Appendix G lists some of the largest Schmidt cameras in the world. Probably because of the aperture limitation, no large Schmidts have been commissioned since 1978.

An important accessory to the Schmidt with a panoramic detector is an **objective prism**. If a prism with an apex angle of a few degrees is mounted just in front of the corrector, the images of stars formed at focus will be small spectra (see Chapter 11). Such images convey a great deal of information about many objects at once, and objective-prism surveys have created important lists of objects with particular characteristics – quasars, emission-line stars, galaxies with high redshifts, or objects with unusual colors, for example.

6.2.7 Other catadioptric telescopes

A **catadioptric** telescope is one that combines full-aperture refracting (dioptric) and reflecting (catoptric) elements. The category includes the Schmidt and several other designs, the most important of which are two-mirror-plus-corrector telescopes. Given the large number of degrees of freedom of such systems, it is possible to design a telescope with greatly minimized aberrations. As an example, we discuss the **Schmidt–Cassegrain** (S–C), illustrated in Figure 6.9.

The S–C telescope is a popular amateur instrument because its tube is short and closed, it is relatively rugged, and it can produce excellent images. Several manufacturers market systems with apertures up to 40 cm, usually at a focal ratio near $f/10$. Such systems generally have a spherical primary and elliptical secondary, with the aperture stop and corrector placed near the focus of the primary (in Figure 6.9, $\alpha \approx 0$ and $m = f/f_1 \approx 5$). They usually have $D = D_P$ and produce fields of view of around $1°$.

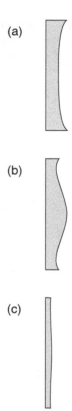

Fig. 6.8 Shapes of the Schmidt corrector: (a) shows a plate whose power decreases radially, starting with zero power at center; (b) shows a plate with positive power at center, decreasing to zero at 0.866 fractional radius; and (c) shows the actual cross-section of a corrector with shape (b) for an $f/3$ system.

___

[2] Prior to 2005, the UK Schmidt, for example, used square glass plates that measured 356 mm (14 inches) on a side and are 1.0 mm thick. Each of these plates covered an area $6.4° \times 6.4°$ on the sky. The camera has subsequently been equipped with several generations of fiber optic spectrographs – these pick off light from images on the prime focus with many individual optical fibers.

Fig. 6.9 The Schmidt–
Cassegrain. The edge of
the corrector plate
determines the aperture.

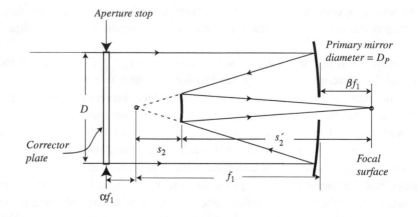

Faster S–Cs (e.g. $f/3$) can have well-corrected fields that rival those of a classical Schmidt, with the added advantages of a shorter tube, accessible focus, and no curvature of field. Their disadvantages are a very large secondary and larger chromatic aberration. You can find discussion of other catadioptric systems in Rutten and van Venrooij (1988). Fast catadioptric systems have found some specialized application in professional astronomy, but, other than Schmidts, no large apertures have been built, mainly because they offer few advantages over the R–C.

6.3 Telescopes in space

> As World War II drew near its end, I was approached by a friend on the staff of
> the RAND Project, an Air Force "think tank." He told me that his group was
> carrying out a secret study of a possible large artificial satellite, to circle the Earth
> a few hundred miles up. "Would you be interested," he asked me, "in writing
> a chapter on how such a satellite might be useful in astronomy?"
>
> – Lyman Spitzer (1914–97), "Dreams, Stars, and Electrons," *Annual Reviews*
> *of Astronomy and Astrophysics*, vol. 27, 1989

In 1946, Lyman Spitzer quickly recognized the usefulness of a space telescope and became a leader in the effort that culminated on April 25, 1990, when the crew of the Space Shuttle *Discovery* placed the Hubble Space Telescope (HST) into Earth orbit.

Was it a good idea? We can judge the excellence of a telescope by at least three criteria: its ability to resolve detail, its ability to detect faint objects, and the angular size of the field over which it can perform these functions. In the first two of these categories, telescopes in space offer obvious advantages over ground-based instruments. The 44-year delay between the conception and actualization of the HST suggests that there are impediments to realizing these advantages.

Several smaller space telescopes preceded and followed the launch of the HST, but to date it remains the astronomical instrument with the largest aperture in space (2.4 m) and has generated results of unprecedented volume: roughly five scientific papers each week since launch have been based on HST data. The HST was designed to operate for a total of 20 years, and its replacement is overdue. NASA, in partnership with the ESA and the Canadian Space Agency, plans to place the 6.5-m James Webb Space Telescope (JWST) at the Sun–Earth Lagrange point, L_2, sometime around the end of 2018.

6.3.1 Advantages of space telescopes

Resolution

In space, the complete absence of the wavefront distortions caused by Earth's atmosphere means a space telescope (if its optics are perfect!) should have diffraction-limited resolution. Specifically, in the absence of atmospheric seeing the "diameter" of the image of a star will be something like the diameter of the central part of the Airy disk in radians (Equation (5.37)):

$$\theta = 2\alpha_A = \frac{2.44\lambda}{D} \approx \frac{2\lambda}{D} \tag{6.10}$$

This equation only applies in the absence of aberrations, and we know that the design and quality of the telescope optics determine image quality and useful field size. The precision and alignment of optical surfaces thus becomes especially critical in space, where seeing will not mask small errors. With excellent optics, the higher resolution of a space telescope produces smaller stellar images and more detail in the images of extended objects like planets and galaxies.

Detection limits

For stellar objects, freedom from wavefront distortions due to the atmosphere also means that a space telescope can detect fainter objects than an identical ground-based telescope, *because the same light can be concentrated in a smaller image*. Consider the simple problem of detecting the presence of a star. Assume that the star produces flux f_λ, that we have a telescope (either in space or on the ground) with aperture, D, with an array of detector pixels. This means that in a narrow wavelength band of width $\Delta\lambda$, the telescope in time, t, will collect a number of photons from the star given by

$$N_* = \text{signal} = (f_\lambda \Delta\lambda t)\left(\frac{\lambda}{hc}\right)\left(\frac{\pi D^2}{4}\right)T_A\varepsilon = \kappa D^2 T_A f_\lambda t \tag{6.11}$$

Here the first term in brackets is the total energy per unit area from the star, the second converts from energy units to number of photons, the third gives the collecting area, ε is total efficiency the system, and T_A is the transparency of the atmosphere ($T_A = 1$ in space). The factor κ just collects all the constants. We will say that the star is just *detectable* if this signal is about the same size as its uncertainty. To estimate uncertainty, we recall (Chapter 3) that our estimate of N_* is the result of counting the all the photons in the star image. Of course, that raw count contains N_* photons from the star plus B photons from the background, so we subtract an estimate of the total background in the image:

$$N_{*est} = (N_* + B) - B_{est} \tag{6.12}$$

If b_λ is the angular surface brightness of the background on the sky and θ is the angular diameter of the image then:

$$B = \kappa \left(\frac{\pi \theta^2}{4} \right) D^2 b_\lambda t \tag{6.13}$$

What is the uncertainty of our estimate of the signal? Propagating the uncertainties for Equation (6.12) and assuming uncertainties arise from shot noise:

$$\sigma^2(N_{*est}) = \sigma^2(N_* + B) + \sigma^2(B_{est}) = N_* + B + \sigma^2(B_{est}) \approx N_* + B \tag{6.14}$$

On the far right of Equation (6.14) we use our assumption that the uncertainty in B_{est} can be made very small by sampling many background pixels. The signal-to-noise ratio is then

$$SNR = \frac{N_*}{\sqrt{N_* + B}} \tag{6.15}$$

For a star that is just detectable, we set the signal-to-noise ratio equal to 1 and use the quadratic formula to find that at this limit:

$$2N_{*lim} = 1 + \sqrt{1 + 4B} \tag{6.16}$$

Now we will assume that the total background count, $B \gg 1$, so that

$$N_{*lim} \cong \sqrt{B} \tag{6.17}$$

Substituting in the above for N_{*lim} from Equation (6.11) and for B from Equation (6.13):

$$f_{limit} = \frac{1}{2} \left(\frac{\pi}{\kappa} \right)^{\frac{1}{2}} \left(\frac{b_\lambda}{t} \right)^{\frac{1}{2}} \frac{\theta}{T_A D} \tag{6.18}$$

Therefore, on the ground, where θ is set by seeing independent of telescope size, the detection threshold decreases only as the first power of D, even though light-gathering power increases as D^2. But in space, a telescope is diffraction limited, so substituting the Airy disk diameter for θ in the previous equation:

$$f_{limit, space} = \left(\frac{\pi}{\kappa} \right)^{\frac{1}{2}} \left(\frac{b_\lambda}{t} \right)^{\frac{1}{2}} \frac{\lambda}{D^2} \tag{6.19}$$

With perfect optics in space, then, the payoff for large apertures is superior, since the detection threshold depends on D^{-2}.

Background

Examining the ratio

$$\frac{f_{\text{limit, grnd}}}{f_{\text{limit, space}}} = \frac{\theta_{\text{see}} D}{\lambda T_A} \left(\frac{b_{\lambda g}}{b_{\lambda s}}\right)^{\frac{1}{2}} \tag{6.20}$$

suggests another advantage to the space environment. The Earth's atmosphere is itself a source of background light, so from space, the value of b_λ is lower than from the ground. At visible and near-infrared (NIR) wavelengths the atmosphere contributes light from several sources: **airglow** (atomic and molecular line emission from the upper atmosphere), plus scattered sunlight, starlight, moon-light, and artificial light. Further in the infrared, the atmosphere and telescope both glow like blackbodies and dominate the background. From space, the main contribution to the background in the visible and NIR comes from sunlight scattered from interplanetary dust (visible from dark sites on the surface as the **zodiacal light**) and in the MIR from blackbody *emission* from zodiacal dust and the telescope itself. In the V band, the darkest background for the HST (near the ecliptic poles) is about 23.3 magnitudes per square arcsec, while at the darkest ground-based site, the sky brightness is about 22.0 magnitudes per square arcsec. In the thermal infrared, the sky from space can be much darker than the sky from the ground because it is possible to keep the telescope quite cold in space. Plans for the JWST suggest the sky at 5 μm should be on the order of 12 magnitudes darker in space than from the ground.

Atmospheric transmission

A fourth advantage of a space telescope is freedom from the absorbing properties of the Earth's atmosphere. This means, of course, that those parts of the electromag-netic spectrum that never reach the surface of the Earth are observable from space, and it is only here that gamma-ray, X-ray, and far-ultraviolet astronomy, for example, are possible. Even in the visible and NIR the atmosphere is not completely transparent, and the effect on detection limits is obvious in Equation (6.18). Moreover, from the ground, a major observational problem arises from *variations* in atmospheric transmission. Not only does the amount of absorbing material vary with zenith distance, but the atmosphere itself is dynamic – clouds form; the concentration of aerosols fluctuates; weather happens. All this variation seriously limits the accuracy one can expect from ground-based astronomical photometry, where astronomers are often pleased to achieve 1% precision. From space, weather never happens, and photometry precise to one part in 10^5 is possible.

Access to sky

A fifth advantage of a space telescope is its improved access to the celestial sphere. From the ground, half the sky is blocked by the Earth at all times, and for more than half the time – daytime and twilight – atmospheric scattering of sunlight makes the sky too bright for most observations. For most locations, part of the celestial sphere is never above the horizon. Even night-time has restrictions, since, for a substantial fraction of each month, scattered moonlight limits the kinds of observation that can be made. In space, a telescope far enough away from the Earth and Moon has access to most of the sky for most of the time. The HST, in a low Earth orbit with a period of about 97 minutes, has somewhat greater restrictions. Many objects are occulted by the Earth once each orbit. Because of scattering by residual atmosphere and zodiacal dust, the telescope cannot point within 50° of the Sun, within about 25° of the illuminated Earth, or within 10° of the illuminated Moon. The JWST (see Section 6.3.4) will be in an orbit that softens some of the HST constraints.

Perturbing forces and environment

A telescope on the ground will experience changing gravitational stresses as it points in different directions and will respond by changing shape – it is impossible for large telescopes to maintain the figures and alignments of optical surfaces without careful and expensive engineering. Stresses induced by wind or by temperature changes generate similar problems. A whole other set of difficulties arises from the toxic environment for telescopes on Earth – optical coatings get covered with dirt and degraded by atmospheric chemicals; abundant oxygen and high humidity promote corrosion of structures. Most of the expense of a large modern telescope is not in the optics, but in the systems needed to move and shelter the optics, while maintaining figure and alignment. A telescope in space is in an ultra-clean environment, in free fall. The forces needed to point, track, and maintain optical integrity can be relatively small, and the large mechanical bearings, massive mounts, and protective buildings of ground-based observatories can be eliminated or downsized.

6.3.2 Disadvantages of space telescopes

It might seem that telescopes belong in space, and that it would be foolish to build any serious ground-based astronomical facilities. This is not the case. At the present time, the total optical/infrared aperture on the ground exceeds that in space by a factor of at least 200, and that factor is likely to increase in the near future. The disadvantages of a space observatory are epitomized by its enormous cost compared to a ground-based observatory of similar aperture. For example, the two 8-m Gemini telescopes had a construction budget of about $100 million per telescope. The 2.4-m HST cost $2000 million to construct and launch, and the NASA construction and launch costs for the JWST are $8700 million.

Part of the great expense is space transportation – boosting a large telescope (or anything else) into orbit requires an enormous technical infrastructure.

The transportation requirements also place severe constraints on telescope design – the instrument needs to be lightweight, but also sturdy enough to survive the trauma of a rocket launch. Once in space, the telescope must function automatically or by remote control, and needs communicate its observational results to the ground. This technical sophistication entails substantial development costs and investment in ground stations and staff.

Although the space environment offers many benefits, it also harbors hazards. Any spacecraft is potentially vulnerable to intense thermal stresses, since its sunward side receives a high heat flux, while its shadowed side sees only cold space. Low Earth orbits exacerbate thermal problems as the spacecraft passes in and out of the Earth's shadow. X-rays and ultraviolet light from the Sun can damage electronic and structural components. Although a spacecraft can be shielded from sunlight, it is impossible to avoid energetic particles, either from cosmic rays or from the solar wind, especially for orbits that encounter particles trapped in the Earth's magnetosphere.

6.3.3 Airborne telescopes

You can gain some of the advantages of a space telescope simply and more inexpensively by flying your telescope at high altitudes. The best current example is **SOFIA (Stratospheric Observatory for Infrared Astronomy)** which mounts a 2.6-m telescope in a wide-body Boeing 747 that had been retired from 18 years of commercial service. Flying at an altitude of about 12 km, above 99% of the water vapor and most of the atmosphere, the low atmospheric turbulence in the stratosphere permits diffraction-limited seeing at wavelengths longer than 15 μm. SOFIA has the tremendous advantage of access to most of the spectrum from 0.3 to 1600 μm and, as a mobile telescope, has some unique capabilities (e.g. in observing Solar System occultations).

6.3.4 The James Webb Space Telescope

This replacement for the HST, built by NASA, the European Space Agency (ESA) and the Canadian Space Agency (CSA), should launch sometime before the year 2019. The design, illustrated in Figure 6.10, provides substantially advanced observational capabilities compared to HST. The main features are:

- **A heliocentric orbit at the L_2 libration point.** This orbit, about 1.5 million kilometers from the Earth, keeps the Sun, Earth, and Moon all in roughly the same direction. It avoids some of the troublesome features of a low Earth orbit: repeated occultation of targets by the Earth, periodic thermal cycling, the terrestrial radiation belts, and dynamic drag and scattering of light by residual atmosphere.
- **A stable low-temperature environment.** A highly reflective sunscreen made up of aluminized plastic film will keep the telescope and instruments in perpetual shadow.

Fig. 6.10. The James
Webb Space Telescope.

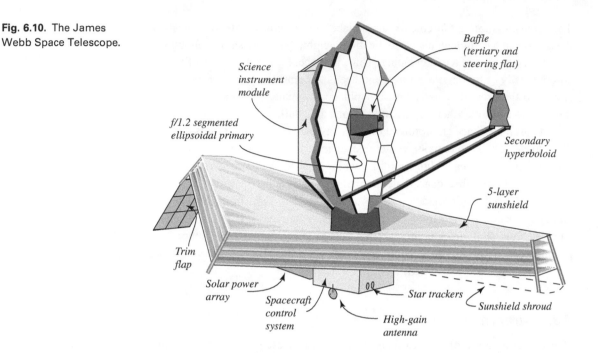

*Science
instrument
module*

*f/1.2 segmented
ellipsoidal primary*

*Baffle
(tertiary and
steering flat)*

*Secondary
hyperboloid*

*5-layer
sunshield*

*Trim
flap*

*Solar power
array*

*Spacecraft
control
system*

*High-gain
antenna*

Star trackers

Sunshield shroud

Five layers separated by vacuum minimize conductive heat transfer through the plastic itself. Under these conditions, the telescope should reach a relatively constant temperature of around 40 K. This substantially reduces the background level in the mid infrared (MIR), so that it is dominated by emission from the interplanetary dust (the ***zodiacal light***) rather than from the telescope.

- **Large aperture and high resolution.** The JWST has a lightweight segmented beryllium primary mirror with a diameter of 6.5 m. The $f/10$ anastigmatic optical system is similar to the Paul–Baker configuration in Figure 5.18, with the tertiary behind the primary. A small flat mirror will redirect the beam from the tertiary into the science instruments. The optics should produce images that are diffraction–limited at a wavelength of 2 μm ($\theta = 150$ mas) over a field of 20 minutes of arc. The limited dimensions of the Ariane 5 launch vehicle's cargo shroud mean that JWST must launch in a "folded" configuration, and deploy the primary mirror, sunshield, and secondary support in space. Once unfolded, a critically important active optics system will align the 18 primary mirror segments and other mirrors and then continuously maintain image quality.

- **Advanced instrumentation.** The telescope will have four instruments: (a) a fine-guidance camera; (b) a near infrared (NIR, 0.6–5 μm) imager, wavefront sensor, and coronagraph; (c) a NIR spectrograph; and (d) a mid-infrared instrument, MIRI, providing imaging and low-resolution ($R = 3000$) spectra in the 5–28 μm range. Because of the large aperture and low background, the JWST should improve detection limits in the 2–28 μm region by factors of over 100 compared to any existing telescope, including Spitzer and HST.

The JWST promises revolutionary advances in observational capabilities. Meanwhile, however, a new species of telescope gives every indication that equally spectacular advances can be expected on the ground.

6.4 The current revolution in ground-based observing

By the end of the nineteenth century, the most successful species of astronomical telescope, the achromatic aplanatic refractor, had evolved to the 1.0-m aperture limit set by the strength of the Earth's gravity and the fluidity of glass. Telescopes long before this time had encountered the limit on resolving power set by the Earth's atmosphere. There was barely a pause, however, before a new species of telescope, the reflector with a massive silvered-glass primary, shattered the aperture limit. The resolving-power limit was more stubborn, but did yield a bit as astronomers realized that atmospheric turbulence was minimized at certain locations.[3] Evolution reached a plateau with the 5-m Hale[4] reflector at Palomar in 1948. Over the next 30 years, no successful larger apertures appeared, and the elements of telescope design remained static: (a) a rigid primary mirror, (b) a rigid support system for the optics with passive adjustments for gravity, and (c) an equatorial mount. Some evolution in seeing quality (improved sites and enclosure architecture) and field width (R–C design) manifested in telescopes like the 4-m Blanco Telescope (1976) on Cerro Tololo in Chile and the 3.6-m Canada–France–Hawaii Telescope on Mauna Kea in Hawaii (1979).

Beginning in the 1980s, however, a spectacular series of technological advances produced a third species of reflecting telescope with 6- to 10-m apertures, capable of HST-quality resolution over narrow fields of view. Appendix G gives the current list of the largest telescopes on Earth. This new ground-based species has by no means evolved to its fundamental limits, and we expect to see greater resolving power on even larger telescopes soon. A 25–39-m class seems almost certain by 2025, with more speculative plans for 60–100-m apertures after that.

---

[3] Until late in the 1800s most of the great observatories of the world used telescopes conveniently located in university towns or near national capitals, e.g. Paris, Greenwich, Potsdam, Chicago. This gradually changed with the realization that better conditions existed at remote mountaintop locations like Mt. Hamilton (Lick Observatory, 1888) and Mt. Wilson (1904), both in California, Lowell Observatory (1894) in Arizona, and Pic du Midi Observatory (first large telescope, 1909) in the French Pyrenees.

[4] George Ellery Hale (1868–1938) an astronomer and extraordinary administrator, founded the Kenwood, Yerkes, Mt. Wilson, and Palomar Observatories, and four times raised funds and supervised the construction of the largest telescopes in the world: the Yerkes 1-m (40-inch) refractor, the Mt. Wilson 1.5-m (60-inch) reflector, the 2.5-m (100-inch) Hooker reflector, and the 5-m (200-inch) Hale reflector.

6.4.1 Large mirrors

The mirror for the Hale 5-m telescope has a finished weight of 14 tons. Made of Pyrex glass, it was designed to be rigid enough to maintain its paraboloidal figure in all orientations, requiring a support structure (tube and mount) that has a moving weight of 530 tons. The Hale mirror set the standard for an entire generation of telescopes. In this classical fabrication method, the mirror-maker pours molten glass into a cylindrical mold with a diameter-to-thickness ratio of about 6:1. The mold sometimes impresses a ribbed pattern on the back of the glass to reduce mirror weight while retaining stiffness, and to provide a method for attaching the mirror support structure. After casting, the mirror must be annealed – gradually cooled at a rate slow enough to avoid thermal gradients that would stress the glass. Improperly annealed glass can chip or shatter during the later stages of figuring, and the first 5-m blank had to be rejected in 1934 because of poor annealing. The second 5-m blank required 10 months to anneal, during which time the lab survived both a flood and an earthquake.

The front surface of the blank is next ground into a spheroid, and then polished into a conic with the desired focal ratio. For the Hale mirror (an $f/3.3$ paraboloid), this process required about 3 months, removed 5 tons of glass, and consumed 10 tons of abrasive.

A mirror larger than the Hale mirror cannot be made rigid enough, even with a massive support structure, to retain its shape in a moving telescope. Moreover, the classical fabrication method becomes very costly with increasing size (cooling and grinding time should scale as the second or third power of the mirror diameter). The advance to larger apertures required a new strategy of telescope design, and new methods of mirror fabrication. The new strategy recognizes that any large mirror will be "floppy," and uses techniques known as **active optics** to adjust and maintain the mirror shape. Since mirrors are expected to be flexible, they can be low mass, which cuts cost, fabrication time, mount bulk, and thermal response time.

To fabricate very large mirrors, three different approaches are currently in use:

> **Honeycombed monolithic mirrors** are an extension of the classical design, but with reduced mass and stiffness. Fabrication of these mirrors is greatly facilitated by a technique known as **spin casting**. The idea here is to rotate the glass mold at a constant rate, so that the centrifugal effect forms the surface of the molten glass into a paraboloid. Spin casting greatly reduces or even eliminates the grinding phase, saving months or years of work, and makes it possible to fabricate very fast ($f/1.2$) surfaces simply by selecting the correct spin rate for the mold. The two 8.4-m mirrors of the Large Binocular Telescope and the LSST primary/ tertiary are spun-cast honeycombed mirrors.

> **Segmented mirrors** are mosaics of several easily fabricated smaller mirrors arranged to produce a single large aperture. The primaries of the two 10-m Keck telescopes on Mauna Kea, for example, each consist of 36 hexagonal

segments. Individual 1.8-m segments are held in place by an active control system that constantly adjusts mirror positions to compensate for misalignments. The JWST will have an 18-segment primary.

Thin meniscus mirrors have a diameter-to-thickness ratio of something like 40:1. They are usually spun-cast in a bowl-shaped mold. Unlike the honeycombed monolith or the individual segments of a mosaic, these mirrors have no ability to retain their shapes unless supported by an active cell. The Gemini 8.4-m telescopes and the four 8.2-m elements of the Very Large Telescope (VLT) use meniscus primaries.

6.4.2 Observatory engineering

We list six important principles that govern modern observatory design. Except for the first, none of these were part of the thinking that produced "classical" 3–5-m telescopes prior to 1980.

1. **The location of an observatory is crucial to its success**. In general, atmospheric stability is greatest at subtropical latitudes. Seeing is substantially better at high altitudes on isolated islands like Hawaii and La Palma, and at the various sites in northern Chile. Remote sites at high altitude have dark skies. Dry climates are important because clouds are so detrimental to optical work and because atmospheric transmission in the infrared is closely linked to total atmospheric water-vapor content.

2. **Lightweight primary mirrors with fast focal ratios are cost effective**. Modern mirrors for large telescopes are lightweight, which means a smaller moving mass, which translates into lower cost and easier control. The moving weight of one of the Keck 10-m telescopes is 300 tons, while the moving weight of the Hale 5-m is 530 tons. Modern primary mirrors have fast focal ratios ($f/3.3$ for Hale, $f/1.75$ for Keck, $f/0.93$ for the E-ELT), which means a shorter telescope length and a smaller building, and again, smaller moving mass. Modern mounts are altazimuth, and occupy less space than the classical equatorial, again producing a smaller (cheaper) enclosure. Although one Keck telescope has four times the light-gathering power of the Hale telescope, the observatory domes are about the same size. Finally, the reduction in mass makes it easier to maintain the telescope at the same temperature as the outside air.

3. **Active optics (ao) are essential**. Computer-controlled active optics systems use motorized push–pull attachments to adjust mirror shape and position on a timescale of seconds or minutes to optimize image quality. Such systems are required for any telescope with a non-rigid mirror. Even a smaller, relatively rigid mirror will experience minor shape and position changes due to shifting gravitational stress and differential thermal expansion and can benefit from an active system.

4. **Local climate control can improve natural seeing**. Appreciable turbulence can exist inside the telescope shelter. A "dome" (in fact, many telescope enclosures are not dome-shaped, but astronomers use the term generically) and its contents will warm up during daylight hours. At night, when the air temperature drops, the dome itself and

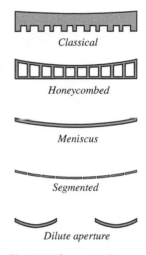

Classical

Honeycombed

Meniscus

Segmented

Dilute aperture

Fig. 6.11 Cross-sections of types of large astronomical mirrors. The dilute aperture option combines the beams of several individual widely separated mirrors, as in the four ESO VLT 8.2-m telescopes and the Large Binocular Telescope.

objects inside – like the telescope mirror, the tube, the mount, the floor – generate turbulence as they cool convectively. Even worse, a poorly designed observatory might have artificial heat sources near the telescope – a leak from another room, or non-essential power equipment in the dome itself. Structures that permit substantial airflow (e.g. fans, louvers, retractable panels) while still protecting the telescope from wind buffeting can often improve seeing appreciably. For example, the WIYN 3.5-m telescope, installed on Kitt Peak in 1994, has a low-mass mirror and active optics. The WIYN enclosure maximizes airflow and minimizes seeing induced by thermal effects from the mirror, mount, and other structures. The WIYN delivered a median 0.7 arcsec seeing during 1995–97. During this same period, the Kitt Peak Mayall 4-m Telescope (dome and telescope both of the classical Palomar design), delivered median seeing of 1.1 arcsec.

5. **Novel focal arrangements can reduce costs of specialized telescopes**. Both 10-m telescopes of the Keck Observatory can combine their beams at a common focus, as can the two 8.4-m mirrors of the Large Binocular telescope. The four 8.2-m unit telescopes of the Very Large Telescope (VLT) at the European Southern Observatory can do the same, producing a light-gathering power equivalent to a 16-m aperture. The Hobby–Eberly Telescope, in another example, has an 11-m diameter spherical primary made of identical 1.0-m hexagons, and is intended for spectroscopic use only. The optical axis is permanently fixed at an elevation angle of 55 degrees. To observe, the entire telescope structure rotates to the desired azimuth and stops. The system tracks during the exposure (limited to a maximum of 2.5 hours, depending on declination) by moving an SA corrector and the detector along the focal surface. A large fraction of the sky is accessible. This design produces substantial savings due to the tiny moving weight during exposures and the invariance of the gravitational load on the primary.

6. **Adaptive optics (AO) can eliminate some of the effects of atmospheric seeing**. Astronomers can build an optically perfect telescope and place it at a very good site, inside a well-designed structure that minimizes local turbulence. Still, uncontrollable turbulence in the upper and lower atmosphere will largely determine image quality. Technology can reduce the effects of seeing by adapting the shape of an optical element on millisecond timescales and thus undo the distortions caused by the atmosphere. Large ground-based telescopes with adaptive optics have in fact attained image resolution approaching 0.02 arcsec over narrow fields of view, and systems are under development to widen the field and further sharpen image quality.

6.4.3 Computers

I wake and feel the fell of dark, not day.
What hours, O what black Hours we have spent
This night! What sights you, heart, saw; ways you went!
And more must, in yet longer light's delay. . ..
— Gerard Manley Hopkins (1844–89),
"I wake and feel the fell of dark, not day"

The advent of inexpensive and powerful digital computers completely transformed the practice of observational astronomy. Without computers to monitor and adjust mirror shape, the large "floppy" mirrors or segmented primaries of modern telescopes are useless. Without computers in control of the fabrication process, it is doubtful these mirrors could be made in the first place. Without computers to manage pointing, tracking, and instrument rotation, an altazimuth mount becomes a very tricky proposition. Without computers to command instruments and gather the data from a modern camera or spectrograph, the flow of information from even the largest telescope would choke off to a trickle. Without computers, elimination of the effects of atmospheric seeing on the ground would be impossible. Without computers, the HST and JWST would be utterly unthinkable.

Astronomers were quick, in the 1950s and 1960s, to utilize early "mainframe" electronic computers for the reduction and analysis of data, and for the construction of theoretical models of astrophysical phenomena. Then, in 1974, the 3.9-m Anglo-Australian Telescope became the first large telescope to use computer-controlled pointing and tracking. As the price-to-power ratio of mini- and micro-computers fell, digital electronics moved into observatories. The advent of CCD detectors in the 1980s meant that computers not only reduced data and moved telescopes but also controlled instruments and acquired data.

In 1870 (or 1670) an observational astronomer woke to spend the night in the cold and dark, eye to ocular through the black hours, making occasional notes or calling measurements to an assistant. By 1970, little had changed, except things were sometimes a bit more gymnastic: still in the cold and dark, the astronomer used the ocular only to guide the telescope, perhaps while exposing a photographic plate. A frantic rush to the (blissfully warm) darkroom provided an occasional interlude when the photograph needed to be developed.

Today, a night at the telescope differs little from a day at the office: the warm room is brightly lit, and the astronomer types an occasional command at a computer console: move the telescope; change the filter; expose a CCD frame; start a pre-programmed sequence of moves, changes, and exposures. Data accumulate in computer storage, and if data flow in slowly, the astronomer can start reducing them as they arrive. Rewards are immediate, right there on the monitor. Often, though, data flood in at a mind-smothering rate, and teams of astronomers have worked for years to build software to digest them.

The telescope is in another room: cold, dark, and open to the sky. That room could be next door, or thousands of kilometers away on an oxygen-poor mountain on another continent. Or there might be no room at all, as the telescope orbits above. The older methods are exhausted, and discovery and adventure come with the new, but the price is reduced acquaintance with the fell of dark, and the exchange of exotic photons from the depths of space for mundane emissions from a monitor screen.

6.5 Atmospheric blur

> Long telescopes may cause Objects to appear brighter and larger than short ones
> can do, but they cannot be so formed as to take away that confusion of Rays
> which arises from the Tremors of the Atmosphere. The only Remedy is a
> most serene and quiet Air, such as may perhaps be found on the tops of the
> highest Mountains above the grosser Clouds.
>
> – Isaac Newton, *Opticks*, 1704

6.5.1 Atmospheric wavefront (WF) distortion

For nearly four centuries, Newton's judgment about astronomical seeing held
true, but there are now technological remedies for unquiet air. We mentioned
wavefront (WF) distortions and atmospheric seeing in Chapter 5 but now wish to
examine these quantitatively and understand methods for removing them. The
index of refraction of air depends on its density (and the wavelength of light –
see Table 5.3). In a perfectly serene and quiet atmosphere, the density and index
will depend only on altitude, and every point at the same height will have the
same index.

In the real and imperfect atmosphere, however, solar heating drives convective cells in the lowest layer of the atmosphere, a region about 10–12 km thick
called the ***troposphere***. Here, one mass of air in contact with the surface can
become warmer and thus more buoyant than its neighbors. That mass rises.
Another moves horizontally to fill its place; cold air from above drops down to
make room for the rising mass and completes the circulation around a cell. Many
cells are established, and the air, especially at the boundaries of the flow, tends to
break up into ever smaller eddies and lumps – this break-up of the flow is
turbulence. See Figure 6.12. Lumps of different temperature (and hence, density) will have different indices of refraction: Near Earth's surface, a temperature
change of 1 °C will cause an index change of about 10^{-6}.

Now consider a wavefront from a distant star that passes through the turbulent atmosphere. The WF arrives as a plane, but each ray from the front will
encounter slightly different patterns in the index of refraction and will traverse
a slightly different total optical path. By the time it reaches the entrance aperture of a telescope, the WF will be crumpled with dents and bumps corresponding to larger or smaller total optical path lengths through the atmosphere.

Fig. 6.12 Circulation and
turbulence in the Earth's
atmosphere. (a) The basic
circulation pattern set up
by convection above a
warm surface. Winds here
have a laminar flow.
(b) The circulation
modified by turbulence –
producing a mix of small
moving warm (H) and cold
(C) air masses. On Earth,
turbulence is usually
strongest within 1 km of
the ground (the boundary
layer), and has a weaker
peak higher up in the
"free atmosphere" near
the tropopause – the
upper limit of the
convective region. Large
local values of wind shear
at any layer produce
appreciable turbulence.

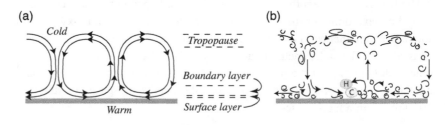

See Figure 6.13. Since the turbulent lumps and eddies at each altitude move at the local wind speed, the shape of the distorted WF changes very quickly.

Consider, now, what happens if you point a large telescope at this star (Figure 6.14). When different segments of its distorted wavefront reach the telescope entrance aperture, rays normal to each segment are traveling in slightly different directions, so each segment is imaged in a slightly different spot in the focal plane. In Figure 6.14a, for example, we can approximate the distorted WF in one dimension (along the y-axis) as N different plane wavefronts each with a diameter of about D/N, each producing an Airy disk of radius $\alpha \sim \lambda/ND$, each disk at a different location in the focal plane. These spots will have different phases, and will create a complex system of criss-crossing interference fringes. The combination of these interference patterns in two dimensions produces a multi-spotted image in the x–y focal plane. This is called a ***speckle pattern*** – the size of each local spot, or speckle, in the pattern turns out to be roughly equal to the diffraction limit of the telescope, $\alpha_A \sim \lambda/D$ (Figure 6.14b). Turbulence, moreover, moves the lumps of air around at high velocity so a particular speckle pattern is evanescent, and a new pattern forms after something like 10 ms, depending on wind velocities. Therefore, in a long exposure, the variation in the speckle pattern blurs into the ***seeing disk*** (Figure 6.14b).

Atmospheric turbulence produces somewhat different phenomena for small apertures. ***Scintillation*** is the intensity variation produced by an overall concave (bright) or convex (faint) WF distortion. Scintillation is very apparent as "twinkling" to the unaided eye (aperture about 7 mm), and as short timescale photometric variability in telescopes. Telescopes with apertures smaller than a single atmospheric refractive cell produce a single Airy pattern, not a speckle image. However, the pattern moves around the focal plane, and produces, on long exposures, the usual seeing disk.

6.5.2 High resolution on short exposures

Speckle interferometry is a set of techniques that analyzes the speckle patterns on multiple very short exposures usually taken with large telescopes, and resolves characteristics of a target object at the telescope diffraction limit. The technique works well for measuring the separation of close binary stars, the angular diameters of the nearest giant and supergiant stars, and the details of relatively bright circumstellar material.

A related technique works under conditions of relatively good seeing, where there is a chance that the image of a bright star will consist of a single speckle. The technique, called ***lucky imaging***, accumulates a series of very short exposures and selects only the very few in which a single speckle predominates. These "lucky" exposures are then shifted and combined to produce a high-resolution longer exposure. Although lucky imaging and speckle interferometry work well

(a)

Plane wavefront

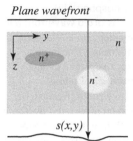

Distorted wavefront

(b)

Plane wavefront

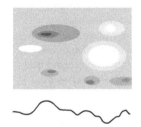

Distorted wavefront

Fig. 6.13 Distortion of wavefronts by a turbulent atmosphere. The figure shows plane wavefronts incident on turbulent layers, and the distortion that results. Darker colors indicate lower temperatures (higher index), light colors, warmer temperatures and lower index. The distortion in (b) is much greater than in (a). Wavefront distortions are greatly exaggerated in the z-direction.

Fig. 6.14 Image formation through a turbulent atmosphere. (a) A distorted wavefront (*D*) incident on a telescope aperture (*T*), can be approximated as a number of plane wavefronts (*P*), each of which is focused by *T* to a different spot in the focal pane (*FP*). The converging wavefront at *C* will retain the original distortions. (b) The instantaneous image in the *x*–*y* focal plane of each parallel wavefront in *P* forms an Airy pattern modified by interference effects – the speckle image, (c). The movement, dissolution, and formation of different speckles produces a seeing disk on a long exposure.

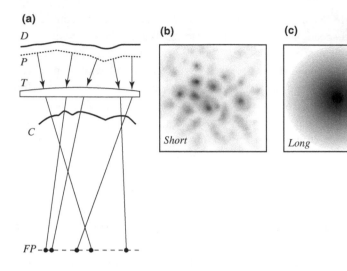

in some circumstances, a broader approach is needed for most astronomical investigations, so we now look more closely at WF distortions.

6.5.3 Quantifying wavefront distortion

Figure 6.15 suggests a method for quantifying the WF distortion in a medium with index *n*. First consider the *y*–*z* plane. Assume that the average tilt of the WF is zero. We start at one end of the WF ($y = 0$, $z = 0$) and fit a straight line $\Delta z(\Delta y)$ to a segment of the front. We note the value of Δy required to make $\Delta z = \lambda/2\pi n$ (about 1/6 of a wavelength), and call this, r_1, the **coherence length** of the first segment. The change in the phase of the wave across the coherence length is less than one radian. Now move along the front, fitting successive segments with straight lines, noting the coherence lengths of each, r_i. The statistical mean of all the r_i values is r_{avg}, the **coherence length of the wavefront**. Each segment has a different slope, so each will propagate in a slightly different direction, and each will focus at a different spot in the image plane of a telescope (review Figure 6.14). The shorter the coherence length, the more speckles in the image, and the greater the spread in their locations

Now we extend the idea of coherence length to two dimensions. Suppose we select a random point on a two-dimensional wavefront, and ask how large a two-dimensional patch of the front we can expect to be coherent. The answer is given by the **Fried parameter**, $r_{0\lambda}$.

$$r_{0\lambda} = \text{the expected diameter over which the root-mean-square}$$
$$\text{optical phase distortion is 1 radian}$$

Note that the Fried parameter is a statistical description of how the optical path length varies across the WF. Nevertheless, you will find it most useful to

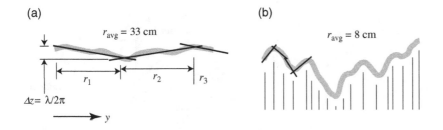

(a)

$r_{avg} = 33$ cm

r_1 r_2 r_3

$\Delta z = \lambda/2\pi$

y

(b)

$r_{avg} = 8$ cm

Fig. 6.15 The coherence length of a wavefront. Within each segment of length r_i, the variation in optical phase is less than one radian. Front (a) has a much larger average coherence length than front (b). The scale is greatly exaggerated in the z-direction.

regard $r_{0\lambda}$ as a measure of how large a segment of the wavefront you can expect to treat as a plane wave. The Fried parameter is a good indicator of image quality. In particular, the full width at half-maximum (FWHM) of the seeing disk is given by

$$\theta(\text{seeing, in arc sec}) \approx 0.2 \frac{\lambda[\mu m]}{r_{0\lambda}[m]} \tag{6.21}$$

If the diameter of a telescope, D, is larger than $r_{0\lambda}$, then Equation (6.21) gives the image size. If $D < r_{0\lambda}$, then the telescope is ***diffraction limited*** and the instantaneous image size is given by the Airy disk, Equation (6.10). The overall tilt of the WF will change over time, so even in a small telescope, the diffraction-limited image will move around in the focal plane, so Equation (6.19) will apply for long exposures.

Since the index of refraction of air depends only weakly on wavelength, the same must be true of the wavelength dependence of the optical path length, $s(x, y)$. The variation in phase produced by a variation in s is therefore inversely proportional to wavelength. The Fried parameter is thus a function of wavelength:

$$r_{0\lambda} = r_0 \left(\frac{\lambda}{0.5 \ \mu m}\right)^{\frac{6}{5}} (\cos \zeta)^{\frac{3}{5}} \tag{6.22}$$

Here ζ is the angle between the direction observed and the zenith. The Fried parameter is usually quantified simply by quoting its value at a wavelength of 500 nm, r_0. Values for r_0 generally range from a few centimeters (poor seeing) to 15 or 20 cm (superb seeing). Whatever the value of r_0 at a site, $r_{0\lambda}$ will be larger (and the size of the seeing disk will be somewhat smaller) for observations made nearer the zenith as well as for observations made at longer wavelengths. The Fried parameter can fluctuate by large factors over time, on scales of seconds or months.

6.6 Adaptive optics

6.6.1 The idea of adaptive optics

The idea of adaptive optics (AO) is simple: remove WF distortions by inserting one or more adjustable optical elements into the path between source and detector to exactly cancel those distortions. In practice, the adjustable elements

Fig. 6.16 Schematic of an adaptive optics system. Actual systems will differ in important details.

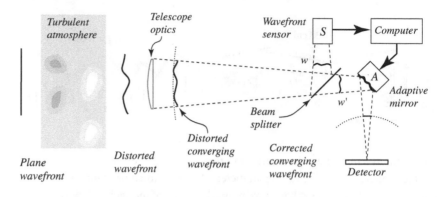

Plane wavefront   *Distorted wavefront*   *Distorted converging wavefront*   *Corrected converging wavefront*   *Detector*

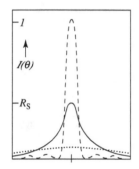

Fig. 6.17 The point-spread function in one dimension. The dashed curve is the Airy function, the dotted curve is the seeing disk, and the solid curve is an AO-compensated PSF with Strehl ratio of R_S

are usually mirrors located between the telescope primary and the focal plane. Figure 6.16 is a schematic representation of the AO concept. In the figure, a partially reflecting mirror splits the distorted wavefront from the telescope into two fronts, w and w'. These fronts have identical distortions. Front w proceeds to a sensor, S, which detects the magnitude of its distortion at some number of locations on the front. The other front, w', is reflected from an adjustable mirror, A, onto the detector. Meanwhile, the computer has read the distortions mapped by S, and has commanded A to adjust the shape of its surface to exactly cancel them. If all goes perfectly well, the compensated image formed at the detector will be diffraction-limited, with all effects due to the atmosphere removed.

Things seldom go perfectly well. One measure of how well an AO system succeeds is the **Strehl ratio**, R_S. For a point source detected by a particular telescope, the run of intensity per unit solid angle with position is called the **point-spread function, PSF(x,y)**. If I_{PSF0} is the peak intensity per unit solid angle of the point-spread function, and I_A is the peak intensity of the Airy function for the same source, then

$$R_S = \frac{I_{PSF0}}{I_A} \tag{6.23}$$

The Strehl ratio of a long-exposure (seeing-imposed) PSF is $(r_0/D)^2$. The hope is that an AO system will transfer intensity from the outer part of the seeing disk to the core, and increase R_S over the uncompensated value. A *perfect* AO system will produce $R_S = 1$, and thus improve the Strehl ratio by a factor of $(D/r_0)^2$.

6.6.2 The Greenwood time delay

The idea of adaptive optics is simple. Its execution is not. Davies and Kasper (2012) provide a good introduction to some practical applications, and the books by Roddier (1999) and Hardy (1998) provide a more complete discussion. We can only examine a few concepts here. One has to do with the time element.

Clearly, any time delay between sensing the wavefront and adjusting the deformable mirror is a problem. The maximum delay that can be tolerated will depend on the size and velocity of each distortion in the atmosphere. Statistically, if $\overline{v}$ is the weighted average velocity of turbulent features, then the **Greenwood time delay**, or the **coherence time** in seconds, before a the wavefront error at some location changes by 1 radian is (λ in µm)

$$\tau = 0.559 \frac{r_0}{\overline{v}} \lambda^{\frac{6}{5}} (\cos \zeta)^{\frac{3}{5}} \tag{6.24}$$

For typical values of the parameters, the Greenwood time delay is several milliseconds at visual wavelengths. Practical AO systems must not only respond within the Greenwood time, but must also update the shape of correcting elements at at least the Greenwood frequency, $1/\tau$. The wavelength dependence of τ is one (of many) reason(s) why AO is a lot easier at longer wavelengths. The *overall tilt* of the wavefront changes on a somewhat longer timescale than the coherence time and is more easily compensated.

6.6.3 Anisoplanatism

Figure 6.18 illustrates another serious limitation of simple AO systems. A turbulent layer is some height, h, above a telescope. Rays from two sources, separated in the sky by angle θ, traverse the turbulence along different paths. The layer introduces different WF distortions for the two sources. For very small values of θ, the distortions will not differ greatly, but if $\theta \geq r_{0\lambda}/h$ the phase distortions will be uncorrelated. The **isoplanatic angle**, θ_i, in radians, is the largest angle for which the expected distortions differ by less than one radian over the whole front:

$$\theta_i = 0.314 \frac{r_{0\lambda} \cos \zeta}{\overline{h}} \tag{6.25}$$

Since turbulence generally occurs at different heights, $\overline{h}$ represents a weighted mean height. For a typical site, the isoplanatic angle ranges from around two

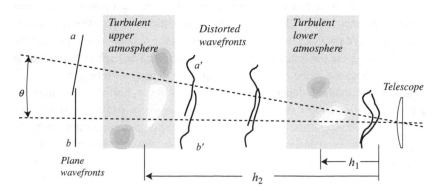

Fig. 6.18 Anisoplanatism. Plane wavefronts *a* and *b* arrive from sources separated by angle θ. They experience different distortions, a' and b'. An AO compensation for a' will be incorrect for b'.

seconds of arc in the blue to around 30 seconds of arc at 10 μm. This means that the compensated field of view – the isoplanatic patch – is very limited. Note also that ground-level turbulence (height h_1 in the figure) does not restrict the size of the compensated field as seriously as motions in the upper atmosphere. Again, the wavelength dependence means that AO systems will be more useful in the infrared than in the optical.

6.6.4 Guide stars

Yet another limitation for AO arises from the obvious condition that AO only works if it can sense the distortions in a wavefront. This sensed WF is usually (but not always) from a point source – the *guide star*. Exposure times for the WF sensor must be less than the Greenwood delay time, so the guide star must be bright ($m_V < 13$–15). If the science source itself is faint, and if there is no *natural guide star (NGS)* of sufficient brightness nearer to it than the isoplanatic angle, then AO cannot compensate the image.

In practice, it is difficult to find suitable natural guide stars, especially away from the Galactic plane. A rescue technique is the *laser guide star (LGS)*: use a laser to illuminate a small spot in the upper atmosphere well above the turbulence layer.

In addition to expense, there are a few drawbacks to using laser guide stars. Any LGS will suffer from the *cone effect* – the wavefront from an LGS samples a conical volume of air, not the cylinder traversed by the science object wavefront. See Figure 6.23. The cone effect produces errors because the dark-colored regions (X in the figure) in turbulent layers near the tropopause (Tr) are not sampled by the LGS wavefront, and because the wavefront distortions in the light-gray areas are stretched by the lateral expansion of the wavefront as it moves downward. Moreover, turbulence tilts the laser beam in its upward passage through the atmosphere as well as in its downward passage. The position of an LGS, unlike that of an NGS, is not useful in determining the tip–tilt correction for the science wavefront: an NGS must be used for tip–tilt correction. Nevertheless, for many cases, the alternative to an LGS is no AO correction.

There are two current methods for implementing an LGS. The first uses a *pulsed* laser to illuminate a narrow column of air and observes the back-scattered light (Rayleigh scattering). The WF sensor observation is also pulsed, so that the altitude of the illuminated spot can be selected by adjusting the delay between pulsing the laser and making the observation. Maximum altitude for Rayleigh laser beacons is about 20 km because of the exponential drop in air density with height.

The second method depends on a curiosity of the Earth's atmosphere – the presence of a 10-km thick layer in the mesosphere (90 km up) with an unusually high concentration of neutral sodium and potassium atoms, probably of meteoritic origin. A laser on the ground near the telescope is tuned to one of the sodium D lines (589.00 or 589.59 nm) and fired to pass through the mesospheric layer at the desired position. The laser light excites sodium atoms, which in turn emit line radiation by spontaneous emission (after about 10^{-8} seconds), with most of the emission concentrated in the sodium layer. For astronomy, sodium beacons are superior to Rayleigh beacons in that their higher altitude permits a more accurate replication of the ray path from the science object.

Laser beacons (especially sodium lasers), though expensive and difficult to maintain, are implemented at most major observatories.

6.6.5 Wavefront correctors

Wavefront correctors in current astronomical AO systems are small to medium-sized *deformable mirrors (DMs)* of several types. All must have surfaces whose shape can quickly (within milliseconds) adjust to compensate for sensed distortions in the incoming wavefront. Early AO systems used segmented mirrors made up of independent flat reflectors, each capable of piston, tip, and tilt motions. The number of segments needed depends on the coherence length of the WF, and should be roughly $(D/r_0)^2$. Segmented mirrors have the disadvantage of diffraction and scattering effects produced by the gaps between segments, as well as requiring a relatively large number of actuators. Continuous-surface DMs eliminate these problems, and have seen the most use in astronomical AO.

An *actuator* converts an electrical signal into a change in position, and a few different types have been used in DMs. Some, for example, rely on the piezo-electric effect: certain polarized ceramic materials respond to an imposed electric field by changing dimension, and have a relatively fast response time. In the pure piezoelectric effect, the change in dimension is directly proportional to the voltage applied in the polarization direction as illustrated in Figure 6.19a.

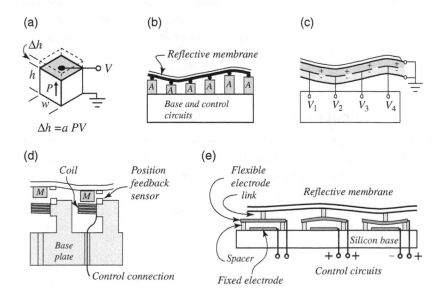

Fig. 6.19 Deformable mirror designs.
(a) A piezoelectric stack, which will change its vertical dimension in response to an applied voltage. Voltage applied opposing the polarization will shrink the stack.
(b) Piezoelectric actuators (*A*) bonded to a thin monolithic reflecting plate and a rigid base.
(c) A bimorph actuator whose curvature is controlled by adjusting the local tension/compression of the bimorph pair. The top surface is coated with reflecting material.
(d) An element of a magnetically controlled DM: permanent magnets (*M*) are bonded to a flexible faceplate. Electromagnetic coils mounted on the reference surface attract or repel the magnets.
(e) A cross-section of a MEMS deformable mirror.

Other actuators rely on electrostatic forces or on magnetic forces between a wire coil and a permanent magnet.

A DM can be the secondary mirror in a reflecting telescope – a design that eliminates one reflection. Designs for these to date rely on permanent magnet–electromagnet interactions to change the shape of a thin reflecting plate relative to a rigid back surface (Figure 6.19d). Capacitive sensor feedback helps verify and control the deformation. DMs large enough to act as secondaries can accommodate the rather bulky coil/magnet actuators, but these mirrors are very, very expensive, and only a few are in use. More often, the deformable element is located behind the focus, often at an image of the primary.

An alternative and somewhat less expensive DM design, probably still the most common, bonds a flexible thin reflecting glass, quartz, or silicon face-sheet to an array of piezo actuators (Figure 6.19b). A second, even less expensive design is the bimorph mirror. This consists of two large piezoelectric disks of opposite polarities with local electrodes sandwiched between them (Figure 6.19c). Voltage applied at a particular location causes local material in one disk to expand while the material in the other disk contracts, producing surface curvature.

Deformable mirrors that are micro-electronic machined systems (MEMS) are potentially very inexpensive. Employing the lithography methods of the electronics industry, MEMS technology fabricates electrodes, spacers, and electrostatic actuators on a silicon chip, and bonds a flexible reflecting membrane to the surface of the device. The primary disadvantage is the limited size of the resulting mirror.

Most practical AO systems correct the wavefront in two stages. One flat mirror corrects the overall *tip–tilt* and focus (push–pull) of the entire wavefront, and a deformable mirror corrects the remaining distortions. Although this arrangement is more complex, it has important advantages, including a reduction in the overall piston motion required by the elements of the deformable mirror. Piston movement in adaptive secondaries can be large, but piezoelectric actuators are limited to about 10 μm and MEMS electrostatic actuators usually cannot exceed about 2 μm piston motion.

6.6.6 Wavefront sensors

A wavefront corrector is only useful if the AO system is able to measure the distortions that need compensation. There are at least a half-dozen techniques for performing this function. (The most important are: pyramidal prism sensors, shearing interferometers, and curvature sensors.) Here, however, we will describe only one, the *Shack–Hartmann sensor*, because it has seen wide application in astronomy, relies on a mature technology, and is relatively easy to understand.

The Shack–Hartmann sensor, illustrated in Figure 6.20, exploits the fact that rays propagate perpendicular to the surface of the wavefront, so that sensing the

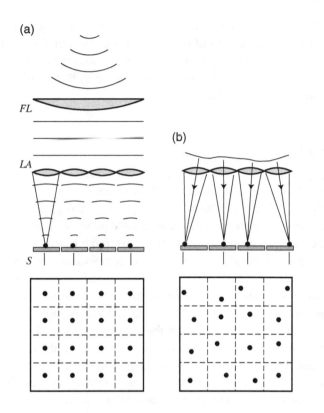

(a)

FL

LA

S

(b)

Fig. 6.20 The Shack–Hartmann sensor. (a) The beam from a perfect point source is rendered afocal by the field lens (*FL*). The resulting plane wave is imaged by the lenslet array (*LA*) – all images on the sensors (*S*) are in the null position. (b) A distorted wavefront and the resulting image displacements from tilted segments.

direction of a particular ray is equivalent to sensing the slope of the local wavefront. In Figure 6.20a, we sketch a Shack–Hartmann sensor operating on a WF that was a perfect plane when it was incident on the aperture. The device, in this case, is located behind the focus, where a ***field lens*** (usually a mirror) converts the diverging spherical wavefront back into a plane. An array of lenses then separates the rays from each segment of the front into isolated bundles, called ***sub-apertures***, each of which is brought to a different focus on a detector array.

Now, in Figure 6.20a, we consider what happens if the original wavefront is distorted. The distortion causes local changes in the wavefront slope and corresponding changes in the directions of the rays. Each bundle of rays on a lenslet now comes to a focus whose location depends on the local slope of the WF. If the detector array has at least 4 pixels in each sub-aperture, a computer can sense the position of the focused spots and determine the WF slope in each sub-aperture. Note that the guide object need not be a point source: the system only requires some image whose relative displacement can be determined in each sub-aperture. Because exposure times must be milliseconds, and the guide images distributed over many ($> 4(D/r_{0\lambda})^2$) pixels, guide star faintness is a serious limitation.

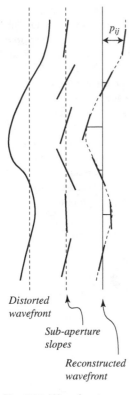

Distorted wavefront

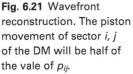

Sub-aperture slopes

Reconstructed wavefront

Fig. 6.21 Wavefront reconstruction. The piston movement of sector *i, j* of the DM will be half of the vale of p_{ij}.

One more step remains before we can command a deformable mirror. Figure 6.21 shows the process of ***wavefront reconstruction***. Our computer must convert the array of slopes for each sub-aperture into the actual shape of the distorted front. For a large number of sub-apertures, this requires a very fast computer and clever algorithms.

6.6.7 A simple AO system

Building a practical AO system is complex and expensive, and astronomers can usually justify one only for the largest of telescopes, where its cost constitutes a reasonably small fraction of the overall budget. Figure 6.22 is a very schematic layout of a practical system. In this case, the AO optics are behind the telescope focus, where a parabolic mirror converts the diverging spherical wavefront from the guide star to an afocal beam before the uncompensated front encounters the deformable mirror.

The corrected beam then separates at the dichroic beam-splitter – this is a special mirror that passes infrared light but reflects the optical. Since optical path-length deviations introduced by the atmosphere depend only weakly on wavelength, the NIR image can be corrected by sensing the distortions in the optical image. This allows all the infrared light to be directed to the image, while all the optical light (from a sodium laser, for example) is used for wavefront sensing. A second paraboloid then refocuses the compensated infrared WF on the detector.

The illustration shows a separate system to correct for errors in the overall position of the image – errors that can arise not only from seeing, but also from telescope drive imperfections. A sensor commands the tip and tilt orientation of a flat mirror to maintain the centroid of the image at the same spot on the detector. This separate tip–tilt correction minimizes the amount of correction (and therefore actuator motion) that the deformable mirror needs to make.

Another important feature of the practical system is the location of the WF sensor. The sensor examines the WF *after* the tip–tilt and the deformable mirrors have corrected it, an arrangement termed ***closed-loop*** operation. The task of the AO computer is to null out only *changes* in the WF errors introduced (or residual) since the last command cycle. This requires that the null point be well calibrated, but has the benefit that – if the adaptive cycle is sufficiently short – the range of motions directed by the feedback loop will be relatively small, and any errors in compensation will tend to be corrected in the next cycle.

6.6.8 Advanced AO systems

An AO system like the one sketched in the previous section is capable of producing Strehl ratios of 0.4–0.7 in the K band (2.2 μm) at good sites.

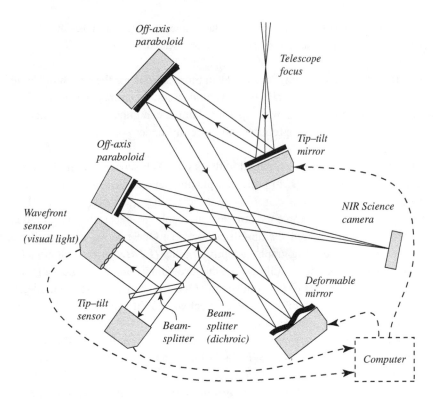

Off-axis paraboloid

Telescope focus

Tip–tilt mirror

Off-axis paraboloid

NIR Science camera

Wavefront sensor (visual light)

Deformable mirror

Tip–tilt sensor

Beam-splitter

Beam-splitter (dichroic)

Computer

Fig. 6.22 Schematic of a practical AO system.

A ground-based telescope can thereby attain angular resolutions approaching those of a space telescope of similar aperture. Of course, an AO system on, say, a 6.5-m telescope is very expensive (telescope plus AO around $50 million) and often difficult to operate, but when compared to the cost ($8700 million) and complexity of the 6.5-m JWST, the ground unit looks like a mighty bargain. It is hardly surprising, then, that as soon as the technology became available in the 1990s, astronomers began finding funds and developing AO on existing telescopes, then migrating these systems (and their experience) to larger telescopes as they were constructed. Currently (2015) most of the world's largest telescopes ($D > 6$ m) can operate with AO. Most systems are restricted to wavelengths longward of 2 μm, but improvements in DMs and processing speed are making some AO at shorter wavelengths possible.

Until around 2011, the productive AO systems could be characterized as *single-conjugate AO (SCAO)* – they use a single guide star and a single DM (Figure 6.23), and have serious restrictions because of anisoplanism and small Greenwood delays:

- Natural guide stars are seldom available, so an LGS system, with the attendant cone effect, is required.
- The point-spread function varies with distance from the guide star, so photometry is difficult.

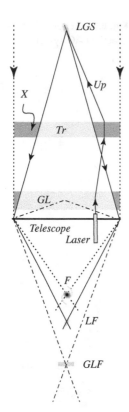

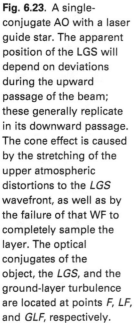

Fig. 6.23. A single-conjugate AO with a laser guide star. The apparent position of the LGS will depend on deviations during the upward passage of the beam; these generally replicate in its downward passage. The cone effect is caused by the stretching of the upper atmospheric distortions to the *LGS* wavefront, as well as by the failure of that WF to completely sample the layer. The optical conjugates of the object, the *LGS*, and the ground-layer turbulence are located at points *F, LF,* and *GLF*, respectively.

• Only a tiny angular field is corrected, étendue is microscopic, and studies of extended objects like clusters of stars or galaxies are poorly served.

These restrictions are limiting, and are even more serious for the extremely large (D = 24–42 m) telescopes *(ELTs)* now under construction: The cone effect becomes more pronounced with aperture, and the number of DM actuators increases as D^2. Several approaches, all currently under some degree of development, offer relief.

Multiple guide stars, each paired with its own WF sensor, are the basis for many of the improvements. In principle, placing several LGS near the circumference of the field of view can mitigate the cone effect because their overlapping cones completely sample the turbulent atmospheric layers. (In addition, one or more NGS are required for tip–tilt correction.) One goal of these multiple lines of sight is laser tomography adaptive optics *(LTAO)*, a computationally intensive real-time modeling (almost) of the three-dimensional turbulence structure at all relevant layers of the atmosphere. One can then collapse the modeled distortions along the path of the science object wavefront and command a single DM, perhaps an adaptive secondary.

A related application of multiple guide stars recognizes the fact that a large fraction (perhaps 50–60%) of WF distortion occurs in the lowest layers of the atmosphere – where the cone effect, wind velocities, and Greenwood frequencies are all low. The idea of *ground-level adaptive optics (GLAO)* is to average the results of the multiple wavefront sensors. Distortions caused by the upper atmosphere will tend to cancel and those caused by the ground layer will remain (all lines of sight pass through it) – and this signal is used to command a single DM. Since blurring by the upper atmosphere is not removed, the result is a modest bump up in the Strehl, but very welcome improvement in seeing ($\approx$ factor of 2) and PSF uniformity over a relatively wide field (1–3 arc minutes).

Successful tomography is the basis for two more complex advances in AO. *Multi-object AO (MOAO)* uses a dense distribution of guide stars + WF sensors to model the three-dimensional atmospheric structure over a wide field (5–10 arc minute radius), and one small, mobile DM for each object of interest. Light from each DM is then conducted to an integrated camera or spectrograph. *Multi-conjugate AO (MCAO)* models WF distortions for N_L layers in the atmosphere, then produces a corrected contiguous field by using N_L separate DMs conjugated one to each layer. Field sizes are similar to GLAO, but more highly corrected.

Finally, we mention *extreme AO*, which is basically enhanced SCAO using a very bright natural guide star, fast correction cycles, an advanced DM, sophisticated data processing, and attention to stability issues. The goal is to produce very high contrast (Strehl > 0.9) images that would be capable – as has been demonstrated in at least one case – of detecting large planets around nearby stars.

The different AO techniques are very much in the state of development. Astronomers have tested all of them on existing large telescopes, and each of the three ELTs under construction will incorporate sophisticated adaptive optics.

6.7 Extremely large telescopes

> Perhaps more excellent things will be discovered in time, either by me or by
> others with the help of a similar instrument, the form and construction of
> which . . .I shall first mention briefly. . ..
>
> – Galileo Galilei, *Sidereus Nuncius*, 1610

If adaptive optics can produce Strehl ratios close to unity over a wide field, then the aperture payoff in detection and resolution limits is roughly governed by the equations for space telescopes (Section 6.3), so, depending on sky brightness, a 40-m telescope should detect stellar sources 100 times fainter than an AO-equipped 4-m, and resolve detail six times finer than the JWST.

As of this writing (2015) three multinational groups worldwide are constructing single-aperture steerable telescopes much larger than the current crop of 6–10-m instruments. All three projects are expected to attain first light in around the years 2022–27, and Table 6.1 gives some details for each. Note that no technology exists for casting monolithic mirrors larger than 8.4 m, so any ELT must use segmented mirrors.

Each of the three ELTs in progress will cost in the range of one billion US dollars – a price that forces multi-institutional, multinational cooperation. Doing ELT astronomy is similar to doing astronomy with space missions like the JWST and quite unlike the historic single astronomer/single telescope practice. In addition to massive funding and wide-ranging cooperation, an ELT's success depends on satisfying three requirements:

- A telescope structure with sufficient stiffness to withstand normal wind loading, as well as the occasional spectacular storm or mild earthquake, with an active optics system that will maintain segment shape, positioning and alignment in the face of inevitable deformations.
- An advanced adaptive optics system that is capable of producing a large Strehl ratio over a large isoplanatic patch.
- A suite of instruments (cameras, spectrographs) capable of utilizing the imaging characteristic of the planned optics.

To these three points we should add a host of particular scale-related financial, mechanical, optical, and logistic difficulties. Just a few examples: (a) The annual operating costs of a telescope are typically 5% of the construction budget (i.e. $50 million/year to maintain an ELT). (b) The drive system must move

Table 6.1 *ELT projects currently under way.*

Project	Effective aperture (m), [# of segments × sub-aperture (m)]. Design and AO	Name, major partners, URL
E-ELT	39, [798 × 1.4] 3-mirror anastigmat MCAO	European Extremely Large Telescope, European Southern Observatory (15 member nations), www.eso.org/sci/facilities/eelt/
GMT	24.5, [7 × 8.4] Aplanatic Gregorian Adaptive segmented secondary LTAO	Giant Magellan Telescope, nine partner institutions from the USA, Australia, Chile, Brazil, and Korea, www.gmto.org/
TMT	30, [492 × 1.4] R–C, MCAO	Thirty Meter Telescope, 11 institutions from USA, India, China Japan, Canada, www.tmt.org/

huge ($1\text{-}3 \times 10^6$ kg) loads without transmitting vibrations to instruments or optics. (c) Cleaning and recoating a primary mirror is typically done once per year. If a telescope has 300–800 segments, one or two of those must be removed, cleaned and recoated every day. (d) Without AO, focal plate scales are poorly matched to typical detector pixel sizes.

Despite difficulties, scaling up and optimizing existing altazimuth telescope structures appears to be a reasonable approach. As discussed above, advances in AO systems are well under way, and all three telescopes utilize designs that integrate AO to some extent, rather than adding it on after construction. As of this writing (2015), the world's enthusiasm for such expensive astronomy seems to be limited, competition for funding is fierce, and construction delays mount. In hindsight, many have questioned the wisdom of trying to build three large telescopes rather than one or two.[5] Nevertheless, most astronomers remain optimistic that all three projects will see first light within the decade.

---

[5] At least some of the fragmentation of effort has historical roots in Hale's fundraising for the Palomar 5-m, which devolved into a long-standing rivalry between Carnegie Observatories and the Caltech/University of California consortium that built the Keck telescopes.

Summary

- Ground-based telescopes usually utilize either an equatorial or an altazimuth mount. Each has advantages. Telescopes in space use rockets and/or reaction wheels to point and guide. Concepts:

  setting circle   *clock drive*   *guide star*

  instrument de-rotator   *reference gyroscope*

- Prime focus and Newtonian optical layouts with a parabolic primary mirror suffer from off-axis coma.

- Cassegrain and Gregorian layouts use two curved mirrors, are compact and convenient, and can be made aplanatic. Simple algebraic relationships govern the design of these telescopes. Concepts:

  back focal distance   *aplanatic Gregorian*   *Ritchey–Chrétien*

  Nasmyth focus   *coudé focus*

- The addition of optical surfaces can remove aberrations besides SA and coma: the LSST is an example of a three-mirror design that produces large étendue.

- Catadioptric telescopes utilize both reflecting and refracting optics. The Schmidt telescope combines a spherical mirror, aperture stop, and a transmitting corrector plate to remove all third-order aberrations except curvature of field. The Schmidt–Cassegrain is a modification popular in the amateur market.

- A space telescope is generally superior to ground-based telescopes of the same aperture in its ability to resolve detail and to detect faint objects. The James Webb Space Telescope should have advanced capability in the IR. Concepts:

  Airy disk   *seeing*   *airglow*

  zodiacal light   *HST*   *JWST*

- Since 1980, ground-based telescopes have achieved very large apertures (6 to 10 -m) and an ability to compensate for atmospheric seeing. Concepts:

  5-m Hale telescope   *floppy mirror*   *active optics*

  honeycombed mirror   *segmented mirror*   *spin casting*

  meniscus mirror   *local climate control*

- Adaptive optics (AO) technology is based on three components: a WF sensor, a deformable mirror, and a computer that quickly interprets the sensor output and adjusts the DM shape to cancel atmospheric distortions in the WF. Concepts:

  speckle pattern   *seeing disk*   *scintillation*

  speckle interferometry   *lucky imaging*   *coherence length*

  Strehl ratio   *Fried parameter*   *point-spread function*

  isoplanatic angle   *Greenwood time*   *natural guide star*

  laser guide star   *closed-loop operation*   *tip–tilt correction*

- Practical deformable mirrors have flexible surfaces. DM actuators may be based on voice coils + magnets, piezoelectric elements, or MEMS technology.

  (continued)

Summary (*cont.*)

- One form of wavefront sensor is the Shack–Hartmann device, which depends on intercepting the guide object's wavefront with multiple apertures.
- Advanced AO includes a variety of approaches:

  SCAO   LTAO   MCAO

  GLAO   MOAO   extreme AO
- Three extremely large telescope projects (GMT, TMT, E-ELT) in the 24–39-m range are expected to see first light before the year 2025.

Herschel focus

Exercises

1. Describe the kind of motion an altazimuth mount must execute to track a star through the zenith. Describe the motion of an instrument rotator at the Cassegrain focus during such a maneuver. Assume an observatory latitude of 45 degrees, and only consider tracking close to the zenith.

2. Investigate and provide an image of, or sketch, the different configurations of the equatorial mount known as (a) German (universal or Fraunhofer), (b) English, (c) horseshoe, and (d) open fork. The book by King (1979) is a good source for older illustrations.

3. Investigate and describe the operation of the telescope mountings known as (a) a siderostat and (b) a heliostat.

4. William Herschel made many of his discoveries with an $f/13$, 0.47-m diameter telescope operated in the "direct view" or "Herschel focus" mode – see figure in margin – in which the focus, F, of a tilted mirror, M, is viewed at a location off the optical axis, A. You may assume that the distance, d, provides sufficient clearance for the head of a gentleman wearing a wig (i.e. about 12 cm). (a) Ignoring possible aberration corrections provided by the ocular, compute which is the better choice for the primary mirror of this telescope, sphere or paraboloid? (b) If the best seeing is 1.5 arcsec, does the shape of the mirror really matter?

5. A prime focus telescope with a parabolic mirror will be used at a site where the best seeing is 0.9 arcsec. What is the limiting radius of the field of view for which the comatic blur is smaller than the best seeing disk? Your answer will depend on the focal ratio of the primary, so compute the radius for $f/2.5$, $f/8$, and $f/10$. Show that for the $f/8$ telescope, the blur due to astigmatism at the edge of this coma-limited field is much smaller than the seeing disk.

6. Compare the tube lengths of the following telescopes, all of which have 1.0-m apertures.

(a) an $f/10$ refractor,

(b) an $f/10$ Schmidt,

(c) an $f/2.5$ Schmidt,

(d) an $f/10$ classical Cassegrain with an $f/3$ primary mirror and final focus that is 20 cm behind the vertex of the primary (length is the distance from the secondary to the final focus),

(e) an $f/10$ Cassegrain with an $f/2$ primary mirror and final focus that is 20 cm behind the vertex of the primary.

7. A Gregorian telescope and a Cassegrain telescope have identical primary mirrors, back focal distances, and final focal lengths (i.e. $\beta_{Cass} = \beta_{Greg}$, $m_{Cass} = -m_{Greg}$) Using the definitions in Equations (6.3) show that the difference between the lengths of two tubes is proportional to

$$\frac{|m|}{m^2 - 1}$$

8. An $f/7.5$ 4-m RC telescope has an $f/2.5$ primary and a final focus 25 cm behind the vertex of the primary. (a) Compute conic constants of the primary and secondary mirrors, and the diameter of the secondary. (b) Compute the angular diameter of the field over which the astigmatic blur is less than 1 arc second.

9. A classical Schmidt telescope is being designed to have a 1-meter diameter aperture stop and an un-vignetted field of view of 10 degrees diameter.

(a) Compute the diameter of the primary mirror, D_p, if the focal ratio of the system is $f/3.5$.

(b) Compute the diameter of the primary if the focal ratio of the system is $f/1.7$.

(c) Ignoring the effect of any central obstruction, show that the image of a star will be dimmer at the edge of the field that at center by a factor of $\cos 5°$.

10. The Palomar Oschin Schmidt telescope has a 1.2-m diameter corrector and an $f/2.5$ primary. The observatory plans to install a 360 megapixel detector (the ZTF camera) at the prime focus. The ZTF detector is a square measuring 37 cm on a side.

(a) What fraction of the beam is obscured by the detector?

(b) Compute the pixel scale and étendue of this system.

(c) Compare your result in (b) with your computed pixel scale and étendue of the Pan-STARRS -1 instrument: a 1.8-m $f/4$ R–C telescope with a 1.4 gigapixel detector measuring 40 cm on a side.

11. The detection threshold of the HST (aperture 2.4 m) for a certain application is $m = 26.0$. What is the magnitude threshold for the same application for the JWST (aperture 6.5 m) at the same wavelength? Assume both telescopes are diffraction limited, detectors and exposure times are identical, and the background for the JWST is 1.0 magnitudes per square arcsec fainter than for the HST. (Caution: recall the relation between magnitude difference and flux ratio.)

12. Assume the detection threshold of the JWST (aperture 6.5 m) for a certain application is $m = 28.0$. What is the magnitude threshold for the same application of for an ELT with a 24-m aperture? Assume the background for the ELT is

2 magnitudes per square arcsec brighter than for the JWST, and that the PSF for the ELT is 4 times as wide (FWHM).

13. What would happen to the Strehl ratio delivered by the JWST if one of the 18 primary mirror segments introduced an optical path length error of $\lambda/2$?

14. "Good" seeing in most major cities can approach 2.5 arcsec at 0.5 μm. What is the Fried parameter of such a site and the Strehl ratio of such an image in a 0.5-m telescope? How many elements would be required in a deformable mirror to compensate for this seeing?

15. Investigate a device known as a Lyot coronagraph. Give an example of an astronomical application of the device. In what ways might AO be helpful with a coronagraph?

16. A particular site has a median Fried parameter of 10 cm during the month of August.
    (a) Estimate the expected FWHM of the uncompensated seeing disk and the Strehl ratio, R_s, in the U band, I band, and K band.
    (b) Above what wavelength will the images in an optically perfect 1-m telescope be unaffected by turbulence at this site (i.e. $R_S = 1$)?
    (c) For an AO system in K band for this situation, compute the Greenwood time delay imposed by ground-layer turbulence alone (average velocity 10 km hr^{-1}) and by tropospheric turbulence alone (average velocity 120 km hr^{-1}).

17. Compute the anticipated plate scale of the Thirty Meter Telescope (D=30-m, f/15) at the Nasmyth focus. If the telescope delivers a 20-arc minute field of view, how many 30-μm pixels would be needed to cover this field?

Chapter 7
Matter and light

Because atomic behavior is so unlike ordinary experience, it is very difficult to get used to, and it appears peculiar and mysterious to everyone – both to the novice and to the experienced physicist. Even experts do not understand it the way they would like to, and it is perfectly reasonable that they should not, because all of direct human experience and of human intuition applies to large objects.

– Richard Feynman, *The Feynman Lectures on Physics*, 1965

Chapter 1 introduced the situations that produce line and continuous spectra as summarized by Kirchhoff's laws of spectrum analysis. This chapter descends to the microscopic level to examine the interaction between photons and atoms. We show how the quantum mechanical view accounts for Kirchhoff's laws, and how atomic and molecular structure determines the line spectra of gases.

To understand modern astronomical detectors, we also turn to a quantum mechanical account – this time of the interaction between light and matter in the solid state. The discussion assumes you have had an introduction to quantum mechanics in a beginning college physics course. We will pay particular attention to some simple configurations of solids: the metal-oxide-semiconductor (MOS) capacitor, the p–n junction, the photo-emissive surface, and the Type 1 superconductor. Each of these form the physical basis for a distinct class of astronomical detector.

7.1 Isolated atoms

7.1.1 Atomic energy levels

A low-density gas produces a line spectrum, either in absorption or emission, depending upon how the gas is illuminated (review Figure 1.7). The formation of lines is easiest to understand in a gas composed of single-atom molecules, like helium or atomic hydrogen. Consider the interaction between a single atom and a single photon. In either the absorption or the emission of a photon, the atom usually changes the state of one of its outermost electrons, which are therefore termed the **optical electrons**. The same electrons are also called the **valence electrons**, since they largely influence an atom's chemical properties by participating in covalent and ionic bonds with other atoms.

Observations of atomic spectra and the theory of quantum mechanics both demonstrate that the energy states available to any bound electron are **quantized**. That is, an electron can only exist in certain permitted energy and angular momentum states. In theory, these permitted states arise because an electron (or any other particle) is completely described by a **wave function**. In the situation in which the electron is bound in the potential well created by the positive charge of an atomic nucleus, the electron's wave function undergoes constructive interference at particular energies, and destructive interference at all others. Since the square of the wave function gives the "probability density" of the electron existing at a certain location and time, the electron cannot have energies that cause the wave function to interfere destructively with itself and go to zero. Physicists call these **forbidden** states. In the isolated atom, most energies are forbidden, and the energies of the rare **permitted** states are sharply defined.

Figure 7.1a illustrates the permitted energy levels for a fictitious atom, which appear as horizontal lines. In the figure, energy, in units of electronvolts, increases vertically. (One electronvolt (eV) is the energy gained by an electron accelerated by a potential difference of one volt. $1 \text{ eV} = 1.6022 \times 10^{-19}$ J.) There are seven **bound states**, labeled $a - g$, in this particular atom. (Real atoms have an infinite number of discrete states.) These different energy levels correspond to different configurations of, and interactions among, the outer electrons. The atom must exist in one of these permitted energy states. The lowest energy state, the one assigned the most negative energy, is called the **ground state** (level a in the figure). This is the configuration in which the electrons are most tightly bound to the nucleus, and would be the state of an undisturbed atom at zero

Fig. 7.1 (a) Permitted energy levels for an optical electron in a hypothetical atom that has seven bound states (*a–g*). The most tightly bound states (lowest energy) correspond to an electron location closer to the nucleus. (b) Absorption or emission of photons. The probabilities of different transitions can be vastly different from one another.

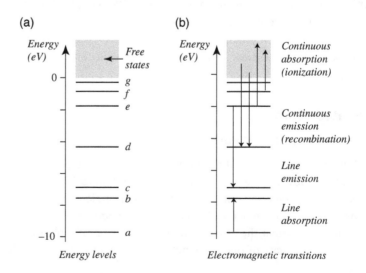

temperature. Above the ground state are all other permitted **excited** states, up to the **ionization level**. The ionization level, conventionally assigned zero energy, corresponds to an atom that has so much internal energy that an electron is just able to escape. In that situation, the free electron is no longer part of the atom, and the remaining positive ion will have internal energy states described by a completely different diagram. The energy of the free electron is not quantized.

You can think of bound states with higher energies as situations in which the optical electrons are *on average* physically further away from the nucleus. Be aware, though, that the vision of electrons orbiting the nucleus like planets in the Solar System (i.e. the early Bohr theory) is limited in its usefulness. The best answer to the question "where is this electron?" is a function that says certain locations are more likely than others, but, unlike the energy situation, a rather broad and sometimes complicated range of positions is possible for each bound state.

7.1.2 Absorption of light by atoms

Even though we can't see the positions of an atom's electrons, we *can* measure their energies when light interacts with atoms. Remember that a photon carries a specific amount of energy that is directly proportional to its frequency, v:

$$E = hv = \frac{hc}{\lambda} \tag{7.1}$$

The atom can make a transition from one bound state to another by either absorbing (the process is called **photo-excitation**) or emitting a photon of the correct frequency or wavelength, as illustrated in Figure 7.1b. In the process of photo-excitation, the photon is truly absorbed and ceases to exist. The figure shows a photo-excitation transition from the ground state (level a, which has energy E_a, to the first excited state level b, which has energy E_b). The photon responsible for this transition must have wavelength

$$\lambda_{ab} = \frac{hc}{|\Delta E_{ab}|} \tag{7.2}$$

where $\Delta E_{ab} = E_b - E_a$.

This explains why a beam of light with a continuous spectrum that passes through an atomic gas will emerge exhibiting an absorption line spectrum. Since only photons with energies corresponding to the energy difference between bound electron states, ΔE_{ij}, can be absorbed, only lines with the corresponding wavelengths of λ_{ij} will be present as absorption features in the spectrum that emerges.

As Figure 7.1b illustrates, photons capable of *ionizing* the atom can have any wavelength, so long as they are energetic enough to move an electron from a bound to a free state. This minimum energy is observed in the spectrum as a feature called an **absorption edge** – an intensity discontinuity in the continuum at the minimum wavelength.

7.1.3 Emission of light by atoms

An isolated hot gas produces an emission-line spectrum. Again, you can understand why by considering the quantized energy levels. In Figure 7.1b, for example, an atom changing from state e to state c must lose energy. It can do so by creating a photon with energy ΔE_{ec}. This process of **de-excitation** by photo-emission can occur spontaneously, or it can be stimulated to occur by an incoming photon of exactly the transition energy. This latter process is the equivalent of negative absorption: one photon collides with the atom and two identical photons emerge. Stimulated emission is the basis for the operation of lasers and masers.

If there are a significant number of free electrons in a hot gas, then the gas will emit continuous radiation along with the usual emission lines. A photon is emitted if a free electron loses energy and recombines with a positive ion, forming the bound state of the neutral atom. The resulting radiation will be continuous since the energy of the free electron is not quantized. There will be an **emission edge**. **Free–free transitions** from one free state to another are also possible, and will also contribute to a continuous spectrum.

7.1.4 Collisions and thermal excitation

Atoms prefer to exist in the lowest possible energy state, the ground state. An isolated atom in any excited state will spontaneously decay to a lower state. The length of time we can expect an atom to remain in a particular excited state depends on the rules of quantum mechanics, but if there is a "permitted" transition to a lower state, the half-life of the excited state usually is on the order of 10^{-8} seconds. How do atoms get into an excited state in the first place? One way, of course, is by absorbing electromagnetic radiation of the proper wavelength. A second path is via collisions with other atoms or particles. A collision can convert kinetic energy into internal energy in the form of optical electrons in excited states. In the very eventful environment of a hot gas, atoms that want to stay in the ground state have little chance of doing so for long, because frequent collisions kick them up into higher states. A hot gas glows as the resulting excited atoms decay back to lower energy levels, emitting photons in the process.

Collisions can transfer energy out of an atom as well as into it. With many collisions, at constant temperature, the population and de-population rates for one level due to all processes are equal, and the expected number of atoms in a particular bound state is well defined. The **Boltzmann distribution** describes the number of atoms in each energy state in such a situation of ***thermodynamic equilibrium***. Consider any two bound states, *i* and *j*, having energies E_i and E_j. The Boltzmann distribution gives the ratio of the number of atoms in these two states as

$$\frac{n_i}{n_j} = \frac{g_i}{g_j} \exp\left\{\frac{E_i - E_j}{kT}\right\} \tag{7.3}$$

Here g_i and g_j are the ***statistical weights*** of each level (g_i is the number of distinct quantum mechanical states in the *i*th energy level – see the next section). Boltzmann's constant, *k*, has the value 1.381×10^{-23} J K^{-1} = 8.62×10^{-5} eV K^{-1}.

7.1.5 Specification of energy levels

In the terminology of quantum mechanics, the state of every bound electron is specified by four ***quantum numbers***:

- ***n***, the principal quantum number, can take on all positive integer values 1, 2, 3, ... This number is associated with the radial distribution of the probability density of the electron as well as with its energy, and in the terminology used by chemists, specifies the ***shell***.
- ***l***, the azimuthal quantum number, can take on values 0, 1, ... ,(*n* − 1). It can be associated with the angular distribution of the probability density, and can have a secondary effect on the energy of the state.
- ***m***, the magnetic quantum number, can take on values 0, ±1, ... , ±*l*. It describes the possible interaction between the electron and an imposed magnetic field. It can have an effect on the energy of the electron only if a field is present.
- ***s***, the electron spin quantum number can have only two values, +1/2 or −1/2. It can affect the electron energy by interacting with the angular momenta of other parts of the atom.

In particle physics, a ***fermion*** is a particle like the electron, proton, or neutron, whose spin quantum number has a half-integer value like ±1/2, ±3/2, etc. Any particle's intrinsic angular momentum has the value $(h/2\pi)\sqrt{s(s+1)}$, where *h* is Planck's constant and *s* is the spin quantum number. Particles with integer spin (0, ±1, etc.) are called ***bosons***.

The ***Pauli exclusion principle*** states that no two identical fermions may occupy the same quantum state. This demands that ***no two electrons bound in an atom may have the same four quantum numbers (n, l, m, s)***. Table 7.1 lists all possible values of the four quantum numbers for electrons in the first few levels. Each of the states listed must be either empty or occupied by a single electron. The ground state of an atom with atomic number *Z* will have the lowest energy configurations occupied, up to the *Z*th available state, and all other states

Matter and light

Table 7.1 *Quantum numbers of the first 30 bound atomic states (up to the ground state of zinc). In the periodic table, the 4 s states are usually filled before the 3 d states, 5 s before 4 d, etc. See Figure 7.2.*

Quantum numbers				Name of configuration	Number of states
n	l	m	s		
1	0	0	$\pm 1/2$	1 s	2
2	0	0	$\pm 1/2$	2 s	2
2	1	-1	$\pm 1/2$		
		0	$\pm 1/2$	2p	6
		$+1$	$\pm 1/2$		
3	0	0	$\pm 1/2$	3 s	2
3	1	-1	$\pm 1/2$		
		0	$\pm 1/2$	3p	6
		$+1$	$\pm 1/2$		
3	2	-2	$\pm 1/2$		
		-1	$\pm 1/2$		
		0	$\pm 1/2$	3 d	10
		$+1$	$\pm 1/2$		
		$+2$	$\pm 1/2$		
4	0	0	$\pm 1/2$	4 s	2

empty. The actual energy of a particular state depends not only on the atomic number and the values of the four quantum numbers for the occupied states, but also on other details like the atomic weight, and interactions between the electron, nucleus, and electrons in other states.

The energy of an electron will depend most strongly upon both n and l quantum numbers. The **configuration** of electrons in an atom is therefore usually described by giving these two numbers plus the number of electrons in that n, l level. The **spectroscopic notation** for a configuration has the form:

$$ny^x$$

where

> n is the principle quantum number
> x is the number of electrons in the level – many electrons can have the same n, l so long as they have different m and/or s values, and
> y codes the value of the l quantum number according to the following scheme:

l	0	1	2	3	4	5	6	7, 8, ...
Designation	s	p	d	f	g	h	i	k, l, etc.

Table 7.2 *Examples of a few electron configurations.*

Element	Atomic number	Ground-state configuration
Hydrogen	1	$1s^1$
Helium	2	$1s^2$
Boron	5	$1s^2\,2s^2\,2p^1$
Neon	10	$1s^2\,2s^2\,2p^6$
Silicon	14	$1s^2\,2s^2\,2p^6\,3s^2\,3p^2$
Argon	18	$1s^2\,2s^2\,2p^6\,3s^2\,3p^6 = [Ar]$
Potassium	19	$1s^2\,2s^2\,2p^6\,3s^2\,3p^6\,4s^1 = [Ar]\,4s^1$
Germanium	32	$[Ar]3d^{10}\,4s^2\,4p^2$
Krypton	36	$[Ar]3d^{10}\,4s^2\,4p^6 = [Kr]$
Rubidium	37	$[Kr]\,5s^1$

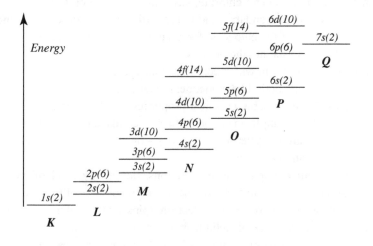

Fig. 7.2 Schematic energy levels of electronic configurations. Each level is labeled with its spectroscopic designation, including the number of electrons needed to fill the configuration in parentheses. Chemical shell designations (K, L, etc.) are at the bottom of each column. The diagram indicates, for example, that the two 5 s states will fill before the ten 4 d states. Energy levels are illustrative only of the general order in which configurations or sub-shells are filled and are not to scale. There are several exceptions to this overall scheme.

Lithium, atomic number 3, for example, has the ground-state configuration $1s^2\,2s^1$; that is, two electrons in the $n = 1$ state, one with quantum numbers $(1, 0, 0, -1/2)$, the other with $(1, 0, 0, 1/2)$. The third lithium electron (this is the valence electron) is in the $n = 2$ level with quantum numbers either $(2, 0, 0, -1/2)$ or $(2, 0, 0, 1/2)$. Table 7.2 gives some further examples of electron configurations.

Figure 7.2 is a schematic energy-level diagram that shows the relative energies of the electron configurations in atoms. As one moves from element to element in order of increasing atomic number, electrons are added from the bottom up in the order suggested by Figure 7.2. (There are minor exceptions.)

The ***periodic table***, one of the triumphs of human learning, summarizes our knowledge of the chemical properties of the elements, and recognizes that chemical behavior is periodic in atomic number. The table is organized according to similarities in optical electron configurations. Each row or period contains elements with identical values of n for outer electrons. In chemical terminology, the valence electrons of atoms in the same period are all in the same *shell*. The atomic properties of elements, like ionization energy, ionic radius, electronegativity, and chemical behavior all trend generally in one direction along the row. Period three, for example, ranges from the reactive metal sodium, through the less reactive metals magnesium and aluminum, the semi-metal silicon, the increasingly reactive non-metals phosphorus, sulfur, and chlorine, and the inert gas, argon. Elements in the same *column* of the table, in contrast, have the same electron configuration in their outer shells, and therefore all have very similar chemical properties. The noble gases (helium, neon, argon, krypton, xenon, and radon – column 18 or group VIIIA), for example, all exhibit chemically inert behavior, and all have a filled outer shell with eight electrons in the s^2p^6 configuration. Similarly, the halogens in column 17, all highly reactive non-metals like fluorine and chlorine, have outer shells with the s^2p^5 configuration. There is also a secondary trend in properties moving down a column: the chemical reactivity of the halogens, for example, decreases steadily from fluorine, the lightest, to astatine, the heaviest.

Because of the order in which configurations are filled (see Figure 7.2) many elements in the same period have identical valence electron configurations, and thus almost indistinguishable chemical properties; but they differ in their inner electron shells. For example, in period 6, the rare-earth elements, or lanthanides – cerium ($Z = 58$) through ytterbium ($Z = 70$) – have identical outer shells ($6s^2$) and nearly identical chemistry.

For atoms with multiple valence electrons, the energy level of an excited configuration may depend not only on the quantum numbers of the electrons, but upon the interactions between the electron spins and angular momenta. For example, the excited state of helium that has configuration $1s^1\ 2p^1$ has *four* possible energies spread over about 0.2 eV. States differ because of different relative orientations of the two electron spins and the $l = 1$ angular momentum of the p electron (directions are quantized and thus limited to four possibilities). The details of how multiple electrons interact are beyond the scope of this book, but for now, it is sufficient to recognize that such interactions can cause the energy level of a configuration to split into multiple values.

7.2 Isolated molecules

The outermost electrons of a molecule see a more complex binding potential due to the presence of two or more positively charged nuclei. Generally, this results in a greater density of electronic energy states. Each electronic state is still

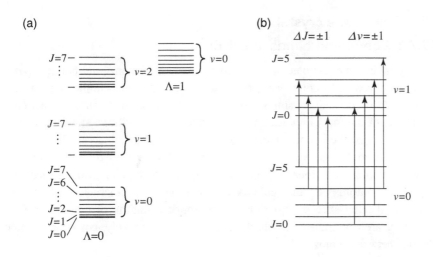

Fig. 7.3 Simplified energy levels and transitions in a diatomic molecule. (a) Right and left columns are different electronic states, as indicated by the total orbital angular momentum quantum number Λ. Quantum numbers J and v specify the rotational and vibrational states, respectively. We show only three rotation states and seven vibration states in the lower electronic level. (b) Permitted vibration–rotation absorption transitions from the $v = 0$ state. Only five rotation states are shown. Absorption lines increase in frequency from left to right and constitute a band.

characterized by four quantum numbers, but in the molecule, the value of the m quantum number has an important effect on the energy level. More importantly, the molecule itself has internal degrees of freedom due to its ability to rotate around its center of mass, as well as its ability to vibrate by oscillating chemical bond lengths and angles. These internal rotational and vibrational modes are quantized as well, and they vastly increase the density of energy states permitted to the molecule.

Quantum mechanical theory approximates the total internal energy of a molecule as the sum of three independent terms:

$$E_{molecule} = E_{electron} + E_{vibration} + E_{rotation} \qquad (7.4)$$

In addition to the quantum numbers specifying its electronic state, a simple diatomic molecule like CO or TiO will have one quantum number, J, to specify its rotational state, and one, v, for its vibrational state. Specification of the vibrational mode for molecules with more than two atoms becomes quite complex.

Figure 7.3 is a schematic energy-level diagram for a fictitious diatomic molecule. The energy levels in the figure are not to scale: Transitions between the ground state and the first excited electronic state are usually in the range 0.5 to 100 eV. Transitions between adjacent vibrational states are about 100 times smaller than this, and between adjacent rotational states, about 10^6 times smaller yet.

The spacing between the rotational levels at different vibrational states is similar. As a result, the spectra of even simple diatomic molecules show a pattern of *bands*, with each ***molecular band*** in the spectrum composed of many closely packed lines. See Figure 7.3.

7.3 Solid-state crystals

7.3.1 Bonds and bands in silicon

A crystal is a mega-molecule in which the pattern of atoms and bonds repeats periodically with location. Many of the detectors we discuss in the next chapter are made of crystalline solids, so we now describe in detail the electronic structure of the most important of these materials. The silicon atom, located in column IVa of the periodic table, has 14 electrons, 4 of which are in the outer shell, with configuration $3s^2 3p^2$. The outer shell will be filled when it contains eight electrons, not four. According to the theory of chemical valence, the component atoms of a molecule try to attain the electron structure of an inert gas (eight outer-shell electrons) by an appropriate sharing or transfer of electrons. Shared or transferred electrons produce, respectively, covalent or ionic bonds between atoms.

Consider the formation of a silicon crystal. Figure 7.4a shows what happens to the energy levels of an isolated silicon atom when a second silicon atom is brought closer and closer to it. As the electron wave functions begin to overlap, the levels split into two, outermost first. The nearer the neighbor, the greater is its influence, and the greater the splitting of levels. The outer electrons of both atoms can enter either of those levels since their wave functions overlap.

If we construct a crystal atom by atom, new energy states appear with each addition. For five atoms in a row, we expect something like Figure 7.4b. As crystal construction continues, more and more electron states become available as more and more atoms are added to the structure. Since even a tiny crystal contains on the order of 10^{17} atoms, each causing a split in the energy levels, the spacing between levels must be on the order of 10^{-17} eV. These levels are so closely spaced that for practical purposes we treated them as a continuous **band** of available energies. If bands do not overlap, they will be separated by energy **band gaps**. An electron anywhere in the crystal lattice is permitted an energy anywhere in a band, and is forbidden an energy anywhere in a gap.

Figure 7.5 shows the energy situation in crystalline silicon. The preferred interatomic spacing between nearest neighbors is R_0 (0.235 nm at room temperature). Note that at this spacing, the 3p and the 3 s energy levels overlap. The result is called a crossover degeneracy, and energies in the crossover region are

Fig. 7.4 (a) Splitting in the electron energy levels in a silicon atom as a second atom is brought into close proximity. (b) The same diagram for the case of five atoms in a linear matrix.

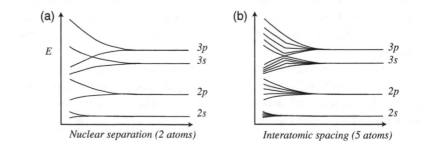

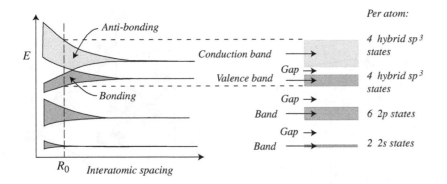

Fig. 7.5 Schematic diagram of the bands in silicon crystals. The diagram on the right shows the bands formed at the preferred interatomic spacing. Dark gray bands are occupied; light-gray are empty but permitted. Energies in the band gaps (white) are forbidden to electrons.

forbidden by quantum mechanics. The permitted states are certain linear combinations of s and p states, called ***sp³-hybrid orbitals*** – not the separate s and p states of the isolated atom. Each silicon atom contributes eight such permitted states to the bands. Four of the sp³ hybrid orbitals, the ones with lowest energy, correspond to an electron having its most probable location midway between the atom and one of its nearest neighbors. The nearest neighbors are at the four vertices of a tetrahedron centered on the nucleus. These four sp³ hybrid orbitals all have energies that lie in the ***valence band*** and constitute the ***bonding states***. Four other sp³ hybrid states have energies in the ***conduction band*** and locations away from the bonding locations. These are the ***anti-bonding states***.

From now on, we will use band diagrams, like the right side of Figure 7.5, to account for all the electrons in the entire crystal. At zero temperature all the anti-bonding states are empty and make up the conduction band. The difference between the energy of the top of the valence band, E_v, and the bottom of the conduction band, E_c, is called the ***band gap energy***:

$$E_G = E_c - E_v$$

In silicon, the band gap is 1.12 eV at room temperature.

In a silicon crystal, one would most likely find two electrons in each of the light-gray regions between the nuclei in Figure 7.6a – one from each atom. Each pair of shared electrons constitutes a ***covalent bond***. Four such bonds are symmetrically placed, and each atom therefore "sees" eight outer electrons – a complete shell. X-Ray diffraction studies confirm that this tetrahedral structure repeats throughout the crystal in a three-dimensional pattern called a ***diamond lattice***, as sketched in Figure 7.6b.

7.3.2 Conductors, semiconductors, and insulators

It is very instructive, although overly simple, to explain the differences between electrical conductors, semiconductors, and insulators as arising from differences

Fig. 7.6 (a) Tetrahedral covalent bonds for sp³ hybridized orbitals for one atom (black) pictured at the center of a cube (solid lines). Its nearest bond-forming neighbors are at four corners of the cube. These define the vertices of a tetrahedron (dashed lines). The electron bonding states are shown as light-gray ellipsoids – the regions where there is the highest probability of finding a valence electron. (b) A stick-and-ball model of the diamond lattice. Each of the two large cubes outlines a unit cell of the crystal. A complete crystal is built by assembling many identical adjoining unit cells in three dimensions.

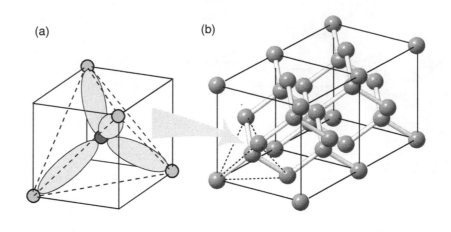

(a)   (b)

in the size of the band gap and in electron populations within the bands. The important principle is that a material will be a good conductor of electricity (or heat) if its electrons can accelerate (i.e. change quantum state) easily in response to an applied electric field.

An analogy may help. Imagine that you are standing in an almost empty room. You are free, in this environment, to respond to a whim to run across the room at top speed. On the other hand, if the same room is packed shoulder-to-shoulder with people, running is out of the question, no matter how strong your desire. Indeed, a sufficiently dense crowd makes moving completely impossible.

Similarly, an electron in relative isolation can help conduct electricity or heat because it can accelerate without obstruction when a field is imposed. In a crystalline solid, however, options are more restricted. The Pauli exclusion principle dictates that an electron can only accelerate (i.e. change quantum states) if it can move into a new state that is (a) permitted and (b) not occupied by another electron. A perfect silicon crystal at zero temperature cannot meet the second condition: Every electron is in the valence band and part of a covalent bond. Electrons occupy every permitted state in the band. They, in effect, are packed shoulder-to-shoulder. Although we have ignored movement of the nuclei (which can oscillate around their mean positions), as well as surface effects, the basic conclusion is: electron crowding makes cold silicon a very poor conductor of electricity, heat, and sound.

There *are* available states at much higher energies – the anti-bonding states in the conduction band. If an electron can acquire at least enough energy to jump the band gap, then it finds itself in the empty conduction band where it is able to move around. In the crowded-room analogy, if you have enough energy to climb up a rope through a trap door to the empty room on the next story, then you are free to run. Silicon conductivity thus improves at higher temperatures, because electrons in a hot crystal can gain enough energy from thermal collisions to reach states in the conduction band.

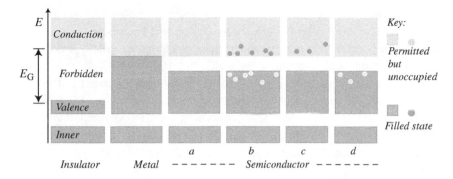

Fig. 7.7 Band structure of insulators, conductors, and semiconductors: (a) an intrinsic semiconductor at zero temperature; (b) the same material at a higher temperature; and (c) and (d) extrinsic semiconductors.

Figure 7.7 shows simplified band diagrams typical of insulators, metals, and semiconductors. In an insulator, the valence band is completely filled. The band gap is large compared with both the thermal energy, kT (at room temperature (300 K), $kT = 0.026$ eV), and with any other energy sources. Because of the large gap, valence electrons cannot reach any permitted states in the conduction band. Since the exclusion principle forbids any electron to move into an already occupied valence state, electrons cannot move at all – the material is a non-conductor.

A metallic conductor, in the second panel of the figure, has unoccupied permitted states immediately adjacent to the occupied valence states. If an electron near the top of the valence band absorbs even a tiny amount of energy, it will move into the conduction band, and from there to virtually anywhere in the material.

Figure 7.7 shows four different views of materials called **semiconductors**. The first, (a) an **intrinsic semiconductor**, looks like an insulator, except it has a small band gap similar to silicon at zero temperature. A valence electron can jump the gap into the conduction band by absorbing a modest amount of energy, either from thermal excitation or from some other energy source. Illustration (b), for example, shows the material in (a) at a high temperature. A few electrons have absorbed sufficient thermal energy to rise to the conduction band. This material will conduct, but the size of the current is limited because only these few electrons are in the conduction band. More electrons, of course, will rise to the conduction band to improve the conductivity if the temperature is increased further, and materials of this kind, in fact, can be used to make temperature gauges (thermistors).

Another thing to notice in Figure 7.7b is that whenever an electron is boosted into the conduction band, it must leave behind an empty state in the valence band. Another valence electron can shift into this vacated state and create a new empty state in the location it vacates. Since yet another electron can now move from a third location to fill this second empty state, it is clear that valence electrons can move through the crystal by occupying and creating empty states. It is easier to concentrate on the motion of the empty states, and to think of these *valence-band holes* as the entities that are moving. *Holes thus behave like mobile positive charges in the valence band*, and will contribute to the overall electrical conductivity. In intrinsic semiconductors, holes are usually less mobile than conduction-band electrons.

The third semiconductor (Figure 7.7c) also has a few electrons in the conduction band, but without any corresponding holes in the valence band. Materials of this kind, called *extrinsic semiconductors*, are extremely important in the construction of most electronic devices. A second class of extrinsic semiconductors (Figure 7.7d) has valence-band holes without corresponding conduction-band electrons.

7.3.3 Intrinsic semiconductors

Semiconductor crystals

A pure silicon crystal forms by linking all atoms with the tetrahedral covalent bond structure pictured in Figure 7.6. This geometry, called the *diamond lattice*, insures that each atom shares eight electrons in four bonds, completely filling its outer shell and producing a chemically stable structure. Indeed, the regularity of the diamond-lattice structure is tightly enforced, even if impurities are present in the silicon.

Elements with similar outer-electron configurations form similar diamond-lattice crystals. These are in column IVA (also called column 14) of the periodic table, and include carbon, germanium, and tin.[1] Similar bonds also form in binary compounds of elements symmetrically placed in the table on either side of column IVA. For binary compounds, the crystal structure is called the "zinc blend" structure, which resembles Figure 7.6b except for alternation of the chemical identity of the nuclei on either end of each bond. Most useful semiconductors exhibit the diamond or zinc blend crystal structure (exceptions include lead sulfide and zinc oxide.) Table 7.3 shows part of the periodic table containing elements that combine to form important semiconductors.

Examples of binary-compound semiconductors are gallium arsenide (GaAs, a III–V compound) and cadmium telluride (CdTe, a II–VI compound). Some ternary compounds, notably $(Hg_{1-x}Cd_x)Te$, and quaternary compounds like $In_xGa_{1-x}As_yP_{1-y}$, also form useful semiconductors. Commercially, silicon is by far the most commonly used semiconductor. Germanium and gallium arsenide also find important commercial applications.

Semiconductor materials generally have a room-temperature resistivity in the range 10^{-2} to 10^9 ohm cm, midway between that of a good conductor (10^{-6}

---

[1] The commonest allotrope of tin, white tin, the familiar metal, has a tetragonal crystal structure. Gray tin, a less common allotrope, crystallizes in the diamond lattice. Lead, the final member of column IVA, crystallizes in a face-centered-cubic lattice.

Table 7.3 *Periodic table of the elements near column IVA. Beryllium (IIA/ 2–1s²2s²) and copper (IB/11 – [Ar]3d¹⁰4s¹) are sometimes used as semiconductor dopants.*

IIB/12 – s^2	IIIA/13 – s^2p^1	IVA/14 – s^2p^2	VA/15 – s^2p^3	VIA/16 – s^2p^4
	B	^{6}C	N	O
	Al	^{14}Si	P	S
Zn	Ga	^{32}Ge	As	Se
Cd	In	^{50}Sn	Sb	Te
Hg	Tl	^{82}Pb	Bi	Po

Table 7.4 *Some common semiconductors. Forbidden band gap energies and cutoff wavelengths at room temperature. A more complete table is in Appendix H2. Data from Sect. 20 of Anderson (1989).*

Material	Symbol	Band gap (eV)	λ_c (µm)
IV			
Diamond	C	5.48	0.23
Silicon	Si	1.12	1.11
Germanium	Ge	0.67	1.85
Silicon carbide	SiC	2.86	0.43
III–V			
Gallium arsenide	GaAs	1.35	0.92
Indium antimonide	InSb	0.18	6.89
II–VI			
Cadmium sulfide	CdS	2.4	0.52
Cadmium selenide	CdSe	1.8	0.69
Mercury cadmium telluride	$Hg_xCd_{1-x}Te$	0.1–0.5 (x = 0.8–0.5)	12.4–2.5
IV–VI			
Lead sulfide	PbS	0.42	2.95

ohm cm) and a good insulator ($> 10^{14}$ ohm cm). As we have already seen, resistivity depends critically on both temperature and the size of the band gap. Table 7.4 lists the band gap energies for several semiconductors. Note that since the lattice spacing in a crystal is likely to change with temperature, so too will the band gap. Carbon in the diamond allotrope is an insulator because its band gap is so large; other carbon allotropes (graphite, carbon nanostructures) are conductors.

Conductivity and temperature

At zero temperature, all the materials in Table 7.4 are non-conductors. As temperature increases, thermal agitation causes *ionizations*: electrons are promoted to the conduction band, free of any one atom; corresponding mobile holes are created in the valence band. The material thus becomes a better conductor with increasing temperature. At constant temperature equilibrium, we expect the rate of electron–hole recombinations to exactly equal the rate of thermal ionizations.

How, exactly, does an electron in a bonding state receive enough energy to jump the band gap? Optical electrons can collide with one another, of course, but it is important to note also that the lattice itself is an oversized molecule that can vibrate by oscillating bond length or angle. Just as with molecules, lattice vibration states are quantized with respect to energy. Solid-state theory often associates each discrete lattice vibration energy with a particle, called a *phonon*, an entity analogous to the photon. Changes in electron state may thus involve the absorption or emission of a phonon. An electron can jump the band gap because it absorbs a phonon of the correct energy, and can lose energy and momentum by creation of, or collision with, a phonon.

At a particular temperature, the density of electrons at any energy within the bands will depend upon the product of two functions, (a) the probability, $P(T, E)$, of an electron having that energy, and (b) the number density of available states at each energy, S:

$$n_e(T, E) = P(T, E)S(E) \tag{7.5}$$

With respect to the probability of a fermion having a particular energy, recall that the exclusion principle causes important restrictions on occupancy. This is certainly the case for the electrons in the bands of a semiconductor, where most of the valence states are fully occupied. In such a situation of *electron degeneracy* the probability per unit energy that an electron has energy, E, is given by the *Fermi–Dirac* distribution:

$$P(T, E) = \frac{1}{1 + \exp\{(E - E_F)/kT\}} \tag{7.6}$$

This expression reduces to the Boltzmann distribution, Equation (7.3), at high temperatures. At the limit of zero temperature, the Fermi–Dirac distribution requires that all of the lowest energy states be occupied, and that all of the higher states (those with energies above E_F) be empty. That is, at $T = 0$,

$$P(E) = \begin{cases} 1, E < E_F \\ 0, E > E_F \end{cases} \tag{7.7}$$

The parameter E_F is called the *Fermi energy*, and might be defined as that energy at which the probability for finding an electron in a permitted state is exactly one-half. According to this definition, the Fermi energy will itself be a function of temperature for some systems at high temperature. However, for all

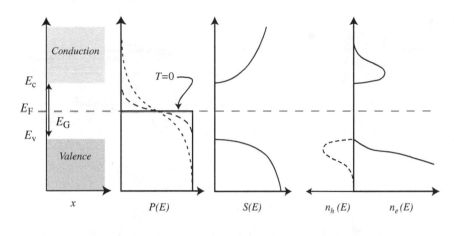

Fig. 7.8 Electron and hole
density in an intrinsic
semiconductor. (a) The
locations of the band
edges and the Fermi level
midway between them. (b)
The probability of finding
an electron in a permitted
state as a function of
energy, *P(E)*. The solid line
shows *P* at zero
temperature, and the
broken lines at two higher
temperatures. (c) The
density of permitted states
as a function of energy, *S
(E)*. (d) The densities of
electrons and holes as
functions of energy (the
product of (b) and (c)) for
the highest temperature
curve in (c)). The
horizontal scale of plot (d)
has been expanded to
show detail.

cases we are concerned with, the Fermi energy can be treated as a constant equal
to the energy of the highest permitted state at $T = 0$.

Figure 7.8a shows the energy bands for silicon at absolute zero, where
electrons will fill all available states in the permitted bands up to the Fermi
level, which falls midway between the valence and conduction bands.
Figure 7.8a plots Equation (7.6) at three different values of temperature.

Figure 7.8c shows a schematic representation of $S(E)$, the **number of permitted states** per unit energy per unit volume for the valence and conduction bands
of silicon. Note that $S(E)$ decreases near the permitted band edges and vanishes
in the band gap.

The product $P(E)S(E)$ gives $n_e(E)$, **the number density of electrons** at energy
E. The **number density of holes** at energy E in the valence band is just
$n_h = [1 - P(E)]S(E)$. Figure 7.8d plots these two functions, n_e and n_h, for a
non-zero temperature. The *total* number densities of charge carriers of each kind
(negative or positive) are given by the integrals of these functions in the
appropriate band –

$$n_N = \int_{E_F}^{\infty} n_e(E)\mathrm{d}E$$
$$n_P = \int_{-\infty}^{E_F} n_h(E)\mathrm{d}E$$

(7.8)

In intrinsic semiconductors, the number density of these two kinds of charge
carriers in equilibrium must be equal, so $n_P = n_N$.

The temperature dependence in Equation (7.8) follows from the Fermi
distribution and has the form

$$n_P = n_N = AT^{\frac{3}{2}}\mathrm{e}^{-\frac{E_G}{2kT}}$$

(7.9)

The exponential term dominates in most practical circumstances (i.e. low
temperatures).

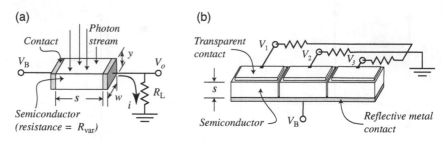

Fig. 7.9 Simple
photoconductors. In (a),
light strikes the exposed
surface of a
semiconductor linked to a
simple circuit by two
metal contacts. Photo-
ionization produces
charge carriers that reduce
semiconductor resistance.
Current through the
device will increase with
increasing illumination,
and output is the voltage
across a load resistor. In
(b), a three-pixel device
registers three different
voltages in response to
local illumination. Here
photons pass through
upper (transparent)
contacts. The lower
contact is reflective so that
photons passing through
the device are redirected
for a second pass.

7.3.4 Intrinsic photo-absorbers

If it is given sufficient energy an electron can leave a covalent bond and jump
the band gap into the conduction band. The required energy could be supplied
by a photon if it has a wavelength less than the **cutoff wavelength** for the
material:

$$\lambda_c = \frac{hc}{E_G} = \frac{1.24\mu m}{E_G[eV]} \tag{7.10}$$

The band gap for silicon, for example, corresponds to a cutoff wavelength λ_c of
1.1 µm. Since the band gap energy is a mild function of temperature, so is the
cutoff wavelength.

Figure 7.9a shows a simple device that utilizes photo-absorption to measure light intensity. Photons absorbed by a block of semiconductor material
produce ionization events – electrons in the valence band promoted to the
conduction band, leaving an equal number of holes. The greater the flux of
incoming photons, the greater the equilibrium concentration of charge carriers, and the greater the conductivity of the detector. If you maintain a
constant voltage, V^+, across the semiconductor, as in the figure, then the
electrical current through the circuit should increase with the number of
photons absorbed per second. Measuring the voltage at V_0 thus monitors
light intensity. Figure 7.9b shows an alternative structure that facilitates a
close-packed array of detectors.

Notice that this **photoconductor** responds to the *number* of photons per
second absorbed, not, strictly, to the rate of *energy* absorbed. Of course, if you
know their spectral distribution, it is an easy matter to compute the energy flux
carried by a given number of photons.

There are at least three reasons why a photon incident on the top of the
device in Figure 7.9a will fail to generate an electron–hole pair. First, we
know that those with frequencies below the band gap frequency, E_G/h, cannot
move an electron from the valence to conduction band, and thus cannot be
detected.

A second failure is due to reflection of photons from the top surface of the device. As we saw in Chapter 5, minimal reflection occurs at normal incidence, and depends on the refractive index of the material:

$$R = \frac{(n_1 - n_2)^2}{(n_1 + n_2)^2} \tag{7.11}$$

The refractive index (and thus reflectivity) for silicon and most other semiconductors is very high in the ultraviolet, decreases through visible wavelengths, and is low (3.5 to 4) in the red and infrared. Reflectivity is also low in the X-ray band. Anti-reflection coatings can considerably reduce reflectivity for a particular wavelength.

A third reason for detection failure is that photons above the band gap frequency might pass completely through the device. Once entering a semiconductor, the distance a photon can travel before being absorbed depends very strongly on its wavelength. If a beam of photons enters material in the z-direction, its intensity at depth z will be

$$I(z) = I_0 e^{-\alpha z} \tag{7.12}$$

where I_0 is the intensity at $z = 0$, and α is the **absorption coefficient**. A large absorption coefficient means light will not travel far before being absorbed. Figure 7.10b shows the absorption coefficient as a function of wavelength for silicon, germanium, and gallium arsenide, and illustrates an important quantum mechanical distinction. Notice that GaAs absorbs strongly right up to the cutoff wavelength, whereas Si and Ge very gradually become more and more transparent approaching that wavelength. Materials with an abrupt cutoff, like GaAs and InSb, are called **direct transition** semiconductors. Materials of the second kind, like Si and Ge, are called **indirect transition** semiconductors.

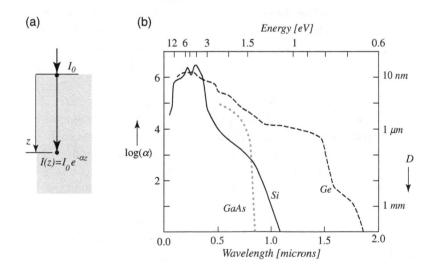

Fig. 7.10 (a) Light incident on a semiconductor with transmitted intensity I_0 at depth $z = 0$, declining to intensity $I(z)$ at depth z. (b) The absorption coefficient, α, measured in m^{-1}, as a function of photon wavelength or energy. The absorption depth, $D = 1/\alpha$, is on the right axis.

For both direct and indirect materials, photo-absorptions almost always produce electron–hole pairs. The exceptions are usually due to flaws in the material. In some cases, a photon can interact with the lattice (particularly defects in the lattice) and deposit its energy as a phonon, not as a photo-ionization. For this reason, light-detecting devices require a semiconductor material that has been crystallized with strict controls to assure chemical purity and lattice integrity.

7.3.5 Extrinsic semiconductors

Crystals inevitably have some chemical impurities and mechanical imperfections. These alter the energies and momenta of the states available near the sites of the defects, usually in undesirable ways. Curiously, though, some of the most useful semiconductor devices are made by intentionally introducing impurity atoms into the lattice.

Figure 7.11a shows a flattened schematic of the positions of the atoms and outer electrons in an intrinsic semiconductor like silicon. Each atom shares eight valence electrons, forming four complete bonds. All atoms and bonds in the lattice are identical. Diatomic semiconductors like GaAs have a similar structure, except the chemical identity of the atoms alternates along rows and columns.

Now we intentionally introduce an impurity into the lattice, as in Figure 7.11b, where a few of the silicon atoms have been replaced by atoms that have only three valence electrons, like boron, gallium, or indium. Each impurity creates a vacancy in the electron structure of the lattice – a "missing" electron in the pattern.

The crystal, in fact, will try to fill in this "missing" electron. The impurity creates what is called an ***acceptor state***. It requires relatively little energy (on the order of the room temperature thermal energy, kT) to move a valence electron from a silicon–silicon bond elsewhere in the lattice into this gap at the impurity site. This creates a hole at the site that donates the electron. Such a hole behaves just like a mobile hole in an intrinsic semiconductor – a ***positive charge carrier*** that increases the conductivity of the material. Semiconductors in which impurities have been added to create positive charge carriers are termed ***p-type extrinsic*** semiconductors.

Figure 7.12a is an energy-band diagram for a p-type semiconductor. At zero temperature, a small number of (unoccupied) acceptor energy states exist within the band gap of the basic material. The energy difference, E_i, between the top of the valence band and the acceptor states is typically on the order of 0.05 eV in silicon (see Table 7.5). At a finite temperature, excitation of electrons from the valence band into these intermediate states creates valence-band holes (Figure 7.12a). Because the electrons in the intermediate states are localized at the impurity sites (they have no available nearby states at about the same energy), they are immobile and cannot contribute to the conductivity. The mobile holes in the valence band, of course, can contribute, and are termed the

(a) Pure crystal

(b) Acceptor impurities

(c) Donor impurities

Fig. 7.11 A schematic of the bond structure in intrinsic and extrinsic semiconductors. In an actual crystal, the bond arrangement is three-dimensional – see Figure 7.5: (a) a pure intrinsic material; (b) an extrinsic material: a lattice with three p-type impurity atoms; and (c) three n-type impurity atoms.

Table 7.5 *Ionization energies, in eV, for different impurity states in silicon and germanium. Data from Kittel (2005) and Rieke (2003).*

Acceptors	Si	Ge
B	0.045	0.0104
Al	0.057	0.0102
Ga	0.065	0.0108
In	0.16	0.0112
Be	0.146	0.023
Donors		
P	0.045	0.0120
As	0.049	0.0127
Sb	0.039	0.0096
Bi	0.069	

(a)

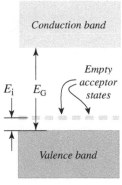

(b)

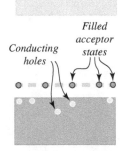

Fig. 7.12 Band structure of p-type extrinsic semiconductors. (a) A p-type material at zero temperature. The energy difference between the top of the valence band and the acceptor states is typically on the order of 0.05 eV. (b) The same material at a higher temperature. Electrons excited into the acceptor states have created valence-band holes.

majority charge carriers. In contrast with intrinsic semiconductors, $n_P > n_N$ in p-type materials. Adding impurities to create an extrinsic semiconductor is called *doping*, and the more heavily doped the material is, the higher is its conductivity. The transparent conductors used as contacts in Figure 7.9b, for example, are often made of highly doped silicon.

Figure 7.11c and Figure 7.13 illustrate the second kind of extrinsic material. Here intrinsic material has been doped with atoms that have five valence electrons, like arsenic or antimony. The result is an *n-type extrinsic* semiconductor. Here, the "extra" electrons from the *donor* impurities are easily ionized into the conduction band. This ionization restores the Si-bond structure (only eight shared outer-shell electrons, not nine) and consequently produces conduction-band electrons that constitute the majority carriers

Extrinsic semiconductors respond to light in nearly the same way as intrinsic material. In fact, because the concentration of impurity atoms is always quite small (typically one part in 10^3 or 10^4), the presence of dopants does not appreciably modify intrinsic photo-absorption *above* the band gap energy. The important difference occurs with photons whose energies lie *below* the intrinsic band gap energy but above the dopant ionization energy, E_i.

Suppose, for example, a sample of boron-doped silicon (usually symbolized as Si:B), a p-type material, is kept so cold that the acceptor states, which lie 0.045 eV above the top of the valence band, are mostly empty. Intrinsic absorption in silicon cuts off at wavelengths longer than 1.12 μm. Shortward of this cutoff wavelength, our sample absorbs as if it were intrinsic silicon. Now, however, photons with wavelengths shorter than $\lambda_i = hc/E_i = 26$ μm can ionize electrons from the valence band into the acceptor states. In effect, extrinsic

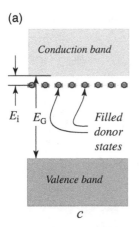

(a)

(b)

Fig. 7.13 Band structure of n-type extrinsic semiconductors. (a) An n-type material at zero temperature; (b) The same material at a higher temperature, where electrons from the donor states have been ionized into the conduction band.

absorption moves the cutoff to the longer wavelength. The implication for the construction of detectors for infrared light is obvious.

The absorption coefficient for extrinsic operation depends upon the dopant concentration. An important difference, then, between intrinsic and extrinsic photo-absorption is that the coefficient for extrinsic absorption can be adjusted in the manufacturing process. However, there are limits to the amount of impurity that can be added, so the absorption coefficient for extrinsic operation will normally be low. Extrinsic detectors therefore tend to be rather thick (1 mm) to provide adequate depth for photo-absorption. Detectors this thick are difficult to fabricate into arrays.

7.4 Photoconductors

7.4.1 Simple photoconductors

Both intrinsic and extrinsic semiconductors, employed in a circuit like the one illustrated in Figure 7.9a, should in principle make excellent light detectors. The output voltage for the circuit is

$$V_0 = \frac{R_L}{(R_L + R_{var})} V_B \tag{7.13}$$

If s, y, and w are the dimensions of the semiconductor in Figure 7.9a, its resistance will be

$$R_{var} = \frac{s}{\sigma yw} = \frac{s}{n_e \mu qyw} \tag{7.14}$$

Here we express the conductivity, σ, as the unit charge, q, times the density of charge carriers, n_e, times their mobility, μ. In a photoconductor at very low temperatures, n_e is directly proportional to N_ϕ, the number of photons per second incident on the device. More precisely,

$$n_e = N_\phi \frac{\eta \tau}{syw} \tag{7.15}$$

Here τ is the expected lifetime of a charge carrier before recombination, which shortens with the concentration of impurities and flaws in the crystal, and lengthens with increasing temperature. Typical values range from 10^{-7} s (InSb) up to 10^{-2} s (germanium) The factor η is the quantum efficiency, that is, number of charge carriers actually produced per incident photon. It depends upon the factors discussed above: surface reflectivity, the absorption coefficient, and the thickness of the sensitive layer. (The fraction of the photons entering a layer of thickness z that are absorbed is just $1 - e^{-\alpha z}$.) The absorption coefficient, in turn, will depend upon wavelength and (for extrinsic materials) impurity concentration.

Substituting (7.15) into (7.14):

$$R_{\text{var}} = \frac{s^2}{q\mu\eta\tau N_\phi} = \frac{b}{N_\phi} \tag{7.16}$$

So the output voltage is (substituting (7.16) into (7.13)):

$$V_0 = V_D R_L \left(R_L + \frac{b}{N_\phi} \right)^{-1} \tag{7.17}$$

This means that the voltage response to light in the simple photoconductor circuit is in general non-linear. However, in the special case of low light levels and low temperatures, $R_{\text{var}} \gg R_L$ and the current is almost entirely due carriers created by photo-ionization, so

$$V_0 \simeq \frac{1}{b} V_B R_L N_\phi = s^{-2} q\mu\eta\tau V_B R_L N_\phi \tag{7.18}$$

and

$$I_{\text{photo}} = s^{-2} q\mu\eta\tau V_B N_\phi \tag{7.19}$$

To build a device, you would want to make V_0/N_ϕ and I_{photo}/N_ϕ as large as possible. You can maximize carrier lifetime by using high-purity materials. You can increase the quantum efficiency by increasing the thickness or the doping in an extrinsic absorber. You can increase V_B/s^2 by increasing the voltage or decreasing electrode spacing. All these strategies have limits: For example, at large electric field strengths, electrons will gain enough kinetic energy to ionize atoms by collision, and create new charge carriers. These secondaries will in turn accelerate, collide with atoms, and produce more carriers. At high enough voltages, in a condition called **breakdown**, the avalanche of charge production becomes constant, destroying the resistance of the material and making it useless as a detector.

7.4.2 The blocked impurity band photoconductor

For infrared detection, extrinsic semiconductors like Si:As and Ge:Ga are very attractive because of their long cutoff wavelengths. Their absorption coefficients are low, however, because dopant concentrations must be low. This is because at high concentrations, the dopant atoms are so close together that their electron's wave functions overlap, producing an **impurity band**. If the states in this band are partially occupied, then the material will be conducting – with charge carriers "hopping" from one impurity state to another, effectively short-circuiting any photoconductive effect.

The **blocked impurity band (BIB) detector**, also called the **impurity band conduction (IBC) detector**, is a device that prevents impurity carrier hopping,

Fig. 7.14 A cross-section (a) and band diagram (b) of a BIB detector. A photon enters from the left, passing through a transparent Si substrate and a very highly doped Si contact ($n{+}{+}$). Charges created in the depletion zone of the IR-sensitive region ($n{+}$) sweep toward the oppositely charged contacts. The high-purity blocking layer (i) is responsible for most of the electrical resistance, and prevents charge flow from the impurity band. A metal contact supplies positive voltage to one side of the blocking layer. Low concentrations of acceptor sites help control the width of the depletion zone.

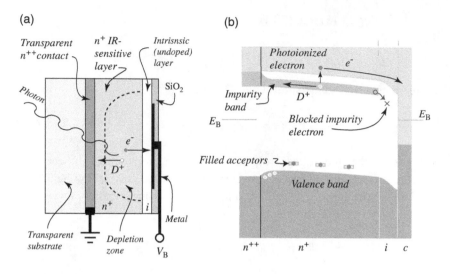

thereby permitting high doping levels. (In Si:As the density of silicon atoms is 5×10^{22} cm^{-3}, and the maximum level acceptable level of arsenic concentration without hopping is around 10^{16}.) Figure 7.14a shows a simple BIB design based on n-type material. A thin layer of highly doped material semiconductor (n^{+}) is bonded to a layer of intrinsic material (i), so that the intrinsic material breaks the continuity (and conductivity) of the impurity band. The band diagram (Figure 7.14b) shows that electrons and holes (D+) created by photo-ionization will contribute to the current, but other electrons in the impurity band are blocked. See Rieke (2003).

7.5 The MOS capacitor

The metal-oxide-semiconductor (MOS) capacitor is the basic element of an important class of astronomical detectors. The device is a three-layer sandwich (Figure 7.15a). In the figure, the left-hand layer is a block of p-type semiconductor, usually doped silicon. The left-hand face of this block is connected to electrical ground. A thin layer of insulator, usually silicon dioxide, forms the middle of the sandwich. The right-hand layer is a thin coating of metal, which is held at a positive voltage. If the insulating layer is not made of SiO$_2$ (silicon nitride, Si$_3$N$_4$, is the usual alternative), then the device is called an ***MIS (metal–insulator–semiconductor) capacitor***.

Figure 7.15b shows the band structure of the device. The positive voltage of the metal layer distorts the energies of the bottom and top of the semiconductor forbidden gap. The tilt of the band reflects the strength of the electric field. In the diagram, the electric field forces electrons to move down and to the right and holes upwards and to the left.

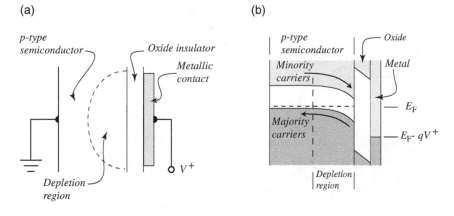

(a)

(b)

Fig. 7.15 (a) Cross-section of the physical structure of a MOS capacitor. Positive voltage (usually a few volts) applied to the metal layer creates a depletion region in the semiconductor. (b) An energy-band diagram for the MOS capacitor. Majority carriers are swept out of the depletion region. Minority carriers are swept toward the boundary with the insulator. Donor sites have been omitted from the diagram.

The large band gap in the insulator prevents minority electrons from crossing into the oxide layer. The flow of majority holes to ground in the valence band, in contrast, is not impeded. The result is that, in equilibrium, a **depletion region** devoid of the majority charge carriers develops in the semiconductor adjacent to the insulator. The minority carriers here are immobile – trapped in the potential well formed by the bottom of the semiconductor valence band and the band gap of the insulator.

The MOS capacitor is especially useful because it will *store* electrons that are generated by ionization. Referring to Figure 7.16a, it is clear that if an electron–hole pair is created in the depletion region, the pair will be swept apart before they can recombine: the electron goes into the well, and the hole leaves the material. Electrons in the well remain there indefinitely, since they sit in a region depleted of holes. Ionizations outside the depletion zone are less likely to produce stored electrons since charges there move by diffusion, and the longer it takes for the electron to reach the depletion zone, the greater are its chances of encountering a hole and recombining.

For ionizations in the depletion zone, however, charge storage can be nearly 100% efficient. Eventually, if enough electrons accumulate in the zone, they will neutralize the effect of the positive voltage and remove the potential well for newly generated electrons. Figure 7.16a illustrates this saturated situation. Saturation destroys the depletion zone and generated charge carriers move only by diffusion, eventually recombining in equilibrium. Newly created electrons are no longer stored. The capacitor has exceeded its **full well capacity**.

Short of saturation, the MOS capacitor is a conceptually simple detector of light. For every photon below the cutoff wavelength absorbed in the depletion zone, the device stores something like one electron. Making a photometric measurement then consists of simply counting these electrons, a wonderful characteristic. It means that a very weak source can be detected by simply exposing the capacitor to light from the source for a time long enough to accumulate a significant number of electrons.

Fig. 7.16 (a) The movement of charge carriers created by ionization in the semiconductor layer of an MOS capacitor. Conduction-band electrons will move into the potential well, while valence-band holes move out of the material to ground. There is a net increase in the negative charge in the semiconductor layer. (b) In a saturated device, there is no longer a potential gradient in the semiconductor, so recombination and ionization will be in equilibrium, and there will be no further gain in stored charge.

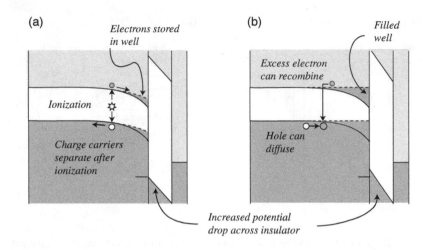

7.6 The p–n junction

Important physics occurs if a p-type material and an n-type material are brought into contact. Junctions of this sort are the basis for many solid-state electronic devices and for some astronomical detectors. Figure 7.17 illustrates the behavior of charge carriers at a ***p–n junction.*** We imagine that a block of n-type material has just been brought into contact with a block of p-type material, and Figure 7.17a shows the non-equilibrium situation immediately after contact.

The majority charge carriers start to flow across the junction. Electrons in the n-side conduction band will move across the junction to fill the available acceptor states on the p side (broken lines). Likewise, mobile holes in the valence band of the p-type material will move across the junction to neutralize any electrons in donor states in the n-type material. Opposite charges build up in the doping sites on either side of the junction – excess negative charge on the p side, excess positive charge on the n side. Electrostatic repulsion eventually halts further transfer of carriers across the junction.

7.6.1 Generation and recombination

Figure 7.17b shows the situation once equilibrium is established. As in the MOS capacitor, a ***depletion region***, constantly swept clear of mobile charge carriers, has formed in the volume surrounding the junction. The lack of charge carriers means this region should have very high electrical resistance. In equilibrium, the energies of the bands change across the depletion region – it requires work to move an electron from the n region to the p region against the electrostatic force. The potential difference across the depletion zone, E_b, is just sufficient to bring the Fermi energy to the same level throughout the crystal. In equilibrium,

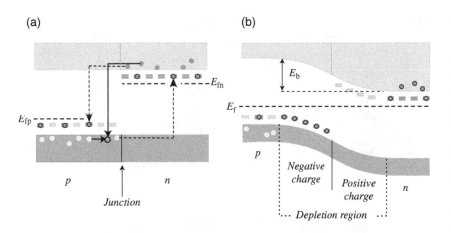

Fig. 7.17 The p–n junction;
(a) shows the flow of
charge carriers
immediately after contact
between the two regions.
Majority carriers (n-side
electrons and p-side
holes) recombine, fill
acceptor sites, and ionize
donor sites. The band
structure in equilibrium is
shown in (b). The
accumulation of charges
near the junction creates a
built-in field, which alters
the energy levels of
available states so that the
Fermi energy is the same
everywhere in the crystal.

charges do move through the depletion region, but the two electric currents here cancel:

$$I_r = -I_g$$

The first current, the ***recombination current***, I_r, is due to the majority carriers that are able to overcome the potential barrier, cross the depletion region, and undergo recombination. This I_r is a positive current that flows from p to n; it has two components: one caused by n-side electrons, the other by p-side holes. Figure 7.18a illustrates the flow of the recombination current, whose magnitude *will depend on the size of the barrier* and on the temperature.

The second current, I_g, the ***generation current***, is due to minority carriers and flows in the opposite direction (from n to p). The minority carriers are thermally ionized conduction-band electrons on the p side and valence-band holes on the n side, which diffuse away from their creation sites. If such a carrier reaches the depletion region, it will be swept across. Diffusion speed outside the depletion region depends on the temperature and the impurity concentration, but is (to first order) *independent* of E_b. Thus, I_g, depends on temperature, but in contrast to I_r is virtually independent of the size of E_b.

7.6.2 p–n junction diodes

The different behaviors of the two currents mean that the p–n junction can function as a ***diode***: it will carry (positive) current in the direction p to n, but not in the reverse direction. Figure 7.19 illustrates the process.

In the condition known as ***forward bias***, a positive voltage connected to the p side of the junction reduces the size of the potential barrier E_b. The recombination current, I_r, will flow more strongly. (That is, more electrons will have energies greater than the barrier, and can move from n to p.) The size of this current will depend in a non-linear fashion on the size of the applied voltage,

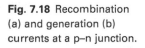

Fig. 7.18 Recombination (a) and generation (b) currents at a p–n junction.

Fig. 7.19 Biased diodes: (a) forward bias reduces the size of the barrier, so the recombination current increases; (b) reverse bias increases the barrier and decreases the recombination current. In both cases, the generation current remains unchanged.

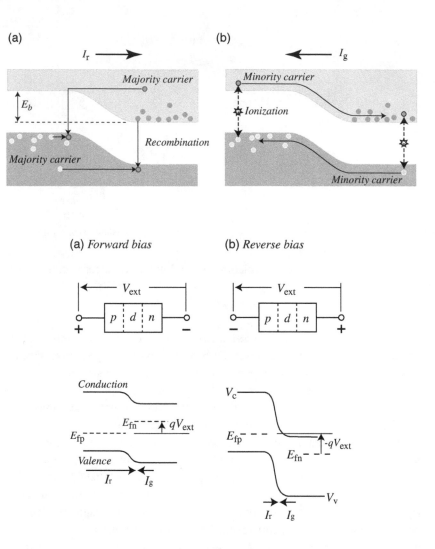

V_{ext}. The applied voltage, however, does not affect the generation current in the opposite direction, I_g, due to minority carriers. The relatively poor conductivity of the depletion region guarantees that almost all of the potential drop will occur here, and the applied voltage will have little influence on the diffusion rate outside the depletion region. Thus, in the forward bias case, $I_r > -I_g$, and current flows from p to n.

A negative voltage connected to the p side of the junction – a condition known as **reverse bias** – increases the size of the potential barrier E_b. This chokes off the flow of majority carriers and lowers I_r from its equilibrium value. Again, the minority carrier current, I_g, remains little changed, so the result of the reverse bias circuit is a very small current in the direction n to p. Boltzmann's law and the above arguments suggests a **diode equation** that gives the voltage–current relationship for an "ideal" diode:

$$I_{\text{TOTAL}} = I_{\text{r}} + I_{\text{g}} = I_{\text{s}}\left(e^{\frac{qV_{\text{ext}}}{kT}} - 1\right) \tag{7.20}$$

Here, q is the electron charge, and current and voltage are assumed to be positive in the p to n direction. You can verify that this formula corresponds to the behavior seen in an actual diode illustrated in Figure 7.20. The formula does not describe the phenomenon of diode breakdown at large reverse biases.

7.6.3 Light detection in diodes

Figure 7.21 illustrates photo-absorption in a p–n diode. Each absorption can create a conduction electron and valence hole. This adds a new contribution to the generation current, this one dependent on ϕ, the number of photons that enter the detector per second. The inclusion of a photocurrent modifies Equation (7.20):

$$I_{\text{TOTAL}} = I_{\text{ph}} + I_{\text{r}} + I_{\text{g}} = -q\phi\eta + I_{\text{s}}\left(e^{\frac{qV_{\text{ext}}}{kT}} - 1\right) \tag{7.21}$$

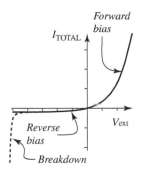

Fig. 7.20 The current–voltage relation for an ideal p–n diode. The solid line is the relation given by Equation (7.20). The dotted line shows the phenomena of breakdown in real diodes, which become conducting at very negative external voltages.

Here η is a factor that depends on the fraction of incident photons absorbed as well as the probability that a generated charge carrier will cross the junction before recombining. Note that charge pairs created outside the depletion zone must move by diffusion to the junction, as discussed above, and have good chance of recombining before crossing the junction. Electron–hole pairs created in the depletion zone, on the other hand, are immediately swept apart by the strong electric field there and have little chance of recombining. Majority carriers will thus tend to accumulate on either side of the depletion zone, and the junction will behave like a charge storage capacitor if an external circuit does not remove the carriers.

There are different strategies for exploiting the light sensitivity of a photodiode. Figure 7.22 plots Equation (7.21) for three different light levels, as well as three different modes of operation: (a) In the *photoconductor* mode, a battery holds the external voltage to a constant value, and the current is a linear function of the incident photon flux. (b) In the *power-cell* mode, the diode is connected to a constant-load resistance, and the power output depends on the incident photon flux. This is the principle of operation for solar power cells. (c) In the *photovoltaic* mode, current from the diode is held at zero (making it a storage capacitor by connecting it to a very high impedance voltmeter, for example), and the voltage across it increases with time and is a non-linear function of the photon flux.

7.6.4 Variations on the junction diode

Some modifications of the simple p–n junction can improve the device's response to light. Several are important in astronomy.

The ***PIN diode*** sandwiches a layer of intrinsic (undoped) silicon between the p-material and the n-material of the junction. This increases the physical size of the depletion zone, and the resulting p–intrinsic–n (PIN) diode has larger

Fig. 7.21 Photo-absorption in a p–n junction diode.

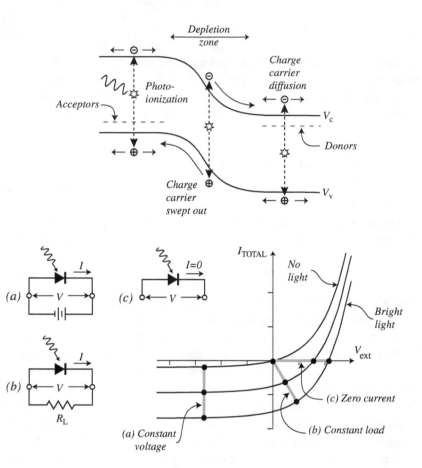

Fig. 7.22 Current–voltage relations for a photodiode at three different levels of incident photon flux. Heavy lines show electrical properties as a function of photon intensity for three different modes of operation.

photosensitive volume, higher breakdown voltage, lower capacitance, and better time response than the simple p–n device.

The *avalanche photodiode* is both a physical modification and a mode of operation. Consider a photodiode (a modified PIN type) that is strongly back-biased at close to its breakdown voltage. Because of the large voltage drop across the intrinsic region, charge carriers created by photo-absorption will accelerate to high kinetic energies – high enough to produce impact ionization of additional charge carriers. These secondaries will in turn accelerate to produce further ioniza-tions. The resulting avalanche of carriers constitutes a current pulse that is easy to detect. At low levels of illumination, counting the pulses is equivalent to counting photons. At higher illuminations, the pulses are too close together to count, but the resulting current, although noisy, is very large and therefore easy to detect.

7.7 The vacuum photoelectric effect

The vacuum photoelectric effect depends on the ejection of electrons from the surface of a solid. Figure 7.23a is a band diagram that illustrates the effect, which

(a)

(b)

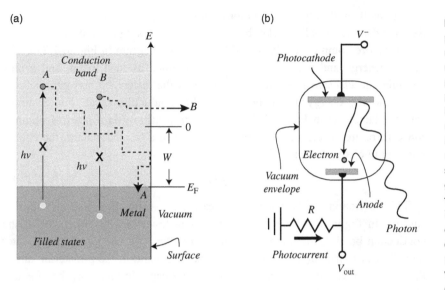

Fig. 7.23 (a) The vacuum photoelectric effect in a metal. Photoelectron *B* reaches the vacuum with positive energy, while photoelectron *A* does not. Both photoelectrons make collisions with the lattice, and execute a random walk to the surface. Photoelectrons gradually become thermalized – if the metal is cold, they tend to lose energy on each lattice collision. (b) A vacuum photodiode. A photon with sufficient energy to eject an electron strikes the photocathode. The photoelectron then accelerates to the anode and flows through the load resistance to ground.

is simplest in metals. A thin slab of the metal cesium, which has a relatively loose hold on its surface electrons, occupies the left side of the figure. The surface of the metal runs vertically. If the potential energy of an electron at rest well away from the metal is zero, then the **work function**, **W**, of the material is the difference between this free electron energy and the Fermi energy of the solid. In the case of cesium, the work function is 2.13 eV.

We would like to use the energy of one photon to move one electron from the metal to the vacuum. This operation has two requirements: the electron must be given a positive energy, and it must be located at the surface. In general, the absorption of a photon with energy $h\nu \geq W$ will take place in the interior of the metal, and will promote an electron there into the conduction band. If, after diffusing to the surface, the electron still has both positive energy and an outward-directed momentum (*case* B in Figure 7.23a) it can move into the vacuum.

A simple device called a **photocell** (or more properly, a **vacuum photodiode**), illustrated in Figure 7.23b, uses this effect to measure the intensity of light. In the diagram, two conductors are sealed in an evacuated cell with a transparent window. One conductor, the **photocathode**, is made from some material (e.g. cesium) that exhibits an efficient vacuum photoelectric effect. The photocathode is held at a negative voltage. The other conductor, the anode, is connected through a load resistor to the ground as illustrated. Illumination of the photocathode ejects electrons into the vacuum. These accelerate to the anode. The result is an output current and voltage across the resistor that is proportional to the photon arrival rate at the cathode.

Metals actually make rather poor photocathodes. They are highly reflective and exhibit large work functions. (Cesium, the metal with one of the smallest

Matter and light

(a)

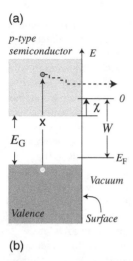

p-type semiconductor

(b)

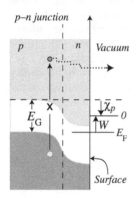

p–n junction

Fig. 7.24 The vacuum photoelectric effect in semiconductors.

values for W, will only detect photons with wavelengths shorter than 580 nm.) Semiconductors usually make better photocathodes but there the effect is slightly more complex, as illustrated by the band diagram in Figure 7.24a. The zero of energy and the work function are defined as in a metal, and a new variable, the ***electron affinity***, χ, is defined as the difference between the zero point and the energy at the bottom of the conduction band. For a simple semiconductor, as in Figure 7.24a, the electron affinity is a positive number. Since there are no electrons at the Fermi level in a semiconductor, the energy required to eject an electron is

$$h\nu \geq E_G + \chi \tag{7.22}$$

This restriction can be relaxed by creating a p–n junction near the emitting surface. In Figure 24b, the junction forces a downward displacement of the conduction band in the n-material. Thus a photon with energy slightly greater than E_G can ionize an electron to the conduction band in the p-material, and if this electron migrates to the n-material, it can escape. In this case, the effective electron affinity of the p-type material is a negative number. The n-type layer is so thin and transparent that it does not detract from the cathode's sensitivity to long wavelengths. Materials of this type, termed ***NEA photocathodes*** (negative electron affinity), are usually fabricated with a III–V semiconductor as the p-type material and oxidized cesium as the n-type material. For example, an NEA photocathode made from p-doped gallium arsenide ($E_G = 1.4$ eV) with a surface layer of n-doped Cs_2O ($E_G = 2.0$ eV, $\chi = 0.6$ eV) is sensitive out to 880 nm and has been important for some astronomical applications.

We have assumed that emitted photoelectrons will leave from the surface that is illuminated. This need not be the case, and many photocathodes are ***semi-transparent***: photons enter on one side and electrons emerge from the opposite side.

7.8 Superconductivity

Superconducting material has an electrical conductivity that falls to zero at and below a critical transition temperature, T_c, where the material also becomes diamagnetic (repels magnetic fields). The simplest superconductors, the first investigated and best-understood, are all metallic elements with very low critical temperatures ($T_c < 10$ K). These are called Type 1 superconductors. Type 2 superconductors are normally insulators at room temperature and less-well understood, but can have high transition temperatures. They are complex materials: alloys, ceramics, and various exotic compounds like $YBa_2Cu_3O_7$ ($T_c = 177$ K). Good evidence exists for superconductivity at room temperature or above under unstable conditions in Type 2 materials.

Type 1 superconductors are the basis of some very promising light detectors in astronomy, so we briefly describe their behavior here. The website superconductors.org or the modern physics text by Harris (1998) gives a more complete introduction, and chapter 10 of Kittel (2005) provides a more advanced treatment, as does Blundell (2009).

7.8.1 The superconductor band gap

Above the critical temperature in a superconducting metal like lead, the Fermi–Dirac formula describes the energy distribution of the electrons. This changes at and below T_c (7.19 K for lead) where a complex lattice-mediated force between electrons makes new energy states available just below the Fermi level – two spatially separated electrons can form a ***Cooper pair*** of exactly cancelling momenta and spins. Each pair has a binding energy well below the thermal energy of the lattice and, with zero spin, behaves in many ways like a boson – the Pauli exclusion principle does not apply to these states, and all pairs have the same momentum (zero, when there is no current). It is the Cooper pair states that are responsible for superconductivity and many resultant behaviors – including perpetual electric currents and ***magnetic field repulsion***. Although a superconductor exhibits no resistance to a direct current, it does exhibit a property called ***kinetic inductance*** – Cooper pairs in a super-current have mass and store kinetic energy, so the superconductor will resist a change in current direction.

Our concern, however, is the manner in which a superconductor interacts with light. Figure 7.25 shows an energy-band diagram for a superconductor. At temperatures below T_c, an unlimited number (no exclusion principle!) of super-conducting states exist at an energy Δ below the Fermi level. Electrons will therefore occupy only states of energy $(E_F - \Delta)$ or lower. The value of Δ is a strong function of temperature, rising from zero at T_c to a maximum value of Δ_m at temperatures below about $0.3T_c$. The value for Δ_m – the binding energy per electron of a Cooper pair – is tiny, 1.4×10^{-3} eV for lead, which is typical.

Consider what must happen for a superconductor to absorb a photon: if the photon has energy larger than 2Δ can it break apart a Cooper pair and promote

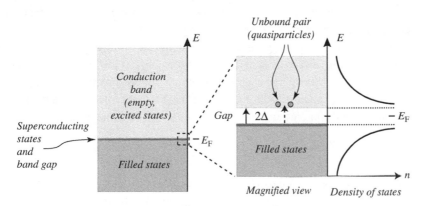

Fig. 7.25 Energy bands in a Type 1 superconductor. The band gap energy is the minimum energy required to break apart two electrons bound in a Cooper pair, placing them in an excited quasiparticle state (dotted arrow). The density of states just below and just above the band gap is very high, although there are no states in the gap itself.

Table 7.6 *Some Type I superconductor characteristics. Data from Kittel (2005).*

Atom	T_c, K	Band gap $2\Delta_m$ units of 10^{-3} eV	Atom	T_c, K	Band gap $2\Delta_m$ units of 10^{-3} eV
^{13}Al	1.14	34	^{72}Hf	0.12	4
^{41}Nb	9.5	305	^{73}Ta	4.48	140

Fig. 7.26 An STJ diode. (a) A cross-section of the physical device. In most practical detectors the three layers and their contacts are deposited as films on a transparent substrate, so a more accurate diagram would extend vertically several page heights. The band structure is shown in (b). Not shown is the possibility that quasiparticles on the right can tunnel back to the left to break apart additional pairs.

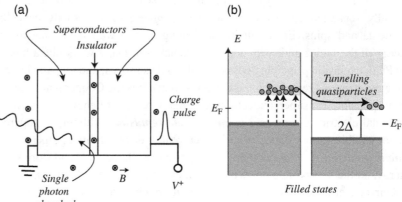

the two electrons to higher energies. Lower energy photons will not be absorbed: the material has an effective band gap of magnitude 2Δ. The electrons promoted to the excited states in the "conduction" band in the superconductor have quantum characteristics that differ from energetic electrons in an ordinary metal, and are therefore termed **quasiparticles**. The number of states available to quasiparticles at energies just above the gap is very large. Table 7.6 lists the gap energies and transition temperatures of a few superconductors that have been useful in astronomical detectors.

The tiny band gap suggests that superconductors have great potential as **energy-resolving detectors** at optical wavelengths, since n_q, the number of quasiparticles produced by a single photon will be directly proportional to $h\nu/\Delta$. (Energetic photons will break apart many Cooper pairs.) The theoretical uncertainty in the energy of a detected photon will be something like:

$$\sigma(h\nu) = \sigma(n_q)\cdot\Delta = \Delta\sqrt{\frac{h\nu}{\Delta}} = \sqrt{h\nu\Delta} \qquad (7.23)$$

More detailed consideration of the physics as well as practicalities of detector construction mean that Equation (7.23) is optimistic by about a factor of 3 or more.

7.8.2 Light detection in an SIS junction

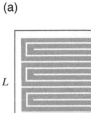

Two superconductors separated by a thin layer of insulator (SIS = superconductor–insulator–superconductor) constitute a ***Josephson junction*** if the insulator is thin enough (around 1 nm) to permit quantum mechanical tunneling. Figure 7.27 shows such a junction arranged as a light-detecting diode: a positive bias voltage less than $2\Delta/q$ is applied to the right-hand superconductor, and a magnetic field is applied parallel to the junction. If the junction is very cold, all excited states are empty. In a normal Josephson junction, it would be possible for the Cooper pairs to tunnel from left to right, but the magnetic field suppresses that current, so the diode does not conduct.

If the left-hand superconductor absorbs a single visible light photon of wavelength λ (energy hc/λ), it receives enough energy to break apart multiple Cooper pairs, promoting a maximum of $n_{q\text{max}} = hc/\lambda\Delta$ electrons into excited states. These quasiparticles are not repelled by the magnetic field and *can* tunnel across the insulator, and those that do produce a current pulse whose total charge is inversely proportional to the wavelength of the exciting photon.

Devices of this kind, called ***superconducting tunnel junctions*** (**STJs**), operated with sufficient time resolution, can count individual incoming photons and determine the wavelength (from X-ray to near infrared) of each. The uncertainty in the wavelength determination of a detected photon can be derived from Equation (7.23). Although still very much in the development stages, a few experimental but practical multi-pixel STJ-based detectors have begun to appear at telescopes. See chapter 4 of Rieke (2003) and the references by Eisenhauer and Raab (2015) and Verhoeve et al. (2004).

Superconducting tunnel junctions promise to be the near-ideal astronomical detector: They can be fashioned into an array that produces an image yielding both spectroscopic information and high time resolution. Especially because they must operate at milli-kelvin temperatures below $0.2T_c$, there are formidable engineering issues in developing them as practical and affordable astronomical tools, but there is no doubt about their potential as detectors.

7.8.3 Light detection in kinetic induction devices

The microwave kinetic induction device (MKIS or KID) also depends on photon-generated quasiparticles in superconducting material. In this case, the relevant physics is the increase in both the resistance and (especially) the inductance of the material due to the quasiparticles. One observes the resonant frequency of an RLC circuit in which the inductor is a superconductor. The material inductance will increase if the superconductor absorbs a photon, and the resonance will shift to a lower frequency, with the shift depending on the energy of the photon. See Figure 7.27 – the resonant frequency (usually in the microwave region) is approximately $1/\sqrt{LC}$ if C_s and R_s are small. In the detector

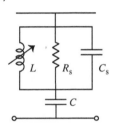

Fig. 7.27 An MKID, after the OLE array design by Mazin et al. (2012) (a) A top view of the physical device. The darker gray inductor (L) and capacitor (C) are etched in a superconducting TiN thin film deposited on a substrate. The light-gray read probe line (P) capacitively couples to the resonant circuit. Photons are concentrated on the inductor by a microlens (not shown). (b) An equivalent circuit of this single MKID pixel.

described by Mazin et al. (2012), each pixel in a 1024-element array contains a titanium nitride (TiN) superconductor film fabricated into a serpentine-shaped inductor connected to a planar capacitor. The resonant frequency is measured by sending a multi-frequency probe signal down the read line. The superconducting circuit will absorb energy at the resonant frequency, and if a photon has been absorbed that frequency will drop and the shift observed. MKIDs, like STJs, are still very much in the development phase, but have demonstrated very real promise as superb detectors.

Summary

- Quantum mechanics accounts for a quantized pattern of permitted states for the energies, angular momenta, magnetic interactions, and spins of electrons bound to an isolated atom. Concepts:

free state	*ground state*	*Pauli exclusion principle*
valence electron	*periodic table*	*quantum number*
fermion	*boson*	*spectroscopic notation*

- The outer (optical) electrons of an atom gain or lose energy by making transitions between permitted states. Concept:

excitation	*photo-emission*	*photo-ionization*
ground state	*photo-absorption*	*absorption edge*
thermal excitation	*Boltzmann distribution*	

- Permitted quantum states of isolated molecules are distinguished by the electronic states of their component atoms, but also by the quantized rotation and vibration states of the molecule. Concept:

  molecular absorption bands

- The energy states for electrons in solid-state crystals typically arrange themselves in continuous bands separated by forbidden band gaps. Concepts:

diamond lattice	sp^3-*hybrid orbitals*	*bonding state*
anti-bonding state	*valence band*	*conduction band*
holes	*semiconductor*	*intrinsic semiconductor*
electron degeneracy	*phonon*	*Fermi–Dirac statistics*
Fermi energy	*band gap energy*	*cutoff wavelength*

- Adding small quantities of a selected impurity can produce desirable properties in a semiconductor. Concepts:

Dopant	*extrinsic semiconductor*	*donor atom*
acceptor atom	*p-type*	*n-type*
impurity band		

- Photoconductors absorb a photon and create an electron–hole pair, thereby increasing the electrical conductivity of the material. Concepts:

> absorption coefficient   breakdown   absorption depth
> BIB photoconductor   impurity band hopping   load resistor

- The MOS capacitor is a block of extrinsic semiconductor separated from a metal electrode by a thin layer of insulation. With the proper voltage across the insulator, the device can store charges produced by photo-absorptions. Concepts:

  > SiO_2 depletion region   potential well
  > full well capacity   saturation

- The p–n junction produces a depletion region where photo-absorptions can generate charge carriers and an electric current. Concepts:

  > p–n junction   recombination current   generation current
  > diode   forward bias   reverse (back) bias
  > breakdown   diode equation   p–n photodiode
  > avalanche photodiode   PIN photodiode

- Electrons can leave the surface of material in a vacuum if they have energies greater than the material's work function. Photons can supply the needed energy, and thus produce an electric current in a vacuum.

  > photocathode   vacuum photodiode   anode
  > electron affinity   NEA photocathode

- The electrons in Type I superconducting materials are joined in Cooper pairs and exhibit no resistance to electric currents. Pairs can be broken by photo-excitation. The unpaired electrons move across a very small band gap into quasiparticle states in the material's conduction band.

  > critical temperature   Cooper pair   quasiparticles
  > kinetic induction

- A superconducting junction diode produces a number of conduction-band electrons that is proportional to the energy of the incoming photon. A kinetic induction device measures a change in the resonant frequency of a superconducting LC circuit due to creation of quasiparticles by photo-absorption. In pulse-counting mode, either device can measure both the intensity and the wavelength distribution of a source. Concepts:

  > tunneling   Josephson junction   resonant circuit
  > kinetic inductance   MKID   STJ

Exercises

> For the things we have to learn before we can do them, we learn by doing them, for example, men become builders by building, and lyre-players become lyre-players by playing the lyre.
>
> – Aristotle, *The Nicomachean Ethics*, Book II, Chapter I, *c.* 340 BCE

1. Using the nl^x notation, write down the electron configuration for the ground state, first excited state, and third excited state of iron (atomic number 26) as suggested by Table 7.2.

2. There are several exceptions to the configuration-filling scheme presented in Table 7.2. The configuration of the ground state of copper is an example. Look up a table of electron configurations in atoms and find at least five other examples.

3. Suppose a certain diatomic molecule has an energy-level diagram similar to Figure 7.3 and consider only transitions within the $\Lambda = 0$ states. Suppose that relative to the ground state, state $(J = 1, v = 0)$ has an energy of 1 eV. Suppose also that, no matter what the rotational state is, the relative energies of the lowest vibrational states are $v(v + 1)d$, where $d = 10^{-5}$ eV and v is the vibrational quantum number. (a) Compute the wavelengths of all permitted emission lines arising between levels $J = 0$ and $J = 1$, and involving vibrational states $v = 0, 1, 2, 3, 4$. The only permitted transitions are the ones in which $\Delta v = \pm 1$. (b) Sketch the emission spectrum for these lines.

4. Compute the relative probability of finding an electron at the bottom of the conduction band relative to the probability of finding an electron at the top of the valence band in a silicon crystal at a temperature of (a) 3 K and (b) 300 K. Use Fermi–Dirac statistics. Compare your answer with the one given by the Boltzmann equation.

5. A block of semiconductor is being used to measure temperature changes in a bolometer. The device is operated at 40 K. Assume the electrical conductivity of the block depends only on the number density of conduction-band electrons. By what fraction does the conductivity change when the temperature increases by 1 K, (a) if the semiconductor material is silicon and (b) if it is germanium?

6. You are designing a photoconductor like the one in Figure 7.9a to operate in the visible (500 nm). What is the minimum thickness, y, needed to achieve a quantum efficiency of 0.4 if the photoconductive material is bare silicon (index of refraction is 4.4 at 500 nm)?

7. Suppose the photoconductive elements in Figure 7.9b are made of a thin film of germanium, with thickness $s = 25$ μm, in a camera operating at a wavelength of 1.5 μm. Discuss whether decreasing the value of s to 15 μm will increase or decrease the photocurrent. Assume the voltage drop across the electrodes remains constant and the reflectivity of the positive contact is 100%.

8. How does an MOS capacitor made of an n-type semiconductor work? Why do you think p-type material is usually preferred for these devices?

9. Derive a relationship between the full well capacity of an MOS capacitor and the maximum possible relative precision that the device can produce in a brightness measurement. What is the risk in planning to achieve this precision with a single measurement?

10. Assume you have a meter that measures electric current with an uncertainty (noise) of 100 picoamps. (One picoamp $= 10^{-12}$ amp $= 10^{-12}$ coulomb s^{-1}.) You employ your meter with a photodiode in a circuit like the one in Figure 7.18a. You have a 2-m telescope at your disposal, and use a filter to limit the light received to those

wavelengths at which the detector is most sensitive. Compute the magnitude of the faintest star you can detect with this system. "Detect" in this case means the signal-to-noise ratio is greater than 3. Assume the photon flux from a zero-magnitude star in the bandpass you are observing is 10^{10} photons m^{-2} s^{-1}. Your photodiode detects 45% of the photons incident, and you may ignore any background signal.

11. In response to an incoming photon, a niobium-based STJ diode detects a pulse of 500 electrons. Assume tunneling operates with 100% efficiency, and the only source of noise is counting statistics. (a) Compute the energy of the incoming photon and its uncertainty. (b) What is the wavelength of the photon and its uncertainty? (c) Compute the spectroscopic resolution ($R = \delta\lambda/\lambda$) of this device as a function of wavelength. (d) Find the equivalent expression for a device in which the superconductor is hafnium instead of niobium.

Chapter 8
Detectors

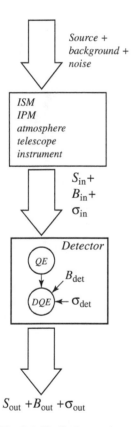

Source +
background +
noise

ISM
IPM
atmosphere
telescope
instrument

$S_{in}+$
$B_{in}+$
σ_{in}

Detector

QE

B_{det}

DQE ← σ_{det}

$S_{out} + B_{out} + \sigma_{out}$

Fig. 8.1 Mediation and detection of a light signal, *S*, (*IGM* = intergalactic medium, *ISM* = interstellar medium) background, *B*, and noise, *σ*. The detection step may fail to record some of the mediated signal, and may introduce additional noise and background.

[Holmes said] Honestly, I cannot congratulate you upon it. Detection is, or ought to be, an exact science, and should be treated in the same cold and unemotional manner. You have attempted to tinge it with romanticism, which produces much the same effect as if you worked a love-story or an elopement into the fifth proposition of Euclid.

"But romance was there," I remonstrated.

– Arthur Conan Doyle, *The Sign of the Four*, 1890

Astronomical detection, even more than the work of Sherlock Holmes, is an exact science. Watson, though, has an equally important point: no astronomer, not even the coldest and most unemotional, is immune to that pleasant, even romantic, thrill that comes when the detector *does* work and the universe *does* seem to be speaking.

An astronomical detector receives photons from a source and produces a corresponding **signal**. The signal characterizes the incoming photons: it may measure their rate of arrival, their energy distribution, or perhaps their wave phase or polarization. Although detecting the signal may be an exact science, its characterization of the source is rarely exact. Photons never pass directly from source to detector without some mediation. They traverse space, the Earth's atmosphere, telescope optics, and other elements of the observing system like filters and spectrograph gratings, all of which alter the stream. Only in the end does the detector do its work. Figure 8.1 illustrates this two-stage process.

An astronomer must understand both mediation and detection if she is to extract meaning from measurement. This chapter describes only the second step in the measurement process, detection. We first outline the qualities an astronomer will generally find important in any detector. Then we examine a few important detectors for the optical and IR in detail: the CCD, a few photoemissive devices, the hybrid array, and the bolometer.

8.1 Detector characterization

Why does an astronomer choose one detector instead of another? Why did optical astronomers in the 1980s largely abandon the photograph, the then-dominant detector for imaging, in favor of solid-state arrays? Why are these same arrays useless for other purposes, such as measuring very rapid changes in brightness? Is there a *perfect* detector? We begin an answer by listing several critical characteristics of any detector.

8.1.1 Detection modes

We can distinguish three distinct modes for detecting light.

Photon detectors produce a signal that depends on the *number* of incident photons. For example, in the last chapter, we saw how photo-ionization in a photoconductor, photocathode, or photodiode can produce a change in the macroscopic electrical properties like conductivity, voltage, or current. Other photon-induced changes in quantum state might produce chemical reactions (as in photography). Photon detectors are particularly suited to shorter wavelengths (infrared and shorter), where the energies of individual photons are large compared to the thermal energies of the electrons in the detector.

Thermal detectors produce a signal that depends on the *energy* deposited by the incoming photon stream. In these devices the signal is the temperature change in the body of the detector. Although thermal detectors are in principle useful at all wavelengths, in practice, thermal detectors, especially a class called *bolometers*, have been fundamentally important in the long-wavelength infrared and microwave regions, as well as very useful in the gamma and X-ray regions.

Wave detectors produce a signal in response to the oscillating electric or magnetic field of the incoming electromagnetic waves, usually by measuring the interference effect the incoming fields have on a wave produced by a local oscillator. In principle, these detectors, unlike photon and thermal detectors, can gauge the phase, intensity, and polarization of the detected wave. Wave detectors are especially useful in the radio and microwave parts of the spectrum.

8.1.2 Efficiency and yield

> Thou shalt not waste photons.
>
> – Anonymous, *c.* 1980

A good detector is efficient. We construct costly telescopes to gather as many photons as possible, and it seems perverse if a detector does not use a large fraction of these expensive photons to construct its signal.

Photography, for example, is relatively inefficient. The photographic detector, the emulsion, consists of a large number of tiny crystals, or *grains*, of silver halide (usually AgBr) suspended in a transparent gelatin matrix. Photons can interact with a grain to eventually turn the entire grain into elemental silver.

The more silver grains present in the emulsion after it has been processed, the stronger is the signal.

Why is the process inefficient? Some photons reflect from the surface of the emulsion and are not detected. Some pass right through the emulsion, while others are absorbed in its inactive parts without contributing to the signal. Nevertheless, silver halide grains absorb something like 40–90% of the incident photons. An absorbed photon breaks an ionic bond, creating a neutral silver atom and a neutral bromine atom. The bromine atom can escape the crystal, either by combining with the gelatin or with another bromine to form a mobile molecule. Most bromines, however, will bond with a silver atom, so few neutral silver atoms survive. It is only after three to six silver atoms drift and clump together at a spot on the grain that the crystal becomes developable. In the end, very few of the incident photons actually have an effect in photography. The process is inefficient.

The *quantum efficiency*, *QE*, is a common measure of detector efficiency. It is usually defined as the fraction of photons incident on the detector that actually contribute to the signal.

$$QE = \frac{N_{detect}}{N_{in}} \tag{8.1}$$

In a perfect detector, every incident photon would be absorbed in a fashion that contributed equally to the signal, and the detector would have a QE of 100%. Photographic emulsions have QE values in the range 0.5–5%.[1] Most of the solid-state devices discussed in the last chapter have QE values in the 20–95% range. Astronomers prefer these devices, in part, because of their high quantum efficiencies.

The quantum efficiency of a particular device is not always easy to measure, since (as in photography) the chain of events from incident photon to output signal may be difficult to describe and quantify. *Absorptive quantum efficiency* is physically more straightforward, but somewhat less informative. It is defined as the photon flux absorbed in the detector divided by the total flux incident on its surface:

$$\eta = \frac{N_{abs}}{N_{in}} \tag{8.2}$$

Because absorbed photons are not necessarily detected, $QE \leq \eta$.

---

[1] Quantum efficiency is a bit of a slippery concept in photography. For example, once a grain has formed a stable clump of three to six silver atoms, absorbed photons can make no further contribution to the signal, even though they create additional silver atoms. The entire grain is either developed or not developed depending only on the presence or absence of the minimum number of atoms. In photography, QE is thus a strong function of signal level – the highest efficiencies only apply if the density of developed grains is relatively low.

The **quantum yield** of a photon detector is the number of detection "events" per incident photon. For example, in silicon photoconductors, the detection event is the production of an electron–hole pair. If an incident photon has energy less than about 5 eV, it can produce at most one electron–hole pair, so the quantum yield is 1. For higher energy photons, a larger number of pairs are produced, around one e–h pair per 3.65 eV of photon energy. What happens in detail is that the first electron produced has so much kinetic energy that it can collide with the lattice to produce phonons that generate additional pairs. A 10-angstrom X-ray, therefore, will yield (on average) 34 photoelectrons. An STJ-based detector, you will recall, is particularly attractive because of its very large, wavelength-sensitive quantum yield.

8.1.3 Noise

> It's whabbit season, and I'm hunting whabbits, so be vewy, vewy quiet!
>
> – Elmer Fudd, *Looney Tunes*, c. 1940

Although efficiency in a detector is important, what really matters in evaluating a measurement is its uncertainty. The uncertainty in the output signal produced by a detector is often called the **noise**, and we are familiar with the use of the **signal-to-noise ratio**, **SNR**, as an indication of the quality of a measurement. It would seem that a perfect detector would produce a signal with zero noise. This is not the case.

You will recall that there is an uncertainty *inherent* in measuring the strength of any incident light ray. For a photon-counting device, this uncertainty arises from the Poisson statistics[2] of photon arrivals, and is just

$$\sigma_{\text{Poisson}} = \sqrt{N} \tag{8.3}$$

where N is the number of photons actually counted. A perfect detector, with QE = 1, faithfully counts all incident photons and will therefore produce

$$(\text{SNR})_{\text{perfect}} = \frac{N_{\text{out}}}{\sigma_{\text{out}}} = \frac{N_{\text{in}}}{\sigma_{\text{in}}} = \sqrt{N_{\text{in}}} \tag{8.4}$$

Real detectors will differ from this perfect detector by either counting fewer photons (reducing the output noise, but also reducing both the output signal and the output SNR) or by exhibiting additional noise sources (also reducing the SNR). The **detective quantum efficiency** (*DQE*) describes this departure of a real

---

[2] Although we have been treating the photon-counting process as if it were perfectly described by Poisson statistics, both theory and experiment show this is not the case. Photon arrivals are not statistically independent – real photons tend to clump together slightly more than Poisson would predict. This makes little practical difference in the computation of uncertainties.

detector from perfection. If a detector is given an input of N_{in} photons and has an output with signal-to-noise ratio $(SNR)_{out}$, then the DQE is defined as a ratio:

$$DQE = \frac{(SNR)^2_{out}}{(SNR)^2_{perfect}} = \frac{N_{out}}{N_{in}} \qquad (8.5)$$

Here N_{out} is a fictitious number of photons, the number that a perfect detector would have to count to produce a signal-to-noise ratio equal to $(SNR)_{out}$. The DQE gives a much better indication of the quality of a detector than does the raw QE, since it measures how much a particular detector degrades the information content of the incoming stream of photons. For a perfect detector, DQE = QE = 1, and it should be clear from Equation (8.2) that DQE $\leq$ QE. If two detectors are identical in all other characteristics, you should choose the detector with the higher DQE.

Returning to the example of the photographic emulsion, the noise in an image is experienced as **granularity**: the microscopic structure of, say, a star image consists in an integral number of developed grains. Statistically, counting grains in an image is a Poisson process, and has an uncertainty and a SNR of $\sqrt{N_{grains}}$. Since it takes something like 10–20 absorbed photons to produce one developed grain, the photographic process clearly degrades SNR. In addition, grains are not uniformly distributed in the emulsion, and some grains not activated by photons will nevertheless get developed to produce a background "fog." Both of these effects contribute noise, and thus reduce the DQE. A typical emulsion might have $\eta = 0.5$, QE = 0.04, and DQE = 0.02. Many solid-state detectors introduce little noise, and their DQE values are close to their QE values – in the range 20–90%.

The DQE generally is a function of the input level. Suppose, for example, a certain QE = 1 detector produces a background level of 100 electrons per second. You observe two sources. The first is bright. You observe it for 1 second, long enough to collect 10 000 photoelectrons (so $SNR_{in} = 100$). For this first source, $SNR_{out} = 10\,000/\sqrt{(10\,100 + 10^2)} = 98$ and DQE = 0.96. The second source is 100 times fainter. You observe it for 100 seconds, and also collect 10 000 photoelectrons. For the second source, $SNR_{out}/\sqrt{20\,000 + 10\,000} = 57.8$, and DQE = 0.33.

8.1.4 Spectral response and discrimination

The QE of a detector is generally a function of the wavelength of the input photons. Pure silicon devices, for example, cannot respond to photons with $\lambda > 1.1$ μm since these photons have energies below the silicon band gap energy. The precise relationship between efficiency and wavelength for a particular detector is an essential characteristic.

One can imagine an ideal detector that measures both the intensity and the wavelength distribution of the incoming beam. An STJ diode, operated in a pulse-counting mode, for example, **discriminates** among photons of different wavelengths.

8.1.5 Linearity and dynamic range

In an ideal detector, the output signal is directly proportional to the input illumination. Departures from this strict linearity are common. Some of these are not very problematic if the functional relation between input and output is well behaved and calibrated. For example, in the range of useful exposures, the density of a developed photograph is directly proportional to the logarithm of the input flux. Figure 8.2 illustrates two very typical departures from linearity. At lower light levels, a detector may not respond at all – it behaves as if there were an input **threshold** below which it cannot provide meaningful information. Often the noise level in the detector sets this threshold. At the extreme of large inputs, a detector can **saturate**, and an upper threshold limits its maximum possible response

The **dynamic range** in input is the ratio of the maximum signal to the minimum signal to which the detector can respond: i.e. the ratio P_{high}/P_{low} in Figure 8.2. Subtle details of the detection process can influence the value S_{max}/S_{min}, which is the manifestation of the dynamic range in the output signal. Most commonly, if the signal is recorded digitally as a 16-bit binary integer, then the smallest possible signal is 1, and the largest is 65 535 ($= 2^{16} - 1$). Thus, even if the input range set by saturation is larger, the range in output is limited by data recording to 1:65 535. Digitization can often set the **sensitivity** of a device, by defining the change in input that produces a one-unit change in output.

8.1.6 Stability

The environment of a detector will change over time, perhaps because of variation in temperature, atmospheric conditions, or orientation with respect to gravity

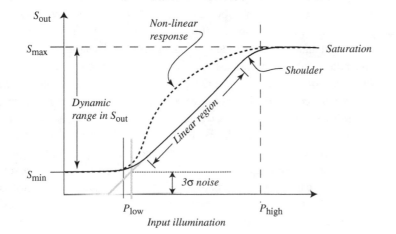

Fig. 8.2 Linear and non-linear regions in a typical detector response curve. The dashed response is completely non-linear. The gray continuation of the solid curve is linear at low light levels, but this signal is masked by detector noise.

or to local magnetic fields. The detector itself may age because of chemical or mechanical deterioration, electrical damage, or radiation and particle exposure. Unrecognized changes can introduce systematic effects and increase uncertainties.

Two general approaches cope with detector instability. The first is to avoid or minimize anticipated changes: e.g. use thermostatic controls to maintain a constant temperature, keep the detector in a vacuum, shield it from radiation, use fiber-optic feeds so that the detector remains motionless. Basically, isolate the detector from the environment. The second approach is to recognize that some changes are unavoidable and calibrate the detector to correct for the instability. For example, if the response of a detector deteriorates with age, make repeated observations of the same standard source so you can compute a correction that compensates for the deterioration.

Hysteresis is a form of detector instability in which the detector response depends on its illumination history. Human vision, for example, exhibits the phenomena of positive and negative afterimage, and some solid-state detectors can continue to report ghost signals from bright objects long after the source has been removed.

8.1.7 Response time

How quickly can the detector make and report a measurement, then make and report the next measurement? Readout procedures for large CCDs, for example, can limit their response time to tens of seconds, photodiodes need time (usually much less than a CCD read time) to charge a readout capacitor, while STJs, MKIDs, and photo-emissive devices have sub-millisecond response times.

8.1.8 Physical size and pixel count

The surface area of the detector is important. One wants to match it to the area on which the light to be measured falls (e.g. the image of an exit pupil, of an object, of a spectrograph slit, of the end of a fiber, etc.). If the detector is too small, it will not intercept all the light from the source; if it is too large, it will intercept unwanted background light and probably produce a higher level of detector noise. Physical size is also related to other properties like dynamic range and response time.

A *single-channel* detector measures one signal at a time, while a *multi-channel* or *multi-detector* device measures several at once. An astronomer might use a simple two-channel device, for example, to simultaneously measure the brightness of a source and the brightness of the nearby background sky. A *linear array* (a string of closely packed detectors arranged in a straight line) might be a good configuration for sensing the output of a spectrograph. A *two-dimensional array* of detectors can record all parts of an astronomical image simultaneously.

Clearly, the physical size of each detector of an array determines how closely spaced its elements, or *pixels* (for *picture element*) can be. Sometimes there must be some inactive area between the sensitive parts of the pixels, sometimes not. Large arrays are more easily manufactured for some types of detectors (e.g. MOS capacitors, photodiodes) than for others (e.g. bolometers and wave detectors). There is an obvious advantage in field of view for detectors with a large number of pixels.

How closely should pixels be spaced? ***Sampling theory*** was originally developed to understand electronic communications in media such as radio broadcasting and music reproduction. The Nyquist theorem states that the optimum sampling frequency of a waveform should be about two times the highest frequency present in the wave. Extending this theorem to the spatial domain means that to preserve maximum detail, pixel-to-pixel spacing should be less than the ***Nyquist spacing***. The Nyquist spacing is one-half the full width at half-maximum (FWHM) of the point-spread function of the telescope. If pixel spacing is larger than the Nyquist value, the resulting ***undersampling*** of the image degrades resolution. (There is little payoff to oversampling an image.)

At the beginning of the CCD era, photographic plates had a clear advantage in pixel number: for a very moderate cost, a photographic plate had a very large area (tens of centimeters on a side), and thus, in effect, contained up to 10^9 pixels. Mosaics of CCD arrays, although quite expensive, now match the size of the largest photographic plates, and (a few) astronomers currently employ mosaics of solid-state arrays of up to 3 billion pixels in the optical, and somewhat smaller mosaics in the NIR. Focal-plane arrays of hundreds of pixels are used on some far-infrared (FIR) and submillimeter telescopes. Radio detectors are almost always single-pixel or few-pixel devices.

8.1.9 Image sampling and degradation

Astronomers go to extremes to improve the resolution of the image produced by a telescope – minimize aberrations, launch the telescope into space, and create active and adaptive optics systems. Two-dimensional detectors like arrays should preserve that resolution, but in practice can often degrade it. Signal can drift from its pixel of origin into a neighboring pixel, or photons can scatter within an array before they are detected. Improper movement of signal from one pixel to another is called ***cross-talk***, and is an important characteristic for an array.

8.2 The CCD

> One morning in October 1969, I was challenged to create a new kind of computer memory. That afternoon, I got together with George Smith and brainstormed for an hour or so. . . . When we had the shops at Bell Labs make up the device, it worked exactly as expected, much to the surprise of our colleagues.
>
> – Willard Boyle, Canada Science and Technology Museum, 2008

When Boyle and Smith (1971) invented the first **charge-coupled devices** at Bell Laboratories in 1969 they quickly recognized the CCD's potential as multi-pixel light detector instead of a computer memory. By 1976, astronomers had recorded the first CCD images of celestial objects.[3] Since that time, the CCD has become a standard component in applications that include scanners, copiers, mass-market still and video cameras, surveillance and medical imagers, industrial robotics, and military weapon systems. This large market has diluted the research and development costs for astronomy. The consequent rapid evolution of the scientific CCD has profoundly revolutionized the practice of optical observational astronomy. This section gives a basic introduction to the principles of operation of the CDD and its characteristics as a detector.

8.2.1 General operation

Recall (Section 7.5) how an MOS capacitor stores photoelectrons in a potential well. A CCD is an array of MOS capacitors (one capacitor per pixel) equipped with circuitry to read out the charge stored in each pixel after a timed exposure. This readout scheme (called "charge-coupling") moves charges from one pixel to a neighboring pixel; pixel-by-pixel shifting is what makes the array a CCD.

The basic idea is simple. Imagine a matrix of MOS capacitors placed behind a shutter in the focal plane of a telescope. To take an image, first empty all the capacitor wells of charge, open the shutter for the exposure time, then close the shutter. While the shutter is open, each pixel accumulates photoelectrons at a rate proportional to the rate of photon arrival on the pixel. At the end of the exposure, the array stores an electronic record of the image.

Figure 8.3 sketches how the CCD changes this stored pattern of electrons into a useful form – numbers in a computer. In part (a) we show the major components of the detector. There is the light-sensitive matrix of MOS capacitors: in this case an array three columns wide by three rows tall. Each capacitor is a pixel. A column of pixels in the light-sensitive array is called a **parallel register**, so the entire light-sensitive array is known collectively as the parallel registers. There is one additional row, called the **serial register**, located at the lower edge of the array and shielded from light. The serial register has one pixel for each column or parallel register (in this case, three pixels). Both the serial and parallel register structures are fabricated onto a single chip of silicon crystal.

Reading the array requires two different charge-shifting operations. The first (Figure 8.3b) shifts pixel content down the columns of the parallel registers by one pixel. In this example, electrons originally stored in row 3 shift to the serial register, those in row 2 move to row 3, electrons in row 1 move to row 2.

[3] The first CCD images reported from a professional telescope were of the planets Jupiter, Saturn, and Uranus, taken in 1976 by Bradford Smith and James Janesick with the LPL 61-inch telescope outside Tucson, Arizona.

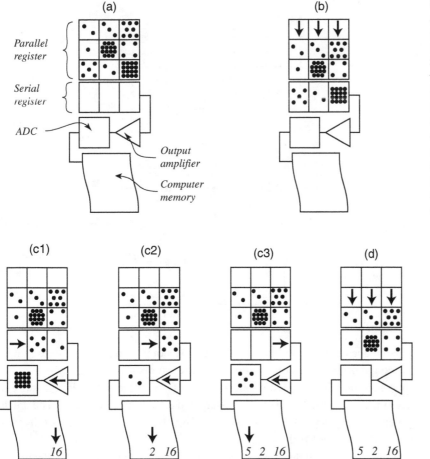

Fig. 8.3 CCD components and readout.
(a) The accumulated photoelectrons in a 3×3 array of capacitors – the parallel register. (b) Shift off the bottom row into the serial register, all remaining rows shift down in the parallel register. (c) Read off the serial register one column at a time. (d) Next row shifts down into the empty parallel register.

The second operation (C1) now reads the newly filled serial register by shifting its contents to the right by one pixel. The electrons in the rightmost pixel shift into a new structure – one or more *output amplifiers* – that ultimately converts the charge to a voltage. This voltage is in turn converted to a binary number by the next structure, the *analog-to-digital converter (ADC)*, and the number is then stored in some form of computer memory. The first stage of the output amplifier is usually fabricated onto the same silicon chip as the registers. The subsequent amplifiers and the ADC are often located in a separate electronics unit. The CCD continues the shift-and-read of the serial register, one pixel at a time (Figures 8.3c2 and 8.3c3), until all serial register pixels have been read.

Now the whole operation repeats for the next row: there is another shift of the parallel register to refill the serial register with the next row (Figure 8.3d); the serial register is in turn read out to memory. The process repeats until the entire array has been read to memory.

Fig. 8.4 Gate structure
in a three-phase CCD.
Two pixels are shown in
cross-section. Collection
and barrier potentials
on the gates isolate the
pixels from each other
during an exposure.
Overlapping gates
produce a gradient
in the barrier region
(dashed curve in lower
figure) that enhances
collection.

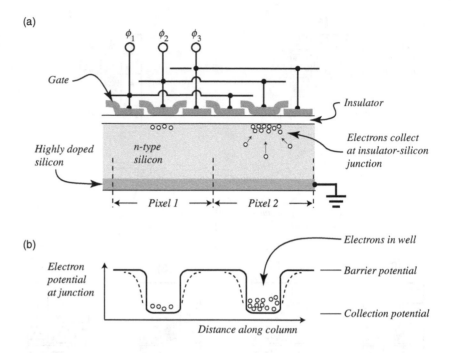

How does the CCD persuade the electrons stored in one capacitor to move to the neighboring capacitor? Many strategies are possible, all of which depend upon manipulating the depth and location of the potential wells that store the electrons. A parallel or serial register is like a bucket brigade. The bucket (potential well) is passed down the line of pixels, so that its contents (electrons) can be dumped out at the end.

Figure 8.4 illustrates one strategy for moving the well. The depth of a potential well depends on the voltage applied to the metal, and is greatest at the Si–SiO$_2$ junction, closest to the metal layer. (See, however Section 8.2.6 on the buried channel CCD.) The figure shows two pixels in the same register of a three-phase CCD. In this device, the metal electrode is separated into three gates, and these are interconnected so that gate 1 of every pixel connects to gate 1 of every other pixel, and likewise for gates 2 and 3. Thus, a single pixel can simultaneously have three separate voltages or phases applied to its front side, producing a corresponding variation in the depth of the potential well, as illustrated in the figure. The interconnection of gates insures that the pattern of well depth is identical in every pixel of the register.

Setting the correct voltages on three separate gates implements both charge-shifting and pixel isolation. For example, during an exposure, phase 2, the voltage on the central metal electrode, can be set to a large positive value (say 15 V), producing what is known as the ***collection potential*** in the semiconductor. The other two phases are set to a smaller positive voltage (say 5 V), which produces the ***barrier potential***. The barrier potential maintains the

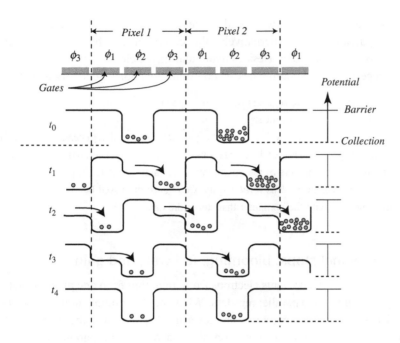

Fig. 8.5 Shifting potential wells in a three-phase CCD. See Figure 8.4 for the corresponding physical structure. Two pixels in the same register (either parallel or serial) are illustrated here. At the end of the shift, electrons stored in pixel 1 have shifted to pixel 2.

depletion region in the silicon, but prevents electrons from drifting across pixel boundaries. Photoelectrons generated in the barrier region of the silicon will diffuse into the nearest deep well under the collection phase and remain there. Each isolated pixel thus stores only charges generated within its boundaries.

Figure 8.5 illustrates the three voltage changes that will shift charges by one pixel. Assume again that the pixels are isolated during an exposure with collection under phase 2 ($\phi_2 = +15$ V) and a barrier under the other phases ($\phi_1 = \phi_3 = +5$ V). At the end of the exposure the time is t_0. Then:

1. At time t_1, gate voltages change so that $\phi_3 = 15$ V and $\phi_2 = 10$ V. The electrons under ϕ_2 will diffuse to the right, and collect under ϕ_3.
2. At time t_2, after a delay that is long enough for all electrons to diffuse to the new location of the deep well, voltages change again, so that $\phi_1 = 15$ V, $\phi_3 = 10$ V, and $\phi_2 = 5$ V. Stored electrons drain from phase 3 of the original pixel to phase 1 of the neighboring pixel.
3. A third cycling of gate voltages ($\phi_1 = 10$ V, $\phi_2 = 15$ V, and $\phi_3 = 5$ V) brings the electrons to the middle of the pixels at time t_3, and the one-pixel shift is complete.

The values of the barrier and collection potentials are somewhat arbitrary, but there are usually some fairly well-defined optimal values. These values, along with the properties of the insulator layer, determine required values of the **clock voltages** (the input values for ϕ_1, ϕ_2, and ϕ_3). An electronic system called the **CCD controller** or **CCD sequencer** sets the clock voltages and manages the very precise timing of their changes. The controller, usually built around a simple microprocessor, is generally housed in the same electronics board as the ADC and output amplifiers. Alternatively, the controller can be a program on a general-purpose computer. Besides manipulating the clock voltages, the controller also performs and coordinates several other functions, generally including:

- clearing the appropriate registers before an exposure or a read;
- opening and closing the shutter;
- controlling the sequence of reads of the parallel and serial registers, including the patterns for special reads (see the discussions of on-chip binning and windowing below);
- controlling the parameters of the output amplifiers and the ADC (in particular, setting two constants called the **bias level** and the **CCD gain** discussed below);
- communicating with the computer that stores the data.

8.2.2 Channel stops, blooming, full well, and gain

The barrier potential prevents electrons from migrating from one pixel to another along a column in the parallel registers. What about migration along a row? In a classical CCD, shifts along a row are never needed, except in the serial register. The CCDs prevent charge migration along a row in the parallel register by implanting (by heavily diffusing a dopant) a very highly conductive strip of silicon between columns. These **channel stops** held, say, at electrical ground, produce a permanent, extra-high barrier potential for stored electrons. Think of a pixel as a square bucket that holds water (or electrons). Two sides of the bucket, those that separate it from the adjacent columns, are maintained by the channel stop and are permanently tall and thin. The other two sides, the ones that separate it from its neighbors on the same column, are not as tall, and can be lowered or moved by "clocking" the gate voltages.

Consider what might happen if a pixel in an array fills with electrons during an exposure. As additional photoelectrons are generated in this saturated pixel, they will be able to spill over the barrier potential into the adjacent wells along their column, but cannot cross the channel stop. This spilling of charge along a column is called **blooming** (see Figure 8.6). Bloomed images are both unattractive and harmful: detection of photons in a pixel with a filled well becomes very non-linear; moreover, blooming from a bright source can ruin the images of other objects that happen to lie on the same CCD column. Nevertheless, in order to optimize the exposure of fainter sources of interest, astronomers will routinely tolerate saturated and bloomed images in the same field.

The maximum number of electrons that can be stored in a single pixel without their energies exceeding the barrier potential is called the CCD's **full well**. The size of the full well depends on the physical dimensions of the pixel, the design of the gates, and the difference between the collecting and barrier potentials. Typical pixels in astronomical CCDs are 8–30 μm on a side and have full well sizes in the range 25 000 to 500 000 electrons.

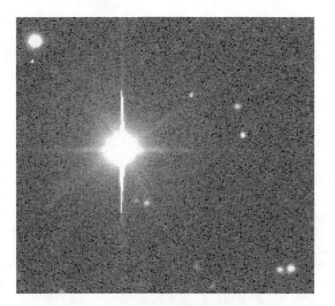

Fig. 8.6 Blooming on a CCD image: the saturated vertical columns are the bloom. The other linear spikes on the bright star image result from diffraction by the vanes supporting the telescope's secondary mirror.

The final output from a scientific CCD is an array of numbers reported by the ADC to the storage computer. The number for a particular pixel is usually called its *pixel content*, and is measured in *ADU*s (analog-to-digital units) or *DN*s (data numbers). Pixel contents are proportional to the voltage the ADC receives from the output amplifier. The *gain* of the CCD is the number of electrons that need to be added to a pixel in order to increase the output contents for that pixel by one ADU.

For example, suppose a particular CCD has a full well of 200 000 electrons, and is equipped with a 16-bit ADC. The ADC is limited to digital outputs between 0 and 65 535 (= $2^{16} - 1$). A reasonable value for the gain might be 200 000/65 535 = 3.05 electrons/ADU. A smaller gain would mean that the CCD is better able to report small differences in pixel content, but would reach *digital saturation* before reaching the electronic full well. One might do this intentionally to avoid the non-linear shoulder in Figure 8.2. At a larger gain, the CCD would reach full well before the output could reach the maximum possible digital signal, so dynamic range would be reduced.

8.2.3 Readout time, read noise, and bias

To maximize DQE, the amplifier and ADC of an astronomical CCD should introduce the smallest possible noise to the output. A technique called *correlated double sampling* (*CDS*) is capable of very low noise operation – only a few electrons per pixel. The noise added by the CDS circuit depends crucially on how quickly it does its job – the faster, the noisier. Another consideration – the time needed for the analog-to-digital conversion – also limits the read time per pixel. Practical times correspond to a pixel sample frequency of 10 to 500 kHz, with higher frequencies producing higher noise. Except for low frequencies, noise added by the amplifier stage is proportional to the square root of the frequency.

The basis of charge-coupled readout is the one-pixel-at-a-time movement of the array contents through a single amplifier, and this is a bottleneck. A low-noise CCD must read out slowly, and the larger the array, the longer the read time. An important difference between scientific-grade CCDs and the commercial-grade CCDs and (especially) CMOS arrays in mobile phones and camcorders is the readout rate – to obtain real-motion video images, an array must read out about 30 times a second. The large read noise that results is usually not objectionable in a consumer device because of the high input level. In contrast, the astronomical input signal is usually painfully low, and a low-noise, *slow-scan* CCD for astronomy may require many tens of seconds to read a single image.

There are some cases in astronomy where the large read noise of a rapid-scan CCD is not objectionable, and in which time resolution is very important – observations of occultations of bright stars or rapid changes in solar features, for example. Also note that a rapid scan is not a problem if no data are being digitized. Thus, reading an array to clear it before an exposure can be done very quickly.

For the usual astronomical tasks, though, it is mainly lengthy readout time that limits the number of pixels in a CCD. (Time spent reading the detector is time wasted at the telescope!) The strategy to shorten read times is to read in parallel. One notion uses multiple amplifiers on a single array. Imagine, as in Figure 8.7a, an array with an amplifier at each corner. The CCD has two serial registers, at the top and bottom. The controller clocks the readout to split the parallel registers – they read out to both ends simultaneously – and does the same with each serial register. Each amplifier reads one-quarter of the array, so the total read time is reduced by the same factor. The image can then be reassembled in software.

A powerful extension of the parallel-read strategy is a mosaic of several very closely spaced but electrically independent CCDs. Figure 8.7a shows an eight-element mosaic read by 16 amplifiers. Gaps between the individual CCDs are of course an issue, but a relatively simple combination of shifted multiple exposures will fill in those parts of an image masked by the gaps on a single exposure. Mosaics have become so important that some modern CCDs are manufactured to be "almost-four-side-buttable" – so that the width of the gaps in a mosaic need be only to 30–100 pixels on all sides. An important early (2000–08) mosaic of this sort was the detector for the *Sloan Digital Sky Survey* (30 devices, each 2k × 2k pixels). A current example is the *HyperSuprimeCam* on the Subaru 8-m telescope (116 2k × 4k devices, 20 sec readout time). The detector for the LSST camera is being constructed (2015) as a mosaic of 189 4k × 4k devices. Each device has 16 separate amplifiers, and a readout time of 2 sec. A major problem with these large-format arrays becomes simple data storage and transfer, since data rates will be on the order of terabytes per night.

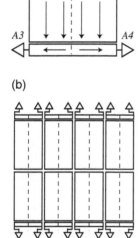

(a)

(b)

Fig. 8.7 Large-format CCD strategies. (a) A large monolithic detector with multiple serial registers and amplifiers (four, in this case). Read time is reduced by a factor equal to the number of amplifiers, and the total CTE is improved. (b) A mosaic of eight arrays butted to form a single large-area detector.

8.2.4 Dark current, cooling, and vacuum enclosures

At room temperature, a CCD is a problematic detector for astronomy. The energy of thermal agitation generates electron–hole pairs in the depletion zone and the resulting steady flow of electrons into the CCD potential wells is called *dark current*. Dark current is bad for two reasons:

1. It adds some number of electrons, N_D, to whatever photoelectrons are produced in a pixel. You must make careful calibrations to subtract N_D from the total. At room temperature, the dark current can saturate a scientific CCD in seconds.
2. Dark current adds not only a background *level*, N_D, but also introduces an associated uncertainty or *noise* to any signal. Since the capture of dark-current electrons into the pixel wells is a random counting process, it is governed by Poisson statistics. The noise associated with N_D dark electrons should be $\sqrt{N_D}$. This noise is more insidious than the background level, since it can never be removed. Dark current *always* degrades SNR.

Lower the temperature of the CCD, and you reduce dark current. The Fermi distribution governs the rate at which dark charges accumulate in a semiconductor pixel, and the exact temperature dependence varies with details of the device, but is roughly of the form:

$$\frac{dN_D}{dt} = A_0 T^b e^{-\frac{E_{eff}}{2kT}} \tag{8.6}$$

Here T is the temperature in kelvin, A_0 is a constant that depends on pixel size and structure, b is a constant between 0 and 3, and E_{eff} is an "effective" band gap energy. A large fraction of dark current in a pixel arises at the Si–SiO_2 interface of the capacitor, where discontinuities in the crystal structure produce many energy states that fall within the forbidden band. Electrons in these interface states have small effective band gaps, and hence produce a large dark current.

A common method for cooling a CCD is to connect the detector to a *cryogen* – a very cold material with a large thermal mass. A very popular cryogen is a bath of *liquid nitrogen (LN2)*, a chemically inert substance that boils at 77 K = −196 °C. Since it is generally a good idea to keep the CCD at a somewhat warmer temperature (around −100 °C), the thermal link between detector and bath is often equipped with a heater and thermostat.

A cold CCD produces difficulties. The band gap energy of silicon increases by about 6% from room temperature to 0 K, and the cutoff wavelength and IR sensitivity of the CCD drop as it cools. The CCD and the LN_2 reservoir must be sealed in a vacuum chamber since a CCD at −100 °C in open air will immediately develop a coating of frost and other volatiles. In addition, the vacuum thermally insulates the LN_2 reservoir from the environment, and prevents the cryogen from boiling away too rapidly. Filling the CCD chamber with an inert gas like argon is a somewhat inferior alternative. Vacuum containers, called *Dewars*, can be complicated devices (see Figure 8.8), but are quite common in large observatories.

Fig. 8.8 A simple Dewar for cooling a detector using liquid nitrogen. This design is common for devices that "look upward," and prevents cryogen from spilling out of the reservoir as the Dewar is tilted at moderate angles.

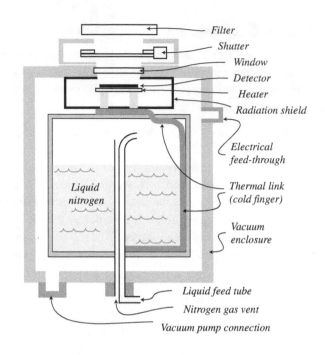

Compact and relatively inexpensive thermoelectric (***Peltier junction***) coolers instead of cryogens require very small Dewar sizes. These solid-state coolers can maintain a detector in the -30 to -50 °C range, where the dark current of an ordinary CCD is still quite high, but where the dark current from an MPP CCD (see Section 8.2.7) is acceptable for many astronomical applications. Such coolers are considerably more convenient to use than cryogens.

At the other extreme, superconducting devices, many small band gap detectors for the mid- and far-infrared, and most bolometers, require extreme cold. ***Liquid helium***, which boils at 4.2 K, is an expensive cryogen that is difficult to handle. Liquid ^{3}He boils at 3.2 K, but is even more difficult and expensive. To avoid the expense of evaporating helium into the air, one option is a ***closed-cycle refrigerator*** that compresses and expands helium fluid in a cycle. If they employ two or three stages, these systems can cool detectors to the 10–60 K range. Special closed systems using helium-3 evaporation can bring small samples to temperatures in the 0.3–3.2 K range.

8.2.5 Charge transfer efficiency

The charge-coupled readout works perfectly only if *all* the electrons in a well shift from pixel to pixel. The fraction of electrons in a pixel that are successfully moved during a one-pixel transfer is the ***charge transfer efficiency***, or ***CTE***. In a single-amplifier CCD, p is the actual number of full pixel transfers needed to read a particular charge packet. If the rows and columns of the parallel registers are numbered from the corner nearest the amplifier, then $p = R + C$, where R and C are the row and column numbers of the pixel in question. The fraction of the original charge packet that remains after p transfers (the total transfer efficiency, or TTE) is just

$$\text{TTE} = (\text{CTE})^p \tag{8.7}$$

The CTE needs to be very close to one. For example, suppose a 350×350 pixel array has a CTE of "three nines" (CTE = 0.999), which in this context is *not* very close to 1. Then the minimum value for p is 2 and the maximum is $p = 350 + 350 = 700$, so the minimum TTE = $(0.999)^{700} = 0.49$; this device will lose over half the charge from the most distant pixel in the array before bringing it to the amplifier. Disaster results. Images will appear streaked along row and column, and photometry becomes inaccurate. Multi-megapixel arrays require CTE values approaching six nines.

What limits CTE? One issue is time – when CCD gate voltages change, electrons need time to diffuse into the new potential minimum. Usually, the required time is shorter than the time needed to complete a low-noise read. However, at very low temperatures, electron velocities can be so small that CTE suffers because of slow diffusion, and so operation below about $-120\ ^\circ$C is inadvisable.

Charge *traps* are a more serious limitation. A trap is any location that will not release electrons during the normal charge transfer process. Traps can result from imperfections in the gates, channel stops, or insulation; from radiation damage, unintended impurity atoms, structural defects in the silicon lattice, or some effects not completely understood. The surface of the silicon layer in contact with the insulator invariably has a large number of charge traps; these are such a serious problem that all modern CCDs are designed so that the potential well excludes the front surface (see Section 8.2.6). Some traps affect only a few electrons each. If scattered throughout the entire body of a CCD, they produce a small decrease in the overall CTE. Other traps can render a pixel non-functional, so that it will not transfer charge in a meaningful way. This compromises the entire column upstream from the trap. Devices with a "bad column" or two are still very useful, but they place additional demands on the observing technique.

Manufacturing defects can also cause a complete failure of charge transfer. The expense of a particular CCD is directly related to the *manufacturing yield* – if many devices in a production run need to be discarded, the cost of a single good device must rise. In the early days of CCD manufacture, yields of acceptable devices of a few percent were not uncommon, but yields have improved as the technology has matured.

8.2.6 The buried channel CCD

The simple MOS/MIS (metal–insulator–semiconductor) capacitor we have been discussing up until now has its minimum electron potential (i.e. the bottom of

Fig. 8.9 A buried channel in a p–n junction capacitor. (a) There is no buried channel in the electron potential when the normal collection phase voltage is applied. If the gate voltage is reduced, as in (b), electrons collect away from the interface. (c) Inverting the voltage on the barrier-phase electrodes pins the surface potential to the channel-stop value and allows a current of holes to flow to neutralize dark-current electrons.

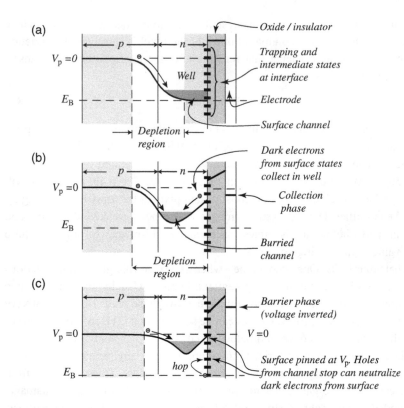

the collection well) at the Si–SiO$_2$ interface. A CCD made of these capacitors is a ***surface-channel*** device, since charge transfer will require movement of electrons close to the interface. The high density of trapping states at that interface makes it impossible to achieve acceptable charge transfer efficiency in a surface-channel CCD. (Values of only 0.99 are typical.) Instead, modern scientific CCDs are ***buried channel CCDs*** (***BCCDs***), in which all electrons collect in a region safely removed from the surface traps, and all charge transfers take place within the unperturbed interior of the semiconductor lattice.

Manufacturers can produce a buried channel by constructing a p–n junction near the semiconductor surface. Figure 8.9a shows the potential energy plot for electrons in an MOS or MIS device in which the semiconductor consists of a thin n-type region (perhaps 300–800 nm thick) layered on top of a much thicker p-type region. Within the semiconductor, the potential exhibits the basic pattern for a junction diode – there is a high-resistivity region depleted of majority charge carriers near the junction, and a potential difference, E_B, across the depletion zone. We connect the p side to electrical ground and the electrode to E_B. In this state, photoelectrons created in or near the depletion zone will be swept into the broad channel in the n-region, where they can still interact with surface traps.

In Figure 8.9b, we create the buried channel by setting the gate voltage to a positive voltage smaller than E_B. This alters the shape of the potential and produces a minimum deeper in the n-region. The required voltage on the gate is the **collection phase**. Note two important features: First, electrons that collect in the well do not contact the surface. This is good – CTE is vastly improved. Second, the capacity of the well is reduced compared to a surface-channel device made from the same material. This is not so good, but worth the trade-off.

8.2.7 The MPP CCD

Interface states at the Si–SiO$_2$ junction remain the major source of dark current in a simple BCCD. Thermal electrons can reach the conduction band by "hopping" from one interface state to another across the forbidden gap. You can eliminate this electron hopping by **pinning** a phase, as in Figure 8.9c. To pin the phase, you set the voltage on the gate to so negative a value that the potential at the interface **inverts**, that is, it reaches the same potential as the back side of the p region, V_p, which is also the same as the potential of the conductive channel stops. Any further reduction in the gate voltage has little effect on the interface potential, since the surface is now held at ground by holes that flood in from the channel stops. The abundance of holes means that thermal electrons are neutralized before they can hop through the inter-face states. Dark current in a pinned phase is reduced by several orders of magnitude.

A **partially inverted** three-phase CCD operates with one non-inverted phase (the collection phase, as in Figure 8.9b), and with the other two phases pinned and serving as the barrier phases, as in Figure 8.9c. Dark current in such a device is about one-third of what it would be in a completely non-inverted mode. If all three phases are pinned, the CCD is a **multi-pinned phase** (*MPP*) device, and dark current less than 1% the rate in non-inverted mode. The obvious difficulty with MPP operation is that there is no collection phase – the buried channel runs the entire length of a column. Multi-pinned phase devices therefore require additional doping under one of the phases to make a permanent collection potential. This is possible because the value of E_B in Figure 8.9 depends on the density of dopants in the semiconductor. In an MPP device, for example, the surface under phase 2 might invert with the collection phase set at −5 V, while the other two (barrier) phases require −7 V for inversion.

With their remarkably low dark currents, MPP CCDs can operate at room temperature for several minutes without saturation. In recent designs, dark rates below 0.1 electron per second are routine at −40 °C, a temperature attainable with inexpensive thermoelectric coolers. An MPP CCD controlled by a standard personal computer is a formidable and inexpensive astronom-ical detector within the financial means of many small observatories, both professional and amateur. As a result, modern observers using telescope apertures below 1 m are making quantitative astronomical measurements of a kind that would have been impossible at the very best observatories in the world in 1975.

The full well capacity of an MPP device is a factor of two or three less than a partially inverted BCCD. Modern MPP devices nevertheless have respectable full wells. Appendix I gives the specifications for a few devices currently on the market. If the larger full well is more important than the reduced dark current, the proper selection of clock voltages makes it possible to run a device designed for MPP operation in a partially inverted mode.

8.2.8 CCD variations

You should be aware of several variants of the basic BCCD that offer some special advantages. Consult the websites of manufacturers (e2v, Hamamatsu, Kodak, Teledyne) and observatories (e.g. pan-STARRS, Keck, ESO, Gemini, Carnegie) for further details.

The *orthogonal-transfer CCD*, or *OTCCD* has a gate structure that permits charge-coupled shifting of pixel contents either along the row or along the column, on either the entire array or on subsections. Orthogonal-transfer CCDs can make small image shifts to compensate for tip–tilt image motion during an exposure, or larger shifts in other applications. The Pan-STARRS project uses OTCCDs in its 1 gigapixel mosaic.

Frame-transfer CCDs drastically shorten the time interval between successive frames. Since it is the amplifier stage that limits the readout rate of a scientific CCD, a FTCCD rapidly reads an acquired frame into a matching set of parallel registers that are shielded from light. The device then reads the shielded frame slowly through the amplifier while the next frame is being acquired.

Electron Multiplication CCDs (EMCCDs) have additional extra-large, deep-well MOS capacitors in a "charge multiplication" extension of the serial register. The device clocks charges from the serial register into these capacitors at a very high voltage, so that the energy of a transferred electron can produce an additional electron–hole pair when it enters a multiplication capacitor. Several hundred multiplication transfers typically produce multiplication gains of 100–1000 before amplification, so read noise is insignificant, permitting rapid readout (1–10 MHz) and true photon-counting at low light levels. EMCCDs are thus often termed **low-light-level CCDs** or *L3CCDs.*

These devices find a ready application in the wavefront sensors of AO systems, where rapid reads are essential and guide star images are faint. Figure 8.10 is a very schematic view of the architecture of an L3CCD that employs frame-transfer, electron multiplication, and multiple amplifiers to speed readout. The sketch is based on the e2V CCD220, which has 240×240 active pixels, a frame rate of 1300 fps, eight amplifiers, and a programmable electron multiplication factor of up to 1000.

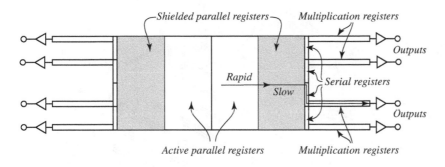

Fig. 8.10 Schematic design for a low-light-level CCD (L3CCD).

8.2.9 CCD sensitivity issues

We need to address the very practical question of getting light absorbed in the depletion region of the CCD pixels.

Frontside options

The most direct pixel design sends light through the metal gates. Since even very thin layers of most metals like copper or aluminum are poor transmitters, the "metal" layer of the CCD is usually made of highly doped *polysilicon*: silicon in a glass-like, amorphous state – a random jumble of microscopic crystals. A thin (about 0.5 micron) layer of doped polysilicon is both a good electrical conductor as well as relatively transparent – but it does, however, absorb green, blue, and (especially) ultraviolet light. An alternative, doped *indium tin oxide* (*ITO*) has better transparency; ITO electrodes are becoming common, but are somewhat harder to fabricate.

There are two strategies for further improving the short-wavelength QE of a front-illuminated CCD. The first is somehow to make the gate structure more transparent by manipulating the geometry – leaving gaps, perhaps coupled with an array of microlenses (*open-electrode architecture*) – or by replacing some phases with dopants implanted in the silicon itself (the *virtual-phase CCD)*.

The second applies a thin coating of *phosphor* on top of the gates. The useful phosphors are organic molecules that absorb a short-wavelength photon, then de-excite by emitting one or more longer-wavelength photons that easily penetrate the gates. Since phosphors emit in all directions, they will slightly degrade image resolution at short wavelengths. Another drawback is that some phosphors tend to evaporate in a vacuum, especially at high temperatures.

Backthinning

A completely different solution sends the light in through the back (from the bottom of Figure 8.11) of the device, avoiding the gates completely. This *backside illumination* has the advantage that light that would be attenuated by the polysilicon or ITO gates instead will pass directly into the silicon. Since these photons have a short absorption depth, they create photoelectrons mainly

Fig. 8.11 Schematic of a thinned, three-phase CCD. In a conventional CCD, insulated gate electrodes usually overlap, while in an open architecture, gaps more closely follow the pixel pattern. This drawing is of a backthinned device. A front-illuminated device would have a much thicker silicon layer, with the AR coating above the gates.

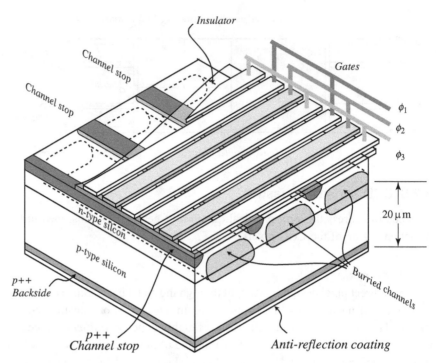

near the back face of the device. This is a serious problem if the photoelectrons cannot diffuse from the back face into the depletion zone without recombining, so the semiconductor layer needs to be very thin (10–20 μm). "Thinning" the silicon will in turn reduce its ability to absorb NIR photons, which have a large absorption depth. The final geometry needs to be something of a compromise. Nevertheless, astronomers have generally embraced backthinned CCDs, since they detect a considerably larger fraction of incident photons of all wavelengths than does any frontside-illuminated device (see Figure 8.12). Their main drawback is that they are difficult to manufacture and therefore expensive.

If red and near-infrared QE is very important (and many times it is), the ***deep-depleted CCD*** offers a considerable improvement over the normal backthinned device. Because the depth of the light-sensitive depletion zone is inversely proportional to the dopant concentration, use of a lightly doped (high-resistivity) silicon layer means that the total layer thickness of the CCD can be increased to about 50–200 μm. The thicker detector has greater long-wavelength ($\lambda >$ 500 nm) sensitivity, but more inter-pixel cross-talk. However, achieving the required resistivity can be difficult, and susceptibility to cosmic rays increases with depletion depth. In a normal, 20-micron backthinned CCD, considerable IR light reflects from the back of the electrodes and can create interference fringes in the background of an image, so an additional advantage of DDCCD's good IR absorption is that fringing is minimized.

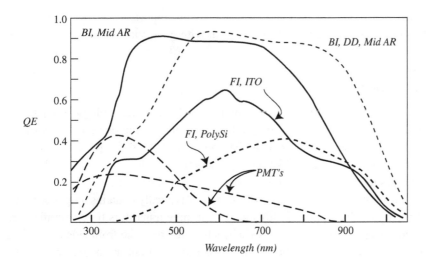

Fig. 8.12 Efficiencies of light detection for various illumination strategies in a CCD, and photocathode choices in a PMT. Curves are representative of the extremes. The abbreviations BI and FI indicate back- and front-illuminated CCDs. The figure shows the QE curves for a normal thinned device with a mid-band AR coating, a deep-depletion (DD) CCD with a near-infrared coating, ITO and polysilicon front-illuminated CCDs. The two photomultiplier tubes (PMTs) are very high-efficiency bi-alkali photocathodes with different spectral sensitivities.

Anti-reflection coatings

An anti-reflection (AR) coating is most effective for light of a particular wavelength, so a CCD designer must choose the coating with the intended use of the detector in mind. (For silicon with index $n = 4$, the usual coating material is HfO_2, $n = 1.7$.) Often CCD manufacturers offer a choice of coatings to enhance either the short-wavelength, mid-wavelength, or NIR response (see Figure 8.12).

8.2.10 Drift scanning and time delayed integration

An important technique in survey astronomy uses the readout characteristics of the CCD to collect data over large areas of the sky with little of the usual array read-time penalty. In its simplest form the idea is to mount the CCD so that columns are oriented east–west and hold the telescope in fixed position (no tracking). Then open the shutter and leave it open all night. Stars, of course, will drift across the detector at the sidereal rate, but if the CCD is also read out at a matching rate, the signal a star deposits in one pixel will shift to the next pixel down the column at the same time the star image does, so there is no trailing of the image, and total integration time for every object will be the time needed to drift across all rows of the detector.

Drift scanning will work in any situation in which there is intentional uniform linear motion of the image relative to the CCD (e.g. the *Gaia* spacecraft). In the literature, **time delayed integration (TDI)** usually refers to the particular drift scan case in which a ground-based telescope is motionless. Besides efficiency, there are additional advantages to drift scans – e.g. pixel sensitivity variations are averaged down the columns. There are difficulties in implementing the technique (e.g. you need to be able to read the array at the right speed, and, except at the equator, the paths of star images are not straight

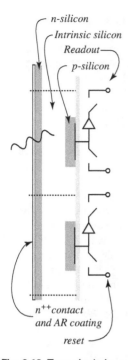

n-silicon
Intrinsic silicon
Readout
p-silicon

*n⁺⁺contact
and AR coating*

reset

Fig. 8.13 Two pixels in a backside-illuminated, monolithic CMOS array. The light-sensitive elements are PIN photodiodes.

lines), but these can be overcome, and many important surveys (e.g. SDSS, Palomar-Quest) have used TDI/drift scan with great success.

8.3 CMOS arrays

Complementary-metal-oxide-semiconductor (***CMOS***) technology is characterized by transistors constructed by diffusing or implanting dopants into a silicon wafer at very specific sites. It is the dominant method for producing modern highly integrated circuits like microprocessors, computer memory, and – what concerns us here – the sensors in consumer digital cameras.

A pixel in a consumer-level CMOS sensor typically contains a light-sensitive element (usually a photodiode) and at least three transistors (one amplifier, one "read" switch, and one "reset" switch) fabricated onto the front side of a silicon base. This design means that the pixels can be read out in parallel, rather than one at a time. Because they use a very mature technology supported by mass production for a huge market, CMOS sensors are less expensive than CCDs of the same size. They consume less power, and also read out more rapidly (around 70 megapixels per second) and more flexibly (random access to a pixel is possible, as are electronic shutters). CMOS sensors have not seen much use in astronomy at optical wavelengths, since historically they had serious inferiorities to CCDs: much higher read noise, dark current, pixel-to-pixel charge diffusion, and (perhaps fatally) lower QE because of the real estate occupied by transistors. However, CMOS sensors are gradually becoming competitive. They are highly amenable to a hybrid architecture (see Section 8.4). Thinning and backside illumination have increased QE, and low noise amplification has reduced read noise to close to CCD levels. This CMOS low-light-level technology is relatively new, so cost is an issue, but CMOS detectors for astronomy could displace CCDs for some applications in the very near future. See Figure 8.13.

8.4 Infrared arrays

Although modern CCDs are usually the astronomical detector of choice in the visual and very near IR, the large size of the forbidden band gap of silicon blinds them to all light with wavelength longer than 1.1 μm. The development of infrared-sensitive arrays of materials with smaller band gaps has faced great technical difficulties, but advances have come quickly. Capable arrays have had an even greater impact in the IR than the CCD has had in the optical. Prior to the CCD, optical astronomers had an excellent, although inefficient, multi-pixel detector – the photographic plate. Prior to the introduction of infrared arrays in the 1980s, infrared astronomers had only single-pixel devices. Different wavelength regions in the infrared place different demands on detector technology.

We first make a brief survey of these differences, then examine the general method of infrared detector fabrication. Chapter 6 of Glass (1999) gives a general qualitative discussion of infrared technology, and chapter 11 of McLean (2008) gives a more technical treatment, as does Rieke (2008).

8.4.1 Detectors at different wavelengths

In general, IR astronomy becomes more difficult at longer wavelengths. Atmospheric transparency decreases and background levels rise. Smaller band gap materials become more difficult to fabricate and the cooling needed to reduce dark current becomes more demanding. Figure 8.14 shows the currently dominant detector choices as a function of wavelength as well as the nomenclature for photometric bands and for different wavelength regions. Astronomy literature tends to distinguish only three regions: near- mid-, and far-IR, with divisions at 5 and 30 µm, but the physics and engineering literature tends to make finer and slightly inconsistent subdivisions (SWIR, ..., etc.). Nomenclature, and obviously, astronomical band locations, are closely related to the transmission characteristics of the atmosphere. When they first appeared, astronomy IR arrays contained only a few pixels (one of the first was a 58×62 InSb array installed on the 3.8-m UKIRT on Mauna Kea in 1986) but, by 2015, manufacturers were producing 4096×4096 pixel 3-, and 4-edge buttable arrays for SWIR, and astronomers were assembling infrared-sensitive mosaics that were only an order of magnitude smaller in size than contemporary CCD-based devices. (The ESO VISTA telescope camera has a 67 Mpixel mosaic sensitive to 2.5 µm).

Silicon IR and SWIR (0.72–1.1 and 0.9–2.5 µm)
The maximum operating temperature needed to hold the dark current of a semiconductor-based detector to an insignificant level (< 0.1 electron s^{-1})

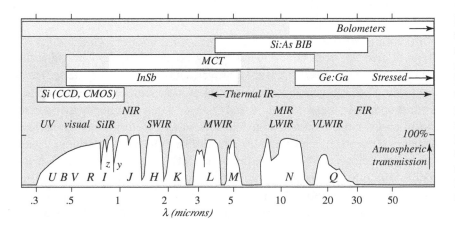

Fig. 8.14 The bottom of the figure gives the atmospheric transmission at a very good site (Mauna Kea), as a function of wavelength, as well as the photometric broadband names. The middle portion shows the names applied to different regions, and the upper parts show the dominant detector types employed.

depends directly on the band gap energy and inversely on the cutoff wavelength (as well as doping levels and applied voltages):

$$T_{max} \simeq \frac{200\,K}{\lambda_c[\mu m]} \qquad\qquad (8.8)$$

The wavelength 2.5 μm constitutes an important threshold for IR astronomy. At wavelengths shorter that this threshold, detectors can be cooled by liquid nitrogen evaporation (77 K), rather than by more expensive cryogens. Moreover, the dominant source of background light shortward of this threshold is emission from excited OH molecules in the upper atmosphere, whereas at longer wavelengths, blackbody radiation from the local environment, including the telescope, dominates.

As we have seen, deep-depleted silicon CCDs are useful in the NIR below about 1.1 micron (the I, and SDSS z and y bands). At longer wavelengths, in the SWIR (J, H, K bands), the most widespread detectors are arrays of junction photodiodes made of either indium antimonide (**InSb** often pronounced "ins-bee") or $Hg_{(1-x)}Cd_xTe$ (mercury cadmium telluride, or **MCT**). The pure form of HgTe is a metal, while pure CdTe has a band gap of 1.6 eV. The band gap of the $Hg_{(1-x)}Cd_xTe$ alloy depends on x, the cadmium telluride fraction, and provides a cutoff wavelength of 2.5 μm at $x = 0.41$ and 100 μm at $x = 0.17$. So far, however, the low-x MCT required for cutoff wavelengths longward of 18–20 μm has been difficult to fabricate. Several modern IR arrays of both InSb and MCT have QE values greater than 70%, read noise below 15 electrons/pixel, and reasonably low dark currents when properly cooled. Both materials are capable of detecting visible light down to 0.4–0.5 μm, although they have not been exploited much there. InSb is more easily fabricated, and 4–16 megapixel Aladdin InSb arrays manufactured by Raytheon are used in many instruments. However, the adjustable λ_c of MCT is very attractive, and liquid N_2-cooled 4- and 16-megapixel arrays in the HAWAII series built Teledyne, with λ_c intentionally set at 2.5 μm, are becoming quite widespread at larger observatories.

Thermal infrared

Progress here has been more modest, partly because high background levels limit ground-based observing at these wavelengths and closed-cycle refrigeration below 77 K is required. Furthermore, at even the best sites, the atmosphere is either opaque or – in the L, M, N, and Q bands – only marginally transparent. Useful observations require a large telescope at an excellent site (Mauna Kea, Atacama, Antarctica), a high-altitude aircraft like SOFIA, or an orbiting observatory like Spitzer or JWST. Both InSb and MCT are capable in the 3–5 μm region. The NIRCam (sensitive to 5.0 μm) on JWST, for example, utilizes assemblies of 4-megapixel MCT arrays. In the N and Q bands, however, the most advanced arrays are of blocked impurity band (BIB/IBC) photoconductors, usually fabricated from silicon doped with antimony (Si:Sb) or arsenic (Si:As). The MIRI instrument on JWST, sensitive in the 5–28 μm range, for example, uses 1024 × 1024 pixel Si:As arrays. Similar devices are appearing on large telescopes at good sites, e.g. the upgrade for the VISIR camera at the ESO VLT.

Beyond 28 μm, the Earth's atmosphere is completely opaque (there is a weak and erratic window at 40 μm at high-altitude sites). In general, extrinsic detector arrays have been the most useful in this region. The Spitzer Space Telescope, for example, carried 32 × 32 pixel arrays of Ge:Ga (cutoff near 115 μm) and a 2 × 20 stressed[4] Ge:Ga array (cutoff near 190 μm). Mosaics of Spitzer-sized devices are under construction. For even longer wavelengths, observers have used small arrays of bolometers (discussed in Section 8.6).

8.4.2 Infrared detector construction

Building an infrared array of photon detectors of any of the types discussed above is different from building a CCD in an important way. Charge-coupled devices are based on a mature technology. Buoyed by the ballooning market in computers and consumer electronics over the past 40 years, manufacturers have refined their skill in the photolithographic fabrication of electronic components (primarily transistors) based on p–n junctions in silicon, silicon dioxide insulation, and metal connections. Expertise with more difficult materials like InSb, MCT, and germanium is limited in comparison. That infrared arrays exist at all is due in large part to their applicability to battlefield imaging, surveillance, and remote sensing. Because building electronics is so much easier in silicon, almost all modern infrared arrays are built as two-layer *hybrids*: one layer is composed of the infrared-sensitive material, the other, made of silicon, provides the CMOS electronics for reading the signal.

Figure 8.15 sketches an example: a schematic cross-section of two pixels of an MCT NIR hybrid array. The top layer is the silicon readout array, which contains several CMOS field-effect transistors (MOSFETs) at each pixel. The lower layer contains the infrared-sensitive material – in this case a p–n photodiode at each pixel. The total thickness of the MCT is quite small, and it is grown or deposited by molecular beam on an inert, IR-transparent substrate that provides mechanical strength, like CdZnTe or sapphire. Initially, the IR-sensitive and CMOS layers are manufactured as separate arrays. A small bump of the soft metal indium is deposited on the output electrode of each photodiode. A matching bump of indium is deposited on the corresponding input electrode

---

[4] Creation of a majority carrier in p-type material requires breaking an atomic bond and remaking it elsewhere (movement of an electron from the valence band to an acceptor state). It is easier to do the bond breaking (it takes less energy) if the crystal is already under mechanical stress. Thus, the cutoff wavelength of a stressed p-type crystal is longer than for an unstressed crystal. Maintaining the proper uniform stress without fracturing the material is a delicate operation.

Fig. 8.15 Cross-section of
two pixels of a hybrid
array of MCT photodiodes.
Each diode connects to
the silicon readout
circuits, which consist of
four to seven transistors,
through an indium
bump conductor.

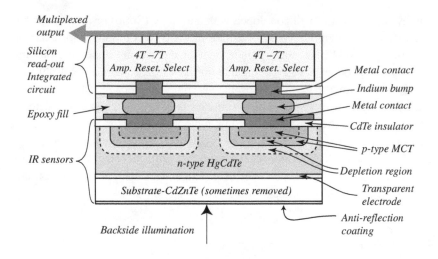

Multiplexed output

Silicon read-out Integrated circuit

Epoxy fill

IR sensors

4T –7T Amp. Reset. Select

4T –7T Amp. Reset. Select

Metal contact

Indium bump

Metal contact

CdTe insulator

p-type MCT

Depletion region

Transparent electrode

Anti-reflection coating

n-type HgCdTe

Substrate-CdZnTe (sometimes removed)

Backside illumination

of each silicon readout circuit. The silicon and the infrared-sensitive arrays
are then matched pixel to pixel and pressed together, so the indium bumps
compress-weld against their mates to make good electrical contact. The spaces
between bumps can then be filled with epoxy to secure the bond, and the
substrate removed to enhance sensitivity to short wavelengths.

There are pitfalls in making arrays using this "bump-bonding" approach –
The high pressure to establish the indium bond can crack the array, some bonds
may fail, and the two layers can de-laminate because of differential thermal
expansion. Nevertheless, the technique is maturing, and the cost, especially of
NIR arrays, has dropped as yields have improved.

Reading an infrared array differs fundamentally from reading a CCD. There
is no pixel-to-pixel charge transfer: each pixel sends output to its individual
readout integrated circuit (***ROIC***) in the silicon layer. Since one of the tasks of
the silicon layer is to organize the multiple signals from all pixels into a single
stream of data from the amplifier, the layer is often called the ***multiplexer*** or
MUX. Many multiplexers, especially in large arrays, read to several (perhaps
four, but sometimes many more) data lines simultaneously. Important differences from CCDs include:

- Since a pixel does not have to (nor is it able to) pass charge to and from its neighbors,
  a "dead" pixel (caused, for example, by a failure in the bump bond) will not kill the
  entire upstream column, as it might in a CCD. Although saturation occurs, there is no
  "blooming" penalty.

- Since readout is separate from sensing, reads can be non-destructive, and the same
  image read several times. Moreover, the array can be read out while the infrared layer is
  still responding to light.

- Very high background levels invariably hamper infrared observations from the ground.
  This forces very short (0.1–10 seconds) integration times to avoid saturation. To cope

with the resulting data rate, controllers often co-add (average) many of the short exposure images and save only that result.

- Many infrared sensors are somewhat non-linear, so calibration for linearity is a much greater concern with an infrared array than it is with a CCD.
- Dark currents in infrared arrays can be a severe problem without proper cooling. Cooling systems become more elaborate with increasing wavelength.

In the thermal IR, careful design of a cold enclosure to shield the detector from the infrared light flooding in from its warm (and therefore glowing) surroundings is essential. These hot surroundings include filters, windows, and the telescope secondary structure and optics. The secondary mirrors of ground-based dedicated infrared telescopes are designed to be as small as possible. Sometimes IR secondaries include a small flat mirror mounted at the center of the secondary, angled to reflect light from the cold sky rather than the warm hole in the primary. IR sky background can vary on short timescales. A "chopping" secondary can cope with this by tilting back and forth between the object position and a clear background position.

8.5 Photo-emissive devices

A few devices based on the vacuum photoelectric effect compete with or enhance CCDs in special circumstances. In this section we examine three of them.

8.5.1 The photomultiplier tube

One disadvantage of the simple vacuum photodiode described in the previous chapter (Figure 7.23) is low signal level. The **photomultiplier tube** (**PMT**) is a vacuum device that increases this signal by several orders of magnitude. Figure 8.16 illustrates its operation. In the figure, a voltage supply holds a semi-transparent photocathode at large negative voltage, usually one or two kilovolts. A photon absorbed by the cathode ejects a single electron. In the vacuum, this electron

Fig. 8.16 A simple photomultiplier tube. The potential of the first dynode accelerates a single photoelectron emitted from the cathode. Its impact releases several secondary electrons, which accelerate and hit dynode 2, releasing another generation of secondaries. After (in this case) eight stages of amplification, a large pulse of electrons flows through the anode and load resistor to ground.

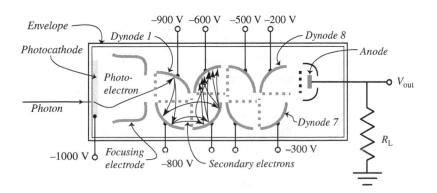

(a)

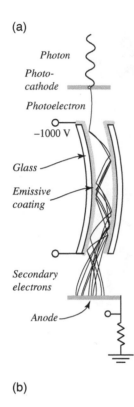

(b)

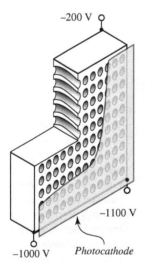

Fig. 8.17 (a) A single microchannel. (b) A closely packed array of channels forming an MCP.

accelerates toward the more positive potential of a nearby electrode called a **dynode**, which is coated with a material (e.g. Cs_3Sb, CsKSb, BeO, GaP) that can easily release electrons to the vacuum if hit by an energetic particle. The original photoelectron impacts the dynode with 100 eV or so of kinetic energy and ejects a number, δ (usually between 2 and 10) of secondary electrons. The group of electrons ejected from the first dynode then accelerates to the second dynode, where each first-dynode electron produces δ second-dynode electrons. The process continues through n dynodes, until pulse of δ^n electrons lands on the anode of the PMT.

The single-channel PMT was the detector of choice for precise astronomical brightness measurements from 1945 until the advent of CCDs in the early 1980s. Photomultipliers have become rare at observatories, and have few advantages over CCDs. One advantage is response time. The temporal spread of a single pulse at the anode limits the shortest interval over which a PMT can sense a meaningful change in signal. Pulse widths are so narrow (5–10 nanosecond) for many PMTs that they can, in principle, detect signal changes as rapid as a few milliseconds. The response time of a CCD, in contrast, is several tens of seconds for a standard slow-scan device, with quicker response possible only with increased noise.

8.5.2 The microchannel plate

Figure 8.17a shows an important variation on the PMT. Take a glass capillary with a diameter between 5 and 25 μm, and a length around 40 times its diameter. Coat the inside surface of this tube with a semiconductor that has good secondary electron-emitting properties, and connect the ends of the channel coating to the voltages as shown. You have created a **microchannel**. Place this microchannel assembly in an evacuated chamber between a photocathode and an anode, and it can serve in place of the dynode chain of a PMT. A photoelectron from the cathode will accelerate toward the upper end of the channel, where it strikes the wall and generates a spray of secondary electrons. These secondary electrons will in turn strike the channel wall further down, multiplying their numbers. After several multiplications, a large pulse of electrons emerges from the end of the microchannel and accelerates to the anode.

A **microchannel plate** (MCP), as illustrated in Figure 8.17b, consists of an array of up to several million microchannels closely packed to form a plate or disk several millimeters in diameter and less than a millimeter thick. The electrical contact that coats the front surface can be made of a metal that has some secondary-electron emission capabilities, so that photoelectrons that do not strike the inside of a channel might still be detected via emission from the contact. You can make a high gain but very compact PMT by sandwiching several MCPs between a photocathode and anode in a vacuum enclosure.

Such *MCP PMTs*, operated as single-channel devices, have an advantage in size, power consumption, response time, and stability in magnetic fields compared to dynode-based devices.

The MCP, however, is most valuable as a component in a two-dimensional detector. Various anode configurations or electron-detection devices can generate an output image that faithfully reproduces the input on the cathode of an MCP. The *multi-anode microchannel array detector* (*MAMA*) is an example. In the MAMA, the anode of the MCP PMT is replaced with two planes of parallel wires that form an x–y grid. A pulse of electrons emerging from a particular microchannel will impact with maximum intensity on one x-wire and one y-wire. Special circuitry then increments the signal count at the corresponding x–y address in the output image.

The MAMA detectors are especially useful at short wavelengths where the DQE of the device can be very high if it is equipped with a "solar-blind" photocathode insensitive to visual and infrared photons. Space astronomy has employed MAMA detectors to great advantage in the detection of X-rays and far-ultraviolet light. Although silicon CCDs are also sensitive at these wavelengths, they suffer from high sky background levels from starlight and scattered sunlight that cannot be completely removed by filtering.

8.5.3 Image intensifiers and the ICCD

An *image intensifier* is not a detector, but a vacuum device that amplifies the brightness of an image. Because military interest in night vision drives the development of intensifiers, the military terminology (Generation I, II, III,[5] etc.) for different designs has become standard. Figure 8.18 shows a Generation II intensifier coupled by optic fibers to a CCD. The intensifier resembles a MCP PMT, but it has a phosphor screen instead of an anode. A photoelectron leaving the cathode enters a stack of MCPs and produces a pulse of high-energy electrons that excites multiple molecules in the phosphor. These then de-excite by emitting photons, usually in the 430–550 nm range. The location of the phosphor emission maps the location of the original photo-absorption on the cathode. A single input photon can generate 10^4 to 10^7 phosphor photons.

As shown in Figure 8.18, the *ICCD* is a device in which the intensifier output phosphor is optically coupled to a CCD. The image on the phosphor is not only brighter than the one that would arrive at non-intensified CCD, it also emits photons of a different wavelength. You can select a solar-blind cathode with a QE of 15% sensitive in the range 180–320 nm, for example, and a phosphor that

---

[5] Generation I devices (now obsolete) used electric or magnetic fields to accelerate photoelectrons from a cathode and then refocus them directly onto the phosphor. Generation II and III devices use an MCP to form the image as described in the text. Generation III devices have advanced photocathodes sensitive in the NIR.

Fig. 8.18 The ICCD. The photocathode is in the image plane of the telescope. The image-intensifier stage produces an image at the phosphor, which is transmitted by a fiber bundle or lens to the CCD. A very thin aluminum film behind the phosphor screen increases forward transmission and reduces the chance that light from the phosphor will feed back to the photocathode.

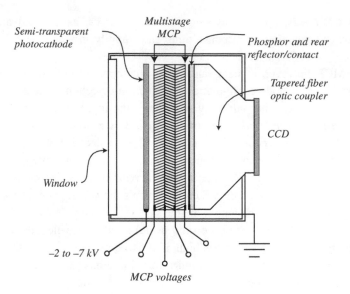

emits near the CCD QE peak in the red. The ICCD thus detects sources that the bare CCD would find absolutely invisible.

It is important to understand, however, that although an intensifier will vastly increase signal strength and decrease exposure times, it will always decrease the *input* SNR for the CCD. For example, consider a source that produces N_i photons at the photocathode of an image intensifier during an integration. If the input is dominated by photon noise (assume background is negligible) then the uncertainty in the input signal is just $\sqrt{N_i}$. The intensifier output at the phosphor is

$$N_{\text{out}} = gN_i \tag{8.9}$$

where g is the gain factor of the intensification. The variance of N_{out} is therefore

$$\sigma_{\text{out}}^2 = g^2\sigma_{\text{in}}^2 + \sigma_g^2 N_i^2 = \left(g^2 + \sigma_g^2 N_i\right)N_i \tag{8.10}$$

Here σ_g is the uncertainty in the gain. Thus, the SNR at the input and output are

$$SNR_{\text{in}} = \sqrt{N_i}$$

$$SNR_{\text{out}} = gN_i\left[\left(g^2 + \sigma_g^2 N_i\right)N_i\right]^{-\frac{1}{2}} = \sqrt{N_i}/\left[1 + \frac{\sigma_g^2 N}{g^2}\right]^{\frac{1}{2}} \leq SNR_{\text{in}} \tag{8.11}$$

So long as intensifier gain is uncertain, intensification will degrade the SNR. *ICCDs* are thus useful in situations where the primary noise source is NOT photon noise in the signal, and/or where rapid signal changes need to be monitored. In such cases (e.g. read noise or dark-current noise dominant), using an intensifier can improve the DQE of the entire device by decreasing the required exposure times.

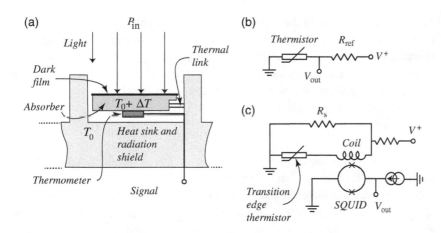

Fig. 8.19 (a) General design for a thermal detector. A thermometer records the increase in the temperature of a light-absorber after it is exposed to a source. A strip of conducting material links the absorber to a large heat sink. The electrical connection to the thermometer may double as the conductive link. (b) A simple circuit to read the thermometer if implemented as a thermistor. (c) A circuit to read a superconducting TES bolometer (adapted from Rieke (2008)).

8.6 Thermal detectors

Thermal detectors work as two-element devices: (1) a thermometer, which senses the temperature increase produced in (2) an absorber, when the latter is exposed to an incoming light beam. The thermal detectors used in astronomy have almost always been **bolometers**, defined as thermal detectors in which the temperature sensor is a **thermistor** (a contraction of the words *thermal* and *resistor*)

Figure 8.19a show a generalized sketch of a bolometer, one that is part of a much larger array. In the figure, the absorber and thermistor are suspended in a shallow well excavated in a heat sink. The heat sink is maintained at temperature T_0. A strip of material with conductance G (in WK^{-1}) provides mechanical support for the absorber and connects it to the heat sink. To make an observation, open the shutter, and the incoming light deposits energy in the absorber at rate P_{in}. After a time, the temperature of the absorber will increase by amount ΔT, a quantity that therefore measures P_{in}.

For example, if the absorber were allowed to reach *equilibrium* with the shutter open, the following condition will apply:

$$P_{in} + P_{elec} = P_{out} = P_{conduct} + P_{radiate} \tag{8.12}$$

$$P_{in} \approx G\Delta T_{eq} + \sigma A\left[\left(T_0 + \Delta T_{eq}\right)^4 - bT_0^4\right] - P_{elec} \tag{8.13}$$

In this equation, A is the effective surface area of the absorbing material, b is the fraction of that area exposed to the heat sink, and σ is Stefan's constant. P_{elec} is the heating produced by the thermistor (e.g. see the circuit in Figure 8.19b), which in general is not negligible. We want to maximize the sensitivity, which is just:

$$\frac{\Delta T_{eq}}{P_{in}} \approx \left\{G + \frac{1}{\Delta T_{eq}}\left[\sigma A(1-b)T_0^4 - P_{elec}\right] + 4\sigma A T_0^3 + \dots\right\}^{-1} \tag{8.14}$$

Thus, it is clear that both the conductance of the link and the temperature of the sink, T_0, need to be kept very low. Indeed, bolometer arrays in the FIR are often operated at liquid helium temperatures or lower.

In making an actual observation, an astronomer also wants the absorber to approach the equilibrium quickly. A short detector *response time* means a large data rate. The time dependence will be given by:

$$\Delta T(t) \approx \Delta T_{eq}[1 - \exp(-tG/C)] \tag{8.15}$$

Here C is the heat capacity of the detector in J K^{-1}. A short time constant requires a small value of C and a large value for G. Choice of the value for G in bolometer design is therefore a compromise between large sensitivity (small G) and large data rate (large G). In general, a thermal detector will employ an absorber that is black or covered with a black film, and having a small heat capacity (and therefore thin). The absorber and the thermometer are often one and the same. The area of the surface ideally would match the expected size of the focal-plane images.

The small signal levels usually characteristic of astronomical measurements have restricted instruments to just a few thermistor materials: extrinsic silicon has been used, but n-type extrinsic germanium (Ge:Ga has a band gap of 0.01 eV) has been the most common choice. As in any semiconductor, the resistance drops as temperature increases. You can monitor this resistance by measuring the voltage drop across the thermistor in a simple circuit, as illustrated in Figure 8.19b: The thermistor is in series with a stable reference resistor whose resistance must be large relative to the thermistor resistance to limit current; some positive heating feedback is helpful ($P_{elec} = I^2R$ in Equation (8.12)) because it speeds the response of the bolometer, but is harmful because it reduces sensitivity.

The situation is somewhat different at temperatures below 5 K, where the mechanism of thermal excitation of electrons across the gap between donor and conduction states becomes unimportant, and amplifier transistors function poorly. A thermometer useful at very low temperatures is the superconducting *Transition Edge Sensor (TES)*, which exploits the very steep drop in conductivity a material undergoes when transitioning from the normal to the superconducting state. Figure 8.19c shows a circuit for reading one pixel of an array of TES bolometers. Because the resistance of the TES is so small, a shunt resistor sets the voltage across the TES to a relatively constant value, and current through the sensor is read by monitoring the magnetic field produced in a coil in series with the sensor. The monitor is a superconducting quantum interference device or SQUID, which produces a voltage proportional to the magnetic field strength surrounding a superconducting current loop.

The 3.5-m Herschel Space Telescope utilizes a 64 × 32 pixel bolometer array in its PACs instrument, sensitive to 130 microns. The array is fabricated in silicon by micromachining techniques and utilizes doped silicon thermistors. The SCUBA-2 array of 10 000 TES pixels operates on the James Clerk Maxwell Telescope on Mauna Kea. It presently operates at 450 and 850 μm (in the "submillimeter" atmospheric windows).

Summary

- An important measure of detector quality is the detective quantum efficiency:

$$\text{DQE} = \frac{(\text{SNR})_{\text{out}}^2}{(\text{SNR})_{\text{perfect}}^2}$$

- Detector have three modes of operation (photon, wave, or thermal). Important detector concepts:

signal	noise	quantum efficiency (QE)
absorptive QE	quantum yield	spectral response
spectral resolution	linearity	saturation
hysteresis	stability	response time
dynamic range	physical size	array dimensions
Nyquist spacing	image sampling	

- The charge-coupled device, or CCD, is usually the preferred astronomical detector at visible wavelengths. Concepts:

parallel registers	serial register	gates
clock voltages	barrier potential	ADU
output amplifier	collection potential CCD	gain
blooming	channel stop	CMOS capacitor
full well	digital saturation	read noise
correlated double sampling	multi-amplifier	dark current
orthogonal-transfer CCD	cryogen	Dewar
buried channel CCD (BCCD)	CTE	traps
multi-pinned phase (MPP)	inverted gate	L3CCD
frontside illumination	backthinned	CCD ITO
deep-depleted CCD	open electrode	microlens
virtual-phase CCD		

- Observational techniques and device performance with infrared arrays are highly dependent on the wavelength region observed. Concepts:

near-, mid-, and far-infrared	InSb	MCT
BIB detectors	Si:Sb	ROIC
MUX	hybrid array	indium bump bond

(continued)

Summary (*cont.*)

- Several important astronomical detectors depend on the vacuum photoelectric effect. Concepts:

  photomultiplier (PMT)   *dynode*   *pulse-counting*
  microchannel plate   *MCP PMT*   *MAMA*
  image intensifier   *ICCD*

- Thermal detector concepts:

  heat sink   *bolometer*   *TES*
  thermistor   *time constant*   *SQUID*

Exercises

> Want to feel better, have more energy and perhaps even live longer? Look no further than exercise.
>
> – Mayo Clinic Website, 2016

1. A photodiode has an overall quantum efficiency of 40% in the wavelength band 500–600 nm. The reflectivity (fraction of photons reflected) at the illuminated face of the detector in this band is measured to be 30%. If this face is treated with AR coatings, its reflectivity can be reduced to 5%. Compute the QE of the same device with the AR coating in place.

2. A certain detector measures the intensity of the light from a stable laboratory black-body source. The signal in three identical trials is 113, 120, and 115 mV. From the blackbody temperature, the experimenter estimates that 10^4 photons were incident on the detector in each trial. Compute an estimate for the DQE of the detector.

3. A photon detector has a QE of q and a quantum yield of y. The uncertainty in y is $\sigma(y)$. Show that DQE = q if $\sigma(y) = 0$, but that DQE < q otherwise.

4. A CCD has pixels whose read noise is 3 electrons and whose dark current is 1 electron per second. The QE of the detector is 0.9. Compute the DQE of a single pixel if 1000 photons are incident in a 1-second exposure. Compute the DQE for the same pixel if the same number of photons is incident in a 400-second exposure.

5. An MOS capacitor observes two sources in the band 400–600 nm. Source A has a spectrum such that the distribution of photons in the 400–600 nm band is given by $n_A(\lambda) = A\lambda^3$. Source B has a distribution of photons given by $n_B(\lambda) = B\lambda^{-2}$ in the same band. If the two sources generate photoelectrons at exactly the same rate, compute their (energy) brightness ratio. You may assume the detector's QE is not a function of wavelength.

6. Construction of a monolithic 8192 × 8192 pixel CCD array is technologically possible. How long would it take to read this array through a single amplifier at a pixel frequency of 25 kHz?

7. The gate structure for four pixels of a certain orthogonal-transfer CCD is sketched at below. Propose a pattern for (a) assigning gate voltages during collection, (b) a method for clocking voltages for a one-pixel shift to the right, and (c) a method for clocking voltages for a one-pixel shift downwards. Gates with the same numbers are wired together.

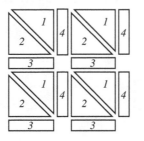

8. Similar to the previous problem – design a CCD with a hexagonal pixel grid. (a) Sketch the gate structure. Propose a pattern for (b) assigning gate voltages during collection, (c) a method for clocking voltages for a one-pixel shift 60° to the upper right, and (d) a method for clocking voltages for a one-pixel shift downwards.

9. Describe the appearance of the CCD image of a star field (a) if the camera shutter malfunctions and does not close until 0.01 seconds after the read has begun. (b) Timing on the serial register malfunctions so that CTE on the serial register (only) becomes very low. (c) Cooling on a CCD with appreciable dark current becomes uneven and one corner is 10 degrees colder than the other three.

10. At an operating temperature of 300 K, a certain CCD exhibits a dark current of 10^5 electrons per second. (a) Estimate the dark rate, in electrons per second, if this CCD is operated at $-40\,^\circ$C (233 K). (b) Compute the operating temperature at which the dark current will be 10 electrons per second. (Assume the constant b in Equation (8.6) is 1.0.)

11. A CCD has a CTE of "three nines" (i.e. CTE = 0.9990). What fraction of the charge stored in the pixel most distant from the amplifier actually reaches the amplifier if the array is (a) 128 pixels on a side or (b) 2048 on a side?

12. A rapid-scan CCD has a read noise of 200 electrons per pixel. You observe a source that produces 400 photoelectrons spread over 25 pixels. Dark current and background are negligible. (a) Compute the SNR for this measurement. (b) Suppose an image intensifier is available with a gain of 10^4 and a gain uncertainty of $\pm 5\%$. Repeat the SNR computation for the intensified CCD. Should you use the bare or the intensified CCD for this measurement?

13. Consider the general situation in which a bare CCD would record N photoelectrons with a total read noise of R electrons in a given exposure time. An intensifier stage has a gain of g and a gain uncertainty of σ_g. If $g \gg 1$, show that the intensifier will improve the overall DQE in the same exposure time if $R^2 g^2 > \sigma_g\, N$.

14. A single-element bolometer operates with a heat sink at 12 K. The thermal link has a conductance of $G' = 5 \times 10^{-7}$ W K^{-1} and a heat capacity of $C = 3 \times 10^{-8}$ J K^{-1}. (a) Compute the time constant and temperature change after 2 seconds of exposure to a source that deposits 10^{-10} W in the bolometer. (b) If the bolometer is a doped germanium thermistor with a resistance of R_0 ohm at 12 K and effective energy gap of $A = 0.02$ eV, compute the fractional change in resistance due to the exposure in (a).

Chapter 9
Digital images from arrays

All the pictures which science now draws of nature and which alone seem capable of according with observational fact are mathematical pictures.

– Sir James Jeans, *The Mysterious Universe*, 1930

Astronomers normally present the output of a sensor array in the form of a digital image, a picture, but a mathematical picture. One appealing characteristic of a digital image is that the astronomer can readily subject it to mathematical manipulation to extract information.

This chapter first presents some general thoughts about array data and some algorithms for image manipulation. We next examine some procedures for removing image flaws introduced by the observing system, as well as some operations that can combine multiple images into a single image. Finally, we examine the important process of *digital photometry*, and derive the *CCD equation*, an expression that describes the quality you can expect from a digital photometric measurement.

9.1 Arrays

Astronomers usually use *panoramic detectors,* like CCDs or IR hybrids, to record two-dimensional images. Unlike a photographic plate (until the 1980s, the panoramic detector of choice), a CCD is an *array* – a grid of spatially discrete but identical light-detecting elements read out electronically as a unit. Although this chapter discusses the CCD specifically, most of its ideas are relevant to images from other kinds of arrays, including, IR hybrid arrays, energy-resolving arrays, ultraviolet-sensitive devices like microchannel plates, and bolometer arrays used in the far infrared and at other wavelengths.

9.1.1 Pixels and pixel response

A telescope forms an image in its focal plane. At each point (x', y') in the focal plane, the image has brightness, $B(x', y')$, measured in W m^{-2}.

Fig. 9.1 Pixels near one
corner of a detector array.
The shaded region
indicates the pixel at [2,
1], which consists of a
photosensitive region
surrounded by an
insensitive border.

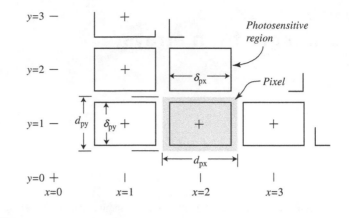

We introduce a panoramic detector (or ***focal-plane array***) to record this image. We assume it is a *rectangular* grid of detectors (other choices are possible), with elements arranged in N_x columns and N_y rows. We denote the location of an individual detector element in this array as $[x, y]$, where it will be convenient to restrict x and y to integer values (running from 1 to N_x and 1 to N_y, respectively). Instead of the phrase "individual detector element," we use the word ***pixel*** (from "picture element"). Pixels exist in the hardware of the array, as well as in the array of data it produces.

Figure 9.1 shows a few hardware pixels of some array. The sensitive area of a single pixel is a rectangle of dimensions δ_{px} by δ_{py}, and the pixels are separated by distances d_{px} horizontally and d_{py} vertically. For most direct-imaging devices, pixels are square ($d_{px} = d_{py} = d_p$) and have sizes in the 5–50 μm range. Linear arrays $\left(N_x \gg N_y \approx 1 \right)$, sometimes used in spectroscopy, are more likely to employ oblong pixel shapes.

If $d_p > \delta_p$ in either direction, each pixel has an insensitive region whose relative importance can be measured by the geometric ***fill factor***,

$$\frac{\delta_{px}\delta_{py}}{d_{px}d_{py}} \tag{9.1}$$

For many CCDs and IR arrays, $\delta_p = d_p$, and the fill factor is unity. Mosaics of arrays will have a reduced overall fill factor. Our detector lies in the focal plane of the telescope, with the x (for the detector) and x' (for the function B) axes aligned. We are free to choose the origin of the primed coordinate system, so can make the center of a pixel with coordinates $[x, y]$ have primed coordinates (we assume square pixels):

$$x' = xd_p, \quad y' = y.d_p \tag{9.2}$$

The light falling on the pixel $[x, y]$ will have a total power, in watts, of

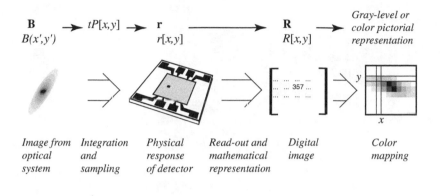

Fig. 9.2 Representations of a digital image.

$$P[x,y] = \int\limits_{(y-\frac{1}{2})\delta_p}^{(y+\frac{1}{2})\delta_p} \int\limits_{(x-\frac{1}{2})\delta_p}^{(x+\frac{1}{2})\delta_p} B(x',y')\mathrm{d}x'\mathrm{d}y' \qquad (9.3)$$

In Equation (9.3), we use square brackets on the left-hand side as a reminder that the detector pixel takes a discrete sample of the continuous image $B(x', y')$ and that x and y can only take on integer values. This **pixelization** or **sampling** produces a loss of image detail if the pixel spacing, d_p, is less than half the resolution of the original image. Such **undersampling** is usually undesirable.

We expose the pixel to power $P[x, y]$ for a time interval, t. It responds by producing a number of photoelectrons or a temperature increase. We call this the photoresponse, $r_0[x,y]$. Note that for photon detectors, including the CCD, $r_0[x,y]$ depends on the *number of incident photons*, not on the energy, so in that case P measures the incident photon flux. To complicate matters, the photoresponse signal usually mixes indistinguishably with that produced by other mechanisms (thermal excitation, light leaks, cosmic-ray impacts, radioactivity, etc.). The pixel gives us, not $r_0[x,y]$, but $r[x,y]$, a total response to all elements of its environment, including $P[x,y]$; see Figure 9.2.

Although it is convenient to think of the CCD response on the microscopic level of individual electrons, this may not be the case for other devices. In some arrays, it will be better to regard $r[x,y]$ as an analog macroscopic property like a change in temperature or conductivity.

9.1.2 Digital images

Our instrument must communicate a quantification of $r[x,y]$ to the outside world. In the case of the CCD, the clock circuits transfer charge carriers through the parallel and serial registers, and one or more amplifiers convert each charge packet to a voltage (the **video signal**). Another circuit, the analog-to-digital converter (ADC), converts the analog video signal to an electronic representation of an integer number, primarily because binary integers are much easier to

Fig. 9.3 A CCD image of the galaxy M51.

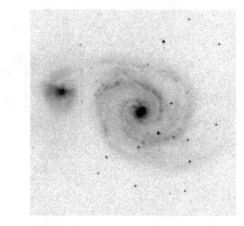

store in a computer. We symbolize the integer output for pixel $[x, y]$ as $R[x, y]$, its *pixel value*.

The entire collection of all $N_x \times N_y$ integers, arranged as a mathematical array to echo the column–row structure of the detector, is **R**, a ***digital image***. Sometimes we call a digital image a ***frame***, or an ***exposure***. We use boldface symbols for an entire array (or image), as in **R**, and the coordinates in square brackets to indicate one element of an array, as in $R[x, y]$. The digital image, **R**, is the digital representation of the detector response, **r**. The relation between **R** and **r** (and especially **B**) may not be simple. Digital images are simply collections of numbers interpreted as images, and often, to help interpret the array, we ***map*** the numbers onto a gray-scale or color-scale and form a pictorial representation.

For example, the "picture" of the nearby galaxy M51 in Figure 9.3 is a representation of a digital image in which a grid of squares is colored according to the corresponding pixel values. Squares colored with 50% gray, for example, correspond to pixel values between 2010 and 2205, while completely black squares correspond to pixel values above 4330. A mapping like Figure 9.3 usually cannot show all the digital information present, since pixel values are often 16-bit integers,[1] while human vision only distinguishes at most a few hundred gray levels, which code as 7- or 8-bit integers.

9.1.3 CCD gain

Quantifying detector response usually means measuring a voltage or current (i.e. an *analog* quantity) and subsequently expressing this as a *digital* quantity.

---

[1] The number of bits (binary digits), n_B, in a computer memory location determines the value of the largest integer that can be stored there. (It is $2^{n_B} - 1$.) Thus, a 16-bit integer can have any value between 0 and 65 535, while an 8-bit integer can have values between 0 and 255.

Hence, each pixel of **R** is said give a count of how many ***analog-to-digital units*** (ADUs) were read from the final detector output. Each pixel value, $R[x, y]$, has "units" of ***ADU***. The terms ***data number*** (DN) and ***counts*** are sometimes used instead of ADU.

The differential change in $r[x, y]$ that produces a change of one ADU in $R[x, y]$ is called the ***gain:***

$$g[x,y] = \text{gain} = \frac{dr[x,y]}{dR[x,y]} \qquad (9.4)$$

In the general case, gain will differ from pixel to pixel, and may even depend on the signal level itself. In the case of the CCD, gain is set primarily by the output amplifier and the ADC, and the astronomer might even set the gain with the controlling software. We expect approximately identical gain for all pixels. Moreover, CCD amplifiers are generally linear, so we usually assume $g[x, y]$ is independent of $r[x, y]$. We often quote a single representative value for CCD gain, a constant with units of *electrons per ADU*:

$$g = \text{CCD gain} = \left\langle \frac{dr[x,y]}{dR[x,y]} \right\rangle [\text{electrons per ADU}], \text{independent of } r, x, \text{ and } y \qquad (9.5)$$

Gain may differ for each amplifier on a multi-amplifier CCD chip, as well as for the components in a mosaic. These variations must be calibrated, as must any pixel-to-pixel gain variations.

9.1.4 Pictures lie

> The world today doesn't make sense, so why should I paint pictures that do?
>
> – Pablo Picasso (1881–1973)

Figure 9.3, the gray-scale map of a CCD image of the galaxy M51, imperfectly represents **R**, the underlying digital image. But even the underlying image is a lie. There are interstellar, atmospheric, and telescopic effects that mask, distort, and destroy information as light travels from M51 to the detector, as well as additions and transformations introduced by the detector itself – all information gains and losses that we would rather not have. Figure 9.4 schematically represents the most obvious elements that might influence the raw digital image, **R**.

For the moment, imagine a "perfect" digital image, **R***. In **R***, the number of ADUs in a pixel is directly proportional either to the energy or number of photons arriving from the source located at the corresponding direction in the sky. The image **R*** is not influenced by any of the elements represented in Figure 9.4. Mathematically, three kinds of processes can cause **R**, the raw image, to differ from **R***, the perfect image:

Fig. 9.4 Additive (ADD), multiplicative (MUL) and non-linear (Non-Lin) effects produce imperfections in detector output. Alterations by optics include intentional restrictions by elements like filters. The local environment may add signal by introducing photons (e.g. light leaks) or by other means (e.g. thermal dark current, electronic interference, cosmic rays).

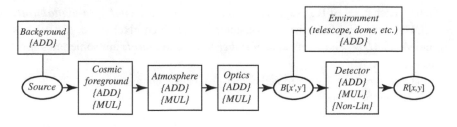

Additive effects contribute or remove ADUs from an output pixel in a way that is independent of the magnitude of $R^*[x, y]$. Examples include:

- background radiation emitted or scattered by the telescope, the Earth's atmosphere, foreground, or background objects;
- impacts of cosmic rays and other energetic particles;
- the ambient thermal energy of the pixel;
- a bias intentionally added to the video signal to optimize amplifier function.

Multiplicative imperfections change $R^*[x, y]$ to a value proportional to its magnitude. Examples include:

- spatial or temporal variations in quantum efficiency or in gain;
- absorption by the Earth's atmosphere;
- absorption, reflection, or interference effects by optical elements like filters, windows, mirrors, and lenses, as well as dirt on any of these.

Non-linear imperfections change $R^*[x, y]$ to a value that depends on a quadratic or higher power of its magnitude. An example would be a quantum efficiency or gain that depends on the magnitude of $R^*[x, y]$. *Saturation*, a decrease in detector sensitivity at high signal levels, is a common non-linear imperfection.

All these imperfections are least troublesome if they are ***flat and stationary***, that is, if they have the same effect on every pixel, every time. Subtracting a spatially uniform background is relatively easy. In contrast, if the imperfection has detail, removing it requires more work. Subtracting the foreground stars from an image of a galaxy, for example, is relatively difficult. Not every imperfection can be removed, and every removal scheme inevitably adds uncertainty. No image ever tells the complete truth.

9.2 Digital image manipulation

> If a man's wit be wandering, let him study the mathematics.
>
> – Francis Bacon (1561–1626)

One of the great benefits of observing with modern arrays is that data take the form of digital images – numbers. Astronomers can employ powerful and sophisticated computing tools to manipulate these numbers to answer questions about objects. We usually first find numerical answers, but eventually construct a narrative answer, some sort of story about the object. Our concern in the remainder of this chapter is to describe some of the computational and observational schemes that can remove the imperfections in astronomical images, and some schemes that can *reduce* those images to concise measurements of astronomically interesting properties. We begin with some simple rules.

9.2.1 Basic image arithmetic

First, some conventions. Boldface letters will symbolize complete digital images: **A**, **B**, **C**, and **D**, for example, are all digital images. Plain-faced letters, like h and k, represent single-valued constants or variables. As introduced earlier, indices in square brackets specify the location of a single pixel, so A [2,75] is the pixel value of the element in column 2, row 75, of image **A**.

If $\{op\}$ is some arithmetic operation, like addition or multiplication, then the notations

$$\mathbf{A} = \mathbf{B}\{op\}\, \mathbf{C}$$
$$\mathbf{A} = k\{op\}\, \mathbf{D}$$

mean that

$$A[x,y] = B[x,y]\,\{op\}\,C[x,y], \text{and}$$
$$A[x,y] = k\,\{op\}\,D[x,y],$$
$$\text{for all indices}, 1 \leq x \leq N_X \text{ and } 1 \leq y \leq N_Y$$

That is, the indicated operation is carried out on a pixel-by-pixel basis over the entire image. Clearly, all images in an equation must have the same size and shape for this to work.

9.2.2 Image dimensions and color

We find it natural to think of digital images as two-dimensional objects – brightness arrayed in rows and columns. But a two-dimensional digital image is just one way to interpret a string of numbers, and there are many cases in which it makes sense to think of images with three or more dimensions. For example, you take a series of 250 images of the same star field to search for the period of a suspected variable. Each image has 512 rows and 512 columns. It makes sense to think of your data as a three-dimensional stack, with dimensions $512 \times 512 \times 250$. You will therefore encounter terms like *data cube* in the astronomical literature. Another common example would be the output of an array of STJ detectors, where spectral distribution would run along the third dimension. Higher dimensions also make sense. Suppose you take 250 images of the field in each of five filters – you then could have a four-dimensional "data-hypercube."

Color images are a special case. Digital color images pervade modern culture, and there are several methods for encoding them, most conforming to the device intended to display the image. For example, each pixel of a color computer monitor contains three light sources: red (R), green (G), and blue (B). The **RGB color model** represents an image as a three-dimensional stack, one two-dimensional digital image for each color. Each pixel value codes how bright the corresponding colored light source should be in that one pixel. The RGB is an additive color model: increasing pixel values increases image brightness.

Subtractive color models are more suited to printing images with ink on a white background. The most common, the **CMYK model**, uses a stack of four two-dimensional images to represent amounts of cyan, magenta, yellow, and black ink in each pixel. In a subtractive model, larger pixel values imply a darker color.

Astronomers *almost never* detect color images directly, but will frequently construct *false color* images as a way of displaying complex data. For example, you might create an RGB image in which the R channel was set by the pixel values of a K-band (i.e. infrared) image, the G channel was set by the pixel values of a V-band (i.e. visual) image and the B channel was set by the pixel values of a far-ultraviolet image. The resulting image would give a sense of the "color" of the object, but at mostly invisible wavelengths.

Astronomers also use *color mapping* to represent the pixel values in a simple digital image. In a color mapping, the computer uses the pixel value to reference a *color look-up table*, and then displays the corresponding color instead of some gray level. Since the eye is better at distinguishing colors than it is at distinguishing levels of gray, a color map can emphasize subtle effects in an image.

9.2.3 Image functions

We expand our notation to include functions of an image. In the following examples, each pixel in image **A** is computed from the pixels with the same location in the images in the right side of the equation:

$$
\begin{aligned}
\mathbf{A} &= -2.5\log(\mathbf{C}) \\
\mathbf{A} &= h(\mathbf{B})^2 + k\sqrt{\mathbf{C}} \\
\mathbf{A} &= \max(\mathbf{B}, \ \mathbf{C}, \ \mathbf{D}) \\
\mathbf{A} &= \mathrm{median}(\mathbf{B}, \ \mathbf{C}, \ \mathbf{D})
\end{aligned}
\tag{9.6}
$$

for each x and y in **A**.

Likewise, the fourth example would compute the median of the three indicated values at each pixel location. You can think of many more examples. We also introduce the idea of a function that operates on an entire image and returns a *single* value. For example, the functions maxPix and medianPix:

$$a = \mathrm{maxPix}(\mathbf{A})$$
$$b = \mathrm{medianPix}(\mathbf{A})$$
(9.7)

will pick out the largest pixel value and the median pixel value in image **A**, respectively. Again, you can think of a number of other examples of functions of this sort.

9.2.4 Image convolution and filtering

The concept of ***digital filtration*** is a bit more complex. Image ***convolution*** is an elementary type of digital filtration. Consider a small image, K, which measures $2V+1$ rows by $2W+1$ columns (i.e. the number of rows and columns are both odd integers). The array **K** is sometimes called the ***kernel*** of the convolution. We define the convolution of **K** on **A** to be a new image, **C**,

$$\mathbf{C} = \mathrm{conv}(\mathbf{K}, \mathbf{A}) = \mathbf{K} \otimes \mathbf{A}$$
(9.8)

Figure 9.5 suggests the relationship between the kernel, the original image, and the result. (1) The center of the kernel is aligned over pixel [x, y] in the original image. (2) The value in each pixel of kernel is multiplied by the value in the image pixel beneath it. (3) The sum of the nine products is stored in pixel [x, y] of the filtered image. (4) Steps (1)–(3) are repeated for all valid values of x and y. Mathematically:

$$C[x,y] = \sum_{i=1}^{(2V+1)} \sum_{j=1}^{(2W+1)} K[i,j] A[(x - V - 1 + i), (y - W - 1 + j)]$$
(9.9)

Note that in convolution, there is a potential problem at the image edges, because Equation (9.2) refers to non-existent pixels in the original image **A**. The usual remedy is artificially to extend the edges of **A** to contain the required

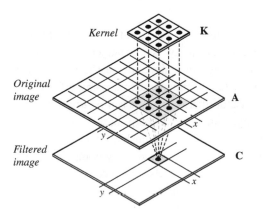

Fig. 9.5 Image convolution operation. The 3 × 3 kernel is aligned over 9 pixels in the original centered on position x, y. The result is the sum of the products of each kernel pixel with the image pixel directly beneath it. The result is stored in pixel x, y of the filtered image.

Kernel — K

Original image — A

Filtered image — C

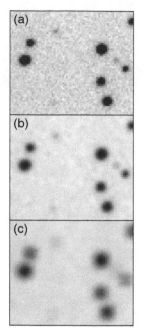

Fig. 9.6 Boxcar filter. (a) Original image of stars in a negative gray-scale map. (b) Convolution of (a) with a 3 × 3 boxcar. (c) Convolution with a 7 × 7 boxcar. Note smoothing effect on the sky.

pixels, typically setting the value of each fictitious pixel to that of the nearest actual pixel.

As an example of a convolution, consider the kernel for the 3 × 3 boxcar filter:

$$\mathbf{B} = \begin{bmatrix} \frac{1}{9} & \frac{1}{9} & \frac{1}{9} \\ \frac{1}{9} & \frac{1}{9} & \frac{1}{9} \\ \frac{1}{9} & \frac{1}{9} & \frac{1}{9} \end{bmatrix} = \frac{1}{9} \begin{bmatrix} 1 & 1 & 1 \\ 1 & 1 & 1 \\ 1 & 1 & 1 \end{bmatrix} \tag{9.10}$$

Figure 9.6 shows an image before and after convolution with a boxcar filter. What happens in the convolution is that every pixel in the original image gets replaced with the average value of the nine pixels in the 3 × 3 square centered on itself. You should verify for yourself that this is what Equation (9.9) specifies. The boxcar is a filter that blurs detail – that is, it reduces the high spatial frequency components of an image. Figure 9.6c shows that a larger-sized boxcar kernel, 7 × 7, has an even greater blurring effect.

Convolutions that blur an image are called *low-pass filters*, and different kernels will blur an image in different ways – a *Gaussian kernel* (whose values are set by a two-dimensional Gaussian function) can simulate some atmospheric seeing effects, for example. Other kernels are *high-pass filters*, and emphasize image detail while suppressing large-scale (low spatial frequency) features. Representative of these is the *Laplacian kernel*. The Laplacian approximates the average value of the second derivative of the intensity map – enhancing pixels that differ from the local trend. The 3 × 3 Laplacian and 5 × 5 Gaussian ($\sigma = 1.25$ pixels) are:

$$\mathbf{L} = \begin{bmatrix} -1 & -1 & -1 \\ -1 & 8 & -1 \\ -1 & -1 & -1 \end{bmatrix}, \ \mathbf{G} = \frac{1}{3.58} \begin{bmatrix} 0.03 & 0.08 & 0.11 & 0.08 & 0.03 \\ 0.08 & 0.21 & 0.29 & 0.21 & 0.08 \\ 0.11 & 0.29 & 0.38 & 0.29 & 0.11 \\ 0.08 & 0.21 & 0.29 & 0.21 & 0.08 \\ 0.03 & 0.08 & 0.11 & 0.08 & 0.03 \end{bmatrix} \tag{9.11}$$

Figure 9.7 shows examples. The factor of 1/3.58 in Equation (9.11) insures that the average value of the array is unchanged after convolution. Other filter kernels can provide edge detection, image sharpening without loss of large-scale features, gradient detection, and embossing effects.

A particularly useful filtering process is *unsharp masking*. The filtered image is the original image minus an "unsharp-mask" image – this mask is a low-pass filtered version of the original. The unsharp mask enhances the high-frequency components and reduces the low-frequency components of the image, emphasizing detail at all brightness levels. Since convolution is distributive, unsharp masking can be accomplished by convolution with a single kernel. For example, convolution with the 5 × 5 identity kernel

$$\mathbf{I} = \begin{bmatrix} 0 & 0 & 0 & 0 & 0 \\ 0 & 0 & 0 & 0 & 0 \\ 0 & 0 & 1 & 0 & 0 \\ 0 & 0 & 0 & 0 & 0 \\ 0 & 0 & 0 & 0 & 0 \end{bmatrix} \quad (9.12)$$

leaves the image unchanged. An unsharp masking filtration operation would be

$$\mathbf{C} = 2(\mathbf{I} \otimes \mathbf{A}) - \mathbf{G} \otimes \mathbf{A} \quad (9.13)$$

or, using the distributive properties of the convolution operation,

$$\mathbf{C} = (2\mathbf{I} - \mathbf{G}) \otimes \mathbf{A} = \mathbf{F} \otimes \mathbf{A} \quad (9.14)$$

So, for example, using the Gaussian kernel in Equation (9.11) an unsharp mask kernel would be:

$$\mathbf{F} = 2\mathbf{I} - \mathbf{G} = \frac{1}{3.58} \begin{bmatrix} 0.03 & 0.08 & 0.11 & 0.08 & 0.03 \\ 0.08 & 0.21 & 0.29 & 0.21 & 0.08 \\ 0.11 & 0.29 & 6.78 & 0.29 & 0.11 \\ 0.08 & 0.21 & 0.29 & 0.21 & 0.08 \\ 0.03 & 0.08 & 0.11 & 0.08 & 0.03 \end{bmatrix} \quad (9.15)$$

Other forms of filtration are not convolutions as defined by Equation (9.9), but do utilize the idea illustrated in Figure 9.5 – the value of a pixel in the filtered image is determined by applying some algorithm to it and its neighboring pixels as described by a kernel. For example, a 3 × 3 *local median filter* sets the filtered pixel value equal to the median of the unfiltered pixel and its eight neighbors.

9.3 Preprocessing array data: bias, linearity, dark, flat, and fringe

When astronomers speak of *data reduction*, they are thinking of discarding and combining data to reduce data volume as well as the amount of information they contain. A single CCD frame might be stored as a few million numbers – a lot of information. An astronomer may only care about the brightness or position of a single object in the frame – information represented by just one or two numbers. Ultimately, he might reduce several hundred of these brightness or position measurements to determine the period, amplitude, and phase of a variable star (just three numbers and their uncertainties) or the parameters of a planet's orbit (six numbers and six uncertainties).

Few astronomers enjoy reducing data, and most of us wish for some automaton that accepts what we produce at the telescope – *raw* data – and gives back digested measurements (magnitudes, colors, positions, velocities, chemical compositions). A great deal of automation is possible, and one characteristic of productive astronomy is a quick, smooth path from telescope to final

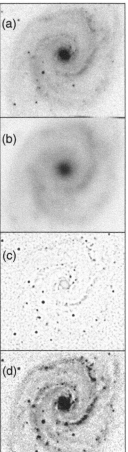

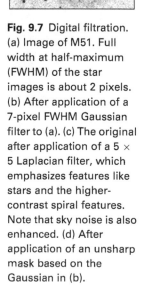

Fig. 9.7 Digital filtration. (a) Image of M51. Full width at half-maximum (FWHM) of the star images is about 2 pixels. (b) After application of a 7-pixel FWHM Gaussian filter to (a). (c) The original after application of a 5 × 5 Laplacian filter, which emphasizes features like stars and the higher-contrast spiral features. Note that sky noise is also enhanced. (d) After application of an unsharp mask based on the Gaussian in (b).

measurement. The path is invariably paved with one or more computer programs working with little human intervention. Indeed, we have seen that the data rates in modern astronomy *require* automation. Eventually, data reduction permits *data analysis* and *interpretation* – for example, what kind of variable star is this, what does that tell us about how stars evolve? The boundaries between reduction, analysis, and interpretation are fuzzy, but each step toward the interpretation stage should become less automatic and more dependent on imagination and creativity.

The first and most automatic steps remove the most obvious imperfections. Data should characterize the astronomical source under investigation, not the detector, telescope, terrestrial atmosphere, scattered light, or any other perturbing element. This section examines the very first steps in reducing array data, and explains reductions that must be made to all CCD data (and most other array data), no matter what final measurements are needed. Other authors sometimes refer to these steps as the *calibration* of the image. I prefer the term *preprocessing*.

Consider, then, a raw image, **R**. Of the many imperfections in **R**, preprocessing attempts to correct for:

- **Bias**. If a detector is exposed to no light at all, and is given no time to respond to anything else in its environment, it may nonetheless produce positive values for a particular pixel, $R[x, y]$, when it is read out. In other words, even when $r[x, y]$, the response of the detector, is zero, $R[x, y]$, *is not*. This positive output from a zero-time exposure is called the bias level, or the *zero level*, and will be present in every frame as a quantity added to the output.

- **Dark response**. If a detector is not exposed to a signal from the telescope, but simply sits in the dark for time t, its response, $r[x, y]$, is not zero. This dark response is the result, at least in part, of thermal effects. In a CCD, electron–hole pairs are created from the energy present in lattice vibrations at a rate that depends on the temperature and the size of the band gap. Like the bias, the dark response adds ADUs to the readout of every frame. Unlike the bias, dark response increases with exposure time.

- **Linearity**. The response of a *linear* detector is directly proportional to incoming signal. All practical detectors have a limited linearity. One of the appealing characteristics of CCDs is the large range of signal over which their response is linear. Even CCDs, however, saturate at large signal levels.

- **Flat field response**. Identical signals generally do not produce identical responses in every pixel of a detector array. This defect can arise because of structural quantum-efficiency differences intrinsic to the array. It can also arise because of vignetting or other imperfections in the optical system like dust, fingerprints, and wildlife (insects turn up in unexpected locations) on filters or windows.

- **Fringing**. Monochromatic light entering the thin layers that compose modern detectors can undergo multiple internal reflections and produce interference patterns. The resulting artificial network of bright fringes perturbs the recorded image background.

As the observer, you want to remove these instrument-dependent characteristics from your images in preprocessing. To do so, you must make some reference observations and appropriate image manipulations. We consider each of the five preprocessing operations in turn. The books by Howell (2006) and by Martinez and Klotz (1998) as well as online instrument manuals treat CCD data reduction in greater detail.

9.3.1 Bias frames

Simply read your array with zero integration time, never exposing it to light (actually, the CCD first clears, then immediately reads out). You have obtained a *bias frame*. The bias frame represents the electronic background present in every frame, no matter how short the integration time. This is uninteresting information; you need to subtract a bias frame from every other frame you plan to use. In practice, one bias frame may well differ systematically from another because of thermal drift in the amplifier stage.

It is good practice to obtain many bias frames. For larger CCD arrays bias readout time may be long enough for much to happen, including cosmic-ray hits, local radioactivity, and electronic interference. For one thing, properly combining several frames will reduce uncertainty about the average level of the bias, as well as minimize the influence of radiation events. For another, the careful observer should monitor the bias level during an observing run, to guard against any drift in the average level, and to make sure any two-dimensional pattern in the bias is stationary.

Assume for the moment that the bias does not change with time, and that you take N bias frames during the run. Call these $\mathbf{Z}_1, \mathbf{Z}_2, \ldots, \mathbf{Z}_N$. How should you combine these frames to compute $\mathbf{Z}_c$, the one representative bias image you will subtract from all the other frames? Here are some possibilities:

(1) **Mean**. Set $\mathbf{Z}_c = \mathrm{mean}(\mathbf{Z}_1, \mathbf{Z}_2, \ldots, \mathbf{Z}_N)$

This is a bad strategy if there are any cosmic-ray hits. Computationally easy, it dilutes the effects of cosmic rays, but does not remove them.

(2) **Median**. Set $\mathbf{Z}_c = \mathrm{median}(\mathbf{Z}_1, \mathbf{Z}_2, \ldots, \mathbf{Z}_N)$

This works well, since the median is relatively insensitive to statistical outliers like the large pixel values generated by cosmic rays. It has the disadvantage that the median is a less robust and stable measure of central value than the mean, and is thus somewhat inferior for those pixel locations not struck by cosmic rays.

(3) **Indiscriminant rejection**. At each $[x, y]$, reject the largest pixel value, then use (1) or (2) on the remaining $(N - 1)$ values.

This removes cosmic rays, but is possibly too drastic, since it skews the central values toward smaller numbers. An alternative is to reject both the largest and the smallest values at each location. This discards two entire images worth of data and skews cosmic-ray pixels to slightly larger numbers.

(4) **Selective rejection**. At each $[x, y]$, reject only those pixels *significantly* larger than the mean, then apply (1) or (2) on the remaining values. To decide whether or not a pixel value is so large that it should be rejected, use a criterion like:

$$Z_i[x, y] > \overline{Z}[x, y] + k\sigma[x, y]$$

where $\overline{Z}$ and σ are the mean and standard deviation of the pixel values, (a) at x, y, or (b) over a segment of the image near x, y, or (c) over the entire image. The value of the constant k determines how selective the rejection will be. For a normal distribution, $k = 3$ will reject 14 legitimate (non-cosmic-ray) pixels out of 10 000.

This is an excellent strategy, but is computationally intensive. Strategy 4b or 4c makes it possible to produce a "clean" $\mathbf{Z}$ from a single frame by replacing the rejected pixel value with the mean or median value of its neighbors. You will undoubtedly think of other advantages or disadvantages to all these strategies, and also be able to compose alternatives. The exact strategy to use depends on circumstance, and we will use the notation

$$\mathbf{Z}_c = \text{combine}(\mathbf{Z}_1, \mathbf{Z}_2, \ldots, \mathbf{Z}_N) \tag{9.16}$$

to indicate some appropriate combination algorithm.

Bias is present in visible and IR photodiode arrays, of course, but these are usually read with a technique called correlated double sampling, in which the output is the difference between a signal near the start of an integration and one near the end. Bias should cancel in the difference, so is never explicitly read.

9.3.2 Overscan and reference pixels

What if the bias drifts over time? The astronomer might compute different $\mathbf{Z}_c$ arrays for different segments of the run, but only if the changes are gradual. A common alternative strategy for CCDs is to use an *overscan*. You produce overscan data by commanding the clocks on the CCD so that each time the serial register is read, the read continues for several pixels *after* the last physical column has been read out.[2] This produces extra columns in the final image, and these contain the responses of "empty," unexposed pixels, elements of the

---

[2] It is also possible to continue to read beyond the last exposed *row*. This means the overscan of extra rows will include the dark charges generated in the parallel registers during the full read time. For arrays operating with significant dark current, this may be significant. Some manufacturers intentionally add extra physical pixels to the serial register to provide overscan data.

serial register that have not been filled with charge carriers from the parallel registers. These extra columns are the overscan region of the image and record the bias level during the read. The usual practice is to read only a few extra columns, and to use the median pixel values in those columns to correct the level of the full two-dimensional $\mathbf{Z}_{l_o}$ image. If ω_i is the overscan portion of image i, and Ω_Z is the overscan portion of the combined $\mathbf{Z}_c$, then the bias frame to apply to image i is

$$\mathbf{Z}_{ci} = \mathbf{Z}_c + (\text{medianP}(\omega_i - \Omega_Z)) \qquad (9.17)$$

Figure 9.8 shows a slightly more complicated application of an overscan. Here the zero level has changed during the read, and shows up in the image most clearly as a change in background in the vertical direction. To correctly remove the bias, the astronomer fit a one-dimensional function (in the y-direction) to the difference $(\omega_i - \Omega_Z)$ and added that function to $\mathbf{Z}$.

9.3.3 Dark current

Even in the absence of illumination, a detector will generate a response during integration time t. This is called the **dark response**. The rate at which the dark response accumulates is the **dark current**. Although primarily a thermal effect, dark current will not be the same for every pixel because of inhomogeneity in fabrication. Some pixels, called "hot" pixels, differ from their neighbors not in temperature, but in efficiency at thermal production of charge carriers.

To calibrate for dark current, you take a long exposure with the shutter closed – this is called a **dark frame**, **D**. In view of the earlier discussion about cosmic-ray hits and uncertainties, it is best to combine several individual dark frames $(\mathbf{D}_1, \mathbf{D}_2, \ldots, \mathbf{D}_M)$ to produce one representative frame:

$$\mathbf{D}_c = \text{combine}(\mathbf{D}_1, \mathbf{D}_2, \ldots, \mathbf{D}_M) \qquad (9.18)$$

The dark frames should be obtained in circumstances (temperature, magnetic environment) as similar as possible to those prevailing for the data frames. If **D** has exposure time t, then you may compute the **dark rate** or **dark current** image as

$$\dot{\mathbf{D}}_i = \frac{\mathbf{D}_i - \mathbf{Z}_c}{t} \qquad (9.19)$$

or, for multiple darks with identical exposure times:

$$\dot{\mathbf{D}}_c = \frac{1}{t} \text{combine}([\mathbf{D}_1 - \mathbf{Z}_{c1}], [\mathbf{D}_2 - \mathbf{Z}_{c2}], \ldots, [\mathbf{D}_M - \mathbf{Z}_{cM}]) \qquad (9.20)$$

If available, you apply an overscan correction for each dark frame as in Equation (9.17). You may then correct for dark current and bias on every data frame by subtraction of the image $(t\dot{\mathbf{D}}_c + \mathbf{Z}_{ci})$. The units for **D** in the above equations are

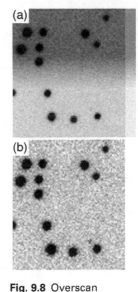

Fig. 9.8 Overscan correction. (a) This frame has a 10-column overscan region on its right edge. The frame in (b) results after the bias frame, corrected for the overscan, is subtracted and the overscan section trimmed from the image. Note that the frame in (b) is displayed with a different gray-scale mapping than in (a).

ADUs. However, dark current for CCDs is usually quoted in units of electrons per second as

$$\dot{\mathbf{d}} = g\dot{\mathbf{D}} \qquad (9.21)$$

where g is the detector gain in electrons per ADU.

Observers routinely cool detectors to reduce dark current and its associated noise. In some cases, the dark rate may be so low that you can omit the correction. In other cases (e.g. thermoelectrically cooled CCDs), dark must be measured. It also appears that the dark current in some multi-pinned phase (MPP) CCDs is somewhat non-linear, which means you must either model the non-linearity or take dark frames whose exposure times match those of the data frames. In the IR, it is usually even more important to match exposure times for dark and data frames.

9.3.4 Detector linearity

All practical devices depart from linearity. If pixels in an array receive photons at rate $P[x, y]$ for time t, the array has a ***linear output*** if

$$\mathbf{R} = \mathbf{Z} + t(\dot{\mathbf{D}} + \mathbf{QP}) = \mathbf{Z} + t\mathbf{E} \qquad (9.22)$$

where $\mathbf{R}$, $\dot{\mathbf{D}}$, and $\mathbf{Q}$ are time-independent arrays: a zero level, dark rate, and efficiency in ADUs per photon, respectively. The typical CCD response resembles the solid curve in Figure 9.9, where the horizontal variable, E, the exposure, is defined in Equation (9.22). The dotted curve is more typical of NIR and MIR photodiode arrays. The CCD in the figure is linear between a threshold exposure, E_T, and an upper limit, E_L. The IR array departs from linearity over a wider range.

The pixels ***saturate*** at response r_s and output R_s in the figure. The threshold effect (see Figure 8.2) for most astronomy arrays is very small, so that $E_T \approx 0$. CCDs' readout circuits are usually set so that digital saturation occurs within the linear region. Non-linearity is an issue for NIR and MIR arrays, but corrections

Fig. 9.9 Linearity: (a) A schematic of the output R, in ADU, and the response, r, in electrons, of a single pixel. The solid curve is a typical CCD response. The dotted curve is typical of an IR array. The gray line is the extrapolated linear response of the IR pixel.

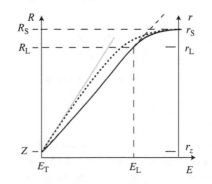

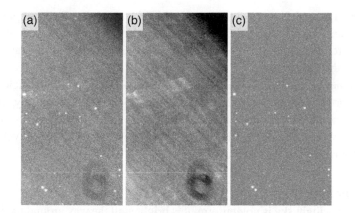

Fig. 9.10 (a) A section of an unprocessed CCD image of a star field. (b) A combined twilight flat for the same section. The two dark rings in the lower right are the shadows of two dust particles on the CCD window. The dark region in the upper right results from vignetting by the filter holder. (c) The section after preprocessing for bias, dark, and flat field.

for HAWAII or RVS MCT and Orion InSb arrays are a few percent when $R[x,y]$ reaches 80% of R_s, as are corrections for the Si:As MIRI array for JWST. When a correction is necessary, one uses a constant source and multiple exposure times to generate a curve like the ones in Figure 9.10. Usually a quadratic fit is sufficient. The corrected response is then

$$\mathbf{R}_L = \text{lin}(\mathbf{R}) = a + b\mathbf{R} + c\mathbf{R}^2 \qquad (9.23)$$

Strictly speaking, the constants a, b, and c could be different for different pixels in an array. In most cases, uncertainties in the pixel-to-pixel variation justify using average values for the entire array.

9.3.5 Flat field

Correcting for pixel-to-pixel variations in device sensitivity is both the most important and the most difficult preprocessing step. Conceptually, the correction procedure is very simple. The astronomer takes an image of a perfectly uniform (or "flat") target with the complete observing system: detector, telescope, and any elements like filters or obstructions that influence the focal-plane image. If the observing system is equally sensitive everywhere, every pixel in this *flat field image*, after correction for bias, dark, and non-linearity, should produce an identical output. Any departure from uniformity in the corrected flat field image will map the sensitivity of the system, in the sense that pixels registering higher counts are more sensitive. Figure 9.10 shows a raw CCD image, a flat field image, and the original image after the flat field correction.

At least three practical difficulties hamper the kind of correction illustrated. First, it is difficult to produce a sufficiently (i.e. 0.5% or better) uniform target. Second, sensitivity variations are in general a function of wavelength. Therefore, the spectrum of the target should match that of the astronomical sources of interest. Spectrum matching becomes especially troublesome with multiple sources of very different colors (stars and background sky, for example) in the same frame. Finally, it is difficult to guarantee that the "complete observing system" remains unchanged between the acquisition of the flat field and acquisition of the data frames.

Commonly, observers employ three different objects as the flat field target: (1) the bright twilight sky, (2) the dark night sky, and (3) a nearby object – usually an illuminated surface inside the observatory dome. Images of these sources are usually termed twilight, dark sky, and dome flats, respectively. Each has advantages and disadvantages.

Twilight flats

The clear twilight sky is not uniform: it is brighter all the way around the horizon than it is near the zenith, and, of course, brighter in the direction of the rising or recently set Sun. By pointing toward the zenith (the exact location of the "flat" spot – usually 5–10 degrees anti-solar from the zenith – is slightly unpredictable), the observer finds a target uniform to about 1% over a one-degree field. It is rare to do better than this. For narrow fields of view, this is acceptable. Clouds usually prohibit good flats.

The advantages of the twilight-sky target are that, for a brief interval, it is the right brightness, and relatively uniform. Moreover, observing in twilight means flat field calibrations do not consume valuable night-time hours. The disadvantages are:

- Large-scale uniformity is limited by the natural gradient in the twilight sky, and small-scale uniformity is limited by the gradual appearance of star images as twilight fades.
- The twilight sky has a spectrum that is quite different from that of most astronomical sources, as well as that of the night sky.
- Twilight brightness and spectrum both change rapidly. The duration of usable twilight is short, and with large arrays (long readout times), or with many filters, it becomes difficult to accumulate sufficient numbers of images.
- Scattered skylight near the zenith has a strong linear polarization, and the flat field of some systems may be polarization sensitive.

Dark-sky flats

The emission from the dark (moonless!) night sky is a tempting source for flat fields. Uniformity is perfect at the zenith and degrades to about 2% per degree at a zenith angle near 70 degrees. Moreover, the spectrum of the night sky is identical to one source of interest: the background that will usually be subtracted from all data frames, an especially important advantage if measuring sources fainter than the background sky. High sky brightness is the rule in the ground-based infrared, where dark-sky flats are a reasonable option.

Offsetting these attractive characteristics are some potent negatives. First, stars are everywhere. Any dark-sky flat will inevitably contain many star images, marring the target's uniformity. The observer can remove star images and construct a good flat with the ***shift-and-stare*** or ***dither*** method. The astronomer takes many deep exposures of the dark sky, while taking care always to "dither" or "nod" or shift the telescope pointing between exposures by many stellar image diameters. He then combines these in a way that rejects the stars. For example, take five dithered images of a dark field. If the density of stars is low, chances are that at any $[x, y]$ location, at most one frame will contain a star image; so computing the median image will produce a flat without stars. More sophisticated combination algorithms can produce an even better rejection of stellar images. The shift-and-stare method should also be employed for twilight flats, since (1) they will usually contain star images and (2) telescope pointing should be shifted back to the flat region near the zenith for each new exposure anyway.

Understand the limitations of shift-and-stare: the scattered-light halos of bright stars can be many tens of seconds of arc in radius and still be no fainter than one percent of the background. Removing such halos, or extended objects like galaxies, can require large shifts and a very large number of exposures.

A second difficulty is that the dark sky is – well – dark. In the visible bands, one typically requires 10^2 to 10^6 times as long to count the same number of photons on a dark-sky frame as on a twilight frame. Sometimes, particularly in broad bands with a fast focal-ratio telescope, this is not a serious drawback, but for most work, it is crucial. Each pixel should accumulate at least 10^4 electrons to guarantee 1% Poisson uncertainty; so dark-sky flats will typically require long exposure times. They are consequently very costly, since time spent looking at blank sky might otherwise be spent observing objects of greater interest.

A modification of shift-and-stare can sometimes help here. It is usually possible to dither and collect many unaligned *data* frames if the objects of interest are small. The median of these unaligned frames is the dark-sky flat, and no time has been "wasted" observing blank sky, since the flat frames also contain the objects of interest.

Dome flats

A source inside the dome is an attractive flat field target, since the astronomer in principle controls both the spectrum and the intensity of the illumination, and observations can be taken during daylight. With very small apertures, it is possible to mount a diffusing light box at the top of the telescope tube, but most telescopes are simply pointed at a white screen on the inside of the dome. In a crowded dome, it is often difficult to set up a projection system that guarantees uniform illumination, the shadow of a secondary may become important in the extrafocal image, and there is an increased possibility of introducing unwanted light sources from leaks or reflections. Nevertheless, dome flats are a very important flat field calibration technique.

Space telescope flats

In space, there is no twilight, no dome, and the sky *really* is dark. Preliminary flat fields obtained on the ground are usually acquired in a simulated space environment. In space, these can be corrected by dithering the same star pattern (e.g. a globular cluster) over the frame to measure relative pixel responses. This is painful, but it works.

Computing simple flats

Assume you have collected N flat field images, all taken through a single filter, using one of the targets discussed above. If $\mathbf{S}_i$ is one if these raw images, then the first step in creating the calibration frame is to remove its bias, dark, and non-linearities:

$$\mathbf{F}'_i = \text{lin}(\mathbf{S}_i) - \mathbf{Z}_i - t_i \dot{\mathbf{D}} \tag{9.24}$$

As before, $\dot{\mathbf{D}}$ is the dark rate, t_i is the exposure time, and $\mathbf{Z}_i$ is the bias. Next, to simplify combining frames, each result should be normalized so that the median pixel has a value of 1.0 ADU:

$$\mathbf{F}_i = \mathbf{F}'_i / \text{medianP}(\mathbf{F}'_i) \tag{9.25}$$

Finally, all normalized frames should be combined to improve statistics, as well as to remove any stars or cosmic-ray events:

$$\mathbf{F}_C = \text{combine}(\mathbf{F}_1, \mathbf{F}_2, \dots, \mathbf{F}_N) \tag{9.26}$$

A different calibration frame must be produced for each observing configuration. Thus, there must be a different flat for each filter used, and a different set of flats whenever the observing system changes (e.g. the detector window is cleaned or the camera rotated).

Compound flats

Given the imperfections of all three ground-based flat fielding techniques, the best strategy sometimes combines more than one technique, applying each where its strengths are greatest. Thus, one uses a dome flat or twilight flat to establish the response of the system on a small spatial scale (i.e. the relative sensitivity of a pixel compared with its immediate neighbors.) Then, one uses a smoothed version of a dark-sky flat to establish the large-scale calibration (e.g. the response of the lower half of the detector relative to the upper half). The idea is to take advantage of the good signal strength and the absence of small-scale non-uniformities (no stars) in the dome or twilight target, but also to utilize the large-scale uniformity of dark-sky targets. To create the compound flat field calibration, assume that $\mathbf{F}_S$ and $\mathbf{F}_L$ are flat frames computed as described in the previous section. Frame $\mathbf{F}_S$ is from a target with good small-scale uniformity, and $\mathbf{F_L}$ from one with good large-scale uniformity. Now compute the ratio image and use a low-pass filter to smooth it:

$$\begin{aligned} \mathbf{C}' &= \mathbf{F_L}/\mathbf{F_S} \\ \mathbf{C} &= \mathrm{conv}\{\mathbf{B}, \mathbf{C}'\} \end{aligned} \tag{9.27}$$

The kernel in the convolution, $\mathbf{B}$, should be chosen to remove all small-scale features from image $\mathbf{C}'$. Image $\mathbf{C}$ is sometimes called an ***illumination correction***. The corrected compound flat is just

$$\mathbf{F} = \mathbf{F_S} \cdot \mathbf{C} \tag{9.28}$$

9.3.6 Preprocessing data frames

Suppose a CCD data frame has output $\mathbf{R_i}$. Preprocessing corrects this image for non-linearity, bias, dark, and flat field:

$$\mathbf{R_p} = \frac{\mathrm{lin}(\mathbf{R}) - \mathbf{Z} - t\dot{\mathbf{D}}}{\mathbf{F}} \tag{9.29}$$

Preprocessing IR array data can differ slightly from the above procedures. For infrared arrays read with double-correlated sampling, the output is the difference between reads at the beginning and the end of an exposure, so bias values cancel and $\mathbf{Z}$ is numerically zero. Also in the infrared, emission from the variable background often dominates the images, so much so that raw images may not even show the location of sources before sky subtraction. A common observing practice then is to "chop" telescope pointing between the object investigated and the nearby (one hopes, blank) sky to monitor its brightness variations. Many infrared-optimized telescopes employ ***chopping secondary mirrors*** that efficiently implement rapid on-source/off-source switching. ***Chopping*** is in this context different from ***nodding*** – manually moving the telescope in the shift-and-stare technique.

In the infrared, then, these high-signal sky frames are usually combined to form the flat field image. A typical preprocessing plan might go like this: use many sky exposures $\mathbf{S_1}, \mathbf{S_2}, \ldots, \mathbf{S_n}$ correct them for non-linearity, then use them to form the combined normalized flat, $\mathbf{F}$, as outlined in Equations (9.24) to (9.26). The processed data frame will be:

$$\mathbf{R_p} = \frac{\mathrm{lin}(\mathbf{R}) - t\dot{\mathbf{D}} - a_{\mathrm{sky}}\mathbf{F}}{\mathbf{F}} \tag{9.30}$$

The constant a in this equation is selected so that $a\mathbf{F}$ is the dark subtracted sky image at the time R was acquired. It could be computed as:

$$a_{\mathrm{sky}} = \left\{ \mathrm{medianP}\left(\frac{\mathrm{lin}(\mathbf{S_i}) - t\dot{\mathbf{D}}}{\mathrm{medianP}(\mathbf{F})} \right) \right\}_{\mathrm{sky}} \tag{9.31}$$

where the median pixel values are computed only over the background clear of sources.

9.3.7 Fringing

Monochromatic light can produce brightness patterns in a CCD image due to reflection and interference within the thin layers of the device. Fringing in direct images is usually due to narrow night-sky emission lines, and if present means that the image of the background sky (only) contains the superimposed fringe pattern. It tends to occur in backthinned CCDs in the far red where night-sky upper atmospheric OH emission is bright and the photon absorption coefficient is low. The fringe pattern is an instrumental artifact and should be removed.

The fringe pattern depends on the wavelengths of the sky emission lines, but its amplitude varies with the ratio of line to continuum intensity in the sky spectrum, which can change, sometimes rapidly, during a night. Fringes will not appear on twilight or dome flats, but will show up on a dark-sky flat produced by the shift-and-stare method – see Figure 9.11.

If fringing is present, you must create a normalized flat F, from either dome or twilight exposures. Then acquire a set of dithered dark-sky images and combine them according to the procedure in Equations (9.24) to (9.26). The result is B, a normalized image of the background that contains no stars but records sky brightness from both continuum and fringes. Flatten this image and set

$$\mathbf{B} = \frac{\mathbf{B}'}{\mathbf{F}} = \mathbf{B}_c + \mathbf{B}_{\text{fringe}} \tag{9.32}$$

Here, $\mathbf{B}_c$ is the part of $\mathbf{B}$ due to the continuum, and $\mathbf{B}_{\text{fringe}}$ the part due to fringes. To separate the two, find the level of the continuum by finding the minimum pixel value in a slightly smoothed version of $\mathbf{B}$: and set every pixel in $\mathbf{B}_c$ equal to that value (the continuum sky should be flat):

$$B_c[x,y] = \text{minP}(\mathbf{G} \otimes \mathbf{B}), \quad \text{for all } x, y \tag{9.33}$$

here $\mathbf{G}$ is a mild Gaussian kernel. Then the fringe part of the background is

Fig. 9.11 (a) A small section of a 300-second I-band exposure on a backthinned CCD. Fringing pattern is apparent in the background. (b) The matching section of an I-band dark-sky flat. (c) A processed version of (a) in which a scaled version of the fringing pattern has been subtracted.

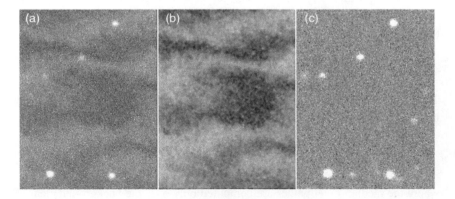

$$\mathbf{B}_f = \mathbf{B} - \mathbf{B}_c \qquad (9.34)$$

Removing the fringes from a processed science image $\mathbf{R}_p$ is then simply a matter of measuring the fringe amplitude on the science image, and subtracting the calibration fringe pattern scaled to match:

$$\mathbf{R}_{pf} = \mathbf{R}_p - a_f \mathbf{B}_{fringe} \qquad (9.35)$$

9.4 Combining images

After preprocessing, astronomers often combine the resulting images. You might, for example, have acquired a dozen images of an extremely fascinating galaxy, and reason (correctly) that adding all of them together digitally will produce a single image with superior signal to noise. The combined image should show features in the galaxy, especially faint features, more clearly than do any of the individual frames. In another example, you may be trying to observe a nebula whose angular size is greater than the field of view of your CCD. You would like to assemble a complete image of the nebula by combining many of your small CCD frames into a large mosaic. Combining images is a tricky business, and this section provides only a brief introduction.

9.4.1 Where is it? The centroid

Suppose you want to combine images $\mathbf{A}$ and $\mathbf{B}$. An obvious requirement is that the pixel location of a source in $\mathbf{A}$ must be the same as its location in $\mathbf{B}$. But what exactly *is* the location of a source? Consider an automated two-step process:

1. Locate those pixels that are part of the source.
2. Compute coordinates for an appropriate center of those pixels.

To complete step 1 in the case of point sources like stars, you can get a good idea of their approximate locations by applying a Laplacian filter (whose size matches the point-spread function – see the next section) to a digital frame and noting the maxima of the filtered image. Different filters could locate extended objects like galaxies. To decide which pixels around these locations are part of the source image and which are not requires some thought. For example, if you ask which pixels in a typical CCD image receive light from a bright star in the center of the frame, the answer, for a typical ground-based point spread function, is: "all of them." A better question might be: "which pixels near the suspected star image receive a signal that is (a) larger than (say) 3σ above the background noise and (b) contiguous with other pixels that pass the same test?"

Figure 9.12 illustrates this approach (there are others) – the bar heights indicate pixel values in a small section of a CCD frame. Although most of the

Fig. 9.12 Bar heights represent pixel values near a faint star image. Darker bars are high enough above the background to qualify as image pixels.

pixels in the area probably registered at least one photon from the star in the center, only those colored dark gray stand out from the background according to the "3σ + contiguous" criterion.

With the "source pixels" identified, you can then compute their ***centroid***. Typically, you consider only that part of the dark gray volume in Figure 9.12 that is above the background level, and compute the (x, y) coordinates (fractional values permitted) of its center of intensity. If $R[x, y]$ is a pixel value and if B is the local background level, then the centroid coordinates are:

$$x_{\text{cen}} = \frac{\sum_x \sum_y x(R[x,y] - B)}{\sum_x \sum_y (R[x,y] - B)}, \qquad y_{\text{cen}} = \frac{\sum_x \sum_y y(R[x,y] - B)}{\sum_x \sum_y (R[x,y] - B)} \qquad (9.36)$$

The sums include only star pixels. Depending upon the signal-to-noise ratio (SNR) in the sums in Equations (9.36) the centroid can locate the image to within a small fraction of a pixel.

9.4.2 Where is it, again? PSF fitting

Finding the centroid of an image is computationally simple, but works well only in cases where images are cleanly isolated. If images blend together the centroid finds the center of the blended object. Even if there is no confusion of images, one object may asymmetrically perturb the background level of another (a galaxy near a star, for example).

In situations like this, you can use knowledge of the ***point-spread function*** (*PSF*) to disentangle blended and biased images. The procedure is to fit each of the *stellar* (only) images on the frame with a two-dimensional PSF, adjusting fits to account for all the flux present. The actual algorithm may be quite complex, and special complications arise if there are non-stellar objects present or if the shape of the PSF varies from place to place due to optical aberrations or to anisoplanatism in AO systems. Despite the difficulties, PSF fitting is neverthe-less essential for astrometry and photometry in crowded fields.

9.4.3 Aligning images: shift, canvas, and trim

Figure 9.13 shows two CCD frames, **A** and **B**, of M33 at different telescope pointings. Each frame has dimensions $x_{max} = 256 \times y_{max} = 256$. We consider the problem of ***aligning*** the two images by applying a ***geometric transformation*** to each – a geometric transformation changes the pixel coordinates of image data elements. In this example, we make the transformation by first measuring the $[x, y]$ coordinates for three stars in the area common to both frames. Suppose that on average, we find for these objects that $x_B - x_A = \Delta x_B = -115$ and that $y_B - y_A = \Delta y_B = 160$. (Assume for now that coordinates are restricted to inte-gers.) There are two possible goals in making the transformation.

First, we might wish to make a new image that contains data from *both* **A** *and* **B**, perhaps to improve the SNR. Do this by creating **A'** and **B'**, two small images that contain only the overlap area from each frame:

$$A'[\xi, \eta] = A[(\xi + \Delta x_B), \eta]$$
$$B'[\xi, \eta] = B[\xi, (\eta + \Delta y_B)] \tag{9.37}$$

The values stored in the pixels of **A'** and **B'** and are the same as the values in **A** and **B**, but they have different coordinates. The ***translation*** operation executed by Equation (9.37) simply slides **B** and **A** until coordinates match. An important step in making the new images discards or ***trims*** any pixels that fall outside the overlap region. Specifically, we trim all pixels except those with coordinates

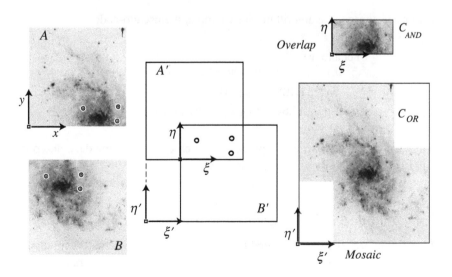

Fig. 9.13 Aligning and combining two images. Alignment and transformation are based on the coordinates of the three marked stars in the overlap region. See text for details.

$1 \leq x_{\max} - \left| \Delta x_B \right|$ and $1 \leq y_{\max} - \left| \Delta y_B \right|$. Both trimmed images thus have the same size, which means we can combine them (add, average, etc.) using image arithmetic. For example:

$$\mathbf{C}_{AND} = \mathbf{A}' + \mathbf{B}' \tag{9.38}$$

Suppose, however, that we wish to combine the two frames to make a wide-angle view, an image that includes *every* valid pixel value from *either* **A** *or* **B**. The procedure is simple: we make two **canvases**, $\mathbf{C_A}$ and $\mathbf{C_B}$, each with dimensions $x_{\max} + \left| \Delta x_B \right|$ by $y_{\max} + \left| \Delta y_B \right|$, large enough to include all pixels.

Then, we "paste" each image in the appropriate section of its canvas, and then combine the large canvases into one large final image. In our example, the canvases have coordinates ξ', η', and the operations that paste the images onto their canvases are:

$$C_A(\xi',,\eta') = \begin{cases} A'(\xi', \eta' - \Delta y_B) = A(\xi', \eta' - \Delta y_B), & 1 \leq \xi' \leq x_{\max}, 1 \leq \eta' \leq y_{\max} + \Delta y_B \\ -10\,000 & \text{otherwise} \end{cases}$$

$$C_B(\xi',,\eta') = \begin{cases} B'(\xi', \eta' - \Delta y_B) = B(\xi' + \Delta x_B, \eta'), & 1 \leq \xi' \leq x_{\max} + \Delta x_B, 1 \leq \eta' \leq y_{\max} \\ -10\,000 & \text{otherwise} \end{cases}$$

$$\tag{9.39}$$

The arbitrary large negative value of $-10\,000$ simply flags those pixels for which there are no data. Any value that cannot be confused with genuine data can serve as a flag. We can combine the two canvases

$$\mathbf{C}_{OR} = \text{combine}(\mathbf{C_A}, \mathbf{C_B}) \tag{9.40}$$

with some appropriate algorithm. For example, the pseudo-code:

> For all x, y :
> $\quad$ IF$\{C_A[x,y] \neq -10,000$ AND $C_B[x,y] \neq -10\,000\}$
> $\quad\quad$ THEN $\quad C_{OR}[x,y] = \dfrac{1}{2}[C_A[x,y] + C_B[x,y]]$
> $\quad\quad$ ELSE $\quad C_{OR}[x,y] = \max[C_A[x,y], C_B[x,y]]$

will compute values for mosaic pixels for which there are some data, and put a flag ($-10\,000$) in those where there is no data.

9.4.4 Aligning images: geometric transformations

Translations are only one of several kinds of geometric transformation. Suppose, for example, you wish to combine images from two different instruments. The instruments have different pixel scales[3] (in seconds of arc per pixel); so one set of images requires a scale change, or **magnification**. The transformation is

$$x = \xi/M_x, \quad y = \eta/M_y \tag{9.41}$$

Again, $[\xi, \eta]$ are the coordinates in the new image, and the equations allow for stretching by different amounts in the x- and y-directions.

Small **rotations** of one image with respect to another might occur if a camera is taken off and remounted on the telescope, or if images from different telescopes need to be combined, or even as the normal result of the telescope mounting (e.g. imperfect polar alignment in an equatorial, an imperfect image rotator in an altazimuth, or certain pointing schemes for a space telescope). If $\mathbf{A}'$ is the image produced when $\mathbf{A}$ is rotated about its origin counterclockwise through angle θ, then $A'[\xi, \eta]$ has the same pixel value as $A[x, y]$ if

$$\begin{aligned} x &= \xi \cos\theta + \eta \sin\theta \\ y &= \eta \cos\theta - \xi \sin\theta \end{aligned} \tag{9.42}$$

For wide fields, optical **distortions** can become significant (e.g. the Seidel pincushion or barrel distortion aberrations). These require relatively complicated transformations.

In creating mosaics from images with different telescope pointings, **projection effects** due to the curvature of the celestial sphere also need to be considered. Such effects have long been an issue in photographic astrometry, and chapter 11 of Birney et al. (2006) outlines a simple treatment of the problem.

To derive any geometric transformation, the general approach is to rely on the locations of objects in the field. In the final transformed or combined image we require that a number of reference objects $(1, 2, 3, \ldots, N)$ have pixel coordinates $(\xi_1, \eta_1), (\xi_2, \eta_2), \ldots, (\xi_N, \eta_N)$. We can call these the **standardized coordinates** – they might be coordinates derived from the known right ascension (RA) and declination (Dec) of the reference objects, or might be taken from the actual pixel coordinates on a single image. Now, suppose one of the images you wish to transform, image $\mathbf{B}$, contains some or all of the reference objects, and these have coordinates

$$(x_{B1}, y_{B1}), (x_{B2}, y_{B2}), \ldots, (x_{BM}, y_{BM}) \quad M \leq N$$

Your task is to find the transformations

$$x = f_B[\xi, \eta], \quad y = g_B[\xi, \eta] \tag{9.43}$$

that will tell you the pixel values in $\mathbf{B}$ that correspond to every pair of standardized coordinates. You specify the forms for the functions from your knowledge

---

[3] Scale differences can have subtle causes: the same CCD–telescope combination can have slightly different scales because of focal-length changes caused by thermal effects on mirrors or chromatic effects in lenses.

of how the images are related. You might, for example, expect that narrow-field images from the same instrument would require just a simple translation, while wide-field images from different instruments might need additional correction for magnification, rotation, distortion, or projection. For a given functional form, the usual approach is to use a least-squares technique to find the best values for the required constants $\Delta x_A, \Delta y_A, \theta, M_x$, etc. Note that some geometric transformations may not **conserve flux** (see Section 9.4.5).

Reducing data from digital arrays very commonly involves a two-step **align and combine** procedure:

(a) apply geometric transforms on a group of images to produce a new set aligned in a common system of coordinates, correcting for flux changes if necessary, then

(b) combine the aligned images with an appropriate algorithm.

This procedure is often termed **shift-and-add**. Basic observational issues make shift-and-add an indispensable technique, and we already discussed some of these in the context of the **shift-and-stare** observing technique for flat field calibration images. (You do shift and *stare* at the telescope, shift and *add* in data reduction.) You will recall (Section 9.3.5) that the aim is to produce a number of equivalent exposures, no two of which are perfectly aligned.

There are many reasons to take several short exposures rather than one long one. For one thing, all arrays saturate, so there may well be an exposure time limit set by the detector. Second, one way to distinguish a pixel illuminated by a cosmic-ray strike from one illuminated by an astronomical object is to take multiple images of the scene. Astronomical objects are present in every image at the same standardized coordinate location; cosmic-ray strikes (and meteor trails and Earth satellites) are not. Similarly, bad pixels, bad columns, and the insensitive regions in array mosaics always have the same pre-transformation coordinates, but different standardized coordinates. When images are aligned, the bad values due to these features in one frame can be filled in with the good values from the others.

9.4.5 Interpolation

Geometric transforms set the values of the pixel at standardized coordinates $\left[\xi_j, \eta_j\right]$ in a new image to those at pixel at $\left(x_j, y_j\right)$ in the original image; see Figure 9.14. Now, ξ_j and η_j must be integers, but x_j and y_j generally contain fractional parts. Therefore, we use round brackets (non-integers permitted) to write, symbolically

$$B'\left[\xi_j, \eta_j\right] = B\left(x_j, y_j\right) = B\left(f_B\left[\xi_j, \eta_j\right], g_B\left[\xi_j, \eta_j\right]\right) \tag{9.44}$$

Since we only know the pixel values for the image **B** at locations where x and y are integers, we must use the pixel values at nearby integer coordinates to *estimate* $B(x_j, y_j)$ – the value a pixel *would* have if it were centered precisely at the non-integer location $\left(x_j, y_j\right)$.

We could, for example, ignore any image changes at the sub-pixel level, and simply round x_j and y_j up or down to the nearest integers, and set $B\left(x_j, y_j\right)$ equal to the value of the **nearest pixel**. This is simple, and largely preserves detail, but will limit the astrometric accuracy of the new image.

Bilinear interpolation often gives a more accurate positional estimate. Figure 9.14 shows the point $\left(x_j, y_j\right)$ relative to the centers of actual pixels in the original image. We compute x_0 and y_0, the values of x_j and y_j *rounded down* to the next lowest integers. Thus, the values of the four pixels nearest the fractional location $\left(x_j, y_j\right)$ are

$$B_{00} = B[x_0, y_0], \qquad B_{10} = B[x_0 + 1, y_0]$$
$$B_{01} = B[x_0, y_0 + 1], \qquad B_{11} = B[x_0 + 1, y_0 + 1] \qquad (9.45)$$

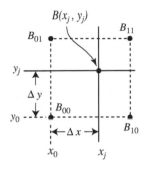

Fig. 9.14 Bilinear interpolation. The method finds the value of the image intensity at point (x_j, y_j), given the nearest pixel values.

As you can see from Equation (9.46) the bilinear procedure takes a weighted average of four pixels – as such, it *smooths* the image:

$$B\left(x_j, y_j\right) \approx (1 - \Delta x)(1 - \Delta y)B_{00} + (\Delta x)(1 - \Delta y)B_{10} + (1 - \Delta x)(\Delta y)B_{01} + (\Delta x)(\Delta y)B_{11}$$

$$(9.46)$$

Bilinear interpolation preserves astrometric precision and affects photometry in predictable ways. Any geometric transformation in which the output grid does not sample the input grid uniformly will change the photometric content of the transformed image. Also, bilinear interpolation chops off peaks and fills in valleys, so an interpolated image is never as sharp as the original. See Figure 9.15. Furthermore, the smoothing effect artificially reduces image noise.

If resolution is of great concern, it is possible to fit the pixels of the original image with a higher-order function that may preserve peaks and valleys. The danger here is that higher-order surfaces may also produce artifacts and photo-metric uncertainties, especially for noisy images. Nevertheless, it is not unusual for astronomers to use higher-order fitting techniques like bicubic interpolation or B-spline surfaces.

9.4.6 Resampling, interlace, and drizzle

Geometric transformations are essential for combining images with the shift-and-add image technique. Transformations, however, require either interpolation (which degrades resolution) or the "nearest-pixel" approximation, which

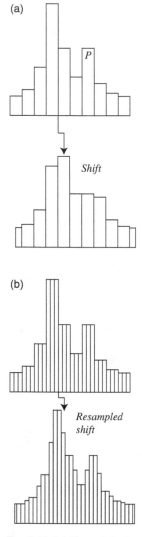

Fig. 9.15 (a) The original pixel values in the top plot are shifted by 0.5 pixels to the right. In the plot, the shift and linear interpolation smooths the original, removing peak P. (b) A 3× resampling of the same pixels preserves more detail after shift and interpolation.

degrades positional information. ***Resampling*** the original image at higher magnification circumvents some of the image degradation that accompanies interpolation, and in some cases can actually *improve* the resolution of the combined image over that of the originals.

The idea is to make the pixels of the output, or transformed, image smaller (in seconds of arc), and thus more closely spaced and numerous, than the pixels of the input image. In other words, the scale (in pixels per arcsec) of the standardized coordinates is larger than the scale of the original input coordinates. We discuss three resampling strategies.

The first is just a modification of the shift-and-add (and interpolate) algorithm. All that is done is to resample each input image by an integral number (e.g. each original pixel becomes nine pixels in the resampled version). After shifting or other transformations, resampling mitigates the smoothing effect produced by interpolation, since this smoothing effect is on the scale of the output pixels. Figure 9.15 shows a one-dimensional example. An image is to be shifted 0.5 pixels to the right from its position in the original. Figure 9.15a shows the result of the shift and linear interpolation without resampling, and Figure 9.15b shows the same result if the output pixels are one-third of the size of the input. Linear interpolation in each case produces some smoothing, but the smoothing is less pronounced with the finer grid. Compared to using the original pixel sizes, aligning multiple images on the finer output grid will of course improve the resolution of their combined image.

The second method, usually called ***interlace***, examines each input pixel (i.e. $B[x, y]$ at only integer coordinates), locates its transformed center inside a particular output pixel in a finer grid (but again, only integer coordinates), and copies the input value to that single output pixel. There is no adjustment for fractional coordinates, nor for the fact that the input pixel may overlap several output pixels. Figure 9.16a gives an example of a shifted and rotated input grid placed on an output grid with smaller pixels. The center of each input pixel is marked with a black dot. Interlacing this single input places values in the output pixels (i.e. the dark-colored pixels), "hit" by the dots, and "no value" or "zero-weight" flags in the other pixels.

Interlace for a single image is a flawed approach. First, it creates a discontinuous image, since only some fraction of the output pixels will score a "hit," and the remainder will have zero weight. Second, we have introduced positional errors because we ignore any fractional coordinates.

Both problems become less significant as more images of the same scene are added to the output. If each addition has a slightly different placement on the output grid, a few additions could well fill in most output pixels with at least one valid value. Moreover, positional information improves as the interlace fills and averaging reduces uncertainty in the brightness distribution.

The combined image is a weighted mean of all the shifted frames, with the weight, $w_i[\xi, \eta]$, of a particular pixel either one (if it is a hit) or zero (if no hit *or*

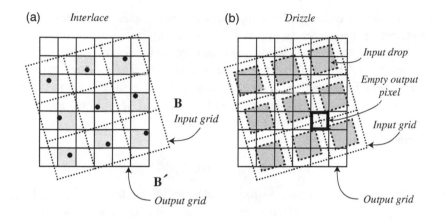

(a) *Interlace*

B
Input grid

B′
Output grid

(b) *Drizzle*

Input drop

Empty output pixel

Input grid

Output grid

Fig. 9.16 Resampling an input grid. The interlace technique (a) regards values in the input grid as if concentrated at points. Grayed pixels on the output copy the values from the input points, white-colored output pixels have no value. The drizzle method (b) assumes values are spread over a square "drop" smaller than an input pixel. Most output pixels overlap one or more input drops, although some, as illustrated, may overlap none.

if we decide the hit is by a cosmic ray or by a bad pixel). Thus, the combined image **C** is

$$C[\xi,\eta] = \frac{1}{\sum_{i=1}^{N} w_i[\xi,\eta]} \left(w_1[\xi,\eta] B'_1[\xi,\eta] + w_2[\xi,\eta] B'_2[\xi,\eta] + \ldots + w_N[\xi,\eta] B'_N[\xi,\eta] \right)$$

(9.47)

We cannot use Equation (9.47) for any pixel in **C** with a combined weight of zero. In this case, the pixel has no valid value. It is possible to interpolate such a missing value from the surrounding output pixels, but this will cause photometric errors unless the "no-value" status is due to masking cosmic rays or bad pixels.

Interlacing shifted images has the potential for actually improving image resolution in the case where the camera resolution is limited by the detector pixel size rather than by the telescopic image itself. Figure 9.17 shows the interlaced result for a one-dimensional example: a double source with a separation of 1.3 input pixels, with each source FWHM = 0.8 pixels. Three dithered input images are shown, none of which shows the double nature of the source, as well as the interlaced combination with 1/3-size output pixels. The combined image resolves the two components.

The interlace technique is powerful, but many telescope controls produce a set of exposures whose grids are dithered randomly, rather than precisely at the sub-pixel level. The ***variable-pixel linear reconstruction*** method, more commonly known as ***drizzle***, can be much more forgiving about input grid placement. Drizzle assumes that the flux in a square input pixel of size (length) d is not spread over the entire pixel, but is uniformly concentrated in a smaller concentric square, called a "drop," whose sides have length fd; see Figure 9.16b, where the drops are the shaded squares. The fractional size of the drops, i.e. the value of f, can be varied to accommodate a particular set of

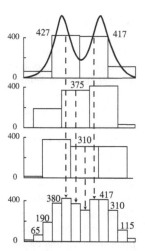

Fig. 9.17 The interlace method in a one-dimensional example. The actual brightness distribution of a double source is sampled with large pixels. In three samples (upper plots) displaced from one another by 1/3 of a pixel, no indication of the double nature of the source appears, yet the combined and interlaced image (bottom plot) does resolve the source.

images. As $f \to 0$ the drizzle method approaches the interlace method, and as $f \to 1$, drizzle approaches resampled shift-and-add.

To examine the drizzle method quantitatively, we introduce a parameter, s, to measure the relative scale of the output pixels: for input pixels of length d, output pixels have length sd. The drizzle algorithm then runs as follows: Input pixel $B_i[x, y]$ in frame i will contribute to output pixel $B'_i[\xi, \eta]$ if any part of the input *drop* overlaps the output pixel. If the area of overlap is $a_i[x,y,\xi,\eta] (fd)^2$, then the contribution will be

$$B_i[x,y]W_i[x,y]a_i[x,y,\xi,\eta]s^2 \tag{9.48}$$

The factor s^2 conserves surface brightness in the final image, and the weighting factor $W_i[x, y]$ accounts for bad pixels and other effects (e.g. exposure time) in the input frame (see Section 9.4.7). Adding all contributions from the input image (up to four input drops can overlap a single output pixel), we assign the output value and weight as

$$B'_i[\xi,\eta] = s^2 \sum_{x,y} B_i[x,y]W_i[x,y]a_i[x,y,\xi,\eta]$$
$$w_i[\xi,\eta] = \sum_{x,y} W_i[x,y]a_i[x,y,\xi,\eta] \tag{9.49}$$

We make the final combination of images by computing the weighted mean of all the input frame contributions to each pixel as in Equation (9.47).

$$C[\xi,\eta] = \frac{\displaystyle\sum_{i=1}^{N} B'_i[\xi,\eta]}{\displaystyle\sum_{i=1}^{N} w_i[\xi,\eta]} \tag{9.50}$$

9.4.7 Cleaning images

Images inevitably have defects caused by bad detector pixels or by radiation events like cosmic-ray impacts or radioactive decays in or near the detector. Most methods for removing such defects require multiple dithered images of the same scene. We are already familiar with shift-and-add and use that as an example. Start with $N > 2$ dithered images $\{\mathbf{R}_1, \mathbf{R}_2, \dots \mathbf{R}_N\}$ whose intensities are *scaled to the same exposure time*. Align them – use a geometric transform to make all astronomical sources coincide. Then combine the transformed images $\{\mathbf{R}'_1, \mathbf{R}'_2, \dots \mathbf{R}'_N,\}$ to form the median image.

The median is relatively insensitive to "no-data" pixels: pixels whose values we want to ignore. These usually differ greatly from the central value (like many radiation events or bad CCD columns), so the median produces a "clean"

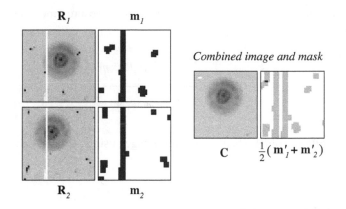

$\mathbf{R}_1$ $\mathbf{m}_1$

Combined image and mask

C $\frac{1}{2}(\mathbf{m}'_1 + \mathbf{m}'_2)$

$\mathbf{R}_2$ $\mathbf{m}_2$

Fig. 9.18 Pixel masks. Two offset images of the planetary nebula NGC 2392 (The Eskimo) are marred by an insensitive column and many cosmic-ray strikes. The mask next to each raw image on the left blocks (black pixels = 0, white = 1) every bad pixel and its immediately adjacent neighbor. The right-hand images show the combined image and mask after alignment. Since there are only two images, the combined image shows noticeably different noise levels in masked and unmasked regions. Two pixels in the upper left are masked in both images and have zero weight. They show as black in the right-hand image of the combined masks.

version of the image. If the detector is a mosaic of smaller arrays, the gaps in the mosaic should have been inserted in the raw image as "no-data" rows and columns. Although simple to execute, the simple "shift-scale-median" does have some shortcomings:

1. At locations where all pixel values are good, the median is not as good an estimator of the central value as is the mean.
2. The median is not *completely* insensitive to deviant values: e.g. the median will be slightly biased toward higher values at the location of cosmic-ray hits.
3. The median will perform poorly in special cases (e.g. if multiple values at the same location are bad).

A modification of shift-and-add can eliminate most of these problems. The idea is to **flag** the defects in the original images by assigning the "no-data" pixels a special value (a large negative number, for example). One way to implement the flags is to generate a special companion image, the **mask**, for each $\mathbf{R}_i$. The mask values (usually either one or zero) indicate whether the corresponding image pixel is to be included or excluded (i.e. flagged) in any subsequent operations; see Figure 9.18.

How can you generate a mask for a particular image? Usually, bad detector pixels or detector mosaic gaps are well documented or are easily discovered on flat field exposures. You can identify radiation events, which occur at random locations and can mimic images of astronomical objects, by their absence from a preliminary median image of the aligned frames. Once the complete mask is generated for an input image, a conservative approach might be to mask all pixels that are adjacent to bad pixels as well, since radiation events tend to spill over. At the end of this process, there will be a separate mask for each input image.

You then geometrically transform all input images, along with their masks, so that all are aligned. The final combination of these aligned images is a weighted mean in which all defective pixels are ignored. That is, if $\mathbf{m}_i$ is the mask for input image i, w_i is the image weight (based, perhaps on exposure time), and $\mathbf{m}'_i$ is the transformed mask:

$$C[\zeta,\eta] = \frac{\sum\limits_{i=1}^{N} w_i m'_i[\zeta,\eta] R'_i[\zeta,\eta]}{\sum\limits_{i=1}^{N} w_i m'_i[\zeta,\eta]} \tag{9.51}$$

Masks are often applied when combining images using the interlace or drizzle methods.

9.5 Digital aperture photometry

We have discussed the preprocessing of individual images (the linearity, dark, bias, flat, and fringe corrections) and the combination of multiple frames to produce a deeper and possibly wider image. As a reminder, we summarize those steps here:

$$\mathbf{R}_{pi} = \frac{\text{lin}(\mathbf{R}_i) - \mathbf{Z} - t\dot{\mathbf{D}}}{\mathbf{F}} - a_f \mathbf{B}_{\text{fringe}} \tag{9.52}$$

$$\mathbf{R}'_i = \text{GXform}(\mathbf{R}_{pi}) \quad i = 1, \ldots, N \tag{9.53}$$

$$\mathbf{C} = \text{combine}(\mathbf{R}'_1, \mathbf{R}'_2, \ldots \mathbf{R}'_N) \tag{9.54}$$

Here the notation GXform(...) indicates the appropriate geometric transform and we understand that combine(...) indicates something like a median image or weighted mean, perhaps utilizing masks and a drizzle or interlace.

The next task in the reduction procedure is often measurement of the brightness of one or more objects. Measuring brightness is at heart a simple task – we did it in the exercises in Chapters 1 and 2. Start with the preprocessed image – an individual frame, $\mathbf{R}_p$, or an aligned/combined accumulation of such frames, $\mathbf{C}$. Then just add up the digital signal from the object of interest, which usually is spread over many pixels. In doing so, remember to remove the background. We will use the terms *sky* and *background* interchangeably for all this unwanted light. Once we have isolated the signal attributable to the source alone, we will need to quantify the uncertainty of the result.

Finally, the signal measured will only be meaningful if it is calibrated – expressed in units like watts per square meter or magnitudes. We consider the calibration process in the next chapter, and confine ourselves here to the tasks of separating signal from background and of estimating the uncertainty of the result.

9.5.1 Digital apertures and PSF fits

Consider a very common situation: from a digital image, you want to determine the brightness of a *point* source – a star, quasar, or small object in the Solar

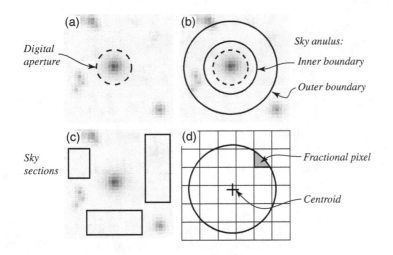

Fig. 9.19 Digital apertures. (a) A circular aperture centered on a point source. (b) An annular aperture for sampling sky emission near a point source. (c) Rectangular apertures for sampling background emission. (d) All curved apertures will require some strategy for dealing with pixels that contain some segment of the boundary.

System. Define a circular area, the ***digital aperture***,[4] that is centered on the centroid of the object (see Figure 9.19). The radius of the digital aperture should include a substantial fraction of the emission from the star. Now make three simple computations:

1. Add all the pixel values inside the aperture. This sum represents the total emission from the aperture – the light from the star plus the light from the background. To deal with fractional pixels (see Figure 9.19) at the edges, multiply every value by $A[x, y]$, that fraction of the signal from the pixel that lies inside the aperture. We approximate this as the fraction of the pixel *area* that lies inside the aperture. (This is a poor approximation for very small apertures.) These sums are understood to extend over the entire x–y extent of the aperture. The number n_{pix} is just the area of the aperture in pixels:

$$\mathrm{Total} = \sum_{x,y} A[x,y]R_{\mathrm{p}}[x,y] \qquad (9.55)$$

$$n_{\mathrm{pix}} = \sum_{x,y} A[x,y] \qquad (9.56)$$

2. Estimate $\langle B \rangle$, the value of the sky emission per pixel, usually from a source-free region near the object of interest (see Section 9.5.2 for details). Compute that part of the emission in the aperture that is due to the background: $sky = n_{\mathrm{pix}} \langle B \rangle$.
3. Subtract the sky emission from the total, and the remainder is the detector response attributable to the source alone; this is the signal in ADUs:

---

[4] Yes, aperture means "opening." The terminology recalls the days of photoelectric photometry, when it was necessary to place an opaque plate with one small clear aperture in the focal plane. This passed only the light from the star and very nearby sky through to the photocathode, and blocked all other sources.

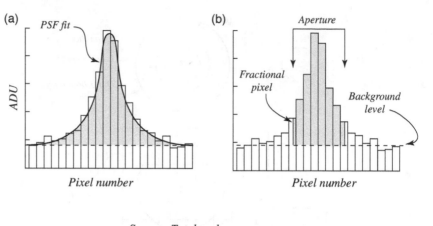

$$S_{\mathrm{ADU}} = Total - sky$$
$$S_{\mathrm{ADU}} = \sum_{x,y} A[x,y]R_{\mathrm{p}}[x,y] - n_{\mathrm{pix}}\langle B \rangle \qquad (9.57)$$

In situations in which star images seriously overlap, digital aperture photometry fails, because it is impossible to estimate easily the polluting star's contribution to the background of the object of interest. We have already discussed
(Section 9.4.2) the idea of fitting a **PSF** to each star image on a frame
(Figure 9.20a). Point-spread-function fitting is required in crowded-field photometry since (at the cost of considerable computational complexity) it can
separate the contributions of individual overlapping images from one another
and from the diffuse background. Once all overlapping images are accounted
for, integration of the PSF fit of the image of interest gives the signal – S_{ADU} in
Equation (9.57).

If he does not use PSF fitting, the astronomer must choose the digital aperture
size; see Figure 9.20b. There are two conflicting considerations: he wants a *large*
aperture because it includes as much light as possible from the star, yet he wants
a *small* aperture because it excludes background light and, especially, its associated noise. An aperture that includes too much sky will decrease the SNR of the
final measurement, as will an aperture that includes too little of the source. The
optimum size varies with the brightness of the star relative to the background.
Since point-source photometry requires the same aperture size for all stars, this
generally means the astronomer chooses the aperture size based on the faintest
star observed. The choice is implemented in software, so it is easy to try a range
of apertures (radius somewhere between 0.75 and 4 times the FWHM of the
image profile) and identify the aperture that yields the best SNR. For most
profiles, a diameter near 0.9–1.5 times the profile FWHM yields the best SNR,
but aperture radius of around 2.5–3 FWHM is needed to collect more than 99%
of the flux.

Note that a digital aperture need not be circular. Indeed, many objects have
decidedly non-circular shapes, and invite equally non-circular apertures.

Photometry via Equation (9.57) applies as well to such shapes. Finally, note that calibrated photometry requires relatively large aperture sizes (see Section 10.6).

9.5.2 Measuring the sky

Both PSF fitting and digital aperture photometry demand an accurate measure of the background emission underlying the source. This, of course, is one area where it is impossible to measure the sky brightness, so we measure the background *near* the source and hope that sky brightness does not change with location. There are some cases where this hope is forlorn. A notorious example is the photometry of supernovae in other galaxies: the background due to the host galaxy changes drastically on the scale of a digital aperture size, so any "nearby" sky measurement is guaranteed to introduce uncertainty. (Fortunately, supernovae are temporary. An image of the galaxy obtained with the same instrument after the supernova has faded can provide the needed background measurement.)

For isolated sources, the nearest possible sample should be the most accurate, and a background sample symmetrically positioned around the source stands a chance of averaging out any trends. Figure 9.19b shows a digital aperture and a *sky annulus*. The annulus is a region between two circles centered on the source. The inner boundary of the annulus is just barely large enough to exclude any appreciable emission from the source. The outer radius of the annulus is less strictly determined, but should be large enough to include a statistically significant number of pixels. If the outer radius is too large, it may sample sky that differs from the sky within the aperture.

The best estimate of the sky value in the annulus is clearly *not* the mean pixel value: the annulus is bound to contain images or halos of other stars. These bias the mean toward larger values. The median is less sensitive to the influence of this kind of pollution, and the mode is even better: the most common value in the annulus certainly sounds like the optimum measurement of the sky. Practical computation of the mode usually requires the construction of a smoothed histogram, with the sky value computed as the mean of the values in the most populous bin of the histogram.

Figure 9.19c illustrates a second approach to measuring the sky value. An astronomer selects one or more relatively star-free sections of the image, and computes the median or modal value. This method avoids the influence of nearby sources on background estimates, and if the field near the source of interest is crowded, this is the only alternative. The disadvantage is that non-local sky samples may not be representative. As explained earlier, in the infrared, one generally obtains sky levels from separate (chopped or dithered) exposures.

9.5.3 Signal and noise in an aperture

Knowing the uncertainty of a measurement is nearly as important as discovering its value. In this section, we develop an equation for the SNR in aperture photometry with a CCD. The general approach, if not the exact equation, will apply for photometry with all digital arrays. For simplicity, we consider only the case of a single exposure, corrected for non-linearity, dark, bias, and flat. From Equation (9.57), we write an expression for N_*, the signal *in electrons* in an aperture as

$$N_* = gS_{\text{ADU}} = \left\{ \sum_{x,y} A[x,y]gR_{\text{p}}[x,y] \right\} - n_{\text{pix}}g\langle B \rangle = \left\{ \sum_{x,y} A[x,y]r_{\text{p}}[x,y] \right\} - n_{\text{pix}}\langle b \rangle$$

(9.58)

Here, the constant g is the gain in electrons/ADU, and the values $r_{\text{p}}[x,y]$ and $\langle b \rangle$ are the preprocessed pixel value and the estimated background value *in electrons*. Our convention will be to replace uppercase array symbols (which stood for values in ADU) with lowercase symbols to indicate measurement in electrons. The noise, or uncertainty in N_*, follows from an application of Equation (2.30) to (9.58). Although it is not always safe to do so, we assume uncertainties in pixel values are not correlated:

$$\sigma_N^2 = \left\{ \sum_{x,y} \{A[x,y]\}^2 \sigma_{r,\text{p}}^2[x,y] \right\} + n_{\text{pix}}^2 \sigma_{\langle b \rangle}^2,$$

(9.59)

To evaluate $\sigma_{r,\text{p}}^2[x,y]$, the uncertainty in a preprocessed pixel value, we write out the preprocessing operation for a single pixel as described for a CCD in Equation (9.52). We ignore fringing.

$$r_{\text{p}}[x,y] = \frac{1}{F[x,y]} \left\{ L(r[x,y])r[x,y] - t\dot{d}[x,y] - z[x,y] \right\}$$

(9.60)

Here F is the (dimensionless) normalized flat field response, and $L(r)$ is a linearity correction expressed, for convenience, as a factor, i.e. $L(r) = Lin(R)/R$. We set, $\dot{d}$ and z as the dark rate and bias levels in electrons. Apply Equation (2.30) to compute the variance of processed pixel value:

$$\sigma_{r,\text{p}}^2[x,y] = \frac{1}{F^2} \left\{ \sigma_r^2 L^2 + \sigma_L^2 r^2 + \sigma_z^2 + t^2 \sigma_{\text{d}}^2 \right\} + \sigma_F^2 \left\{ \frac{Lr - z - t\dot{d}}{F^2} \right\}^2$$

(9.61)

To simplify the notation, we have omitted the $[x, y]$ coordinate references for all the terms on the right-hand side. We can clean up this expression further by noting that $F \approx 1$ and $L \approx 1$. We will make two further assumptions that are harder to justify. First, assume the linearity correction is well known and has negligible uncertainty ($\sigma_L^2 \approx 0$) – this is the case for the CCD, and for many, but not all, IR arrays. Second, assume the flat field correction is well known and has negligible uncertainty ($\sigma_F^2 \approx 0$). This is less often the case, and it is well to keep in mind that we will consequently underestimate the total uncertainty in what follows. Nevertheless, the result we are working toward will prove extremely useful. With these assumptions, Equation (9.61) becomes:

$$\sigma_{r,p}^2[x,y] = \sigma_r^2 + \sigma_z^2 + t^2\sigma_{\dot{d}}^2 \qquad (9.62)$$

We will examine each of the terms on the right-hand side in turn. The first term is the variance in the raw pixel value itself. This value is just

$$r[x,y] = \{n_* + b + t\dot{d}\} + z \qquad (9.63)$$

Here n_* is the signal from the star. Recognizing that the quantity in braces is simply a count governed by Poisson statistics, we can write the variance in the raw pixel value as:

$$\sigma_r^2[x,y] = \{n_* + \langle b \rangle + t\dot{d}\} + \rho^2 \qquad (9.64)$$

The actual values for the background $b[x,y]$ and dark rate $\dot{d}[x,y]$ in a particular pixel are unknown, and we will simply use estimated values: the dark rate is estimated from dark frames (or assumed to be zero if the detector is sufficiently cold) and the background is estimated from nearby sky pixels. The variance of the bias level, ρ^2, is the square of the **read noise**, which does not obey Poisson statistics, but depends on details in the amplifier and readout circuits such as the frequency of the reads, number of samples, output capacitance, temperature, and round-off in the analog-to-digital conversion.

The second term in Equation (9.62) is the squared uncertainty in the "estimated" bias level (different from the read noise in a single pixel!). This estimate is usually computed by averaging a number of calibration frames. If the bias drifts, then σ_z might be large. If we obtain p_z bias frames, the best we can hope for is that the minimum variance of the mean bias at pixel $[x,y]$ is given by

$$\sigma_z^2 = \frac{\rho^2}{p_z} \qquad (9.65)$$

If the bias is obtained from an overscan, and the base bias *pattern* is very well determined, then p_z is the number of columns in the overscan.

The third term in Equation (9.62) is the variance in the estimated dark rate. How do we estimate the dark rate? Take p_d dark frames, each exposed for time t_d, subtract the estimated bias from each and then average and divide by t_d. So the, variance in this *mean* dark rate is:

$$\sigma_{\dot{d}}^2 = \mathrm{var}\left(\frac{1}{p_d t_d}\sum[t_d\dot{d} - <z>]\right) = \frac{1}{p_d^2 t_d^2}\left[p_d t_d \dot{d} + \frac{p_d}{p_z}\rho^2\right] = \frac{\dot{d}}{p_d t_d} + \frac{\rho^2}{p_z p_d t_d^2} \qquad (9.66)$$

Substituting Equations (9.64), (9.65), and (9.66) into Equation (9.62) gives the variance in a single processed pixel:

$$\sigma_{r,p}^2 = n_* + \langle b \rangle + a_d t \dot{d} + a_z \rho^2 \tag{9.67}$$

where

$$a_d = 1 + \frac{1}{p_d t_d} \tag{9.68}$$

$$a_z = 1 + \frac{1}{p_z} + \frac{1}{p_z p_d t_d^2} \tag{9.69}$$

Note that the final term in Equation (9.69) is only present if a correction for dark current has been made. Now return to Equation (9.59). We require a value for the uncertainty in the estimated background. We usually estimate the background by averaging $r_p[x,y]$ in a region of p_b pixels (e.g. the sky annulus) in which $n_*[x,y]$ is zero. That is,

$$\sigma_{\langle b \rangle}^2 = \mathrm{var}\left(\frac{1}{p_b} \sum_{x,y}^{sky\ section} r_p[x,y] \right) = \frac{1}{p_b^2} \sum_{x,y}^{sky\ section} \sigma_{r,p}^2 \tag{9.70}$$

But we have just worked out $\sigma_{r,p}^2$, the variance of a *single* preprocessed pixel, so substituting Equation (9.67) for the case $n_*[x,y] = 0$, the expression above becomes:

$$\sigma_{\langle b \rangle}^2 = \frac{1}{p_b^2} \sum^{sky\ section} \left(b + a_d t \dot{d} + a_z \rho^2 \right) = \frac{1}{p_b}\left(\langle b \rangle + a_d t \dot{d} + a_z \rho^2 \right) \tag{9.71}$$

Now we can finally return to Equation (9.59). Substituting Equations (9.67) and (9.71) into (9.59) we have:

$$\sigma_N^2 = \left(\sum \{A[x,y]\}^2 n_* \right) + n_{pix} a_b \left(\langle b \rangle + a_d t \dot{d} + a_z \rho^2 \right) \tag{9.72}$$

where

$$a_b = \frac{1}{n_{pix}} \left(\sum \{A[x,y]\}^2 + \frac{n_{pix}^2}{p_b} \right) \approx \left(1 + \frac{n_{pix}}{p_b} \right) \tag{9.73}$$

Equation (9.72) does not include some sources of uncertainty that could be important in a specific array, like uncertainties in corrections for charge transfer inefficiency. We have also specifically excluded terms due to the uncertainties in the linearity and flat field corrections. If such effects can be well-modeled, one could in principle represent them with additional terms. Moreover, as a tool for evaluating photometric uncertainty, the most serious problem with Equation (9.72) is its failure to account for systematic effects like drift in bias or dark, color differences between sky, star, and flat, or variations in atmospheric transparency. You should *not* use this equation to evaluate the uncertainty in your digital photometry. *As always, the primary information about the uncertainty of your photometry comes from the scatter in repeated observations and the disagreement of your results with those of others*.

But Equation (9.72) is far from useless. It gives you a way to compare the expected random error with the actual scatter in your data – if you get something unexpected, think hard to understand why. The equation is also a very important tool for *planning* observations, for answering questions like: "how many minutes at the telescope will I need if I hope to measure the brightness of my $V = 22.5$ quasar with a precision of 1%?"

9.5.4 The CCD equation

We now apply Equation (9.72) for the routine case of aperture photometry of a star. For most reasonable apertures, the counts due to the star are very small at the edge of the aperture where the partial pixels are located. In that case, we will not be far off in making the approximation (which applies exactly if partial pixels are not employed):

$$\sum_{x,y} A^2[x,y]n_*[x,y] \approx \sum_{x,y} A[x,y]n_*[x,y] = N_* \tag{9.74}$$

The SNR then implied by Equation (9.72) is

$$\text{SNR} = \frac{N_*}{\left\{ N_* + n_{\text{pix}}a_b\left(\langle b \rangle + a_d\left(t\dot{d} + a_z\rho^2 \right) \right) \right\}^{\frac{1}{2}}} \tag{9.75}$$

This equation, in various approximations, is known as the **CCD equation**. The usual approach is to simplify this expression by assuming good preprocessing practices as well as good fortune: The system remains stable and the observer collects a very large number of bias and (if needed) long-integration dark frames, so that $a_d \approx a_z \approx 1$. If we also make the role of integration time explicit by setting $N_* = \dot{N}_* t$ and $\langle b \rangle = \dot{b}t$, Equation (9.75) becomes

$$\text{SNR} = \frac{\dot{N}_* t}{\left\{ \left[\dot{N}_* + n_{\text{pix}}a_b\left(\dot{b} + \dot{d} \right) \right] t + n_{\text{pix}}a_b\rho^2 \right)}\right\}^{\frac{1}{2}}} \tag{9.76}$$

Using the quadratic formula to solve for exposure time:

$$t = \frac{B + \left(B^2 + 4AC \right)^{\frac{1}{2}}}{2A} \tag{9.77}$$

where

$$\begin{aligned} A &= \frac{\dot{N}_*^2}{(\text{SNR})^2} \\ B &= \dot{N}_* + n_{\text{pix}}a_b\left(\dot{b} + \dot{d} \right) \\ C &= n_{\text{pix}}a_b\rho^2 \end{aligned} \tag{9.78}$$

Figure 9.21 illustrates predictions based on the CCD equation in three different situations.

Fig. 9.21 The CCD
equation. The plot shows
the required time to reach
a specified signal-to-noise
ratio in the three limiting
cases discussed in the
text. Note the
logarithmic scale.

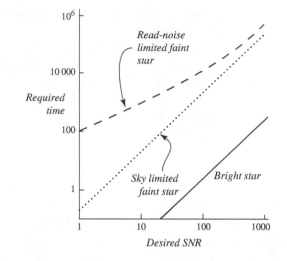

The **bright-star** or **photon-noise limited** case, where the counts from the source, $\dot{N}_* t$, greatly exceeds all other terms in the denominator of Equation (9.76) so

$$SNR = \sqrt{\dot{N}_*}\sqrt{t} \tag{9.79}$$

In the bright-star case, the SNR improves as the square root of the exposure time, and the observer willing to devote sufficient time can produce measurements of arbitrarily high precision. Actual precision attained, of course, will be eventually be limited by one of the processes we ignored in deriving Equation (9.76) (e.g. flat field uncertainties or systematic effect like the atmosphere). In the **background-limited** case, the read noise is small, but the background term, $n_{\text{pix}}a_b(\dot{b} + \dot{d})$, is significant compared to $\dot{N}_*$. In this case, the SNR ratio still increases as the square root of the integration time, but now there is a penalty:

$$\text{SNR} = \frac{\dot{N}_*}{[\dot{N}_* + n_{\text{pix}}a_b(\dot{b} + \dot{d})]^{\frac{1}{2}}}\sqrt{t} = \frac{\sqrt{\dot{N}_*}}{\left[1 + n_{\text{pix}}a_b\left(\frac{\dot{b}+\dot{d}}{\dot{N}_*}\right)\right]^{\frac{1}{2}}}\sqrt{t} = \sqrt{\frac{\dot{N}}{1 + \dot{s}}}\sqrt{t} < \sqrt{\dot{N}_*}\sqrt{t} \tag{9.80}$$

The penalty factor $1 + \dot{s}$ becomes large under any of the following conditions: low source brightness, high sky brightness, high dark rate, large digital aperture, or small sky sample. A frequent ground-based situation is the combination of a faint source and a bright sky, perhaps exacerbated by poor seeing (large n_{pix}).

The final case is the one in which the read noise is large. Here the SNR initially increases linearly with time, but eventually reaches the $\sqrt{t}$ dependence of either the bright-star or the sky-limited case.

Summary

- Digital images are ordered sets of numbers that can represent the output of an array of sensors or other data. Concepts:

pixel	*fill factor*	*undersampling*
pixel value	*gray-scale map*	*detector response*
ADU	*DN*	*CCD gain*

- An important advantage of digital images is that they can be mathematically manipulated to remove defects and extract information. Concepts:

image arithmetic	*data cube*	*RGB color model*
CMYK	*false color*	*color look-up table*
digital filtration	*image convolution*	*kernel*
Gaussian kernel	*Laplacian kernel*	*boxcar*
unsharp mask	*local median filter*	

- Digital images from a CCD can be processed to remove the effects of the detector and telescope. Concepts:

raw image	*bias frame*	*rejection algorithm*
overscan	*dark response*	*dark rate*
reference pixel	*linearity correction*	

- The flat field correction very often limits photometric precision of a detector. Concepts:

flat field image	*twilight flat*	*dark-sky flat*
dome flat	*compound flats*	*shift-and-stare*
dither	*illumination correction*	*space telescope flats*

- Preprocessing images from an array requires subtraction of the dark signal and bias, then division by the normalized flat. Treatment of data from infrared arrays is slightly different because of the strong and variable sky background.

- Fringing is a variation in the sky background intensity due to interference effects in thin layers of a detector. Fringes can be removed with proper calibration.

- Combining images requires alignment, which requires both identification of feature coordinates and transformation of images. Concepts:

centroid	*point-spread function*	*PSF fitting*
image alignment	*translation*	*rotation*
trim	*canvas*	*image mosaic*
magnification	*distortion*	*shift-and-add*
projection effects	*flux conservation*	

- Special methods for combining images can compensate for the loss of resolution due to interpolation, and can compensate for bad pixels. Concepts:

nearest pixel	*resampling*	*bilinear interpolation*
interlace	*pixel flag*	*drizzle*
image mask	*clean image*	

(continued)

Summary (*cont.*)

- Digital aperture photometry is a technique for measuring apparent brightness from a digital image. Concepts:
  digital aperture   sky annulus   PSF fitting
- The CCD equation gives the theoretical relation between the exposure time and expected SNR in digital aperture photometry, given source and sky brightness and detector and telescope characteristics. Concepts:
  read noise   read-noise limited   background-limited
  photon-noise limited

Exercises

1. Exposures to a constant source produce the bias-corrected output from a single-channel infrared detector indicated in the table below. (a) Use these data to derive values for the coefficients a, b, and c in Equation (9.23) for the linearity correction for this detector.

Exptime (sec)	0	1	20	40	60	80
DN	0	500	9878	18 955	26 390	32 267

2. Using this detector, an astronomer measures the magnitude difference between two stars to be 1.25 magnitudes. He used the same integration time for each star, but failed to make any correction for the non-linearity of the detector. Compute the systematic error he made in the magnitude difference if the brighter star produced a signal of 30 000 DN.

3. The table at left below gives the coordinates and pixel values near a faint star on an array image. The small array at right is a sample of the nearby background. Find the x, y coordinates of the centroid of the star image using the criteria outlined in Section 9.4.1 of the text. Use a spreadsheet.

$y\backslash x$	1	2	3	4	5	6	7		
8	23	20	17	19	18	17	23		
7	18	25	20	18	26	18	19	16	19
6	20	27	33	30	27	23	18	14	16
5	19	31	40	34	28	22	25	13	11
4	26	29	53	51	28	28	21	21	18
3	22	26	40	32	33	18	24	16	17
2	23	30	26	24	26	23	14	20	18
1	16	19	20	18	18	17	16		

4. Suggest a strategy, similar to that in the latter part of Section 9.4.3, for combining N unaligned images to create a single mosaic image, $\mathbf{C}_{OR}$, that contains the combined data for every observed location in the collection. Assume that the observed area is contiguous, but that there is no area common to all images.

5. On a 20-second exposure, a star with magnitude $B = 15$ produces an SNR $= 100$ signal with a small telescope/CCD combination. Assuming this is a photon-noise limited case, how long an exposure should be required to produce the same SNR for star with $B = 13.6$?

6. A star with $V = 21.0$ is known to produce a count rate of 10 electrons per second for a certain telescope/detector combination. The detector read noise is 4 electrons per pixel, and the dark rate is zero. Compute the exposure time needed to reach a SNR $=$ 10 under the following conditions:

   (a) dark sky and good seeing: aperture radius $=$ 3.5 pixels, sky brightness $=$ 1.4 electrons per pixel per second;

   (b) moonlit sky and poor seeing: aperture radius $=$ 5.0 pixels, sky brightness $=$ 4 electrons per pixel per second.

7. A certain CCD has a gain of 2.4 electrons per ADU, a read noise of 7 electrons per pixel, and a dark current of 2.5 ADU per pixel per second. In the V filter, the sky brightness averages 8 ADU per second. An astronomer wishes to observe a nebula whose average brightness is expected to be 7 ADU per pixel per second over a digital aperture area of 100 pixels. Compute the expected SNR for measurements of the nebula's brightness on exposures of (a) 1 second, (b) 10 seconds, and (c) 100 seconds.

Chapter 10
Photometry

> The classification of the stars of the celestial sphere, according to different orders of magnitude, was made by ancient astronomers in an arbitrary manner, without any pretension to accuracy. From the nature of things, this vagueness has been continued in the modern catalogs.
>
> – François Arago, *Popular Astronomy*, Vol. I, 1851

Astronomers have measured apparent brightness since ancient times, and, as is usual in science, technology has acutely influenced their success. Prior to the 1860s, observers estimated brightness using only their eyes, expressing the results in the uncannily persistent magnitude system that Ptolemy introduced in the second century.[1] As Arago notes, the results were not satisfactory.

In this chapter, after a brief summary of the history of photometry, we will examine in detail the surprisingly complex process for answering the question: how bright is that object? To do so, we will first introduce the notion of a defined bandpass and its quantitative description, as well as the use of such bandpasses in the creation of standard photometric systems. Photometry is most useful if it represents the unadulterated light from the object of interest, so we will take some pain to describe how various effects might alter that light: spectrum shifts, absorption by interstellar material, and the characteristics of the observing system. We will pay particular attention, however, to the heavy burden of the ground-based photometrist: the influence of the terrestrial atmosphere and the techniques that might remove it.

---

[1] The magnitude system may very well pre-date Ptolemy. Ptolemy's catalog in the Almagest (*c.* 137 CE) may be based substantially on the earlier catalog of Hipparchus (*c.* 130 BCE), which has not been preserved. It is unclear which astronomer – Ptolemy, Hipparchus, or another – actually introduced the scale. Moreover, Ptolemy is largely silent on the practical method used to establish the magnitudes he recorded. Although Ptolemy tends to assign stars integral magnitudes, he designates 156 stars (out of 1028) as slightly (one third of a magnitude?) brighter or fainter than an integral value.

10.1 Introduction: a short history

The history of photometry is brief compared to that of astrometry, due to the symbiotic absences of scientific interest and appropriate instrumentation. John B. Hearnshaw (1996) provides a book-length history of astronomical photometry up to 1970. Harold Weaver (1946) gives a shorter and more technical account of developments up through World War II. A definitive history of the charge-coupled device (CCD) era remains unwritten.

To what degree will two stars assigned the same magnitude by a naked-eye observer actually have the same brightness? Modern measurements show pre-telescopic catalogs (e.g. Ptolemy and Tycho, both of whom were more interested in positions than in brightness) have an internal precision of about 0.5 magnitudes. Even the most skilled naked-eye observer can do little better: Al-Sufi in the ninth century devoted great attention to the problem and achieved a precision near 0.4 magnitudes. At the eyepiece of a telescope, several observers (e.g. the Herschels and, less successfully, the *Bonner Durchmusterung* observers Argelander and Schonfeld) produced better results (0.1–0.3 magnitudes) with a method of careful comparison to linked *sequences* of brightness standards.

After a suggestion by the French physicist François Arago (1786–1853), Karl Friedrich Zöllner (1834–82) built the first optical/mechanical system for astronomical photometry in 1861. Many similar instruments soon followed. An observer using one of these *visual photometers* either adjusts the brightness of a comparison until it matches that of the unknown star, or dims the telescopic brightness of the unknown star until it disappears. Zöllner's instrument, for example, used crossed polarizers to adjust the image of an artificial star produced by a kerosene lamp.

Because it avoids the need for a standard sequence near the unknown in the sky, the visual photometer was efficient. Moreover, these devices were *precise*, because brains are much better at judging equality (or complete extinction) than at making interpolations, especially interpolations based on memory of a sequence. Finally, the visual photometer was more *accurate* since making a mechanical adjustment gives a quantifiable measure fairly independent of a particular astronomer's eye and brain.

Astronomers got busy. Edward Pickering, at Harvard, for example, built a two-telescope "meridian photometer," which used crossed polarizers to equalize the images of two real stars. Between 1879 and 1902, Harvard visual photometrists measured the magnitudes of about 47 000 stars with a precision of about 0.08 magnitudes, and with an accuracy (based on modern measurements) of better than 0.25 magnitudes. Astronomers could now confidently examine the mathematical relationship between brightness and the ancient magnitude scale. Although several fits were proposed, by 1900 everyone had settled on the now familiar "Pogson normal scale":

$$\Delta m = -2.5 \log(b_1/b_2)$$

where b_1 and b_2 are the brightness of objects 1 and 2. The ancient scale turned out to be quite non-uniform in the logarithm: for example, the average brightness ratio between Ptolemy's magnitude 1.0 and 2.0 stars is 3.6, but between his 5.0 and 6.0 stars it is 1.3. The telescopic scales (e.g. Argelander) are closer to Pogson normal.

While the Harvard visual work progressed, photography matured. In 1850, William Cranch Bond and John Whipple, also at Harvard, photographed a few of the brightest stars. The invention of dry photographic plates (1871) increased convenience and sensitivity; eventually (around 1881) stars were recorded that were too faint to be seen by eye in any telescope. Many influential astronomers appreciated the vast potential of this new panoramic detector and with virtually unprecedented international cooperation launched the *Carte du Ciel* project to photograph the entire sky and measure the brightness of every star below magnitude 11.0. Astronomers soon learned to appreciate the difficulties in using photographs for quantitative photometric work, and it was not until the period 1900–10 that several workers (notably Schwarzschild, Wirtz, Wilkins, and Kapteyn) established the first reliable *photographic magnitude scales*. After the introduction (1910–20) of physical photometers for objectively measuring images on plates, photography could yield magnitudes with uncertainties in the range 0.015–0.03 magnitudes. Such precision required very great care in the preparation, processing, and reduction of plate material, and could usually only be achieved in differential measurements among stars on the same plate.

In the first sustained photoelectric work, Joel Stebbins and his students at Illinois and Wisconsin performed extensive and precise photometry, first with selenium cells (1907), but soon with the vacuum photocell. Poor sensitivity at first limited the observations to very bright stars, but in 1932, when Albert Whitford and Stebbins added a vacuum-tube amplifier to the detector circuit, detection limits on their 0.5-m telescope improved from 11th to 13th magnitude. The real revolution occurred in the 1940s, when the *photomultiplier tube* (*PMT*), developed for the military during World War II, became the astronomical instrument of choice for most precision work. It had very good sensitivity and produced uncertainties on the order 0.005 magnitudes in relative brightness.

The years from 1950 to 1980 were immensely productive for ground-based photoelectric work. Harold Johnson pioneered in this era, first using the RCA 1P21 photomultiplier to define the UBV system, and later using red-sensitive photomultipliers to define an extended broadband system through the visual and near-infrared atmospheric windows.

Although astronomers still use photomultipliers for specialized work today, the CCD, IR arrays, and other modern solid-state detectors have superseded them. In the optical, CCDs have superior efficiency, better stability, and a huge multiplex advantage (i.e. they can record many objects simultaneously). For ground-based differential work, CCD photometric precision on bright sources is generally set by photon-counting statistics (e.g. Equation (9.25)) or by uncertainties in calibration. For all-sky photometry and infrared work, the atmosphere imposes more serious limitations – 0.01 magnitude uncertainty is often regarded as routine. Photometry from spacecraft, on the other hand, offers the potential of superb precision in both differential and all-sky work. For example, the *Kepler* space mission (primary mission 2009–15) detects occultations by extrasolar planets, achieving precisions of around 100 µmag for stars brighter than 12th magnitude (each measurement is a combination of 270 6-second exposures). Observations from space are very, very costly, however, so ground-based photometry continues to be a central astronomical activity.

10.2 The photometric response function

A photometric device is sensitive over a restricted range of wavelengths called its **bandpass**. We distinguish three general cases of bandpass photometry to fit three different scientific questions.

10.2.1 Types of photometry

Single-band photometry

Suppose, for example, you suspect an extrasolar planet will move in front of a certain star, and you are interested in the occultation's duration and the fraction of the star's light blocked. You need only use a single band, since a geometric effect like the occultation of a uniform source will be identical at every wavelength. You would probably make a sequence of monitoring observations called a *time series*, a tabulation of brightness as a function of time, and you would tend to choose a wide band (e.g. Kepler uses 430–860 nm) to maximize signal and minimize the required exposure time and telescope size.

Broadband multicolor photometry

On the other hand, you might want to know not just the brightness of a source, but also the general shape of its spectrum. Broadband multicolor photometry measures an ultra-low-resolution spectrum by sampling the brightness in several different bands. A "broad" band is generally taken to mean that the width of the band, $\Delta\lambda$, divided by its central wavelength, λ_c, is greater than 7–10%, or, equivalently, the spectroscopic resolving power $R = \lambda_c/\Delta\lambda < 10$–15. The idea is to admit the maximum amount of light while still providing valuable astrophysical information. For example, the $UBVR_cI_c$ system, the most common broadband system in the optical, uses bandwidths in the range 65–160 nm ($R = 4$–7). It provides information on surface temperature for a wide variety of stars, and more limited information on luminosity, metal content, and interstellar reddening.

The terminology recognizes each band as a "color," so "two-color photometry" measures magnitudes in two separate bands: B and V, for example. For both historical and practical reasons, one traditionally reports the results of n-color photometric measurements by giving one magnitude and $(n-1)$ color indices. The magnitude tells the apparent brightness, and the indices tell about other astrophysical variables like surface temperature. The term "color," as shorthand for "color index" has thus come to have a second meaning – *color is the difference between two magnitudes*. So for example, the results of "*two*-color photometry" in B and V will be reported as a V magnitude and *one* $(B - V)$ color.

Narrow- and intermediate-band photometry

Although multicolor narrow-band photometry (roughly $R > 50$) can provide information about the shape of the spectrum, its intent is usually to isolate a specific line, molecular band, or other feature. The strategy here exchanges the large signal of the broadband system for a weaker signal with more detailed spectroscopic information. Common applications include the measurement of the strength of absorption features like Balmer-alpha or sodium D, or of the ratio of the intensities of emission lines in gaseous nebulae. Intermediate-band photometry ($15 < R < 50$) measures spectroscopic features that cannot be resolved with broader bands, but avoids the severe light loss of the very narrow bands. Examples of such features include discontinuities in spectra (for example, the "Balmer discontinuity" due to the onset of continuous absorption by hydrogen in stellar atmospheres at a wavelength of 364.6 nm), or very broad absorption features due to blended lines or molecular bands (for example, the band due to TiO in the spectra of M stars that extends from 705 to 730 nm).

10.2.2 Magnitudes

Recall that for some band (call it P), the *apparent magnitude* of the source is

$$m_\mathrm{P} = -2.5\log(F_\mathrm{P}) + C_\mathrm{P} = -2.5\log \int_0^\infty R_\mathrm{P}(\lambda) f_\lambda \, \mathrm{d}\lambda + C_\mathrm{P} \qquad (10.1)$$

Here m_P is the bandpass magnitude; F_P is the energy flux (the irradiance) within the band; f_λ is the monochromatic flux (also called the flux density or the monochromatic irradiance – it has units of watts per square meter of area per unit wavelength, or W m^{-3}). We choose the constant C_P to conform to some standard scale (e.g. the magnitude of Vega is zero in the visual system). The function $R_\mathrm{P}(\lambda)$ describes the **response** of the entire observing system to the incident flux: it is the fraction of the energy of wavelength λ that will register on the photometer. We usually assume that f_λ is measured *outside* the Earth's atmosphere.

Photon detectors count photons, rather than measure energy directly. Recall that the **monochromatic photon flux** $\phi(\lambda)$ (number of photons per second per square meter of area per unit wavelength) is related to f_λ:

$$\phi(\lambda) = \frac{\lambda}{hc} f_\lambda \tag{10.2}$$

Photon detectors do *not* directly measure the quantity Γ_P in Equation (10.1) but report a signal proportional to the **photon flux within the band**:

$$\Phi_P = \int_0^\infty R_{PP}(\lambda)\phi(\lambda)\mathrm{d}\lambda = \frac{1}{hc}\int_0^\infty R_{PP}(\lambda)f_\lambda\lambda\mathrm{d}\lambda \tag{10.3}$$

Here $R_{PP}(\lambda)$ is the **photon response**: the fraction of photons of wavelength λ detected by the system. This suggests that photon-counting detectors and energy-measuring detectors will measure on the same magnitude scale if

$$m_P = -2.5\log(\Phi_P) + C_{PP} = -2.5\log(F_P) + C_P \tag{10.4}$$

which requires

$$R_{PP}(\lambda) \propto \frac{R_P(\lambda)}{\lambda} \tag{10.5}$$

Although directly *measured* magnitudes are bandpass magnitudes, it makes perfect sense to talk about and compute a **monochromatic magnitude**. This is defined from the monochromatic flux:

$$m_\lambda = -2.5\log(f_\lambda) + C'(\lambda) = -2.5\log\frac{hc\phi(\lambda)}{\lambda} + C'(\lambda) \tag{10.6}$$

Here again, the value of the function $C'(\lambda)$ is arbitrary, but is often chosen so that the monochromatic magnitude of Vega or some other (perhaps fictitious) standard is a constant at every wavelength. In this case, $C'(\lambda)$ is a strong function of wavelength. Sometimes, however, the function $C'(\lambda)$ is taken to be a constant, and the monochromatic magnitude reflects the spectrum in energy units. You can think of the monochromatic magnitude as the magnitude measured with an infinitesimally narrow band. Conversely, you can think of intermediate or broadband photometry as yielding a value for m_λ at the effective wavelengths of the bands, so long as you recognize the energy distribution referenced is one of very low spectroscopic resolution.

10.2.3 Response function implementation

How is a band response implemented? Both practical constraints and intended controls can determine the functional form of the responses $R_P(\lambda)$ or $R_{PP}(\lambda)$.

The *sensitivity of the detector* clearly limits the range of wavelengths accessible. In some cases, detector response alone sets the bandpass. Ptolemy, for example, based his magnitude system simply on the response of dark-adapted human vision, sensitive in the band 460–550 nm. In other cases, the detector response defines only one edge of the band. Early photographic magnitudes, for example, had a bandpass whose long-wavelength cutoff was set by the insensitivity of the photographic emulsion longward of 450 nm.

A *filter* – an element placed in the optical path to restrict transmission – is the usual method for intentionally delimiting a band. A *bandpass filter* defines both ends of the band by blocking all wavelengths except for those in a specific range. A filter can serve as a *high-pass* or *low-pass* element by defining only the lower or upper cutoff of a band. Filters that limit the transmission of all wavelengths equally are termed *neutral-density filters*.

Another strategy for photometry is to use a dispersing element to create a spectrum. Sampling discrete segments of a spectrum with one or more photodetectors is equivalent to multi-band photometry. Such instruments are termed *spectrophotometers*. A spectrophotometer (see Figure 10.1) generally defines bandpasses by using apertures, slots, or detector pixels of the proper size to select the desired segment of the spectrum. Multi-pixel solid-state detectors like CCDs blur the distinction between a spectrophotometer and a spectrograph: taking a CCD image of a spectrum is equivalent to letting each pixel act as an aperture that defines a band.

For ground-based observations, *atmospheric transmission*, $S_{atm}(\lambda)$, limits the wavelengths that are accessible, and may completely or partially define a response function. Absorption in the Earth's atmosphere set the short-wavelength cutoff of early photographic photometry at 320 nm, for example. In the infrared, absorption by water vapor is significant and variable. Figure 10.2 shows the approximate atmospheric transmission in the near infrared from 0.8 to 2.6 μm expected at a high elevation site. Also marked on the plot are the half-widths of the Johnson J and K bands as defined by filter transmission only. In these bands the atmosphere will set the long cutoff of J and the short cutoff of the K band, and variations in the atmosphere may change the shape of the overall photometric response function. Normally, however, magnitudes are defined outside the Earth's atmosphere, and an astronomer usually removes atmospheric effects during data reduction.

Fig. 10.1 A spectro-photometer. Each aperture defines the range of wavelengths that pass to its detector. It is possible to alter the wavelengths sampled by rotating the dispersing element or translating the apertures. In this case, the instrument is known as a spectrum scanner.

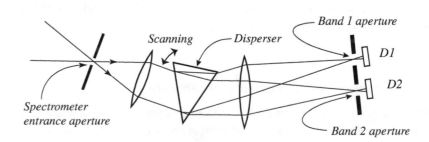

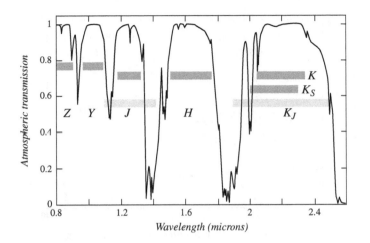

Fig. 10.2 Atmospheric transmission in the near infrared. Transmission curve is based on a model of the atmosphere at an elevation of 2.0 km, and will change with changes in water-vapor content. Light-gray lines locate the Johnson J and K photometric band-filter sensitivity (FWHM). Dark gray lines show the sensitivity of the MKO filters for J, H, and K. The Johnson band definitions are much more susceptible to water-vapor variation than are the MKO definitions. Also shown are the UKIRT Z and Y bands.

As an example of response definition, Figure 10.3 shows how four different factors interact to produce the response of the Johnson U band:

1. The transmissions of the filter – Corning glass number 9863 in Johnson's original definition.
2. The quantum efficiency (QE) of the detector as a function of wavelength. In this case, the detector was a particular photomultiplier, the RCA 1P21, now obsolete. The glass window of early tubes was later replaced with fused quartz, changing the short-wavelength transmission.
3. The transmission of the atmosphere, $S_{atm}(\lambda)$. Photometry in this band assumes that the object is at the zenith, and that the ozone partial pressure is 3 mm. Changes in ozone concentration or zenith angle change the shape of $R_U(\lambda)$. For a PMT with a quartz window, the atmosphere sets the short-wavelength cutoff, a troublesome feature.
4. Transmission of the telescope optics. This is not plotted in the figure, since the reflectivity of freshly deposited aluminum is nearly constant in this region, with a value of around 0.92. Use of glass lenses, windows, or silver surfaces would change the shape of the response function.

10.2.4 Response function description

You will encounter various terms describing the response function. For example, for most responses, there will be a single maximum value, R_{max}, which occurs at the **peak wavelength** λ_{peak}. Likewise, there are usually (only) two half-maximum points. These can be taken as indications of the wavelengths at which transmission begins and ends, λ_{low} and λ_{high}:

$$R(\lambda_{low}) = R(\lambda_{high}) = R_{max}/2 \qquad (10.7)$$

Given the half maxima, we can then define one measure for the width of the response by computing the **full width at half-maximum**:

Fig. 10.3 Response function (shaded) for the Johnson U band. The function $R(\lambda)$ is the product of (1) the filter transmission, (2) the detector quantum efficiency with either a quartz or a glass window, and (3) the transmission of the atmosphere (two extremes, 4 mm and 2 mm of O_3, are indicated). The telescope reflective optics do not affect the shape of $R(\lambda)$.

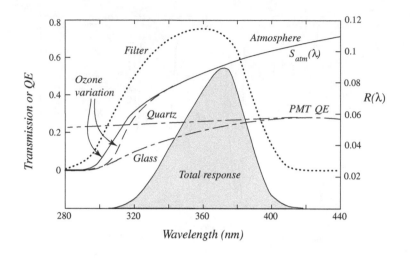

$$\text{FWHM} = \lambda_{\text{high}} - \lambda_{\text{low}} \tag{10.8}$$

The half-maximum points also determine the ***central wavelength*** of the band:

$$\lambda_{\text{cen}} = \left(\lambda_{low} + \lambda_{\text{high}}\right)/2 \tag{10.9}$$

A somewhat more sophisticated and possibly more useful measure of the width of a particular response function is the ***bandwidth***:

$$W_0 = \frac{1}{R_{\text{max}}} \int R(\lambda)\mathrm{d}\lambda \tag{10.10}$$

Likewise, a somewhat more sophisticated measure of the center of a band is its ***mean wavelength***, which is just:

$$\lambda_0 = \frac{\int \lambda \cdot R(\lambda)\mathrm{d}\lambda}{\int R(\lambda)\mathrm{d}\lambda} \tag{10.11}$$

Figure 10.4 illustrates these relations. For a symmetric bandpass, $\lambda_{\text{peak}} = \lambda_{\text{cen}} = \lambda_0$.

Perhaps even more informative is the ***effective wavelength*** of the response to a particular source. The effective wavelength is a weighted mean wavelength (weighted by the source flux) and indicates which photons most influence a particular measurement:

$$\lambda_{\text{eff}} = \frac{\int \lambda \cdot f_\lambda \cdot R(\lambda)\mathrm{d}\lambda}{\int f_\lambda \cdot R(\lambda)\mathrm{d}\lambda} \tag{10.12}$$

Figure 10.5 illustrates that different sources will in general have different effective wavelengths.

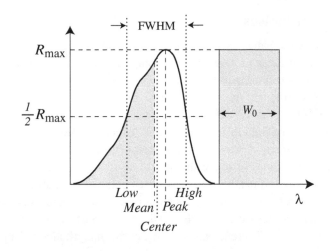

Fig. 10.4 Definitions of the middle and width of a band. The curve shows the function $R(\lambda)$. The mean wavelength divides the area under the curve into two equal parts (shaded and unshaded). The dark gray rectangle has a width equal to the bandwidth and an area equal to the area under the curve.

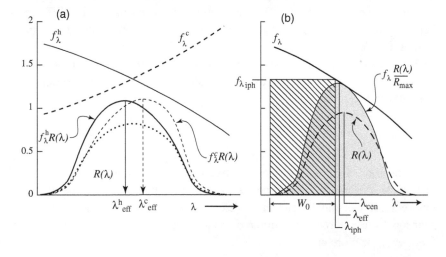

Fig. 10.5 (a) Effective wavelengths for two different sources in the same band. The dotted curve is $R(\lambda)$. The solid curves apply to a hot source, and the dashed curves apply to a cool source with the same magnitude in the band. (b) Definition of the isophotal wavelength: the area of the hatched rectangle is the same as the shaded area under the curve. The dashed curve is the response function.

You might think of any bandpass measurement as equivalent to a measurement of the monochromatic flux at wavelength λ_{eff} multiplied by the bandwidth, W_0. This is nearly correct in practice, and for broadband photometry of stars (provided spectra are sufficiently smoothed) using this equivalence produces an error of a percent or less. To be even more accurate with such an equivalence, we can introduce yet another definition for the "middle" of the band. This one is called the **isophotal wavelength**, λ_{iph}. The isophotal wavelength is the one for which we have

$$W_0 R_{max} \cdot f_{\lambda_{iph}} = \int f_\lambda \cdot R(\lambda) d\lambda = F \qquad (10.13)$$

As with the effective wavelength, the exact value of the isophotal wavelength will depend on the spectrum of the source.

10.2.5 Color indices

If you think of the bands as sampling the monochromatic flux of a smoothed spectrum at their mean or isophotal wavelengths, then you can see that multi-band photometry can measure the shape of an object's spectrum. For example, Figure 10.6 shows the spectra of several blackbodies whose temperatures range from 1600 K to 12 000 K. The vertical scale of the figure shows the monochromatic magnitude in a system in which the constant in Equation (10.6) is independent of wavelength. Remember, this is *not* the usual case in astronomical photometry, where the spectrum of some standard object (e.g. Vega, which is similar to a blackbody with temperature of 9500 K), would be a horizontal line in a plot of m_λ as a function of λ. In the figure, we assume two bands, one with a mean wavelength at 0.4 μm, the other at 0.8 μm. It is clear that the arithmetical difference between these two magnitudes for a particular spectrum depends on the average slope of the spectrum, which in turn depends on the source's temperature. The convention is to speak of the difference between any two bandpass magnitudes used to sample the slope of the spectrum as a ***color index***.

For blackbodies, at least, the color index is not just useful, but definitive – its value uniquely measures the body's temperature. By convention, you compute the index in the sense:

$$\text{index} = m(\text{shorter } \lambda) - m(\text{longer } \lambda) \qquad (10.14)$$

As mentioned earlier, astronomers usually symbolize the color index as the magnitude difference, sometimes enclosed in parenthesis. In the case of Figure 10.6, we might write the index as $(m_{0.4} - m_{0.8})$. In the case of the Johnson–Cousins red and infrared bands, the index would be written $(m_R - m_I)$, or more commonly, $R - I$.

Fig. 10.6 Color indices for blackbodies. Curves are generated by taking the logarithm of the Planck function. Note that monochromatic magnitudes increase downwards. Spectra have been shifted vertically by arbitrary amounts for clarity. In this figure, $\Delta C = 0$.

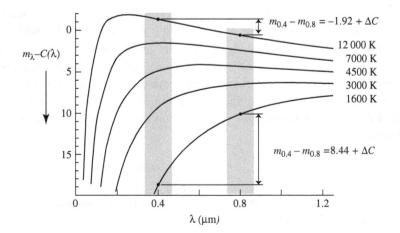

The behavior of the color index at the long and short wavelength extremes of the Planck function is interesting. In the Rayleigh–Jeans region (i.e. where $\lambda kT \gg hc$) you can show that

$$m_\lambda = \log T + C(\lambda) \tag{10.15}$$

so that the color index becomes,

$$(m_{\lambda_1} - m_{\lambda_2}) = C(\lambda_1) - C(\lambda_2) = \Delta C \tag{10.16}$$

a constant independent of temperature. For example, in the Johnson broadband system, a blackbody of infinite temperature has color indices:

$$(U - B) = -1.33, (B - V) = -0.46 \tag{10.17}$$

At short wavelengths, the **Wien approximation** for the surface brightness of a blackbody holds:

$$B(\lambda, T) \approx \frac{2hc^2}{\lambda^5} \exp\left(-\frac{hc}{\lambda kT}\right) \tag{10.18}$$

So the color index is

$$(m_{\lambda_1} - m_{\lambda_2}) = \frac{a}{T}\left(\frac{1}{\lambda_1} - \frac{1}{\lambda_2}\right) + \Delta C \tag{10.19}$$

10.2.6 Line and feature indices

Real objects almost always have more complex spectra than do blackbodies, with features of astrophysical significance that may include absorption and emission lines, bands, and various discontinuities. Multi-band photometric indices can measure the strength of such features.

Two bands often suffice to measure the size of a discontinuity or the strength of a line, for example. In Figure 10.7a, bands C and D sample the continuum on the short- and long-wavelength sides of a sharp break in a spectrum. The index $(C - D)$ will be sensitive to the size of the break – but note two features of the index:

First, the actual relation between the size of the break and the numerical value of the $(C - D)$ index depends on the constants employed in the definition of the bandpass magnitudes in Equation (10.1). It might be convenient to have $(C - D) = 0$ when the break vanishes, but this may violate the convention that all indices should be zero for the spectrum of some standard object. (Examine Figure 1.5 – Vega has several non-zero spectrum discontinuities, yet all its indices are zero in some systems.)

Second, positioning the bands is important. The sensitivity of the index to the size of the break will diminish if either bandpass response includes light from the opposite side of the break. Likewise, if a band is located too far away

Fig. 10.7 Definition of
indices to measure the
strength of (a) a spectrum
discontinuity, and (b) an
absorption line.
Monochromatic
magnitudes are defined
so that the constant in
Equation (10.6) is
independent of
wavelength.

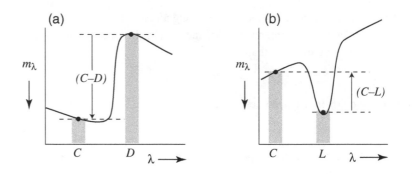

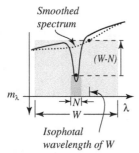

Fig. 10.8 A line index
computed from wide and
narrow bands centered on
the same absorption line.

from the break, unrelated features in the spectrum can affect the index. Obviously, it will be easier to position narrow bands than wide bands, but narrow bands give weaker signals.

A similar index can measure the intensity of an absorption or emission line (Figure 10.7b). Here one narrow band is centered on the feature, and the other on the nearby continuum. The magnitude difference measures the line strength. This strategy is common in detecting and mapping objects with a strong emission line in their spectra: for example, the astronomer takes two CCD exposures – one through each filter. Digital division of the two registered and properly scaled images produces a third image sensitive to the relative strength of the line.

Figure 10.8 illustrates an alternative strategy for measuring a line index. Two bands – one broad, the other narrow – are both centered on the line. The narrow band is quite sensitive to the strength of the line, while the broad band is relatively insensitive, since most of the light it measures comes from the continuum. The index

$$\text{line index} = m_{\text{narrow}} - m_{\text{wide}} \qquad (10.20)$$

tracks the strength of the absorption or emission, in the sense that it becomes more positive with stronger absorption. One widely used line index of this sort is the β index, which measures the strength of the Balmer beta line of hydrogen, useful for luminosity or temperature classification of stars.

Finally, consider a third kind of index. Three bands can measure the *curvature* (i.e. the second derivative, rather than the first) of a spectrum. Curvature can arise on a relatively small scale because of a sharp absorption or emission line, or on a large scale because of broad or diffuse features (molecular bands in gases, or absorption features in the reflection spectra of crystalline solids, for example). Figure 10.9 illustrates two situations with three (a) equally and (b) unequally spaced bands at a short, central, and long wavelength (S, C, and L). If we consider just the monochromatic magnitudes, and if the bands are equally spaced as in Figure 10.9a, the index

$$\text{curvature} = (m_{\text{S}} - m_{\text{C}}) - (m_{\text{C}} - m_{\text{L}}) = S + L - 2C \qquad (10.21)$$

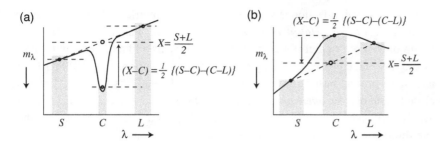

Fig. 10.9 Three bands can measure the curvature of the spectrum. In both (a) and (b), the index $2(X - C)$ tracks the monochromatic magnitude's departure from linearity.

will be zero if the logarithmic spectrum is linear, and positive if the central band contains an absorption feature. *The curvature index depends on the difference between two color indices.* In practical systems the index will still track curvature even if bands are not equally spaced, and even if $C'(\lambda)$ in Equation (10.6) is *not* a constant.

10.3 The idea of a photometric system

The term ***photometric system*** implies at least two specifications:

1. The wavelength response for each band – that is, the shape of the function $R_P(\lambda)$ in Equation (10.1).
2. Some method for *standardizing* measurements made in those bands. This is important for two reasons:
   - Each observer needs to know the value for the constant C in Equation (10.1) that will insure that her magnitudes agree with those of all other observers.
   - The differing hardware make perfect replication of $R_P(\lambda)$ unlikely, so standardization allows correction of the inevitable systematic effects.

The first specification, that of $R_P(\lambda)$, determines the ***instrumental*** or ***natural system***. The first and second together determine the ***standard system***. Observations in the natural system alone can be quite useful (e.g. determining the period of a variable star), but only by placing magnitudes on the standard system can two astronomers confidently compare independent measurements.

Standardization could involve observations of laboratory sources, e.g. a blackbody of known temperature and therefore known absolute flux in $\mathrm{W\,m^{-2}}$. Usually, though, a single astronomical object or set of objects is a much more practical standardizing source. Almost all standard systems today rely upon some network of constant-brightness standard objects distributed around the sky. If everyone agrees on a list of stars and their corresponding magnitudes, anyone can calibrate measurements made in their instrumental system by observing the standards and the unknowns with the same procedures. Because systematic differences will most likely arise if the spectrum of the star observed is different from the spectrum of the standard star, most systems strive to define a set of standards that includes a wide variety of spectral types.

Because standardization is so essential to a photometric system, some astronomers have devised *closed photometric systems*, in which a relatively small group of observers carefully controls the instruments and data reduction, maximizing internal consistency. Many space-based observations (e.g. HIPPARCOS) and many ground-based surveys (e.g. the Sloan Digital Sky Survey) constitute closed systems. An *open photometric system*, in contrast, is one in which all astronomers are encouraged to duplicate the defined natural system as best they can through reference to a published list of standard stars.

10.4 Common photometric systems

Astronomers have introduced several hundred photometric systems. Bessell (2005) gives an extensive review of the most common systems and the Asiago Database http://ulisse.pd.astro.it/Astro/ADPS/enter2.html gives a more extensive catalog. Here we examine only a few of the most widely used as an introduction to the operation of most.

10.4.1 Visual and photographic systems

The dark-adapted human eye determines the band of the *visual photometric system*. In the earliest days of astronomy, the standardization procedure required that magnitudes measured in the system be consistent with the ancient catalogs (e.g. Ptolemy, Al-Sufi, and Bayer). The introduction of optical/mechanical visual photometers led to the establishment of *standard sequences* of stars, including (initially) the *north polar sequence* and (later) many secondary sequences (the 48 Harvard standard regions and the 115 Kapteyn selected areas were perhaps the best studied).

In the early twentieth century, astronomers defined two bands based on the properties of the photographic emulsion (Table 10.1). The poor properties of the emulsion as a photometric detector and lack of very specific definitions limited the success of this system. The *international photographic band* is sensitive in the near ultraviolet–blue region. The response of the *international photovisual band*, somewhat fortuitously, roughly corresponds to that of the visual band (i.e. the human eye, sensitive to green–yellow). The IAU in 1922 set the zero point of both magnitudes so that 6th magnitude A0 V stars[2] in the north polar sequence would have (roughly) the same values as on the old Harvard visual system. This meant that the color index,

---

[2] A0 V is the spectral type of Vega, which is *not* in the north polar sequence. Because of the early decision to keep visual magnitudes roughly consistent with the ancient catalogs, the photographic and photovisual magnitudes of Vega turn out to be close to zero. The importance of Vega stems in part from its brightness, which makes it a good candidate for absolute (i.e. watts per square meter per meter of wavelength) measurement of specific irradiance.

Table 10.1 *Bandpasses of historical importance.*

Band	Symbol	Band definition	λ_{peak}, nm	FWHM
Visual	m_{vis}	Mesotopic* human eye	515 (550)	82 (106)
International photographic	m_{pg}, IPg	Untreated photographic emulsion + atmosphere	400	170
International photovisual	m_{pv}, IPv	Orthochromatic emulsion + yellow filter	550	100

* Visual photometry of stars uses a mixture of photopic (color, or cone) and scotopic (rod) vision, with the shift from cones to rods occurring with decreasing levels of illumination. The effective wavelength of the eye thus shifts to the blue as light levels decrease (the Purkinje effect); see Appendix B3.

$$CI = m_{pg} - m_{pv} \qquad (10.22)$$

should be zero for A0 stars, negative for hotter stars, and positive for cooler stars.

The photovisual magnitude originally depended on "orthochromatic" plates, which were made by treating the emulsion with a dye to extend its sensitivity to about 610 nm. Other dyes eventually became available to extend photographic sensitivity to various cutoffs ranging through the visible and into the near infrared. Twentieth-century astronomers devised many filter–emulsion combinations and set up standard star sequences in a variety of photography-based systems. All these are mainly of historic interest.

10.4.2 The UBVRI system

By far the most widely used ground-based photometric system prior to the present has been the Johnson–Cousins UBVRI system (Table 10.2 and Figure 10.10). Johnson and Harris (1954) defined the UBV portion first, based on the response of the RCA 1P21 photomultiplier, a set of colored glass filters, and a list of magnitudes for a relatively small number of standard stars scattered around the celestial sphere. The V band closely corresponds to the international photovisual band and its zero point was set so that $V = m_{pv}$ for the standards in the north polar sequence. The U and B bands correspond to short- and long-wavelength segments of the photographic band, and to be consistent with the international system, their zero points are set so that the colors $U - B$ and $B - V$ are zero for A0 V stars.

Table 10.2 *The Johnson–Cousins UBVRI system. The R_J and I_J data are from Colina et al. (1996). All other widths are from Bessell (1992). Effective wavelengths and monochromatic fluxes for a zero-magnitude, zero-color star are from the absolute calibration of Vega and Sirius by Bessell et al. (1998). Vega has V = 0.03 on this system. Units are $10^{-12} \, Wm^{-2} \, nm^{-1}$.*

	U	B	V	R_C	R_J	I_C	I_J
λ_{eff}, nm	366	436	545	641	685	798	864
FWHM	66	94	88	138	174	149	197
f_λ at λ_{eff}	41.7	63.2	37.4	22.6	19.2	11.4	9.39

After some early work at longer wavelengths by Stebbins, Kron, and Whitford, Harold Johnson and his collaborators in the period 1960–65 extended the UBV system to include bands in the red (R_J) and near infrared (I_J), as well as the longer infrared bands (JHKLMNQ) discussed in the next section. Modern work with CCDs, however, has tended to replace the R_J and I_J with the R_C and I_C bands specified by Cousins and his collaborators (see Table 10.2). In current practice, the lists of Arlo Landolt (1983, 1992) and Menzies et al. (1989, 1991) define the standard stars for the UBV(RI)$_C$ system.

Modern CCD observers sometimes have difficulty replicating the original photomultiplier-based instrumental system. A complicating factor is the great variation in CCD spectral response due to differing surface treatments, gate material, gate structure, backside illumination, etc. Close matches are possible with a good knowledge of the individual CCD response and a careful choice of filters. For details, see Bessell (1990).

The UBV(RI)$_C$ system was designed with the rough spectral classification of stars in mind. Figure 10.10 shows the responses of the normalized bandpasses superimposed on spectra of an A0 and a G2 dwarf (i.e. matching, respectively, Vega and the Sun). The $U - B$ index is clearly sensitive to the Balmer discontinuity, present very obviously in the A star at 370 nm, and much reduced in the G star. The discontinuity – and hence the $U - B$ index – depends upon luminosity, at least for hot stars. The other indices are primarily sensitive to temperature (and therefore spectral type). The $B - V$ color is more sensitive to metal abundance (In astrophysics, a "metal" is any element other than hydrogen or helium.) than are $V - R$ or $R - I$, and fails as a useful index for M stars because of molecular-band absorption. Because of its long baseline and relative insensitivity to chemical abundances, the $V - I$ index is the most purely temperature-sensitive index in this system ($V - K$ is even better, for the same reason). Appendix K tabulates the colors of various spectral types. The system is useful for measuring the photometric properties of objects besides normal stars: Solar System bodies, supernovae, galaxies, and quasars have all been extensively observed.

Table 10.3 *The bands for the SDSS five-color system, plus the y band. The monochromatic flux for an A0 star of magnitude V = 0 is give in units of 10^{-10} $Wm^{-2}\mu m^{-1}$. Values for the y band for LSST have not yet (2015) been set.*

Fsd	u	g	r	i	z	y
λ_{eff}, nm	354	467	616	747	892	(1000)
FWHM	57	139	137	153	95	(110)
f_λ (λ_{eff})	366	541	250	139	83	(60)

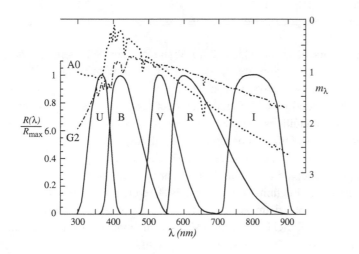

Fig. 10.10 Normalized response functions for the UBVRI system. Also shown are the monochromatic magnitudes for a representative A0 and G2 dwarf. Note the importance of the Balmer discontinuity near 370 nm in the A0 spectrum, and the break due to metal absorption near 400 nm in the G2 spectrum.

10.4.3 The SDSS *ugriz(y)* system

The Sloan Digital Sky Survey (SDSS), the automated ground-based program, used CCD drift scans on a 2.5-m telescope to produce photometry for over 10^8 stellar and non-stellar objects. The project introduced a five-color system (see Table 10.5) designed to extract astrophysical information while making optimal use of silicon CCD sensitivity. The SDSS database is larger than all the PMT-based UBVRI observations accumulated since the 1950s. The LSST will extend this system by adding a filter in the *y* band (1.09 mm). LSST will deploy a deep-depleted CCD camera with enormous étendue, so the volume of the SDSS database is shortly due to explode. Since the SDSS colors give as good or better astrophysical information, the SDSS may (or may not) displace UBVRI as the dominant broadband system in the visual.

10.4.4 The broadband infrared systems: ZYJHKLMNQ

The broadband infrared system (Table 10.4) might be regarded as an extension of the UBVRI system to longer wavelengths and shares a common zero point

Table 10.4 *The broadband infrared system. Z and Y from Hewett et al. (2006) (UKIDSS system), N and Q from Rieke et al. (1985). And the remainder from Tokunaga et al. (2002) (MKO system). Central wavelengths of L′ and M′ are significantly different from earlier L and M, hence the renaming. Monochromatic flux is given in units of $10^{-11}\ W^{-2}\ \mu m^{-1}$ for V = 0, B − V = 0 at the effective wavelength of the band.*

	X	Y	J	H	K	K_S	L′	M′	N	Q
λ_{eff}, μm for A0 stars	0.88	1.03	1.25	1.65	2.20	2.15	3.77	4.67	10.6	21
FWHM	0.08	0.11	0.11	0.16	0.29	0.29	0.34	0.70	3–6	6–10
f_λ at λ_{eff}	859	571	315	114	39.6	45.7	7.1	2.2	0.96	0.0064

(so the colors of an unreddened A0 V star are zero). Except for the Z band, detectors in this region are not silicon CCDs, but infrared arrays or single-channel infrared-sensitive devices.

For the ground-based infrared, bandpass definitions can depend critically on atmospheric conditions (mainly the amount of water vapor encountered along the line of sight). Different observatories with identical hardware can experience different infrared window sizes and shapes, and the same observatory can experience bandpass alterations due to changing humidity. There is thus some variation in band definition, but Table 10.4 represents an effective standard for several bands, based on systems in use at the world's largest IR telescopes (VISTA, UKIRT, IRTF), and the IAU recommendation for JHK in the Mauna Kea Observatory (MKO) near-infrared system.

10.4.5 The intermediate-band Strömgren system: uvbyβ

Bengt Strömgren introduced this intermediate-band system in the late 1950s, and David Crawford and many others developed it observationally in the 1960s and 1970s. The system avoids many of the shortcomings of the UBV system and aims to classify stars according to three characteristics: temperature, luminosity, and metal abundance. Classification works well for stars of spectral types B, A, F, and G, provided the photometry is sufficiently accurate. Photometrists frequently supplement the four intermediate-band colors, *uvby*, with a narrow-band index, β, which tracks the strength of absorption in the Balmer beta line. The β index greatly improves the luminosity classification for hotter stars, and is a good temperature indicator for cooler stars.

Emission in all of the four intermediate bands depends on temperature, but in addition, emission in the u and v bands is depressed by the presence of metals in a star's atmosphere. The u band is further depressed by the Balmer discontinuity, a temperature- and luminosity-dependent feature. To best represent astrophysical information, then, Strömgren photometry is generally presented as a y magnitude, a $(b - y)$ color, and two curvature indices. The $(b - y)$ color closely tracks temperature in the same way as the Johnson $B - V$ (in fact, $b - y \approx 0.68(B - V)$ over a large range of stellar types), but $(b - y)$ is somewhat less sensitive to abundance effects and is more useful at lower effective temperatures than is $B - V$. The two curvature indices are

$$c_1 = (u - v) - (v - b) \tag{10.23}$$

$$m_1 = (v - b) - (b - y) \tag{10.24}$$

The c_1 index measures the strength of the Balmer discontinuity, and in combination with temperature from $(b - y)$ yields information about luminosity. The m_1 index measures metal abundance. The precise relationships between the indices and the astrophysical parameters are more complex than suggested here, but they have been well calibrated for spectral types hotter than K0.

10.4.6 Other systems

Many other photometric systems find less widespread use than those just described, and it is helpful to describe a few examples.

Photometry from space need not contend with any of the atmospheric and many of the background issues that complicate photometry from the ground. Within the parameters of a given detector, space observatories permit much greater freedom to base bandpass design on purely astrophysical considerations. The NICMOS2 NIR camera on the Hubble Space Telescope (HST), for example, carried about 30 filters, many centered at bands completely inaccessible from the ground.

It is nevertheless very important to be able to tie space observations to ground-based measurements. The HIPPARCOS space mission, for example, used a two-filter broadband system closely related to B and V, while some of the HST NICMOS filters correspond to the JKLMN bands. The primary CCD camera for the HST (the WFPC/WFPC2), had slots for 48 filters, but those most commonly used closely matched the UBVRI system. Gaia is using a very broad band for astrometry, but records very low-resolution spectra to obtain multichannel photometry.

The HST standard magnitudes, incidentally, are defined on the **STMAG system** so that a source with constant f_λ has zero colors. Similarly, a star with constant f_ν has zero colors in the **Oke AB_ν magnitude system**, which defines monochromatic magnitudes as

$$m_\nu(AB) = -2.5\log(f_\nu) - 66.10 \tag{10.25}$$

The constant was chosen to so that $m_\nu(AB) = V$ at the center of the V band if f_ν is measured in $Wm^{-2}Hz^{-1}$. Either of these schemes permits a more direct connection to actual flux levels, and is especially appealing for use with spectrophotometers.

10.5 Absorption by the atmosphere

> ... Slowly the Emperor returned –
> Behind him Moscow! Its onion domes still burned ...
> Yesterday the Grand Army, today its dregs!
> ... They went to sleep ten thousand, woke up four.
>
> – *Victor Hugo*, "Russia 1812," trans. Robert Lowell

A grand army of photons leaves a source, but many are lost on their march to our telescope. This section follows one regiment of that army to consider its fortunes in detail. The goal of photometric reduction will be to reconstruct the original regiment from its dregs – to account for all losses and transformations during its long journey in the cold.

At least three different effects can alter the photons on their way to the telescope:

- wavelength shifts
- absorption in space outside the atmosphere
- atmospheric absorption

Because the last of these is both the most obvious and most easily accounted for, we will consider it first.

10.5.1 Atmospheric windows

The Earth's atmosphere removes photons from the stream directed at our telescope. Absorption (in which the photon ceases to exist) and scattering (in which the photon changes direction) are physically distinct processes, but they have the same effect on the regiment of photons headed toward our telescope – they remove photons from the beam. It is common to refer to both processes simply as "absorption." Atmospheric absorption will both reduce the apparent brightness of the source spectrum as well as alter its shape. We therefore distinguish the processes of *atmospheric extinction* and *atmospheric reddening*. The atmosphere also introduces some sharper features in the spectrum, the *telluric lines and bands*.

Extinction is a strong function of wavelength. At sea level, three opaque regions define two transmitting windows. Rayleigh scattering and absorption by atoms and molecules cause a complete loss of transparency at all wavelengths shorter than about 300 nm. This sets the short end of the *optical–infrared window*. The second opaque region, caused by absorption in molecular bands (primarily H_2O and CO_2), begins at around 0.94 μm, has a few breaks in the NIR and MIR (see Figures 8.14, 10.2, and 10.14), then extends from 30 μm until a few breaks appear in the submillimeter region (beginning at 0.35 mm at only the very best sites). Molecular absorption ends at the start of the *microwave–radio window* at around 0.6 cm. In the third opaque region, the ionosphere absorbs or reflects all waves longer than 10–20 m.

Atmospheric extinction has a profound influence on life. The infrared opacity prevents the Earth's surface from radiating directly into space and cooling efficiently. This so-called greenhouse effect is responsible for maintaining the average surface temperature at about 30 K higher than it would be without the atmosphere. Short-wavelength electromagnetic radiation is quite detrimental to biological systems, and none of the forms of life presently on Earth could survive if exposed to the solar gamma-ray, X-ray, and shortwave-ultraviolet radiation that is presently blocked by the atmosphere. Had life here originated and evolved to cope with an environment of either low temperatures or hard radiation, we would not be us.

The wavelength dependence of extinction has an equally profound effect on astronomical life. Astronomy began by peering out at the universe through the narrow visual window and evolved over many centuries to do a better and better job in that restricted region of the spectrum. Astronomy only discovered the radio window in the middle of the twentieth century. Yet later in that century, spacecraft (and aircraft) finally provided access to the entire spectrum. Only with the introduction of decent infrared arrays in the 1980s could astronomers take advantage of the gaps in the near-infrared atmospheric absorption available at dry high-altitude sites. Atmospheric absorption has made optical astronomy old, radio astronomy middle-aged, and gamma-ray, X-ray, and infrared astronomy young.

10.5.2 Absorption by a plane-parallel slab

Figure 10.11a shows a simple model of the atmosphere. A stream of photons traverses a horizontal slab of air toward our telescope at angle z with respect to the vertical. We assume that the density and absorbing properties of the material change with h, the depth measured from the top of the atmosphere, but are independent of the other coordinates. We assume that if a flux of $\phi'(\lambda, h)$ travels over a path of length ds, the material will absorb a certain fraction of the photons. We write the amount absorbed as

$$d\phi(\lambda, h) = -\alpha(\lambda, h)\phi(\lambda, h)ds \qquad (10.26)$$

or

$$\frac{d\phi}{\phi} = -\sec(z)\alpha(\lambda, h)dh \qquad (10.27)$$

Here we introduce the function $\alpha(\lambda, h)$ to describe the fractional absorption per unit distance, and identify z as the zenith angle of the source. However, the geometric and optical properties of the real, spherical atmosphere mean that $z = z$

Fig. 10.11 Absorption geometries. (a) A plane-parallel slab. We assume the top of the atmosphere is at $h = 0$. Note that $ds = \sec(z)dh$, where z is the local zenith angle. Figure (b) illustrates the fact that lower layers are more important in a spherical atmosphere. Figure (c) shows that the angle z increases with depth in a spherical shell. Refraction effects have been ignored.

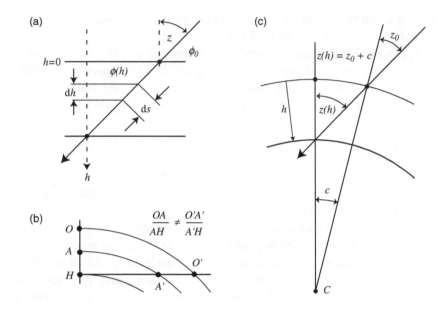

(h). There are two effects involved. First, as you can see from Figures 10.11b and c, because the atmosphere has spherical rather than plane symmetry, the angle z is not a constant, but is an increasing function of h. Second, the actual angle at any height will be even greater than that given by the spherical model because of atmospheric refraction. Taking both effects into account and assuming we have an observatory at depth H in the atmosphere, the solution to Equation (10.27) is

$$\phi(\lambda, H, X) = \phi(\lambda, 0, 0)\exp\left(-\int_o^H \sec(z(h))\alpha(h)dh\right) \equiv \phi(\lambda, 0, 0)e^{-\tau(\lambda, H)X} \quad (10.28)$$

Here $\phi(\lambda, H, X)$ and $\phi(\lambda, 0, 0)$ are the monochromatic photon fluxes inside and outside the atmosphere, respectively. We have introduced two new functions on the right-hand side of Equation (10.28). First, the **optical depth at the zenith**:

$$\tau(\lambda, H) = \int_0^H \alpha(h)dh \quad (10.29)$$

Physically, this definition implies that the monochromatic brightness at the zenith diminishes by the factor $\exp(-\tau)$ due to absorption. The second definition is for **air mass**:

$$X(\lambda, z) = \frac{1}{\tau(\lambda)}\int_0^H \alpha(h)\sec(z(h))dh \approx \sec(z(H)) \quad (10.30)$$

The air mass along a particular line of sight is a dimensionless quantity. It tells how much more absorbing material lies along that line than lies toward the zenith. The approximation $X = \sec(z(H))$ is good for small zenith angles.

(The error is less than 1% for $z < 70°$, corresponding to an air mass of less than 3.) For larger zenith distances, the formula

$$X(z_0) = \sec(z_0)\left[1 - 0.0012\left(\sec^2 z_0 - 1\right)\right] \qquad (10.31)$$

is a much better approximation. Here z_0 is the "true" zenith angle – the angle, z ($h = 0$), between the observer's vertical and the optical path outside the atmosphere, which can be computed from the object coordinates and the sidereal time. We discussed this geometry in Section 3.1.7.

10.5.3 Bouguer's law

From Equation (10.28), we can represent the monochromatic magnitude on the instrumental scale as

$$
\begin{aligned}
m_\lambda(H,X) &= -2.5\log\left[\frac{hc}{\lambda}\phi(\lambda,H,X)\right] + C \\
&= -2.5\log\left[\frac{hc}{\lambda}\phi(\lambda,0,0)\right] + 2.5X\tau(\lambda)\log(e) + C \\
&= m_\lambda(0) + 1.086\tau(\lambda)X
\end{aligned} \qquad (10.32)
$$

$m_\lambda(H,X)$ is the magnitude as observed inside the atmosphere, and $m_\lambda(0)$ is the magnitude in the same system observed outside the atmosphere. Finally, we define the ***monochromatic extinction coefficient***, $k_\lambda = 1.086\tau(\lambda)$, and rewrite the previous equation as

$$m_\lambda(H,X) = m_\lambda(0) + k_\lambda X \qquad (10.33)$$

This expression, which states that the apparent magnitude is a linear function of air mass, is known as **Bouguer's law** (or sometimes, Lambert's law).[3] Bouguer's law suggests the method for determining the value of the extinction coefficient, and thus a method for converting apparent magnitudes inside the atmosphere to magnitudes outside the atmosphere. The astronomer simply measures the brightness of some steady source (the ***extinction source***) at at least two different air masses – then, in a plot of magnitude as a function of air mass, Equation (10.33) tells her that the slope of the straight-line fit is k_λ and the y-intercept is $m_\lambda(0,0)$ see Figure 10.12. Once she knows k_λ, the astronomer can compute outside-the-atmosphere magnitudes for any other star by making a

---

[3] Pierre Bouguer (1698–1758), a French Academician, was celebrated in his day for leading an expedition to Peru in 1735 to measure the length of a degree of latitude. The expedition conclusively demonstrated Newton's hypothesis that the Earth was oblate. Bouguer derived his law for atmospheric absorption by investigation of the general problem of light transmission through a medium. He also holds the distinction of being the first quantitative photometrist in astronomy – in 1725 he measured the relative brightnesses of the Sun and Moon by comparison to a candle flame.

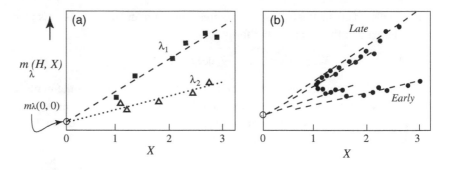

Fig. 10.12 Bouguer's law. (a) A linear fit to extinction star observations gives a measure of the extinction coefficient (slope) and the outside-the-atmosphere instrumental magnitude (intercept). Different observed wavelengths (squares vs. triangles) will give different extinctions. (b) Multiple observations of a single extinction star during a time in which the extinction coefficient is increasing.

single observation and applying Bouguer's law. With one powerful and elegant stroke, the astronomer has removed the absorbing effects of the atmosphere.

Bouguer's law depends on the persistence of two conditions during the time over which observations are made:

(1) that k_λ is stationary – does not change over time,
(2) that k_λ is isotropic – does not change with location in the sky.

If both these conditions hold, observers will say that the atmosphere is ***photometric*** and feel powerful and capable of elegance. If the conditions are violated (visible clouds are one good indication), observers will recognize that certain kinds of photometry are simply impossible. There are intermediate situations – Figure 10.12b shows observations in which condition (1) is violated – extinction here changes gradually over several hours. As long as the changes are carefully monitored, the astronomer can still hope to recover the outside-the-atmosphere data via Bouguer's law.

Condition (2) is always violated to some extent because of the spherical nature of the atmosphere: absorption by lower layers becomes relatively more important at large zenith angles (Figure 10.11b), and total extinction as well the extinction versus wavelength function will change. This effect is not significant at smaller ($<$ 3) air masses, so usually can be (and is) ignored. In general, it is a good idea to avoid *any* observations at very large air masses – the likelihood of encountering non-uniformities is greatly increased, as are all other atmospheric effects like seeing distortions, differential refraction, and background brightness levels.

10.5.4 Sources of extinction

Figure 10.13 plots k_λ for a typical clear (cloud-free) sky in the 0.3–1.4 μm region. As illustrated, the value is the sum of contributions from four different processes, each of which has a characteristic spectral dependence. The processes are:

Rayleigh scattering by molecules
In this process, a photon encounters a molecule of air and is redirected. The probability of scattering is much greater for short-wavelength photons – for pure

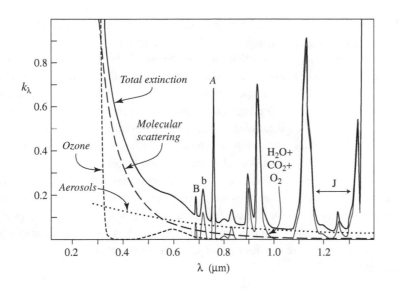

Fig. 10.13 A model for the contributions to the extinction coefficient. Aerosol and water-vapor absorption are highly variable. This is a low-resolution plot, so band structure is smoothed. Letters mark strong telluric Fraunhofer features and the photometric J-band window.

Rayleigh scattering, extinction is proportional to λ^{-4}. Molecular scattering explains why the sky is blue, since multiply scattered photons from the Sun will tend to be those of shortest wavelength. Molecular scattering is stable over time, and its magnitude scales directly with the atmospheric pressure – higher altitudes will have more transparent skies.

Absorption by ozone

Continuous absorption by the O_3 molecule in the ultraviolet essentially cuts off transmission shortward of 320 nm. Ozone also absorbs less strongly in the Chappuis bands in the visible near 600 nm. Ozone abundance is subject to seasonal and global variations, but does not appear to vary on short timescales. Since the gas is concentrated near the stratopause at around 48 km above sea level, the benefit of a high-altitude observatory is minor for ozone absorption.

Scattering by aerosols

Aerosols are suspensions of small solid or liquid particles (particulates) in air. Particulates range in diameter from perhaps 50 µm down to molecular size. Aerosol particulates differ from water cloud drops by their much longer natural residence time in the atmosphere. In fact, the way most aerosols are removed is by reaction with water droplets in clouds[4] and subsequent precipitation. Several different processes inject particulates into the atmosphere. Sea spray and bursting bubbles introduce salt. Winds over deserts introduce dust. Volcanoes inject

---

[4] Aerosol particles are crucial to the *formation* of water clouds: water vapor condenses into droplets much more readily if aerosols provide the "seed" surfaces on which condensation can proceed. Without such seeds, very clean air can supersaturate and reach a relative humidity well over 100%.

ash and sulfur dioxide, a gas that interacts with water vapor to form drops of sulfuric acid. Burning fossil fuel and biomass introduce ash, soot, smoke, and more sulfur dioxide. The wavelength dependence of aerosol scattering depends largely on the size of the particle, and the typical wide range of sizes present (salt particles tend to be large; smoke particles, small) usually produces a relatively "gray" extinction (a λ^{-1} dependence is typical). A pale-blue (rather than deep-blue) sky indicates high aerosol extinction. Sometimes aerosols can produce striking color effects, including the lurid twilight colors from stratospheric volcanic ash and the "green sky" phenomenon due to Gobi Desert dust. Aerosol scattering can be quite variable, even on a short timescale, and different components reside at different atmospheric levels. Although salt, dust, and industrial pollution mainly stay in the lower layers (a scale height of 1.5 km is representative), some volcanic eruptions and intense forest fires can inject aerosols into the stratosphere, where the absence of rain allows them to persist for weeks or even years.

Molecular-band absorption

The main molecular *absorbers* are water vapor and carbon dioxide, although oxygen has a few relatively narrow features, and we have already discussed the ozone bands. Water vapor and CO_2 bands demarcate the relatively transparent windows in the near and middle infrared. Carbon dioxide is well mixed with altitude in the atmosphere, but water vapor is concentrated near the surface and varies with temperature, time, and location. At sea level, the average amount of vapor in one air mass corresponds to about 10 mm of liquid. On Mauna Kea, one of the best conventional sites, the average is about 1 mm of precipitable water. Similar sites have been investigated in the Atacama Plateau, Greenland, and the Himalayas. At the South Pole, which benefits from both high elevation and low temperature, values approach 0.15 mm. Stratospheric observatories carried by balloons or SOFIA enjoy even lower values.

10.5.5 Heterochromatic extinction

The previous discussion strictly applies only for monochromatic magnitudes. For bandpass magnitudes, we must rewrite Equation (10.32) as

$$m_P(H,X) = -2.5\log \int R_p(\lambda)\frac{hc}{\lambda}\phi(\lambda,0,0)\exp\left(\frac{-k_\lambda X}{1.086}\right)d\lambda + C \tag{10.34}$$

so that

$$m_P(H,X) - m_P(0) = -2.5\log\left\{\frac{\int R_p f_\lambda(\lambda,0,0)\exp\left[\frac{-1}{1.086}k_\lambda X\right]d\lambda}{\int R_p f_\lambda(\lambda,0,0)d\lambda}\right\} \tag{10.35}$$

Now, it would be convenient if the right-hand side of Equation (10.35) reduces to $k_P X$ as in Bouguer's law, where k_P is the ***extinction coefficient for band P***. Unfortunately, the function on the right-hand side of Equation (10.35) is *not* linear in X, and will also depend strongly on the shape of the function $f_\lambda(\lambda, 0, 0)$. We should expect, therefore, that $k_P = k_P(X, SED)$, where the variable SED indicates the spectral energy distribution of the object observed.

Think of the variation in k_P as due to two different but related effects. First, as we had seen earlier, the effective and isophotal wavelengths of a bandpass depend on the spectrum of the source. We should expect that the extinction measured, say, for a red star will differ from the extinction measured for a blue star, since the center of the bandpass is different for the two. Because of this effect, the Bouguer plot of apparent magnitude versus air mass will give straight lines of different slopes for stars of different spectral shapes (as we saw in in Figure 10.12 for different monochromatic wavelengths).

The second, more invidious problem arises because atmospheric extinction itself changes the shape of the spectrum that reaches the telescope. This effect – called the ***Forbes effect*** – means you actually observe different spectra for the same star at different air masses. (Alternatively, you can think of the effect as changing the shape of the bandpass response as air mass changes.) A Bouguer plot of apparent magnitude versus air mass will therefore give a *curved* line. The Forbes effect, as illustrated in Figure 10.14, is particularly problematic if strong atmospheric absorption affects some parts of the photometric band more than others. This is the case for Johnson U and many of the wider infrared bands (review Figures 10.2 and 10.3). In such cases, the magnitude change in going from $X = 2$ to $X = 1$ can be considerably less than in going from $X = 1$ to $X = 0$. In some cases, both the width of the band and its effective wavelength can change dramatically at the smaller air masses. Use of outside-the-atmosphere magnitudes when the Forbes effect is present is a little tricky, and sometimes depends on having a good model of the response function, the atmosphere, and the unknown source. For precise work, therefore, it is best to use bands that exclude strong atmospheric absorptions (e.g. the MKO near-infrared system).

10.5.6 Compensating for extinction: theory

An evaluation of the right-hand side of Equation (10.35) would be possible if we could make a good approximation of the monochromatic extinction function k_λ and the shapes of the response function, $R_P(\lambda)$, and spectrum f_λ, either numerically, or perhaps with a Taylor series expansion of each function. The required functions or their derivatives are rarely known with precision, but are easily approximated. The usual approach assumes that the photometric color, which gives some sense of the shape of the source spectrum, accounts for most

Fig. 10.14 The Forbes effect. (a) A model source spectrum, bandpass transmission, and extinction coefficient. The extinction is due entirely to a strong feature near the blue edge of the band. (b) The flux actually detected at the telescope is shown as a function of wavelength for four different air masses. Note the relatively small change between $X = 1$ and $X = 2$ (shaded regions) compared to the change between $X = 0$ and $X = 0.5$. (c) The Bouguer diagram for the data in (b), illustrating the non-linear relationship and the difference between the actual extra-atmospheric magnitude (filled circle) and the intercept of a linear fit to observable data (open circle).

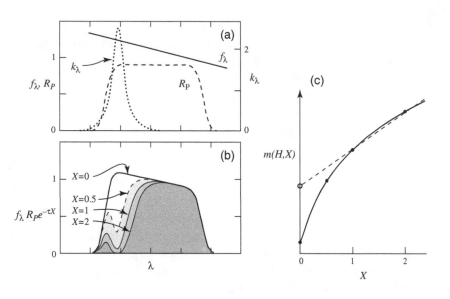

variations in the broadband extinction, and approximate knowledge of the extinction and response functions can account for the Forbes effect. We write:

$$m_P(H,X) = m_P(0) + \left\{ k_P' + (P-Q) \cdot k_P'' \right\} X + k_P''' (SED) \cdot X^2 \qquad (10.36)$$

where $(P - Q)$ represents some "appropriate" photometric index, and k'' is called the ***second-order extinction coefficient***. The coefficient k''' that accounts for the Forbes effect "curvature" is generally on the order of 1 mmag (with important exceptions) and is frequently ignored. For example, the Bouguer relation for the instrumental Johnson V band might be written as:

$$v(H,X) = v(0) + \left(k_V' + (B-V) \cdot k_V'' \right) X \qquad (10.37)$$

Note the convention that lower case letters (b,v) represent instrumental magnitudes and uppercase letters (B,V) represent standard magnitudes. If you use an equation like (10.37), you have not accounted for the Forbes effect well, so if it is severe, you must employ Equation (10.36) and use a value for $k_P''' (SED)$ derived from a model (similar to Figure 10.14) of how the flux in the bandpass changes with air mass.

10.5.7 Compensating for extinction: practice

Is it necessary to remove the effects of atmospheric extinction on ground-based photometry? It depends. If you are performing ***differential photometry***, measuring the brightness ratio of two sources of *identical color* on the same CCD frame, say, the extinction on each should be virtually identical, and their magnitude difference unaffected by the atmosphere. If colors are different, then there will be a small change in the ratio with changing air mass. If you are doing *relative photometry*, or all-sky photometry, measuring the magnitude difference between your target and a standard star on a different image frame, then compensating for extinction is essential. A third type of photometry, **absolute photometry**, measures brightness in physical units (Wm^{-2}). Absolute photometry requires that you remove atmospheric effects, but also that you compare the source with a laboratory standard like a blackbody of known temperature. Absolute photometry is important, but rarely needs repeating in astronomy. Ground-based astronomers use a variety of techniques for removing the effects of atmospheric extinction. We look at a few cases here.

Case 1: assume a mean extinction

At high-altitude sites, the extinction at Vis-UV wavelengths is often due almost entirely to Rayleigh scattering, and is therefore stable. Under these conditions, it may be safe simply to use the average (or better still, the median) extinction coefficients determined by other observers for the same site over the past years. This is a particularly reasonable approach if one is doing differential photometry, or if standard and program stars are observed at nearly the same air mass at nearly the same time. (See Section 10.6 for the relevance of observing standard stars.) It is also true that the second-order extinction coefficient k'' is harder to determine, smaller and less variable than k', so one might assume a mean value for k'' and use one of the methods below (e.g. Case 3) to determine k'.

Case 2: use known outside-the-atmosphere magnitudes

If you wish to determine the extinction yourself from the Bouguer law, this extremely simple but risky method applies – *if* you happen to know $m_P(0)$, the magnitudes outside the atmosphere of several stars in the same field that have differing SEDs. If you have a good model for k_P''' (*SED*) or can assume it is zero, then just point your telescope at the star field, note the air mass, X, and take an image. You measure $m(H,X)$, you know $m(0)$, you have a variety of color indices, so just solve Equation (10.36) for k' and k'' using a least-squares method. But be careful! The apertures you use to measure $m_P(X)$ must capture exactly the same fraction of the PSF as the apertures used to measure $m_P(0)$.

Case 3: draw the Bouguer line from observations

If values for $m(0)$ are not known, then you need to generate a Bouguer plot. Take multiple exposures of the same field of stars over a wide range of air masses. This, of course, requires waiting for the zenith distance of your field to change. Many observers record the extinction field every 90 minutes or so. You plot the resulting instrumental magnitudes as a function of air mass, and if the night is **photometric**, you obtain a plot like Figure 10.12a.

Choose your extinction field so that you have several stars that yield good SNRs on a short exposure. If you are trying to determine the second-order coefficient, these "extinction stars" should have a range of colors. Each star will produce a separate Bouguer line, with slope k. A plot of k as a function of star color will produce values for k' and k''. It will be economical if you can make the extinction field identical to the program field or to a field containing standard stars.

Case 4: variable extinction

What if the extinction changes? If you observe over many consecutive nights, change is likely, mostly due to aerosol response to variations in water-vapor content. If you are very fortunate, the extinction will change slowly over time and uniformly over the entire sky. In this case, some modification of the previous method will yield k' and k''. For example, if you are sure of the outside-the-atmosphere magnitudes of some of your constant stars, then simply monitoring the constant stars will give instantaneous values of the extinction coefficients.

If you do not have the instrumental outside-the-atmosphere magnitudes, there is still hope: if you have the *standard* magnitudes in two fields with very different air masses, observing the two fields in quick succession measures k' and k'' at the time of observation. The practice is to take one frame near the meridian (the "D" frame), and then immediately take a frame containing the second standard field at large air mass in the east (the "M" frame). The stars in field M have standard magnitudes $\{m_{MStd1}, m_{MStd2}, \ldots m_{MStdi}, \ldots\}$ and observed instrumental magnitudes $\{m_{M1}(X_M), m_{M2}(X_M), \ldots m_{Mi}(X_M), \ldots\}$. Similarly, magnitudes for stars in the D field will be designated $\{\ldots m_{DStdj}, \ldots\}$ and $\{\ldots m_{Di}(X_D), \ldots\}$. Provided that the extinction coefficient is the same in all directions, Bouguer's law gives the difference between the instrumental magnitudes of star i in the M frame and star j in the D frame, measured inside the atmosphere, as

$$\Delta m_{ij} = m_{Mi}(X_M) - m_{Dj}(X_D) = m_{Mi}(0) - m_{Dj}(0) + \left(k' + k'' \Delta(ci)\right)(X_M - X_D)$$

(10.38)

where (ci) is an appropriate color index. Now if we choose star M_i and star D_j to have approximately the same color, we assume that

$$m_{Mi}(0) - m_{Dj}(0) = m_{MStdi} - m_{DStdj}$$

(10.39)

and can rearrange Equation (10.38) to give;

$$k' = \frac{m_{MStdi} - m_{DStdj}}{(X_M - X_D)}$$

(10.40)

All the quantities on the right-hand side of Equation (10.40) are either known or observed. Once k' is known, reapplication of (10.40) to stars of different standard colors yields a value for k''. Because of the possibility of extinction change, many photometrists adopt the strategy of observing "MD" pairs every few hours through the night.

Case 5: use all the data

The most general methods make use of all available information, and include data for all nights in which sources are observed in common. Every frame taken during the run of several nights is affected by extinction and therefore contains information *about* extinction. One approach might work as follows: derive values for k' and k'' from the best nights – those for which it is possible to make good linear fits to the extinction data – and compute the outside-the-atmosphere magnitudes for *every* constant star (not just those used for the fits). You then should have a large set of extra-atmospheric magnitudes that you can use to find the extinction as a function of time for the more marginal nights. Cloudy nights in which the extinction changes rapidly or anisotropically will be suitable only for differential work. The extinction problem is well suited to a least-squares solution with constraints imposed by standard stars. See the discussion in chapter 10 of Sterken and Manfroid (1992) for a good introduction.

10.5.8 Indices or magnitudes?

The traditional method for reporting *n*-color photometric data is to give one magnitude and $n - 1$ indices. It has also been traditional to make the extinction computations not for *n* magnitudes, but for one magnitude and $n - 1$ indices.

Suppose, for example, we observe in the Johnson B and V bands. If we have a B frame and a V frame taken at about the same air mass, we can write Equation (10.37) once for each band, and then subtract the equations, yielding

$$(b - v)_X = (b - v)_{X=0} + \left(k'_B - k'_V\right) + \left(k''_{B,BV} - k''_{V,BV}\right) \cdot (B - V)X \qquad (10.41)$$

Combining the coefficients, we have

$$(b - v)_X = (b - v)_{X=0} + \left\{k'_{B-V} + k''_{B-V} \cdot (B - V)\right\}X \qquad (10.42)$$

Here the new first- and second-order coefficients describe the effects of extinction *on the index*. There is an objective reason for analyzing extinction data via Equation (10.42) rather than via Equation (10.37): With a single-channel device, observations were usually performed in a symmetric sequence; e.g. B–V–V–B. If either instrument sensitivity or atmospheric aerosol extinction drifts during an observing run, the effect on the observed color indices will be minor compared to the effect on the individual magnitudes.

This reasoning is less compelling in modern observing. For one thing, CCDs and infrared arrays are much less prone to sensitivity drift than photomultipliers.

Moreover, the requirement that both bands be observed at the same air mass is somewhat restrictive, and may prevent the use of all the available extinction data.

10.6 Transformation to a standard system

Assume we have observed an outside-the-atmosphere instrumental magnitude for some source, either by correcting a ground-based observation for extinction or by direct observation with a space telescope. If we wish to compare our results with those of other observers, we must all use the same photometric system. Thus, the basic idea of relative photometry: the transformation from the instrumental to the standard system. This *transformation* will depend on (1) differences between the response of our instrumental system and that of the standard system and (2) the shape of the spectrum of the source.

It is possible to derive the transformation synthetically, by comparing your own system response with that of the standard system for a particular observed SED, and this is practical if your own system's response function is well known and stable. Indeed, your instrumental response may define the system, in which case no standards are necessary. More often, however, the observer (or observatory) derives transformations by direct observation of a system's *standard stars*.

For monochromatic magnitudes, standard spectral energy distributions for a variety of spectra types have been established. These monochromatic *spectrophotometric standards* are often based on the flux calibration (in $Wm^{-2}nm^{-1}$ or $Wm^{-2}Hz^{-1}$) of physical stars, sometimes supplemented with model stellar atmospheres to provide higher resolution. Several major observatories maintain lists of these standards online, for example:

> www.eso.org/sci/observing/tools/standards/spectra/stanlis.html
> www.gemini.edu/sciops/instruments/gmos/calibration/spectroscopic-stds

Observations of these standards are especially suitable for the calibration of spectra and for synthetic photometry (computing bandpass magnitudes from an input $R(\lambda)$).

For direct standardization of bandpass magnitudes, the de facto standards for the UBVRI system are the sets at declinations $-50°$, $0°$, and $+50°$ produced by Landolt (2013, and references therein), and for the SDSS *ugriz* system, a set produced by Smith et al. (2002). Again, lists are available at major observatory websites.

You establish bandpass transformations as follows: Select a set of N standard stars that have a large a range of colors. These have standard magnitudes $\{m_{Std,1}, m_{Std,2} \ldots m_{Std,i} \ldots\}$ and standard colors $\{C_{Std,1}, C_{Std,2} \ldots C_{Std,i} \ldots\}$. Observe these stars, preferably at low air mass, and correct your measurements for atmospheric extinction. You then have a set of outside-the-atmosphere instrumental magnitudes, $\{m_1, m_2 \ldots m_i \ldots\}$. Because your bandpass width

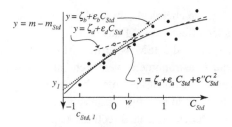

$y = m - m_{Std}$, as a function of color for several stars. Some stars were observed multiple times during the night. Over limited color ranges (e.g. $C < w$), linear fits may be preferable to a quadratic.

Fig. 10.15 Determination of the transformation coefficients from least-squares fits to a plot of the difference between the instrumental and standard magnitude,

and isophotal wavelength may differ from the standard, you assume that your instrumental magnitudes are related to the standards by an equation of the form:

$$m_i = \zeta + \mu m_{Std,i} + \varepsilon C_{Std,i} + \varepsilon' C^2_{Std,i} \qquad (10.43)$$

Here, ζ is the **zero-point** of the transformation, and μ should be equal to 1.0; ε and ε' are the first and second-order **color terms**. Given the N data points, you then use a least-squares technique to find the best-fit values for the three constants, ζ, ε, and ε'. See Figure 10.15. If your instrumental system response is close to the standard, the second-order color term should be unnecessary. For poorly matched systems, the transformation for stars highly reddened by interstellar absorption may differ from the one for unreddened stars of the same color.

To minimize systematic effects, everything about the observing and reduction system (detector temperature, seeing, air mass, flat field) should be as nearly identical as possible for the observations of the standards and program stars. Digital aperture size is a potential hazard. If there are seeing variations use an aperture radius of several PSF FWHM. The most precise transformations to the standard system require that you use the same aperture size that was used in establishing the standards (e.g. Landolt usually used 14 arcsec diameter apertures). Using a different aperture can introduce subtle errors like inclusion or exclusion of background objects.

Many astronomers prefer to transform color indices rather than magnitudes. Subtracting two sets of equations like (10.43) suggests a color transformation like:

$$C_{ij} = \zeta_{i-j} + \varepsilon_{i-j} C_{Std,ij} + \varepsilon'_{i-j} C^2_{Std,iij} \qquad (10.44)$$

where $C_{Std,ij}$ is again the color index in the standard system.

10.7 Absorption outside the atmosphere

10.7.1 The interstellar medium

Space is not empty. Interstellar gas and dust in our own Galaxy (the **interstellar medium**, or **ISM**) absorb, scatter, and emit light. Although a great deal of effort has been to puzzle out the chemical, physical, and structural nature of the ISM, our concern here is the effect the ISM has on our observations of other objects.

Unlike the atmosphere, the ISM is difficult to model. Indeed, prior to the early twentieth century, astronomers were generally unaware of its presence. Trumpler convincingly demonstrated its absorbing properties in 1930 and thereby accounted for several puzzling observational phenomena like the **"zone of avoidance"** (no galaxies seen in directions near the Galactic equator), the existence of stars far too red for their spectral type, and star clusters that were too dim for their apparent sizes.

Subsequent work has shown the ISM to be highly non-uniform, and the properties of both dust and gas to change drastically with temperature and density. The ISM near us in the Milky Way contains at least four more or less distinct components, although a precise three-dimensional map is still uncertain:

1. The *diffuse clouds* are sheets and filaments of cold gas (mostly atomic hydrogen) and dust – with temperatures below 100 K and densities in the range of 0.1 to 100 hydrogen atoms cm^{-3}.
2. The *molecular clouds* (molecules of H_2, CO, and many others) are short-lived, very cold, very dense structures connected with star formation. Their outer regions grade into the diffuse cloud medium.
3. The *warm intercloud medium,* with temperatures of several thousand kelvin and low density, generally embeds the diffuse clouds and has its own complex structure.
4. The *coronal phase* consists of very hot (10^5 K), very low-density bubbles and tubes of highly ionized material that gives the intercloud medium its foam-like structure.

Clarifying the chemical, physical, and structural nature of the ISM is a major research area. However, our concern here is the effect the ISM has on our observations of other objects.

10.7.2 Interstellar absorption and reddening

In the UV, optical, and NIR, diffuse and molecular clouds produce most of the observed effects of the ISM. The dust component in these not only reduces the overall number of photons that arrive at the telescope from a source – an effect called *interstellar extinction* – but it also alters the shape of the arriving spectrum. In the region 0.22–10 μm, dust scatters short-wavelength photons more strongly than long-wavelength photons, so the resulting change in the shape of the spectrum is termed *interstellar reddening*. The dust also adds a few very broad absorption features.

The gas component of the diffuse clouds produces *interstellar absorption lines and bands*. In the optical, the sodium D doublet is usually the strongest interstellar line, and other lines due to atomic Ca, K, Fe Ti, and the molecules CH, CH+ CN, C_2, and NH are present. In the ultraviolet, the Lyman-alpha line at 121.6 nm is usually strongest, with strong lines due to atomic C, N, O, Mg, Si, and Fe, as well as molecular absorptions due to the molecules mentioned above, as well as CO, CH_2, N_2, NO, OH, H_2O, and many others. Especially at short wavelengths, gas produces continuous absorption and absorption edges due to ionization. A strong feature at 91.2 nm due to ionization of atomic hydrogen is very prominent, for example.

When not silhouetted against a continuous source, of course, the gas component of these clouds produces an emission spectrum characteristic of their temperatures and pressures. See the next chapter for details – e.g. atomic recombination lines (like the Balmer, Lyman, etc. series in hydrogen).

10.7.3 The interstellar reddening law

One of the earliest discoveries in astrophysics was that the spectral type of a star is an excellent indicator of its temperature and absolute magnitude. Knowledge of the absolute magnitude permits determination of distance. You recall the monochromatic distance modulus equation:

$$m_{\lambda 0} - M_\lambda = 5\log r - 5 \tag{10.45}$$

But this equation ignores interstellar absorption, so we should really expect that what an observer measures is a dimmer apparent magnitude given by:

$$m_\lambda = m_{\lambda 0} + A_\lambda = \{M_\lambda + 5\log r - 5\} + A_\lambda \tag{10.46}$$

where A_λ is the interstellar extinction in magnitudes, and $m_{0\lambda}$ is the apparent magnitude the star would have in the absence of extinction. Note that for any star whose spectral type and distance are known (e.g. from parallax) the quantity in braces is also known, and A_λ can be computed. Even more simply, an astronomer can select a pair of stars with identical spectral types (O and B types usually work best), one nearby, the other heavily reddened. He observes both at a very long wavelength and assumes extinction at that wavelength is 0 (see Figure 10.16). The ratio of the two spectra normalized at that wavelength and implied distance then gives A_λ as a function of wavelength.

The value of A_λ for a particular source depends critically on the line of sight. In the visual band, observations near us have roughly $A_V = 0.7$–1.1 magnitude kpc^{-1} in the plane of the Galaxy, and $A_V = 0$ perpendicular to the plane.

Apply Equation (10.46) two different wavelengths, $\lambda_2 > \lambda_1$, and subtract them to compute the **reddening excess,** the change in color index produced by the ISM:

$$E(m_1 - m_2) = A_{\lambda_1} - A_{\lambda_2} = m_1 - m_2 - (m_1 - m_2)_0 = C_{12} - (C_{12})_0 \tag{10.47}$$

If the optical properties of the dust grains are the same everywhere, then we expect this reddening effect to be proportional to the amount of extinction. For example, if we consider magnitudes in the Johnson B and V bands, the parameter,

Fig. 10.16 Broadband interstellar extinction as a function of wavelength, normalized to the V band. The two lower curves are from data given by Mathis (1990) for diffuse and molecular clouds. The curve for the Small Magellanic Cloud bar is from Gordon et al. (2003).

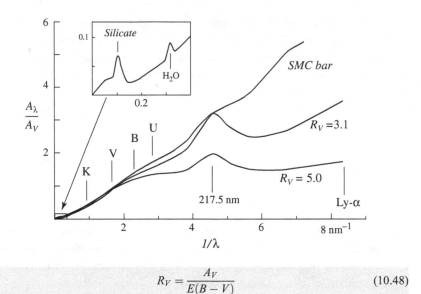

$$R_V = \frac{A_V}{E(B-V)} \tag{10.48}$$

known as "the ratio of the total-to-selective extinction," should be constant. Unfortunately, it isn't. For most lines of sight through diffuse clouds, $R_V \simeq 3.1$, but it varies somewhat unpredictably, and generally becomes large ($R_V \simeq 4-6$) for lines of sight passing through denser regions. To get a useful picture of the variation of A_λ with wavelength, we can examine the **normalized extinctions** that is, the function A_λ/A_V, which we plot in Figure 10.16. At wavelengths longer than 0.9 mm, $A_\lambda \approx A_V\lambda/0.55\,\mu\text{m}$, where $\alpha \approx -1$ for most regions; but it can approach -2 in dense clouds. This behavior at long wavelengths means that heavily obscured regions like the center of our Galaxy or the interiors of star-forming regions are much more easily investigated in the infrared and radio than in the optical. The extinction curves diverge strongly for different values of R_V at shorter wavelengths, with pronounced flattening for large R_V. Most lines of sight show a broad bump at 217.5 nm, probably due to some form of carbon on the surfaces of dust grains.

10.7.4 Spectroscopic parallax

This method uses Equation (10.46) to compute the distance of a star of known spectral type and therefore of expected color, $(B - V)_0$, and absolute magnitude, M_V. If we can assume $R_V = 3.1$, then combining Equations (10.48), (10.47), and (10.46) gives

$$5\log r = m_V - M_V + 5 - 3.1 \cdot \{(B - V) - (B - V)_0\} \tag{10.49}$$

Spectral types are not always available for faint stars, but there still is a sort of spectroscopic parallax available. Reddening produces changes that are similar to temperature changes – but the changes are not identical – so it is sometimes possible to observe a purely photometric **reddening-free parameter** that indicates a star's spectral type. For example, the definition of reddening excess (10.47) implies that the parameter

$$Q_{1234} = C_{12} - \frac{E(m_1 - m_2)}{E(m_3 - m_4)} C_{34} = (C_{12})_0 - \frac{E(m_1 - m_2)}{E(m_3 - m_4)} (C_{34})_0 \qquad (10.50)$$

is independent of reddening as long as the ratio $E(m_1 - m_2)/E(m_3 - m_4)$ is a constant. In the UBV system, the quantity

$$Q = (U - B) - (0.70 \pm 0.03) \cdot (B - V) \qquad (10.51)$$

is both independent of reddening in diffuse clouds and a good indicator of spectral type for O and B stars (only). Other reddening-free parameters for a wider range of types can be defined in intermediate-band systems.

10.8 Wavelength changes

10.8.1 Redshift and photometry

So far, we have described how the number of photons in a telescope-bound stream is changed by absorption and scattering by the terrestrial atmosphere and by the interstellar medium. We now consider photometric changes produced by phenomena that alter a photon's wavelength without removing it from the stream. A wavelength change might be caused by the Doppler effect, or by the expansion of the universe, or by various local relativistic effects (e.g. photons leaving a collapsed object like a white dwarf undergo a gravitational redshift). We write the wavelength of a photon that on arrivals at the top of the atmosphere as

$$\lambda_o = (1 + z)\lambda_E \qquad (10.52)$$

where λ_E is the wavelength of the photon when it as emitted by the source and z is the **redshift parameter** of the source $(z = (\lambda_o - \lambda_E)/\lambda_E)$. Because of this wavelength change, the photons emitted into wavelength interval $\delta\lambda_E$ will arrive spread out over wavelength interval $\delta\lambda_o = (1 + z)\delta\lambda_E$. Now consider an observer (frame O) measuring the flux from a source with a large redshift, z. The spectrum in the source's rest frame (frame E) is $f_E(\lambda)$, but the observer sees spectrum $f_o(\lambda)$, with both functions measured in $J^{-1}s^{-1}m^{-2}nm^{-1}$. Refer to Figure 10.17, and note that the integrated flux in wavelength interval $\delta\lambda_o$ that arrives in observed band A_o originated in wavelength interval $\delta\lambda_E$ (band A) in frame E.

What is the relationship between $f_o(\lambda)$ and $f_E(\lambda)$? Suppose that in frame E, the flux in band A consists of N photons of wavelength λ_E, emitted in time interval δt_E. So that

Fig. 10.17 Wavelength shifts and photometry. The upper panel shows an unshifted spectrum, and the lower panel shows the same spectrum shifted in wavelength by a redshift parameter $z = 0.2$. Energy originating in area A arrives in (equal) area B. Energy measured in band R originates in approximately in bandpass Q.

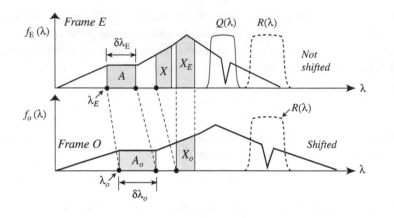

$$N = f_E(\lambda_E)\frac{\lambda_E}{hc}\delta t_E \delta\lambda_E \tag{10.53}$$

These same photons arrive in the observer's frame with a longer wavelength, $\lambda_o = (1+z)\lambda_E$ spread over a larger band, $\delta\lambda_o = (1+z)\delta\lambda_E$ and arrive during a longer time interval $\delta t_o = (1+z)\delta t_E$ (the last photons in the stream have a further distance to travel) so, the observer counts these photons as $N_o = N_E = N$:

$$N_o = f_o(\lambda_o)\frac{\lambda_o}{hc}\delta t_o \delta\lambda_o = (1+z)^3 f_o(\lambda_o)\frac{\lambda_E}{hc}\delta t_E \delta\lambda_E = N_E = f_E(\lambda_E)\frac{\lambda_E}{hc}\delta t_o \delta\lambda_E \tag{10.54}$$

Therefore:

$$(1+z)^3 f_0(\lambda_0) = f_E(\lambda_E)$$

$$f_0(\lambda_0) = (1+z)^{-3}f_E(\lambda_0/(1+z)) \tag{10.55}$$

Now consider a similar case, shown in Figure 10.17 as the areas X, X_o, and X_E. An observer at rest with respect to the source (i.e. in frame E) measures its bandpass magnitude (area X_E) with a photometer whose response is $R_X(\lambda)$, he will obtain the usual result:

$$m_{RE} = -2.5\log\int R_X(\lambda)f_E(\lambda)d\lambda + C_R \tag{10.56}$$

$$C_R = -2.5\log\int R_X(\lambda)g(\lambda)d\lambda \tag{10.57}$$

where $g(\lambda)$ is the spectrum of a photometric standard of magnitude zero, at rest in frame E. If the observer in frame O performs the same operation on the red-shifted spectrum with the same system, again referencing a zero-magnitude standard at rest, he gets:

$$m_{RO} = -2.5\log\left\{(1+z)^{-3}\int R_X(\lambda)f_E(\lambda/(1+z))d\lambda\right\} + C_R \tag{10.58}$$

Clearly, if observer O measures the apparent magnitude of the source in this band, it conveys very different information from the same photometry performed in the reference frame of the source. Note that in the $X_o \leftrightarrow X_E$ case in the figure, the redshift is so large that Observer O's photometry in fixed band X_o actually measures flux that originates in a totally different wavelength region, region X in the figure. If $z = 0.25$, for example, flux emitted in the B band is primarily received in the V band.

In the case of large redshifts, it would be important for an observer in frame O, who performs photometry in a standard band with response $R(\lambda)$, to recognize that his measurement really contains information from the source contained from a different, shorter wavelength, (perhaps standard) band with response $Q(\lambda)$. Again, refer to Figure 10.17.

Wavelength shifts will also affect the colors of objects (e.g. very distant galaxies) with large z. Application of Equation (10.58) for two or more different bands gives an expression for the color change as a function of z. Having observed a color, you can in principle use that expression to discover z, so long as you can approximate the spectrum of a galaxy. These **photometric redshifts** from observed colors are valuable estimators of galaxy distance because they do not require observationally difficult spectroscopy of very faint objects. Indeed, one of the first products of the Sloan survey was the photometric estimate of galaxy distances from *ugriz* photometry.

10.8.2 The *K* correction

Issues of photometry of large redshift objects arise primarily in cosmology. In an expanding, possibly spatially curved universe, distance is a concept that can take on several meanings. The parameter d_L, for example, is the **luminosity distance**, that is, the distance one would compute based on the observed brightness of a source of known luminosity. It gets defined this way: Let $L_E(\lambda_E)d\lambda_E$ be the luminosity of a source measured in a narrow band, $d\lambda_E$, in the emitted frame. This luminosity produces an observed a flux of

$$f_o(\lambda_o)d\lambda_o = \frac{1}{4\pi d_L^2} L_E(\lambda_E)d\lambda_E \qquad (10.59)$$

or, from Equation (10.52):

$$L_E(\lambda_E) = 4\pi d_L^2(1+z)f_o(\lambda_E(1+z)) \qquad (10.60)$$

In an expanding universe, redshift increases with distance according to Hubble's law. (Review Section 3.4.3.) The traditional approach in cosmology is to recognize the sort of spectrum-shifting through the bands noted in Figure 11.17, and to write the measured apparent magnitude of an object as

$$m_{RO} = M_{QE} + \mu_Z + K_{RQ} \qquad (10.61)$$

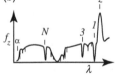

Fig. 10.18 Cosmological redshifts and spectral features. (a) The line of sight from a distant quasar passes through a number of clouds of absorbing gas, each at a distinct redshift or range of redshifts. (b) A portion of the observed quasar spectrum shortward of the redshifted Ly-α emission from the quasar (labeled z.) Some lines in the Lyman-alpha forest are identified with individual galaxies, others with the intracluster medium.

Here M_{QE} is the absolute magnitude in band Q, measured in the rest frame of the object, and μ_z is the luminosity distance modulus: $\mu_z = 5\log(d_L) + 5$. K_{RQ} is called the **K correction**, and accounts for the possible mismatch of the bands. We can write down the K correction by substituting into Equation (10.61), noting first that the absolute magnitude in the Q band is just:

$$
\begin{aligned}
M_{QE} &= -2.5\log\left\{\int \frac{L_E(\lambda_E)Q(\lambda_E)}{4\pi(10pc)^2}d\lambda_E\right\} + C_{QE} \\
&= -2.5\log\left\{\int \frac{d_L^2(1+z)}{(10pc)^2}f_o(\lambda_E(1+z))Q(\lambda_E)d\lambda_E\right\} + C_{QE}
\end{aligned}
\tag{10.62}
$$

And so

$$
\begin{aligned}
K_{RQ} &= -2.5\log\left\{\int f_o(\lambda)R(\lambda)d\lambda\right\} - M_{QE} - 5\log d_L + 5 + C_{RQ} \\
&= 2.5\log(1+z) - 2.5\log\left\{\frac{\int f_o(\lambda)R(\lambda)d\lambda}{\int f_o(\lambda_E(1+z))Q(\lambda_E)d\lambda_E}\right\} + C_{RQ}
\end{aligned}
\tag{10.63}
$$

Here the constant C_{RQ} accounts for possible zero-point differences in the Q and R bands. Equation (10.63) specifies precise K corrections, provided the entire redshifted spectrum, $f_o(\lambda)$, is known. The correction is essential when using photometry of very distant objects to investigate the structure and evolution of the universe at large. Although different authors define the K correction in slightly different ways, the above derivation follows closely the one given by Hogg et al. (2002).

10.8.3 Absorption outside our Galaxy

Absorbing material outside the Milky Way is concentrated in other galaxies and in the intracluster medium in clusters of galaxies. The cosmological redshift makes the situation more complicated – see Figure 10.18. Photons that arrive with wavelength λ will have had wavelength $\lambda/(1+z_1)$ when they passed through galaxy 1 in the figure. Indeed, the photons of interest will have had a different wavelength for each encounter on their path. Although the general problem of the transfer of radiation in an expanding universe is mathematically complex, astronomers use observations of very distant sources to probe the distribution of the intervening material, since it is usually concentrated at distinct redshifts. For example, in Figure 10.18, A quasar at redshift z emits a continuous spectrum, and (usually) a set of very strong emission lines, including one at Lyman-alpha (see Appendix H). Neutral hydrogen in galaxy 1 will produce a Lyman-alpha absorption line at the observed wavelength $121.5 \cdot (1+z_1)$ nm, and every other galaxy and cluster along the line of sight will produce a Ly-α absorption at the wavelength determined by its redshift. Analysis of the resulting ***Lyman-alpha forest*** in spectra can map the often otherwise invisible structure in the gas at cosmologically significant distances.

Summary

- The history of photometry has bequeathed the magnitude scale and the definitions of several important broadband photometric systems.
- Photometric bandpass response functions are generally categorized as broad-, intermediate-, or narrow-band. A response can be implemented by some combination of filters, detector sensitivity, and atmospheric transmission. Concepts:

resolving power	*response function*	*photon response function*
high-pass filter	*peak wavelength*	*central wavelength*
mean wavelength	*effective wavelength*	*isophotal wavelength*
FWHM	*bandwidth*	*photon flux*
zero point	*bandpass magnitude*	

- Photometric indices, which are linear combinations of bandpass magnitudes, quantify characteristics of an object's spectrum. Concepts:

color index	*blackbody spectrum*	*monochromatic magnitude*
line index	*curvature index*	*feature index*

- A standard photometric system specifies both the response functions of its bands as well as some method for standardizing measurements. Concepts:

open system	*closed system*	*instrumental system*
visual magnitude	*standard sequence*	*north polar sequence*
B-V	*international system*	*photovisual magnitude*
UBVRI	*Cousins system*	*JHKLMNQ*
MKO filters	*Strömgren system*	*uvbyβ*
c_1 and m_1	*SDSS ugriz system*	

- Absorption by material inside the atmosphere can produce both reddening and telluric absorption lines and bands. Concepts:

optical–infrared window	*microwave–radio window*
instrumental magnitude	*magnitude outside the atmosphere*

- Photometric data reduction proceeds in steps: (1) preprocessing, (2) digital photometry, (3) atmospheric extinction correction, (4) transformation to a standard system, and (5) further corrections and analysis.
- Bouguer's law is the basis for the correction for atmospheric extinction:

  $$m_\lambda(H, X) = m_\lambda(0) + k_\lambda X$$

Concepts:

(continued)

Summary (*cont.*)

optical depth	*air mass*	*extinction coefficient*
ozone bands	*molecular bands*	*Rayleigh scattering*
aerosols	*Bouguer line*	*mean extinction*
second-order	*monochromatic*	*heterochromatic*
extinction	*extinction*	*extinction*
Forbes effect	*MD pairs*	

- Transformation to the standard system requires observation of standard objects using instruments identical to those used for the unknowns. Concepts:

  zero-point constant   standard star/extinction star
  color coefficient   second-order color coefficient

- Absorption by material outside the atmosphere can produce both reddening and absorption lines and bands in a spectrum. The normalized extinction function depends strongly on the ratio of total-to-selective absorption. Concepts:

  ISM   diffuse interstellar clouds   redenning excess
  Molecular clouds   spectroscopic parallax   redenning-free parameter

- A wavelength shift in an object's spectrum caused by the Doppler effect or cosmological expansion will produce important photometric and spectroscopic effects. Concepts:

  K correction   photometric parallax   Lyman-alpha forest

Exercises

1. Show that for a response function with a boxcar or triangular profile, the bandwidth = FWHM, but that for a Gaussian, the bandwidth < FWHM.

2. The table below gives the response function for a photometric bandpass, as well as the flux distributions for two sources, A and B. Characterize this system by computing (use a spreadsheet) all of the following:

   wavelength at peak transmission   mean wavelength
   the FWHM bandwidth   effective wavelength for each source
                                                           isophotal wavelength for each source

λ(nm)	$R_{BP}(\lambda)$	$f_{A\lambda}$	$f_{B\lambda}$	λ(nm)	$R_{BP}(\lambda)$	$f_{A\lambda}$	$f_{B\lambda}$
500	0	1.70	0.37	540	0.88	0.84	1.24
505	0.04	1.56	0.47	545	0.96	0.77	1.37
510	0.24	1.43	0.57	550	0.99	0.70	1.50
515	0.4	1.31	0.67	555	1 0	64	1.64
520	0.5	1.20	0.78	560	0.8	0.57	1.78
525	0.55	1.10	0.89	565	0.5	0.52	1.92
530	0.64	1.00	1.00	570	0	0.46	2.07
535	0.77	0.92	1.12				

3. A photometer on a spacecraft has an infrared-imaging camera.

   (a) Assume the detector is a perfect bolometer, so that 50% of the energy between 1.0 and 3.0 μm is detected, independent of wavelength. In other words, R_{BP} is a "boxcar" with mean wavelength 2.0 μm and bandwidth 2.0 μm. Compute the effective wavelength of this band for a hot star with $f_\lambda = A\lambda^{-4}$.

   (b) Now assume the detector is replaced with an infrared photon detector with uniform quantum efficiency such that 50% of the incident *photons* at each wavelength in the band are detected. Again, compute the effective wavelength of this band for a hot star with $f_\lambda = A\lambda^{-4}$. Note that you will need to devise an expression for the energy response function of the system.

4. Show that Equations (10.15) and (10.19) follow from the Rayleigh–Jeans and Wien approximations to the Planck law.

5. An MOS capacitor observes two sources in the band 400–600 nm. Source A has a spectrum such that the distribution of photons in the 400–600 nm band is given by $n_A(\lambda) = A\lambda^3$. Source B has a distribution of photons given by $n_B(\lambda) = B\lambda^{-2}$ in the same band. If the two sources generate photoelectrons at exactly the same rate, compute their brightness ratio. You may assume the detector's quantum efficiency is not a function of wavelength.

6. Gabriel very carefully constructs a filter for his CCD photometer so that the response function matches the standard bandpass of the Johnson V color very precisely. He observes two very well-established standard stars whose catalog data are given below. Gabriel discovers that with his CCD, no matter how carefully he observes, he always finds one star is brighter than the other: its image always contains more total analog-to-digital units (ADUs) on the CCD. Liz suggests to him that this is because the CCD is a photon-counting device. (a) Explain her reasoning. (b) If Liz is correct, which star should be the brighter on the CCD and why?

	V	B– V
Star 1	9.874	0.058
Star 2	9.874	0.861

7. Investigate the website for the Sloan Digital Sky Survey. In what ways is the SDSS five-color system superior to the UBVRI system? List the major projects that the SDSS has completed.

8. An astronomy student obtains two images of a galaxy, one in the B band, the other in the V band. Outline the image arithmetic operations the student would execute in order to produce a map of the $(B - V)$ color index for the galaxy. Failure to subtract the constant background sky for each image would cause problems in the map. For which parts of the map would these problems be most serious? On the other hand, would subtracting the background sky introduce any problems in the map? If so, which parts, and why?

9. Speculate, in terms of the Forbes effect, why it might be useful to define the standard magnitude as one measured at 1 air mass, rather than at zero air mass. What difficulties might be inherent in this choice?

10. An observer uses the B and V filters to obtain four exposures of the same field at different air masses: two B exposures at air masses 1.05 and 2.13, and two V exposures at air masses 1.10 and 2.48. Four stars in this field are photometric standards. Their standard magnitudes are given in the table below, as are the instrumental magnitudes in each frame.

	(B − V)	V	b(1)	b(2)	v(1)	v(2)
Air mass			1.05	2.13	1.10	2.48
Star A	−0.07	12.01	9.853	10.687	8.778	9.427
Star B	0.36	12.44	10.693	11.479	9.160	9.739
Star C	0.69	12.19	10.759	11.462	8.873	9.425
Star D	1.15	12.89	11.898	12.547	9.522	10.002

(a) Compute the outside-the-atmosphere magnitudes $v(0)$ and $b(0)$, as well as the extinction coefficients:, k'_B, k''_B, k'_V and k''_V. Hint: at each air mass, Equations (10.37) and (10.40) hold. Write an equation for the difference between the magnitudes at the two air masses (e.g. an equation for $b(2) - b(1)$). You may find it helpful to enter the data from the table into a spreadsheet in performing the computations.

(b) Compute the transformation coefficients, $\zeta_V, \varepsilon_{V,\text{B-V}}, \varepsilon'_{V,\text{B-V}}, \varepsilon_{\text{B-V}},$ and ε'_{B-V} using the method outlined in Section 10.6.

11. Using an adaptive optics camera, a small cluster of stars has been observed closely orbiting Sag A*, the black hole at the center of the Milky Way Galaxy. The extinction law for the ISM between Earth and the Galactic center is

$$A_\lambda / A_{\lambda_0} = \left(\frac{\lambda}{\lambda_0} \right)^\alpha$$

where $\alpha \cong -2.1$, for wavelengths between 1.0 and 8.0 μm. Assuming $a \cong -1$ for wavelengths shorter than 1 μm, compute the expected V magnitude for the brightest of these cluster stars, which has an apparent magnitude in the K band of K = 14.9. Total extinction to the Galactic center in the K band has been measured to be 2.8 magnitudes.

12. A photometric bandpass whose response function is shown in curve A below measures the strength of the emission feature shown in curve B. In the figure, the source has zero velocity. Compute the change in the brightness measurement, in magnitudes, that would result if the source were given a radial velocity of +300 km s^{-1}.

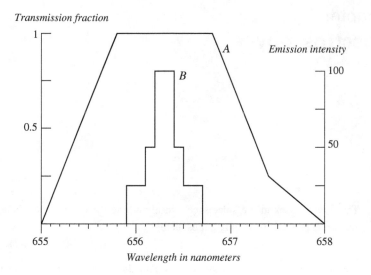

13. The primary low-resolution feature in the visible spectrum of elliptical galaxies is "the G-band break" between 390 and 450 nm (see Figure 11.31), and an important feature in the ultraviolet is the intensity drop due to the Ly-α forest. Considering only these features, over what range of z would you expect SDSS *ugriz* photometry to yield estimates of the redshift?

Chapter 11
Spectroscopy

The dark D lines in the solar spectrum allow one therefore to conclude, that
sodium is present in the solar atmosphere.

– Gustav Kirchhoff, 1862

This news [Kirchhoff's explication of the Fraunhofer solar spectrum] was to me
like the coming upon a spring of water in a dry and thirsty land. Here at last
presented itself the very order of work for which in an indefinite way I was
looking – namely to extend his novel methods of research upon the sun to the
other heavenly bodies.

– William Huggins, 1897

Beginning in 1862, Huggins used a spectroscope to probe the chemical nature of
stars and nebulae. Since then, spectrometry has been the tool for the observational
investigation of almost every important astrophysical question, through direct or
indirect measurement of temperature, chemical abundance, gas pressure, wavelength shift, and magnetic field strength. The book by Hearnshaw (1986), from
which the above quotes were taken, provides a history of astronomical spectroscopy
prior to 1965. Since 1965, the importance of spectroscopy has only increased. This
chapter introduces some basic ideas about spectrometer design and use. Kitchin
(1995, 2008) and Schroeder (1987) give a more complete and advanced treatment,
and Hearnshaw (2009) provides a history of the actual instruments.

Literally, a ***spectroscope*** is an instrument to look through visually, a ***spectrometer*** measures a spectrum in some fashion, and a ***spectrograph*** records the
spectrum. Astronomers are sometimes particular about such distinctions, but
very often use the terms interchangeably. This chapter introduces the basics of
the design and operation of spectrometers in astronomy. We confine our discussion to the class of instruments that use dispersive elements, and examine those
elements in detail: prisms, surface-relief gratings, and volumetric phase holographic gratings. We will direct most of our attention to the important design
parameters for the slit/fiber spectrometer, especially as applicable to the most
productive modern instruments; the faint-object, multiple-object, and integral
field spectrographs.

We will discuss the process of turning raw spectroscopic data, usually in the form of a digital image, into useful astrophysical information. Finally, we will examine a very small sample of the results produced by astronomical spectroscopy.

11.1 Dispersive spectrometry

We note at the outset that there are two methods for generating spectra: dispersing light of different wavelengths into different directions, and analyzing the wavelength distribution of light without such dispersion. We will not treat non-dispersive spectroscopy, but refer to the discussion of Fabry–Perot and Michelson interferometers in chapter 4 of Kitchin (2008) for a quantitative discussion.

Figure 11.1a shows a rudimentary dispersive spectrometer. An abstract telescope–spectrograph combination – represented by a featureless box – accepts a heterochromatic ray, which we assume to be on the optical axis, and disperses it so that rays of wavelength λ are sent in direction θ, while those of wavelength $\lambda + d\lambda$ are sent in direction $\theta + d\theta$. The **angular dispersion** (a concept introduced for prisms in Section 5.2.8) is simply $d\theta/d\lambda$. Figure 11.1b indicates that a useful spectrometer must bring all the rays of wavelength λ to the same point, P, at object distance s_c, where a detector can measure their intensity. Waves of wavelength $\lambda + d\lambda$ will focus at a different spot, a distance dx away on the detector, and the **linear dispersion** is defined as

$$\frac{dx}{d\lambda} = s_c \frac{d\theta}{d\lambda} \qquad (11.1)$$

Astronomers often find the reciprocal of the above quantity to be more intuitive, and may say "dispersion" when the number they quote is actually the **reciprocal linear dispersion**, or **plate factor**, p:

$$p = \frac{d\lambda}{dx} = \left[s_c \frac{d\theta}{d\lambda} \right]^{-1} \qquad (11.2)$$

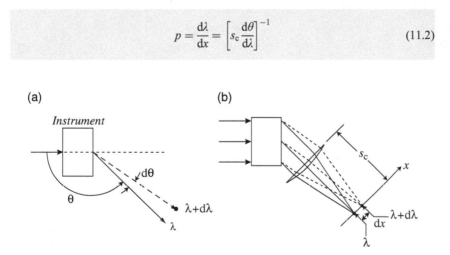

(a)   (b)

Fig. 11.1 (a) Angular dispersion: light from a distant source enters a telescope–spectrograph which disperses different wavelengths to different directions. (b) Linear dispersion: after dispersion all rays of the same wavelength are brought to a focus at image distance, s_c. Images of wavelength λ and $\lambda + d\lambda$ are separated by distance dx in the focal plane.

A "high dispersion spectrometer" is one in which the linear dispersion is large and p is small. The units of p are usually nanometers or Ångstroms (of wavelength) per millimeter of distance in the focal plane.

The size of the image produced by perfectly monochromatic rays that focus at point P cannot be indefinitely small. The image will be smeared out in the x-direction to a linear width, w_0, a distance that depends on the angular size of the source, on the geometric details of the optics, as well as on other processes that limit optical resolution: diffraction, atmospheric seeing, optical aberrations, and errors in the dispersion process. The width w_0 corresponds a range of wavelengths, $\delta\lambda_0$. This quantity, called the **spectral purity of the optics**, measures the spectrometer's ability to resolve details in the spectrum. However, the **effective spectral purity** of the complete instrument also depends on the detector's ability to resolve linear detail. If w_d = width of ≈ 2 detector elements, then

$$\delta\lambda = \text{spectral purity} = \text{the larger of}: \begin{cases} \delta\lambda_0 = & w_0 \dfrac{d\lambda}{dx} \\ \delta\lambda_0 = & w_d \dfrac{d\lambda}{dx} \end{cases} \qquad (11.3)$$

If two emission lines, for example, are closer together than $\delta\lambda$, they will overlap so much that their separate identities cannot be discerned. If we take $\delta\lambda$ to be the full width at half-maximum (FWHM) of the monochromatic instrumental profile, then the definition of resolution rests on the Rayleigh criterion introduced in Chapter 5. Figure 11.2 sketches the **instrumental profiles** (a plot of intensity vs. x or θ) of two monochromatic spectral features and illustrates the distinction between spectral purity and dispersion. Astronomers commonly refer to the spectral purity as the **resolution**. The related dimensionless parameter, the **resolving power** (a concept you may recall from Section 3.4.3), also measures the spectrometer's ability to record detail in the spectrum:

$$R = \frac{\lambda}{\delta\lambda} \qquad (11.4)$$

Fig. 11.2 Instrumental profiles of two emission lines observed at different spectral purities and slit image widths.

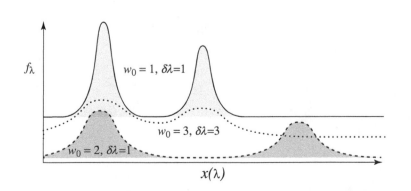

Astronomical spectrometers range in resolving power from very low ($R = $ a few tens) to very high ($R = 100\,000$ or more).

11.2 Dispersing optical elements

11.2.1 Prisms

We discussed the angular dispersion of prisms in Section 5.2.8 where we saw that for a prism

$$\frac{\partial \theta}{\partial \lambda} \cong -\frac{4K_2 \sin (A/2)}{\lambda^3 \cos \alpha} \tag{11.5}$$

Here A is the apex angle of the prism, α is the angle of incidence in the minimum deviation configuration, and K_2 is a constant ranging from about 0.003 to about 0.017 μm^{-2}, for various types of glass.

Prisms find use in astronomical spectrographs, but seldom (except in the near infrared) as the primary disperser. Their weight and relative expense are minor disadvantages. More serious are their low transmission in ultraviolet, low angular dispersion at long wavelengths, and highly non-linear variation of angular dispersion with wavelength.

11.2.2 The diffraction grating

Despite the name, this disperser depends on the interference of diffracted light waves to produce dispersion. Its simplest form, the **amplitude grating**, is a set of closely spaced parallel lines uniformly arranged across a flat surface (see Figure 11.3). This pattern can be either a series of slit-like openings – a **transmission grating** – or a series of separate, tall, but very narrow, mirrors or facets – a **reflection grating**. To produce appreciable diffraction effects, the widths of the slits or mirrors should be on the order of a few wavelengths, so in the optical, astronomical gratings typically have between 100 and 3000 lines per millimeter.

Figure 11.3 illustrates the principles of the grating. Adjacent grating facets are a distance σ apart – the **groove spacing** or **grating constant**. Gratings are conventionally described by giving the reciprocal of σ, the **groove frequency**, usually quoted in lines per millimeter. In Figure 11.3a, which shows a reflection grating, a plane wave strikes the grating at angle α and reflects at angle θ. All angles are measured counterclockwise from the grating normal. Although there will be a preference for reflected rays to travel in the direction specified by geometrical optics, $\theta = -\alpha$, diffraction effects allow rays to spread out from the narrow facets in all directions.

Wave interference means that rays of a particular wavelength, λ, will only diffract without interference from the grating at particular angles. To see this, consider the two rays in Figure 11.3a that strike the centers of adjacent facets.

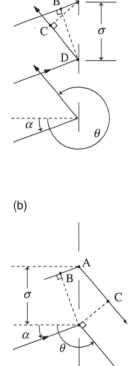

(a)

(b)

Fig. 11.3 (a) A reflection grating. Grating facets are tall, narrow mirrors extending perpendicular to the plane of the paper. Light striking between the facets is not reflected. The figure traces three parallel rays that strike the centers of adjacent facets. (b) A transmission grating consisting of opaque stripes separating clear slits. See the text for further explanation.

Both rays strike at angle α and diffract at angle θ. Upon leaving the grating, the optical path lengths of the two rays differ by the amount $\Delta\tau = \overline{AB} - \overline{CD}$. The two waves will constructively interfere only if $\Delta\tau$ is some integral multiple of the wavelength:

$$\Delta\tau = m\lambda, \quad m = 0, \pm 1, \pm 2, \dots \tag{11.6}$$

From the figure, we have $\overline{AB} = \sigma \sin \alpha$ and $\overline{CD} = \sigma \sin (2\pi - \theta) = -\sigma \sin \theta$, so for constructive interference we require

$$\sin \theta + \sin \alpha = \frac{m\lambda}{\sigma} \tag{11.7}$$

This rather general result also applies to the transmission grating, and is known as the **grating equation**. The integer m is called the **order**. For a large number of facets, interference effects suppress all diffracted rays that fail to satisfy the grating equation. Unlike the dispersing prism, gratings cause red light to deviate more than blue. Differentiating the grating equation tells us that the angular dispersion of a grating is

$$\frac{d\theta}{d\lambda} = \frac{m}{\sigma \cos \theta} \tag{11.8}$$

Since $\cos \theta$ changes only slowly with λ, the angular dispersion of a grating is roughly constant with wavelength, again in contrast to the behavior of prisms. From Equation (11.8), it is clear that high angular dispersions can be achieved by either selecting higher orders or by increasing the number of lines per millimeter on the grating.

An important characteristic of diffraction gratings (and a disadvantage relative to prisms) is dispersion into multiple orders. An analysis of grating efficiency (see the discussion in Kitchen, 2008, pp. 367–72, or Schroeder, 1987, pp. 243–247) shows that most of the light from a simple amplitude grating gets diffracted into the $m = 0$ order, where $\theta = -\alpha$, and where there is no dispersion. Figure 11.4 illustrates an additional problem: the non-zero

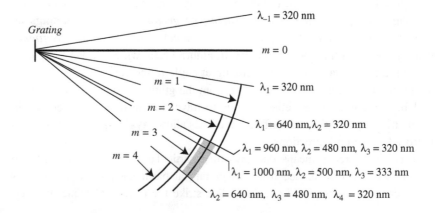

Fig. 11.4 The angular overlap of grating orders. Positions of the blue edges (taken to be at 320 nm) of orders −1 through +4 are shown. The thick gray arc shows the free spectral range of the second order, assuming $\lambda_{max} = 640$ nm.

orders overlap. A particular value of θ corresponds to a different wavelength for each order, i.e. if λ_m is the wavelength in direction θ produced by order m, then $\lambda_1 = 2\lambda_2 = 3\lambda_3 = n\lambda_n$, or

$$m\lambda_m = (m+1)\lambda_{m+1} \tag{11.9}$$

The overlapping of multiple orders means that some method – usually a blocking filter – must be used to eliminate unwanted orders. For example, suppose our detector responds to all wavelengths shorter than some value, $\lambda_{\max}$ (for example, a silicon detector might have $\lambda_{\max} = 1100$ nm). If we record the spectrum in order m with this detector, the spectrum from order $m + 1$ overlaps $\lambda_{\max}$, and at the same value of θ, deposits photons of wavelength $m\lambda_{\max}/(m+1)$. For example, first-order light at 1100 nm mixes with second-order light at 550 nm. We therefore insert a filter to block all light with wavelengths shorter than $m\lambda_{\max}/(m+1)$ to eliminate the overlap. A quantity called the *free spectral range* quantifies the resulting restriction. The free spectral range is just the range of wavelengths not blocked, that is

$$\Delta\lambda_{\text{FSR}} = \lambda_{\max} - \frac{m}{m+1}\lambda_{\max} = \frac{\lambda_{\max}}{m+1} \tag{11.10}$$

11.2.3 Blazed gratings

Amplitude gratings, pictured in Figure 11.3, operate by blocking (reducing the amplitude of) waves whose phases would destroy the constructive interference produced by diffraction in the periodic structure. For example, if the two rays pictured constructively interfere in first order, then the (blocked) ray that strikes between the facets would destructively interfere with them. In any particular order, then, amplitude gratings are inefficient, due to (1) this necessary blocking by the grating as well as (2) diffraction of most light into zeroth as well as other orders. *Phase gratings* produce dispersion effects similar to amplitude gratings, but operate by periodically adjusting the phase of the diffracted waves. They minimize both disadvantages of the amplitude grating.

The *blazed reflection grating* is a common example of a phase grating. As illustrated in Figure 11.5a, the surface of the blazed grating has a sawtooth-shaped profile, with each diffracting facet tilted at an angle ε, the *blaze angle*, measured counterclockwise with respect to the plane of the grating. This is sometimes called an *echellette grating* (from the French *échelle* – stair or ladder). The goal is to arrange the tilt so that all rays diffracted from a single facet are in phase. This will occur if there is specular reflection from the facet, i.e. if the angles of incidence and reflection with respect to the facet normal are equal: $\beta_1 = -\beta_2 = \beta$. This condition means that

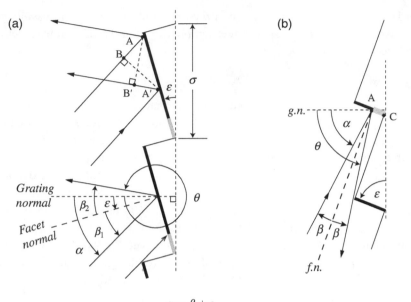

$$\alpha = \beta + \varepsilon$$
$$\theta = 2\pi + \varepsilon - \beta \qquad (11.11)$$
$$\alpha + \theta = 2\varepsilon$$

From the figure, it is clear that the conditions for constructive interference of rays diffracted from adjacent facets are identical to those we had for the amplitude grating, so we can apply the grating equation, (11.2), but substitute $\beta + \varepsilon = \alpha$ and $(\varepsilon - \beta) = \theta$:

$$\sin(\beta + \varepsilon) + \sin(\varepsilon - \alpha) = \frac{m\lambda}{\sigma} \qquad (11.12)$$

which reduces to

$$\sin(\varepsilon) = \frac{m\lambda}{2\sigma \cos(\beta)} \qquad (11.13)$$

or

$$\lambda_{\rm b} = \frac{2\sigma}{m}\sin(\varepsilon)\cos(\alpha - \varepsilon) \qquad (11.14)$$

Once a grating is manufactured with a particular blaze angle, Equation (11.14) suggests an associated **blaze wavelength** in order m, which depends slightly on the angle of incidence. A more complete analysis (e.g. Möller, 1988, chapter 3) shows that the result of blazing is to shift the maximum efficiency of the grating from order 0 to order m. At the blaze wavelength, the grating will be completely illuminated (and most efficient) when mounted in the **Littrow configuration**, with $\beta = 0$ and $\alpha = \varepsilon$, so that incoming light is dispersed back on itself (not always the most convenient arrangement). Except for echelles (see below)

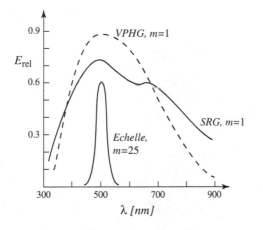

Fig. 11.6 Representative grating efficiencies of three different phase gratings with a blaze wavelength of λ_b = 500 nm. SR = surface relief. VPH = volume phase holographic.

blazed gratings are usually designed to work in order $m = \pm 1$, where their efficiency function (fraction of incoming light dispersed into the order as a function of wavelength) has a somewhat asymmetrical shape; see Figure 11.6.

Blazed transmission gratings also have sawtooth-shaped surface profiles, and achieve a phase shift along a facet by virtue of the changing optical path length in the high-index material. At the blaze wavelength, the transmission is in the direction given by Snell's law for refraction, and light is shifted from zeroth order to the design order. The blaze angle for a transmission grating is:

$$\tan \varepsilon = \frac{\sin \theta_b}{n - \cos \theta_b} \tag{11.15}$$

The blaze wavelength is given by the grating equation.

11.2.4 Echelles

To produce a large angular dispersion, Equation (11.8) suggests operating at high order (m large) and with dispersed rays nearly parallel to the grating surface ($\cos \theta$ near 0). For a blazed grating, where $\theta = \varepsilon - \beta$, this suggests a design like the one in Figure 11.5b, where the blaze angle is very large and the incident and diffracted rays are nearly perpendicular to the reflecting facet (i.e. β small). Echelle gratings used in astronomy are thus very coarse. They typically have groove frequencies between 10 and 100 lines per millimeter, and operate in orders in the range 25–150.

Equation (11.10) means that the free spectral range of any echelle order is very small. Rather than use filters to isolate this tiny range for a single high-m order, the strategy is to separate overlapping orders with a *cross-disperser,* and record all of them. The cross-disperser can be a grating of low dispersing power or a prism; in either case, this second element disperses the echelle output (or, sometimes, input) in a direction perpendicular to the echelle dispersion.

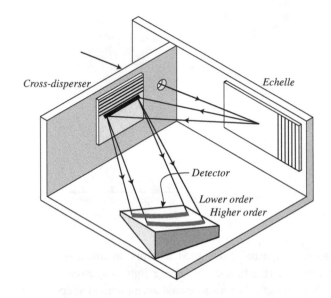

Figure 11.7 shows a representative set-up. With a camera (omitted for clarity in the figure) in place, the echelle-plus-cross-disperser combination produces multiple short spectra, one for each order, and stacked on the output perpendicular to the dispersion direction. Figure 11.8 is an example of the detector output. Most high-resolution astronomical spectrometers are based on echelles.

11.2.5 Volumetric phase gratings

So far we have been discussing phase gratings that rely on surface relief (SR) to produce periodic phase shifts. The ***volume phase holographic (VPH) grating*** produces phase shifts through spatial variations in the index of refraction, rather than surface relief. Researchers have produced VPH gratings in a number of configurations, including both transmitting and reflecting devices, but the form most useful in astronomical applications is the transmission grating illustrated in Figure 11.9. The grating is a thin slab or film of transparent material in which the "lines" are actually parallel planes of higher and lower index of refraction – holographic techniques are used to imprint the index modulation throughout the volume of the slab. In most current devices, the plane-to-plane index variation is roughly sinusoidal with an amplitude of 0.1 or less in the index, and with a peak-to-peak frequency of 200–6000 planes per millimeter.

The VPH gratings used in astronomy are usually in the "normal" or "Littrow" configuration illustrated in the figure, where the planes of enhanced index run perpendicular to the surface of incidence. The film thickness, d, typically ranges from a few microns up to 0.2 mm. If the spacing between fringe maxima is σ, then diffraction by the periodic index structure and the resulting interference

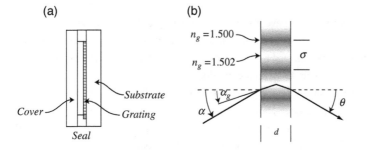

Fig. 11.8 A cross-dispersed echelle spectrogram. Higher orders (shorter wavelengths) are at the top of the figure. The top panel shows the entire image, approximately 90 orders, with the lowest orders (shortest wavelengths) toward the top. The lower panel is an enlargement of the central region of the top panel (CCD image courtesy of Allison Sheffield).

(a)

(b)

$n_g = 1.500$

$n_g = 1.502$

σ

Substrate

Cover

Grating

Seal

α_g

α

θ

d

Fig. 11.9 (a) A VPH grating assembly showing the film-like grating deposited on a glass substrate and sealed against humidity. (b) The path of a ray at the Bragg wavelength and angle of incidence ($\alpha = \theta$) in first order for a normal, or Littrow transmission VPH grating. In typical VPH gratings the refractive index varies sinusoidally, as represented by lighter and darker grays in the figure. In non-Littrow gratings, the planes of constant index are at an angle to the grating normal. We ignore refraction effects of the substrate and cover glass.

effects disperse light exactly as described by the basic grating Equation (11.7):
$$\sin \theta + \sin \alpha = m\lambda/\sigma.$$

However, as in the blazed SR gratings, there is now an additional condition – this one imposed by the fact that the grating extends through the volume of the film. The result is that efficiency will be enhanced by ***Bragg diffraction*** (refer to any introductory physics text) by the planes of constant index. Bragg diffraction occurs in a medium of index n_g if a ray of wavelength λ is incident on the grating fringe at internal angle α_g, defined as

$$n_g \sin \alpha_g = \frac{m\lambda}{2\sigma} \tag{11.16}$$

where we have applied Snell's law at the grating surface to get the external angle of incidence. The efficiency of the VPH grating is usually a maximum at the **Bragg angle** α_B for the wavelength λ_B, which, from the grating equation, diffracts into direction $\theta = \alpha_B$ for first order.

$$\sin \alpha_B = \frac{m\lambda_B}{2\sigma} \tag{11.17}$$

Incident light at other wavelengths does not satisfy the Bragg condition and is diffracted with lower efficiency. The actual efficiency curve depends in a complex fashion on σ, on d, and on the shape of and amplitude, Δn, of the index modulation. For many astronomically relevant gratings, efficiency will be maximum if

$$d \approx \frac{\lambda_B}{2\Delta n} \tag{11.18}$$

Since present technology limits Δn to less than 0.1, this requires d to be relatively large for high efficiencies. However, the width of the efficiency profile depends strongly on the ratio σ/d, so grating design is a compromise between high efficiency (large d) and broad spectral coverage (small d).

Volumetric phase holographic gratings have a number of advantages over their SR counterparts:

- A VPH grating can have a very high efficiency at the Bragg wavelength.
- The **"superblaze"** property: the blaze wavelength (i.e. the Bragg wavelength) of a VPH grating can be selected by tilting the grating to the appropriate Bragg angle.
- Because of their holographic production method, VPH gratings are less liable to imprecision in line spacing and parallelism, can have smaller values of σ, and can be physically quite large compared to ruled gratings.
- Current VPH films are encapsulated in a rugged glass assembly, so unlike SR gratings, can be cleaned and treated with anti-reflection coatings.

The important negative aspects of VPH gratings are:

- The wavelength bandpass can get narrow for high line-density VPH gratings.
- It is currently impossible to produce the VPH equivalent of an echelle – operation at high orders is not well understood.
- A VPH grating costs more than an equivalent SR grating.
- A spectrograph must have special design to take advantage of the superblaze capability of VPH gratings.

11.2.6 Grating manufacture

Palmer (2002) gives a complete treatment of the operation and construction of surface-relief gratings.

Manufacturers produce ***ruled gratings*** by drawing a diamond-tipped cutting tool across the optically flat surface of a soft metal blank. The shape of the diamond controls the shape of the grooves, and the engine that moves the tool (or blank, or both) controls groove depth, spacing, and parallelism. Current engines maintain precision through interferometric methods. The capacity of the engine, and, more critically, the wear of the diamond cutter, limit grating size, so very large ruled gratings are not possible. The ruled metal blank is usually used as a master to mold replica gratings – these are resin imprints of the master, subsequently mounted on a rigid substrate and aluminized. Transmission gratings are not aluminized, but do demand resins that cure to high optical quality.

Production of ***holographic gratings*** is perhaps more elegant. The manufacturer uses a laser to create two or more monochromatic plane waves whose interference creates a pattern of light and dark lines on a flat surface. The surface is coated with photoresist and the development of the photoresist, etching, and reflection coating produces an SR grating. Holographic gratings generally have greater precision, lower surface scattering, and potentially larger size than ruled gratings. The simple holographic technique produces a grating surface with a sinusoidal cross-section, so these devices are not strongly blazed and usually have a lower efficiency than ruled gratings. However, ***ion etching*** or more sophisticated holography can produce an echellette surface pattern on a holographic grating. Very coarse holographic gratings are difficult to produce.

Mosaics of ruled or holographic gratings are also difficult to produce, but feasible, often by bonding multiple replicas on a monolithic surface. Several examples are in operation at large telescopes, for example, a 0.8×0.2-m echelle mosaic is in operation on the HARPS-N spectrograph on the 3.6-m Galileo telescope on Las Palmas.

It is possible to deposit a holographic grating on a curved surface, and ***concave holographic gratings*** find some use in astronomy. It is even possible to vary groove spacing and parallelism to remove optical aberrations. A ***flat field concave grating*** will image a spectrum on a plane surface without additional optics.

As mentioned above, VPH gratings also rely on laser interference effects for their production. In this case, the manufacturer illuminates parallel bright and dark planes in the interior volume of a transparent material. Several materials are under investigation, but current devices use a suspension of ammonia or metal dichromate in gelatin (dichromated gelatin, or DCG). When chemically processed, DCG exhibits a change in index due to differential shrinkage that depends on its exposure to light. Processed DCG must be of limited thickness, and its density enhancements are degraded by humidity, so VPH gratings are usually mechanically stabilized and sealed between glass plates.

11.3 Spectrometers without slits

11.3.1 The objective prism

The objective prism spectrograph mode (probably first employed by Fraunhofer) is conceptually simple. Just mount a prism directly in front of a telescope objective, as illustrated in Figure 11.10. The prism disperses the parallel rays from distant objects and the objective focuses the result. A completely monochromatic source produces a simple image in the focal plane, but a polychromatic source produces a spectrum – a different image at each wavelength. A two-dimensional detector can thus record, in a single exposure, a separate spectrum for every object in the field – a large *multiplex advantage*. For this reason, astronomers frequently have mounted objective prisms on wide-field telescopes like Schmidts. Objective *gratings* are seldom employed because of the expense of large gratings as well as the confusion and light loss produced by multiple orders. Figure 11.11 is an example of an objective prism view of the Pleiades star cluster.

Resolving power and the instrumental profile will depend on both the prism and telescope parameters. A telescope that normally produces monochromatic stellar images of angular diameter $\delta\theta$ should, when equipped with an objective prism, have spectral purity

$$\delta\lambda_0 = \delta\theta\frac{\mathrm{d}\lambda}{\mathrm{d}\theta} \cong \delta\theta\frac{\lambda^3\cos\alpha}{4K_2\sin(A/2)} \simeq \frac{\delta\theta}{0.02A}\lambda^3[\mu m] \qquad (11.19)$$

The numerical values on the right-hand side assume a thin prism of flint glass.

For example, assume a 1-m $f/3$ Schmidt is equipped with a prism with apex angle 1°. According to Equation (11.19), the spectral purity expected in 1 arcsec seeing at 500 nm is $\delta\lambda_0$ = (1 arcsec)(0.125 μm^3)/[(0.02 μm^2)(3600 arcsec)] = 1.7×10^{-3} = 1.7 nm, which improves to 0.85 nm at 400 nm. The optical resolving power is R = 500/1.7 = 300 at 500 nm and 470 at 400 nm. The plate factor at 500 nm will be p = $(1/f)(\mathrm{d}\lambda/\mathrm{d}\theta)$ = 113 nm mm^{-1}. Seeing sets the resolving power in this case. If this same spectrometer were placed in space, stellar image sizes should be around 0.12 arcsec, and we would find R = 2400 at 500 nm.

Remember that for any spectrometer the effective resolving power may not be optics- or seeing-limited, but could be set by the detector resolution. The pixel size of most astronomical charge-coupled devices (CCDs) is in the range 13–26 μm, and sampling theory requires a minimum of 2 pixels per resolution element. In the above example, for instance, if we had 13 μm pixels then $\delta\lambda_d$ = 2.3 × (13 × 10^{-3}) × [p], or 3.4 nm at 500 nm, nearly twice the value for $\delta\lambda_0$ in the seeing-limited case. Since we realize the detector now limits the resolution, we should expect R = 150 at 500 nm.

Spectra from objective prisms are generally very low resolution, but are well suited for survey work that requires rough spectral classification, color determination, or identification of peculiar spectra in a field that contains many objects.

11.3.2 The non-objective prism and grism

Objective prisms must be the same size as the telescope aperture, an expensive requirement, and an impossible one for apertures larger than around 1 m. A prism placed near the focal plane (the "non-objective" configuration) in the converging beam of a telescope will be smaller, but generally produces spectra severely compromised by coma and other optical aberrations.

The *grism* substantially reduces the aberrations of non-objective spectra. (The grism is a grating–prism combination. A similar device, incorporating a grating on a lens surface, is called a *grens*.) The grism usually consists of an SR transmission grating mounted on the hypotenuse of a right-angle prism. The apex angle of the prism is generally chosen so that rays at the grating blaze wavelength, λ_B, converge on the optic axis (Figure 11.10b):

$$\sin A = \frac{\lambda_B}{\sigma(n-1)} \qquad (11.20)$$

As illustrated, the focal plane of the grism can be slightly tilted from the Gaussian plane. A telescopic camera can quickly convert into a low-resolving-power ($R < 2000$) spectroscope by the insertion of a grism in the beam, and many large telescopes, including space telescopes, routinely provide grisms mounted on filter wheels. Volumetric phase holographic grisms, which are capable of relatively high angular dispersions, have begun to appear.

11.4 Basic slit and fiber spectrometers

Slitless spectrometers have serious disadvantages: overlapping spectra in crowded fields, contamination by background light, resolution limits imposed by seeing, limited use for extended objects, and lack of convenient wavelength calibration. The slit or fiber-fed spectrometer addresses all these issues, but at a cost. The basic notion is to restrict the light that reaches the dispersing element to only that light from the small angular area of interest on the sky. This improves resolving power and reduces background, but loses the starlight excluded by the slit or fiber, and, without elaborate design, loses the multiplex advantage.

Figure 11.12 sketches the layout of a simple slit spectrometer with all-transmitting optics (because it is clearer: most astronomical spectrometers use at least some reflecting optics). The rectangular slit has width w_s in the plane of the diagram and height h, and is located at the focus of a telescope of aperture

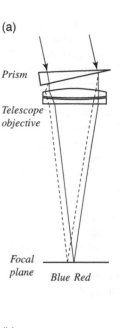

(a)

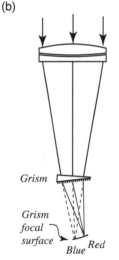

(b)

Fig. 11.10 (a) An objective prism on a refracting telescope. On Schmidt telescopes, the prism would be mounted just in front of the corrector plate. (b) A non-objective grism. In practice, the relative distance between the grism and the focal plane is usually much smaller than illustrated here.

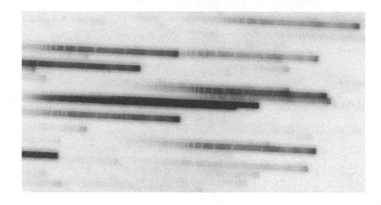

Fig. 11.11 An objective prism spectrogram of the Pleiades star cluster. This illustration is a small section (about 5%) of an original photographic plate taken with a Schmidt telescope and is of relatively high dispersion ($p = 10$ nm mm^{-1} at 400 nm). Note over- and under-exposed spectra, and overlapping spectra; and also the change in linear dispersion and projected spectrum brightness with wavelength (red is toward the right).

D_{TEL} and effective focal length f_{TEL}. The slit limits light entering the spectrometer to the narrow rectangle on the sky defined by the slit, and (as long as the angular slit width is smaller than the seeing disk), the slit width limits the spectral purity of the spectrometer. The angular size of the slit on the sky is

$$\phi_{\mathrm{s}} = \frac{w_{\mathrm{s}}}{f_{\mathrm{TEL}}} \tag{11.21}$$

Alternatively, an optical fiber with one end in the focal plane of the telescope could bring light to the spectrometer. In this case, w_{s} becomes the diameter of the fiber core. Following the light through the spectrometer on the figure, note that rays at the slit (or fiber end) emerge at the focus of the **collimator** lens, which converts them to a bundle of parallel rays. To avoid either losing light off the edge of the collimator or making the collimator larger than needed, we require it to have a focal ratio

$$\mathfrak{R}_{\mathrm{COL}} = \frac{f_{\mathrm{COL}}}{D_{\mathrm{COL}}} = g_{\mathrm{f}}\mathfrak{R}_{\mathrm{TEL}} = g_{\mathrm{f}}\frac{f_{\mathrm{TEL}}}{D_{\mathrm{TEL}}} \tag{11.22}$$

The factor g_{f} here accounts for the fact that an optical fiber can degrade the focal ratio of the beam it transmits. In the case of a slit, $g_{\mathrm{f}} = 1$, but in the case of a fiber-fed system g_{f} may be somewhat smaller, depending on fiber length and quality. The general import of Equation (11.22) is that *the focal ratio of the collimator should match that of the telescope/fiber.*

The **throughput** of a spectrometer is an important measure of its quality. Throughput is just the fraction of light incident in the focal plane of the telescope that actually reaches the spectrometer's detector. Scattered light or light lost in discarded orders, or at the edge of a slit, fiber, grating, or collimator decreases throughput.

Continuing through Figure 11.12, the **collimated beam** (i.e. parallel rays) strikes the dispersing element at angle α. The beam disperses to angle $\theta(\lambda)$, and a camera lens (diameter D_{CAM} and focal length f_{CAM}) finally focuses the rays as a

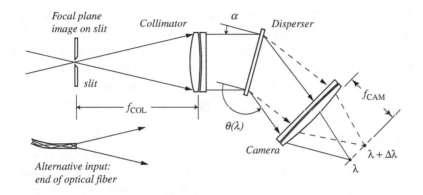

Fig. 11.12 A simple, all-transmission slit spectrometer. An alternative input is the end of an optical fiber, which would replace the slit at the focus of the collimator. The opening of the slit or fiber is w_s, measured in the vertical direction in this diagram.

spectrum on the detector. If the disperser is a grating (the usual case in astronomy) the reciprocal linear dispersion is

$$p = \frac{d\lambda}{dx} = \frac{1}{f_{CAM}}\frac{d\lambda}{d\theta} = \frac{\sigma\cos\theta}{m f_{CAM}} \qquad (11.23)$$

For a prismatic spectrometer, one would substitute an expression like Equation (11.5)) for $d\theta/d\lambda$ above. The spectral purity depends on w_0, the projected size of the slit or fiber on the detector. Without a disperser present (see Figure 11.13), $w_0 = w_s f_{CAM}/f_{COL}$. Inserting the disperser will modify the slit image in the dispersion direction by a projection effect that changes the image size by a factor known as the ***anamorphic magnification***, $r_{an} = d\theta/d\alpha = \cos\alpha/\cos\theta$. Thus, the width of the slit image on the detector will be

$$w_0 = r_{an}w_s\frac{f_{CAM}}{f_{COL}} = r_{an}\phi_s\frac{f_{TEL}}{f_{COL}}f_{CAM} = r_{an}\phi_s\frac{D_{TEL}}{D_{COL}}f_{CAM} \qquad (11.24)$$

Here we have substituted the input slit size in angular units, $\phi_s = w_s/f_{TEL}$, and have assumed the relation in Equation (11.22). The optical limit on spectral purity for the grating spectrograph is therefore

$$\delta\lambda_0 = w_0 p = r_{an}\phi_s\frac{D_{TEL}}{D_{COL}}\frac{\sigma\cos\theta}{m} \qquad (11.25)$$

and the resolving power of the spectrometer is

$$R = \frac{\lambda}{\delta\lambda_0} = \frac{\lambda}{r_{an}\phi_s}\frac{D_{COL}}{D_{TEL}}\frac{m}{\sigma\cos\theta} \qquad (11.26)$$

We have used only first-order geometric optics to derive Equations (11.25) and (11.26), so this result ignores diffraction effects and aberrations. Nevertheless, this expression is central for the practical design of spectrometers. There are three important comments:

First, for ground-based stellar work, you frequently achieve large R by setting the angular width of the slit on the sky, ϕ_s, to less than the full width at

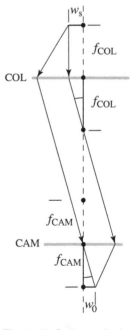

Fig. 11.13 Ray trace in the absence of a disperser, showing the width of the image of the spectrograph slit or fiber in the focal plane of the camera.

half-maximum (FWHM) of the seeing disk, ϕ_{see}. In the cases where the seeing disk does not over-fill the slit, however, we must replace ϕ_s with ϕ_{see} in the above expressions. In this case, the ***seeing-limited resolving power*** is greater than the slit-limited value. Fiber inputs enjoy no such gain in R due to improved seeing, since the entire output end of the fiber is illuminated no matter what the size of the input image. In the usual case where $\phi_s < \phi_{see}$ light will be lost off the edges of the slit or fiber, and throughput reduced. In the case of the slit, at least, there are devices – ***image slicers*** – that redirect this light back through the slit. The direct connection between telescopic image size and spectrograph resolution means that ***adaptive optics*** or any other steps that improve seeing will benefit spectrometer resolution. For space telescopes and adaptive optics systems, the absolute limit, on the resolving power, R_0, is set by the FWHM of the Airy disk: $\phi_0 = 1.22\lambda/D_{TEL}$. Substituting this into Equation (11.26), and making use of the definition of r_{an} we have

$$R_0 = \frac{mD_{COL}}{1.22\sigma\cos\alpha} \approx \frac{mW}{\sigma} = mN \qquad (11.27)$$

where W is the length of the grating and N is the number of lines ruled on it.

Second, if d_{px} is the pixel spacing in the direction of the dispersion, then the projected width of the slit, fiber, or seeing disk must satisfy the condition $d_{px} \leq w_0/2$. Otherwise, the image is under-sampled and R is reduced. Making w_0 very much larger than 2 pixels wastes detector length and usually increases noise.

Finally, suppose you have designed a very successful spectrometer for a 1.0-m telescope. Equations (11.25) and (11.26) say that to apply the same design (ϕ, m, σ, and θ all held constant) to a larger telescope, *the entire instrument must be scaled up in proportion to the size of the primary*. The diameter of the collimator must increase in direct proportion to D_{TEL}, and its focal ratio must match that of the new telescope. The length of the grating increases with the diameter of the collimator, as does the diameter of the camera. On the larger telescope, one would try to reduce the focal length (reduce the focal ratio) of the camera to preserve plate scale and avoid gigantic detectors and over-sampled spectra.

Thus, on a 30-m telescope, the same spectrometer design needs to either be something like 30 times larger (and something like $30^3 = 27\,000$ times more massive), or suffer from lower resolving power. A number of novel strategies (see, for example, Dekker et al., 2000) can reduce the effective slit width and "fold" the optics to reduce mass and soften this scaling rule. For example, the Planet Finder Spectrograph on the 6.5-m Magellan telescopes reflects the beam from the 0.4-m grating back on itself (i.e. near-Littrow configuration) and uses the same transmitting optics as both collimator and camera. Nevertheless, the rule explains the attractiveness of large gratings, as well as part of the popularity of echelles (large m/σ) for even moderate-sized telescopes.

11.5 Single-object spectrometer design for astronomy

11.5.1 An example configuration

Special constraints that may be unimportant in a laboratory environment are crucial in the design of astronomical spectrometers. For example, telescope optics determine the ideal focal ratio of the spectrometer collimator; the need to study faint objects sets a high premium on throughput; and the desire to mount the instrument on the telescope favors compact and rigid designs. Off-the-shelf instruments can be a poor choice for astronomy, and, especially for larger systems, astronomical spectrometers tend to be one-of-a kind devices built for a specific telescope.

Designing any astronomical spectrometer properly begins with careful consideration of the scientific questions the device will address. This sets the range of wavelengths to be investigated as well as the minimum value of R required, with an inevitable trade-off between the value of R and the faintness of the objects to be investigated. We reformulate Equation (11.26) to guide further decisions:

$$R = \frac{1}{D_{\text{TEL}}\phi_s} \frac{\lambda m}{\sigma} W = \frac{1}{D_{\text{TEL}}\phi_s} (\sin\alpha + \sin\theta) W \qquad (11.28)$$

Here we have used the length of the disperser, $W = D_{\text{COL}}/\cos\alpha$, instead of the diameter of the collimator, and in the rightmost expression, have made use of the grating equation.

The astronomer has no control over the seeing at his site or the diameter of his telescope. The remaining design parameters in the middle expression in Equation (11.28) all have to do with the choice of the dispersing element. Until the advent of VPH gratings, modern astronomical spectrometers tended to employ blazed reflection gratings (either echelles or single gratings) because of the difficulty in producing efficient transmission gratings and the drawbacks of prisms. The grating will exert the strongest constraints on the design, because there is a limited choice of sizes and manufacturers. The choice of echelle or single grating is a fundamental one. This choice appears in Equation (11.28) as the value of the dimensionless expression $\lambda m/\sigma$ (which at 500 nm equals 0.1 to 1.5 for a conventional first-order grating, slightly higher than this for a VPH grating, and even larger (1 to 3) for an echelle).

For a given resolution, the echelle requires a smaller grating and therefore produces a more compact spectrometer. The echelle also has the advantage that its coverage of the spectrum, in the form of a one-for-each-order stack of strips, makes efficient use of the square format of most CCDs. The echelle has disadvantages as well – besides requiring a more complex alignment process, the additional reflection/dispersion by the cross-disperser creates more scattered light and lowers overall efficiency. Data reduction, which requires extraction of the orders and reassembly to form a complete spectrum, can become complicated.

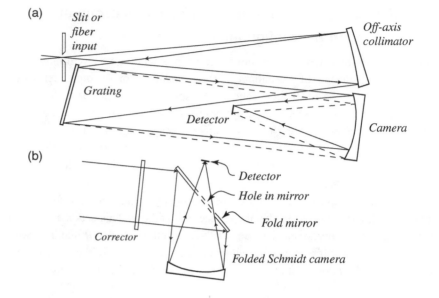

A single grating spectrograph has a few advantages: simplicity makes it relatively versatile and inexpensive compared with the echelle, because it is easy to swap out one grating for another. The somewhat better efficiency favors the single grating for observing faint objects.

The available detectors are a fundamental consideration in spectrometer design. The overall length of the detector limits the range of wavelengths recorded. Detector pixel size limits resolving power.

In the case of both the single grating and the echelle, the projected width of the grating determines the aperture of the camera: D_{CAM} must be at least as large as $W\cos\theta$, plus an allowance for the spread in θ due to dispersion. The camera focal length determines the plate scale, and generally one wants the shortest possible camera focal length consistent with adequate sampling. For large systems, this usually means a very fast camera, and because many rays will be off-axis, a Schmidt camera is a popular choice. Figure 11.14 is a schematic of a simple but representative all-reflecting spectrometer. The collimator is an off-axis paraboloid with its focus at the slit. The blaze angle and tilt of the grating will determine the central wavelength of the spectrum.

The basic layout in Figure 11.14 is subject to vast variation. Cost, as always, is a major consideration: the lower cost of widely available optical components in standard sizes is attractive, but restricts the design options. Spherical mirrors (if focal ratios are slow) can replace paraboloids. A commercial photographic lens is often a useful option for the camera in a small spectrometer if only visual wavelengths are of interest.

For large telescopes, however, the benefits of an optimized and versatile design outweigh its cost. It is not uncommon, for example, to split the collimated beam with a mirror or dichroic filter and construct separate grating/camera arms optimized for short and long wavelengths. You can consult the ESO, Keck, Subaru, AAO, and Gemini websites to get a sense of both the variety and complexity of spectrograph designs for large telescopes.

Spectroscopy in the thermal infrared has significant additional requirements. All mid-infrared spectrographs are cooled, beginning with the slit (a cold slit greatly reduces sky background), and completely encapsulated in a chamber that is either evacuated or filled with an inert gas. Special observing techniques are required to remove the bright background due to the telescope mirrors and sky.

11.5.2 Slit orientation and spectrum widening

Dispersion by the atmosphere means that star images are actually tiny spectra with the red end oriented toward the horizon (see Section 5.2.6). If the telescope is not equipped with an atmospheric dispersion corrector, it is best to orient the slit in the direction of atmospheric dispersion to avoid systematic loss of some parts of the spectrum, as well as some loss of resolution; see Figure 11.15a.

In the earliest days of spectroscopy, astronomers examined spectra visually, and quickly noticed that it was much easier to recognize emission and absorption features if the spectrum was widened in the direction perpendicular to the dispersion (typical widening might be ten times w_0). You can widen spectra through telescope motion or optical scanning that trails the image along the height of the slit during the exposure, or through insertion of a cylindrical lens that widens the image in one dimension only. See Figure 11.15b. In a **widened spectrum** the monochromatic image of the slit really does look like a "line," and in the photographic recording of spectra, sufficient widening was crucial for the visual recognition of features. With digital detectors, spectrum widening is far less important. Widening, in fact, will degrade the signal-to-noise ratio (SNR) except in special cases (e.g. to average out flat field uncertainties).

11.5.3 Getting light in

The astronomer needs to verify that the object of interest is indeed sending light into the slit or fiber aperture of the spectrometer for the duration of the exposure. For a single object, he might use an arrangement like the one sketched in Figure 11.16. In the figure, the jaws of the slit that face the telescope are reflective and tilt away from the normal to the optical axis, so that pre-slit optics let the astronomer view (possibly with a small video camera) the focal plane and slit. He can position the telescope to center the object on the slit and can **guide** telescope tracking during the exposure: either manually positioning the light reflected from the edges of the slit or engaging an autoguider. Image acquisition

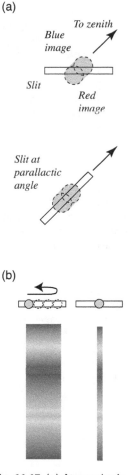

Fig. 11.15 (a) Atmospheric refraction and slit input. In the right-hand panel, the slit is not in the direction of atmospheric dispersion, so light lost at the slit is systematically from the extreme red and blue ends of the spectrum. In the left-hand panel, with the slit at the parallactic angle, light lost is less overall and not systematic with wavelength. (b) Widened (left) and unwidened (right) spectra – the trailed version requires a longer exposure, but is easier to examine visually.

Fig. 11.16 Some strategies for viewing the slit and for acquiring comparison spectra. The optics in the lower part of the figure view reflections from the front of the slit. In the upper part of the figure, a periscope can slide down to cover the slit, and light from either a gas discharge (emission-line spectrum) or solid filament (continuous spectrum) is fed in via an optical fiber and/or relay optics. The left side of the figure shows a cell with transparent windows. Gas in this cell imposes an absorption line spectrum on any light traveling from the telescope to the slit.

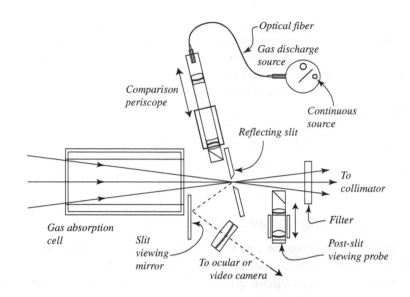

and guiding are more complicated with multi-object and integral field spectrometers, especially with altazimuth mounts.

In addition to light from astronomical objects, spectrometers need to accept light from calibration sources. ***Wavelength calibration*** (see Section 11.8.3) often requires that light from a gas-discharge or spark lamp with a well-understood emission-line spectrum can enter the slit on the same optical path as objects in the sky. ***Flat field*** calibrations require the same for a continuous source, and the retracting periscope sketched in Figure 11.16 is one way to deliver the light from calibration sources. There are many others, some rather easily implemented with fiber optics. It is sometimes useful to provide a moveable probe to examine an image of the slit from the collimator side, and a method for inserting an order-separation filter behind the slit is normal for single-disperser spectrometers.

11.6 Multiplexed spectrometers

The slit/fiber spectrometers described in the previous two sections can only record data from one object at a time. We have seen that the objective prism and the non-objective grism overcome this disadvantage, but at the cost of poor signal-to-noise imposed by background, poor spectroscopic resolution imposed by seeing, and confusion imposed by overlapping spectra. Fortunately, there are other approaches to multiplexing spectra.

11.6.1 Spectra without dispersion – energy-resolving detectors

An astronomer's "dream detector" would not only count the photons incident on a pixel but would also record their wavelength distribution. This detector would produce a data cube of the sort illustrated in Figure 11.17. The x,y coordinates fix location on the sky and each slice of the cube at a particular λ coordinate is an image of the scene in a different (hopefully narrow) wavelength band. Three different classes of this kind of detector show promise at optical-IR wavelengths:

1. Arrays of superconducting tunneling junctions (STJs) – see Section 7.8.2 – have been in use on the 4.2-m Herschel telescope on Las Palmas. The most recent (2004), S-CAM3, a 10×12 pixel device, has a resolution of $R = 10$ at 500 nm.

2. Arrays of superconducting kinetic induction devices (KIDs) – see Section 7.8.3 – have been tested on the Palomar Hale and other telescopes. The most recent (ACORNS) employs 46×44 pixels at a resolution of $R = 8$ at 500 nm, and the design shows great potential for scaling to megapixel size.

3. A 6×6 array of TES micro-bolometers – see Section 8.6 – was tested at McDonald Observatory, and demonstrated $R = 15$ at 500 nm.

All these devices operate in a pulse-counting mode, in which the array is scanned at a rate faster than the incoming photons-per-pixel rate. The data cube is populated as events occur in a specific pixel within a specific range of photon energies. All of these devices demonstrate the feasibility of the dream. Each has also demonstrated the difficulties in attaining it. All require maintaining sub-kelvin temperatures while sending signals, and sometimes bias voltages (on heat-conducting wires), into and out of the device. All must read out at high rates to avoid multi-photon events and all are susceptible to background.

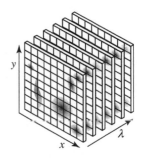

Fig. 11.17 The data cube. The cube can be the output from a single observation from an energy-resolving detector or an imaging spectrograph. It could also be built by collecting multiple observations during a scan in one dimension, as in long-slit spectroscopy.

11.6.2 Long-slit spectrometers

Another approach to creating an (x, y, λ) data cube is to take repeated spectra of an object with a ***long-slit spectrometer***. The slit image (the y-direction on the sky) is oriented parallel to the columns of an ordinary CCD or IR array. The λ-dimension then falls along the rows of the array in the dispersion direction. See Figure 11.18a. A complete data cube can be built up by stepping the slit across the object in the x-direction in successive exposures. This method requires no special equipment other than spectrometer optics capable of forming an excellent image over the entire y-dimension of the slit. Historically, long-slit spectroscopic images have been responsible for important astronomical results, including the confirmation of dark matter as the major material constituent of galaxies. A particularly nice feature of even a single long-slit spectrum of a compact object is that the nearby sky background spectrum gets recorded at the same time. This is especially helpful in the IR, where background is both large and variable. Building up a data cube from long-slit spectra is, of course, time consuming, and subject to errors due to changing atmospheric conditions while stepping in the x-direction.

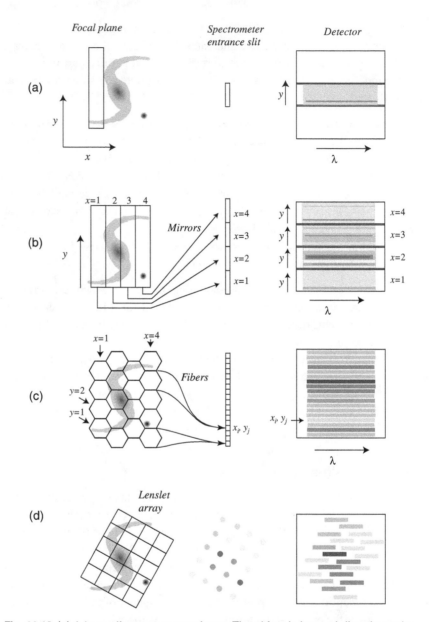

Fig. 11.18 (a) A long-slit spectrometer input. The object is imaged directly on the slit, and the single spectrum produced records any variation in the *y*-direction. (b) An image slicer. Multiple mirrors intercept different parts of the image, each acting like a long slit. Reflections from each mirror are reassembled along a single very long slit at the spectrometer entrance. (c) An integral field spectrometer in which the telescope image plane is tiled with lenslets that feed individual fibers. The fiber output is distributed along the spectrometer entrance slit, (d) an array of lenslets that re-image the focal plane at the input focal plane of the collimator.

11.6.3 Integral field spectrometers

Several types of instrument utilize dispersion spectroscopy to simultaneously record the spectrum of *every contiguous pixel* in the image of an object of interest. These operate by inserting an ***integral field unit (IFU)*** to serve as input to a dispersion-based spectrograph. The instrument that results is called an integral field spectrometer ***(IFS)***. The IFS is a sub-category of ***imaging spectrometer***.

Image slicers

Imagine that we optimize the optics in a conventional spectrograph like the one in Figure 11.12 to produce an aberration-free image of a slit whose length matches the height of the entire detector (i.e. several minutes of arc on the sky). Now, instead of a conventional slit, we project the telescope focal-plane image of the object (a cluster of galaxies, say) onto a stack of very narrow, slit-shaped mirrors, each tilted at slightly different angle – "slicing" the image into many slit-shaped pieces. We direct each slice to its own, somewhat larger, mirror which in turn directs the piece into a small section of the very long slit of our optimized spectrograph. A single exposure then captures a complete (x, y, λ) data cube – a bit scrambled, but easily unscrambled. See Figures 11.18b and 11.19.

We choose the value the angular slit width. ϕ_s, by setting the width of the small mirrors in the first slicer. If these are at the focus of the spectrograph collimator, the very long slit in Figure 11.19 is *not* the physical object that defines the spectrograph resolution, but a pseudo-slit. Alternatively, in a somewhat more difficult design, the object in focus at the small slicer can be re-imaged at the long slit. We can do this either by curving the surfaces of the small slicer mirrors or by inserting small lenses in front of the long slit. Image slicers have benefited

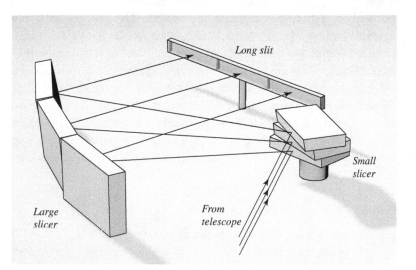

Fig. 11.19 An image slicer. Small mirrors redirect each individual section of the image to a corresponding large mirror. The figure shows three "slices." Each large mirror redirects or refocuses its slice on a segment of the long slit at the spectrometer entrance.

from recent advances in optical fabrication – particularly in the construction of the delicate mirror edges of the small slicer. Several image-slicer based IFUs are in use at large telescopes. Alignment of these devices is non-trivial, but they have high transmittance and make efficient use of detector real estate.

Fiber mosaics

An alternative IFU is bundle of n optical fibers. One end of the bundle is tightly packed in an array in the focal plane of the telescope. The other end is arranged in a line along the long slit of an imaging spectrograph. The result is a stack of n spectra, each corresponding to the (x, y) coordinate of the input aperture. See Figure 11.18c. Advanced systems use an array of lenslets to transfer the focal-plane image to the fiber cores, and this is suggested by the hexagonal grid in Figure 11.18c – otherwise, much light is lost due to fiber cladding and packing gaps. Lenslet optics can also couple the fiber outputs to the spectrograph slit. Blank sky samples and feeds from wavelength-calibration sources are easily implemented with additional dedicated fibers. Fiber mosaic IFUs are less efficient than image slicers because of the light lost due to fiber coupling and fiber f-number degradation, as well as the need to separate individual spectra to prevent overlap on the detector. Compared to the images slicer, fiber bundles can cover a somewhat larger area of the sky, and the technique is considerably easier to implement.

Lenslet arrays

An array of n lenslets behind the focal plane of the telescope can re-image the object of interest as an array of n spots with space between them. These spots, placed at the focus of the collimator, produce a matching array of spectra on the detector. See Figure 11.18d. With care in the orientation of the lenslet array, these spectra will appear staggered on the output. Dispersion is limited because of the need to restrict the length of the spectra. The need to keep spectra separated in both spatial dimensions means that lenslet arrays do not utilize the same fraction of available pixels as do image slicer or fiber-bundle IFUs.

Integral field spectrometers at major observatories are capable of recording up to 10 000 spectra simultaneously on large modern detector arrays. Allowing two spatial samples per angular resolution element, their angular fields of view are therefore about 50 times the width of the point-spread function, i.e. 0.2–1.0 minutes of arc. Additional optics can be useful to re-form an enlarged or reduced image of the focal plane onto the IFU to better match available fiber or lenslet diameters.

11.6.4 Multi-object spectrometers

Integral field spectrometers excel at studies of extended objects in the 0.1–1.0 arc minute size range, like compact gaseous nebula, star clusters, and individual galaxies. ***Multi-object spectrometers,*** in contrast, allow simultaneous recording of

(a) (b)

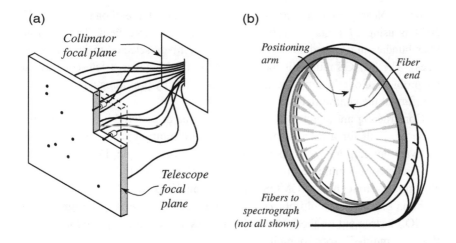

Fig. 11.20 Multi-object spectrometer heads. (a) Fiber ends inserted into openings in a mask, with output arranged along a slit. (b) Fibers connected to right-angle prisms carried at the ends of extensible arms, so that a programmed list of $r.\theta$ positions can be observed simultaneously.

the slit spectra of many small objects scattered over a wide field. Suppose, for example, 50 galaxies are spread over an area 20 arc minutes in diameter in a cluster. An astronomer wants to measure the red shift of the nucleus of every one of them. To avoid taking 50 exposures for a project like this, an instrument-builder modifies the concept of the fiber-bundle IFU: She places the output of 50 different optical fibers in a linear stack along the long slit of the spectrograph, but she positions the input end of each fiber at the location of each galaxy image in the focal plane of the telescope. Each fiber captures the light from a different galaxy, and a single exposure produces 50 spectra stacked vertically (to measure background, it is useful to have some additional fibers at blank sky locations). Saving a factor of 50 in telescope time justifies the tedious job of positioning the fibers. In the early days, multi-object inputs were fabricated by drilling holes in a metal plate at positions matching an image of the field of interest, and then gluing or otherwise positioning a fiber into or behind each hole. Conveniently, many large modern multi-object spectrographs now provide automated fiber positioners, whose robotic arms can place the fiber ends at whatever locations are needed in front of a mask, with no cutting or drilling required. See Figures 11.20a and b.

Alternatively, no fibers are used: a programmable laser milling machine cuts slits in a mask at the desired locations, and the mask is placed at the collimator focus. The output of such a multi-slit spectrometer then resembles an objective prism image. See Figure 11.21.

11.6.5 Refinements

Many of the designs for imaging spectroscopy can be executed on an AO-corrected field. For example, the OSIRIS instrument on the Keck II telescope uses a 3000-element lenslet IFU on AO-corrected images. It is also natural to combine the idea of the IFU with the multi-object concept. MaNGA, the

(a)

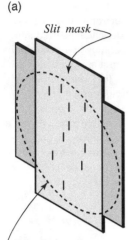

Slit mask

Telescope and collimator focal plane

(b)

Fig. 11.21 Multi-slit spectrometer heads. (a) Multiple slits inserted in the telescope focal plane are also the collimator input. (b) Output spectra.

Mapping Nearby Galaxies at APO project begun at the Sloan telescope in 2014 is using 17 IFUs moveable over a 7-square degree field. The IFUs are fiber bundles in assorted sizes (from 19 to 127 fibers selected to match galaxy sizes). The IMACS multi-slit spectrograph on the Magellan 1 telescope avoids the problem of overlapping spectra in crowed fields by providing an image slicer to reformat the central field onto a slit mask.

Constructing an imaging spectrograph is a major undertaking, and although some are built for a specific purpose, most are multipurpose, and economics motivate building a very adaptable instrument. For example, IMACS provides two cameras and multiple grating and grism choices, including an echelle. By changing its front end, the GMOS instrument on the Gemini North telescope can function as an IFS, a multi-object spectrograph, or a long-slit spectrograph. The VIMOS instrument on the VLT can be used as a direct imager, a 6400-element IFS, or a multi-slit spectrograph.

11.7 Spectrometer stability and mounting

We have seen that although much might be done to reduce the overall dimensions and weight of a spectrometer, its resolving power is directly proportional to the length of the grating employed, so some spectroscopic projects will always require large, heavy instruments.

In general, then, you will find spectrometers suitable for the study of the faintest objects at relatively low resolving power mounted at the Cassegrain or even prime focus, where the fast f-number and limited number of pre-slit reflections encourages a compact design of limited weight but high efficiency. These spectrometers move with the telescope, so their parts experience variable gravitational stresses, and differential motion can produce systematic errors. If uncompensated, these will appear as variations in the focus or intensity distribution of the spectrum of the same object, or as differential shifts between the wavelength calibration and the object spectra.

Spectrometers mounted at the coudé focus are motionless with respect to the Earth, and an altazimuth mount moves those mounted at the Nasmyth focus only in the horizontal plane (if the Nasmyth focus lacks an optical image rotator, the spectrograph must rotate around the optical axis). Such instruments can employ large gratings and camera optics and are correspondingly massive. Their locations permit more robust thermal stability and isolation from mechanical stress and vibrational disturbance. In cases where extreme stability is required, extreme measures are taken: the HARPS-N spectrograph mentioned in Section 11.2.6 is a fiber-fed echelle dedicated to the detection of tiny (0.3 ms^{-1}) stellar radial velocity changes due to orbiting exoplanets. Its critical optics are housed in a vacuum chamber which provides thermal stability to 0.001 K and insensitivity to barometric pressure variations.

Fiber optics facilitate the mechanical and thermal stability of a motionless spectrometer without the large f-number and multiple reflections of the coudé focus. There is some penalty in using fibers, since some light is lost in transmission and fibers will generally degrade the focal ratio (e.g. $f/8$ at fiber entrance might become $f/6$ at exit).

11.8 Data acquisition and reduction

The detailed strategy for collecting and reducing spectrometric data depends on the overall scientific goal. Very precise radial velocity studies have stringent stability and wavelength-calibration requirements. Survey work is more concerned with efficiency. Spectrophotometry requires not only instrument stability, but also precise flux calibration and atmospheric extinction correction. Work on faint sources requires exact background subtraction.

11.8.1 Observing practices

As in direct imaging, the preferred spectroscopic detector is usually an array. For single-object spectrographs, an instrument designer can save some weight and expense by using a linear array, or one with reduced pixel count or resolution in the direction normal to the dispersion. However, given that an observatory is quite likely already to own one or more rectangular, square-pixel detectors for direct imaging, the design will often utilize a CCD or infrared array or a mosaic of arrays. Large-area detectors are indispensable for integral field, echelle, and multi-object spectrographs.

Spectroscopic observations are inherently more complex than direct imaging. The astronomer must oversee more details: grating selection and tilt, focus of the collimator and camera, and slit alignment at the parallactic angle. Multi-object spectrographs require careful positioning of entrance apertures at object images, plus verification and maintenance of this alignment during an exposure. The *bias*, *dark*, and *linearity* exposures for array detectors proceed very much as in direct imaging.

The *flat field* calibration, however, needs special consideration. A precise flat field correction for slitless spectra is extraordinarily difficult, since the wavelength of light falling on a particular pixel in the program image cannot be predicted or repeated in the flat. In twilight (with very few stars) or projection flats each pixel receives the thoroughly blended source spectrum, and that may be the best one can do.

For slit or fiber spectra, the situation is better. We require a source whose image is as uniform as possible along the slit, slits, or integral fields, and whose spectrum is continuous – a quartz halogen lamp and some sort of projection screen is the usual choice. The dark or twilight flats are unsuitable. Slit length,

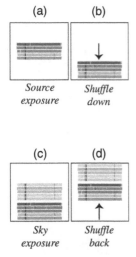

Fig. 11.22 Nod and shuffle with a multi-object spectrometer. (a) Acquire object spectra in the central 1/3 of the CCD. (b) Shuffle spectra down columns by 1/3 to on-chip "storage." (c) Nod the telescope and acquire background spectra. (d) Shuffle spectra up columns by 1/3 to store sky spectra. Now nod the telescope back to the original position and repeat, adding signal to the stored spectra.

multi-slit positioning on the detector, grating tilt, and any other relevant parameter should be the same in the flat as in the object spectrum, so that each pixel in the flat receives light of the appropriate wavelength. The flat field image will depend on the spectrum of the lamp (usually unknown), its projected uniformity along and across the slit, the overall transmission efficiency of the spectrometer in both the wavelength and spatial dimensions, and the quantum efficiency of each pixel. A fiber feed can be engineered to scramble transmitted light so that no matter what occurs at the input end, the fiber output is uniform, which is a help. In many cases, the astronomer can expect to extract the pixel-to-pixel sensitivity in the dispersion direction from the flat, but calibrating variations on a large wavelength or spatial scale is more difficult. The flat fielding problem for echelle spectra is similar to that for single- object spectra, with the complication that an appreciable subtractive correction may be needed for scattered light.

Lamp frames from an emission-line source are usually necessary to establish wavelength calibration and line curvature. These are generally acquired as near in time to the science frames as possible. If isolated background sky is not present on the science images (e.g. single-fiber input), separate ***background frames*** may be required for sky subtraction. Usually the telescope is "nodded" to an area of blank sky near the source. This is especially the practice with bright background in the IR, where a four-frame sequence sky–source–source–sky is the typical observing pattern. With a CCD, a clever ***nod and shuffle*** technique is possible that minimizes sky subtraction error due to flat field and line curvature. See Figure 11.22.

11.8.2 Spectrum extraction

Figure 11.23 is a sketch of a typical detector output spectrum produced by a single star. Echelle, multi-object, and IFS data will contain many such spectra on a single frame. We assume the slit is long enough to include a portion of the sky not significantly contaminated by the star's spectrum, and that the dark, bias, linearity, and flat field corrections have been completed. The figure indicates a few night-sky emission features that show the orientation of the slit.

The slit image is nearly parallel to the y-axis of the detector. This will simplify reduction, but in practice, other orientations can occur. The slit image may or may not have a bit of curvature. With the slit image roughly in the y-direction, the dispersion direction should lie roughly along the x-axis. However, camera distortions (less apparent over the short length of the slit) can tilt and curve the spectrum, as can atmospheric refraction and grating tilt on the axis normal to the grooves. Each order in an echelle spectrogram often exhibits considerable curvature and tilt.

To reduce the data in Figure 11.23, you trace, with a smooth curve, the y-position of the centroid of the spectrum at each x-value, $Y(x)$, and average the

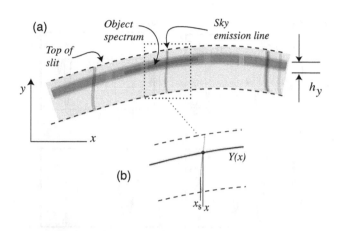

Fig. 11.23 (a) Tracing the object and background spectra in the plane of the detector. (b) Note that a curved slit image means that sky values at $Y(x)$ should be measured at values $x_s(y)$.

pixel values over the height, h_y, of the stellar spectrum at each x-location. It would be appropriate to use a weighting scheme that accounts for the lower statistical values of less-well-exposed pixels at the edge of the image (or discards them). Likewise, use rejection algorithms that remove cosmic rays and bad pixels. The result is a function, $A(x)$, the average intensity of the star plus sky spectrum as a function of x, measured in ADUs. If you want to subtract the background (the usual case), then you do a similar trace of the sky portion or portions of the image, and compute $B(x, Y)$, the average background intensity *at the location x, Y*. Line curvature, if present can complicate this computation, and will require care (see the figure). Finally, the ***extracted spectrum***, $I(x)$, is just the difference, $A(x) - B(x)$.

Different situations will demand some alterations in this basic procedure. For an extended object like a nebula that occupies the entire height of the slit, or for a single-fiber spectrograph, you must "nod" the telescope to a region of blank sky near the object and collect a sky spectrum on a separate exposure. In the infrared, background levels are so high and variable that a very rapid "chopping" to nearby sky and automated subtraction is often part of data acquisition. For multiple-slit mask, multi-object, and integral field spectrographs, you should plan dedicated slits or fibers for the needed sky spectra. Alternatively, data acquired with the nod and shuffle technique simply require matching each sky and object spectrum before subtraction.

The function $I(x)$ is the object spectrum, but modified by atmospheric and optics absorption, the grating efficiency, and detector sensitivity. For some purposes (e.g. a survey for particular spectral features) $I(x)$ is the final data product, but for others (e.g. spectral classification, measurement of relative line strengths) a ***continuum-normalized*** version of the spectrum is more useful. To produce the normalized version, you specify the continuum sections (regions with no absorption or emission features) of $I(x)$, and fit a smooth curve, $C(x)$, to only those parts of the spectrum. The normalized spectrum is just $I_N(x) = I(x)/C(x)$. See Figure 11.24.

Fig. 11.24 (a) An extracted spectrum. The dashed-line curve is a smooth fit to the continuum. (b) A normalized version of the same spectrum. The dashed line has value unity. (c) The flux-calibrated spectrum obtained by dividing the observed spectrum by the response function derived from a standard star. (d) The profile of a single absorption line in a wavelength-calibrated, continuum-normalized spectrum. The equivalent width, W, of the line measures the width of the shaded rectangle. The area of the rectangle equals the area between the continuum and the line profile.

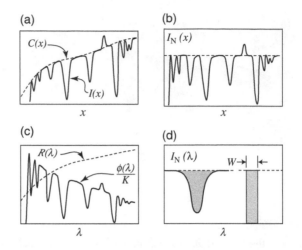

11.8.3 Wavelength calibration

For most purposes, you will want to present the spectrum not as a function of CCD position, x, but as a function of wavelength. The usual – but not the only – method for doing so is to obtain the spectrum of an emission-line source (a gas-discharge tube or "arc") on a separate exposure (the lamp exposure). A popular lamp material is a mixture of ***thorium and argon vapor***, which has many strong emission lines over a wide range in the optical and NIR. For the precise calibration, the light from this ***comparison source*** must follow the same path through the spectrometer as the light from the object. If you suspect any shifts during the object exposure, you should take comparison exposures both before and after the object exposure. It is also possible to check calibrations by measuring the wavelengths of night-sky emission lines on the object spectra. Some spectrographs place the comparison spectrum above and below the object spectrum on the same exposure through dedicated fibers, or do so on a double exposure using a moveable mask called a dekker.

You extract the comparison spectrum as you did the object spectrum. You can then pair the known wavelength of each emission line with the observed x-location of the center of that line in your spectrum. A functional fit to these data (usually a polynomial of some kind) is the ***dispersion solution***, $\lambda(x)$. If you have before-and-after comparisons, extract both and average them. You now can associate a wavelength with every x-value in your spectrum. The next step depends on your scientific goals. You might measure the x values of features of interest and use the dispersion solution to compute their observed wavelengths. For other applications, you may wish to ***linearize*** the spectrum by interpolating pixel values so that there is a constant wavelength change from each pixel to the next. This produces the function $I(\lambda)$ in a convenient form.

One noteworthy departure from the above procedure is the use of a gas-absorption cell for wavelength calibration. The light from the object passes through a low-pressure cell placed immediately in front of the slit (Figure 11.16). Iodine vapor, I_2, maintained at constant temperature and pressure is a popular choice for the gas. Iodine requires a cell only a few centimeters long and produces on the order of 10 lines per nanometer in the red (500–620 nm). This arrangement has the advantage that the light paths through the spectrograph are identical for the object and comparison, and the very narrow iodine lines allow correction for changes in the slit profile due to guiding errors (the center of brightness of the illuminated slit varies if the star is not kept precisely centered) and for mechanical shifts during the exposure. Butler et al. (1996) describe the technique in detail as applied to the detection of extrasolar planets through the minute Doppler shifts a planet induces in its host star. Reducing I_2-calibrated data can be quite complex.

There is no universally preferred emission lamp material for calibration of IR spectra. However, at many wavelengths, one can use the terrestrial atmosphere as a sort of gas cell, and superimposed atmospheric methane and N_2O absorption lines are quite narrow and abundant, permitting rather precise (10–20 m s^{-1}) calibrations in line-rich regions.

11.8.4 Flux calibration

The wavelength-calibrated spectrum $I(\lambda)$ has units of analog-to-digital unit (ADU) per wavelength interval and reflects both the spectral energy distribution of the object and the detection efficiency of the atmosphere–telescope–spectrograph combination. If you wish to transform your data into either absolute or relative flux units, you must observe *photometric standard stars*. There are several sets of such standards (Section 10.6). The rationale is identical to the one described in Chapter 10 for the monochromatic case in photometry, with the simplifying condition that each pixel is a detector in which the central wavelengths of the standard and the program object are identical. Recalling the notation and results of Section 10.5, the observed, inside-the-atmosphere ADU count for a particular pixel will be:

$$I(\lambda) = \phi(\lambda, 0)\Delta\lambda A_{\text{TEL}}Q(\lambda)\exp(-\tau_\lambda X)t \tag{11.29}$$

$$I_{Std}(\lambda) = \phi_{Std}(\lambda, 0)\Delta\lambda A_{\text{TEL}}Q_{Std}(\lambda)\exp(-\tau_\lambda X_{Std})t_{Std} \tag{11.30}$$

The second equation, with subscripted variables, is for the observed spectrum of the standard star. In these equations, $\phi(\lambda, 0)$ is the actual outside-the-atmosphere photon flux, A_{TEL} is the effective light-gathering area of the telescope, $\Delta\lambda$ is the wavelength interval intercepted by the pixel, $Q(\lambda)$ is number of ADUs generated in the detector by one photon of wavelength λ arriving at the telescope aperture. You will also recall that τ_λ is the optical depth at the zenith, X is the air mass, and t is the exposure time. Combining Equations (11.29) and (11.30):

$$\phi(\lambda, 0) = I(\lambda) \left\{ \frac{\phi_{Std}(\lambda, 0)}{I_{Std}(\lambda)} \right\} \frac{Q_{Std}(\lambda)}{Q(\lambda)} \frac{t_{Std}}{t} e^{\tau_\lambda(X_{Std} - X)} \qquad (11.31)$$

This reduces, to:

$$\phi(\lambda, 0) = I(\lambda) \left\{ \frac{\phi_{Std}(\lambda, 0)}{I_{Std}(\lambda)} \right\} K(\lambda) = I(\lambda) \frac{K(\lambda)}{R(\lambda)} \qquad (11.32)$$

You can compute the quantity in the braces in both equations from standard star data and your own observations. We also realize that exact value of the function K is unknown. This is because we usually cannot assure $Q_{Std}(\lambda) = Q(\lambda)$, since that would require that exactly the same fraction of starlight entered the slit or fiber in both the object and standard exposures. Usually guiding errors and seeing variations make that unlikely. Fortunately, as long as a narrow slit is aligned perpendicular to the horizon (see Figure 11.15a) variation in $Q(\lambda)$ due to image motion should be *gray* – independent of wavelength. Therefore Equation (11.31) means that ***relative spectrophotometry***, which determines the shape but not the energy scale of the spectrum (that is, a function proportional to $\phi(\lambda, 0)$), is possible with a narrow slit.

   Flux-calibrated spectrophotometry of stars, in which the observed spectral intensity can be converted to physical units (the function $hc\phi(\lambda, 0)/\lambda$, measured in $Wm^{-2}nm^{-1}$), therefore usually requires a spectrometer slit or fiber opening several (e.g. six) times larger than the FWHM of the seeing disk to capture all significant light from standards and program objects. Alternatively, standard broad or narrow-band photometry (see Table 10.2) applied to a relative spectrum can establish such a calibration. We require ***absolute spectrophotometry,*** in which the measurement is compared directly to a laboratory source like a blackbody, to initially calibrate spectrophotometric standards in physical units.

11.8.5 Other calibrations

Astronomers assign a program star its temperature and luminosity spectral class by comparing the program's spectrum with that of a ***spectroscopic standard star***, noting in particular class-defining features like relative absorption line strengths. The actual appearance of features, and the quantitative relationships among them, depends on the instrumental resolution and its sensitivity function. Precise classification therefore requires observations of spectroscopic standard stars with the same apparatus as the program stars.

   You can determine radial velocities and cosmological redshifts by measuring wavelength shifts in spectra relative to some reference, usually the spectrum of a comparison lamp. Any effects that produce different angles of incidence on the grating for the program and comparison sources will cause systematic errors in your measurements. Similar systematic effects will result if the effective wavelength of a feature changes to an unknown value because of different line blending at different spectrograph resolutions. One precaution against systematic effects for stellar work is to observe a *radial velocity standard* of the same spectral class as the program object. The standards have been cataloged with well-determined velocities in the heliocentric reference frame. If you observe programs and standards under identical conditions, then systematic effects should be apparent in the measured velocities of the standards, and you can therefore eliminate them in the program objects.

11.9 Interpreting spectra

Extracting useful information from the spectrum you have just reduced can require a great deal of analysis, much of it based on astrophysical theory well beyond the scope of our concerns here. However, the analysis almost always begins with some quantitative measurements of the spectrum, a few of which we now examine.

11.9.1 Classification of stellar spectra

> ... in the mid 1930s ... I decided to create a two-dimensional spectral classification that would be determined completely by the appearance of spectral lines, bands, and blends. It was to be autonomous; that is, it would be defined completely by the spectra of standard stars, without having to appeal to any theoretical picture.
>
> – W. W. Morgan, *Annual Reviews of Astronomy and Astrophysics*, Vol. 26, 1988

The introduction of the photographic process in 1830s eventually precipitated a revolution in astronomy. Successful recordings of the solar spectrum appeared as early as 1842, but spectroscopy of fainter objects only became possible after the introduction of the dry emulsion in 1870. Harvard astronomer and photography enthusiast Henry Draper (1837–82) became the first ever to photograph the spectrum of a star (Vega, in 1872). While the international astronomical community embraced photography by initiating, in 1887, a program to record the entire sky to 14th magnitude in the ill-fated astrographic *Carte du Ciel* project, Edward Charles Pickering, the Director of the Harvard College Observatory, began a long-term project aimed at classifying the photographic spectra of stars brighter than $m = 10$. Henry Draper's widow, Anna Palmer Draper (1839–1914), established a memorial fund of several hundred thousand dollars to support Pickering's project.

The primary workers in this monumental undertaking were Williamina Fleming, Antonia Maury, Annie Jump Cannon, and over a dozen other women. It was Maury, Draper's niece, who first recognized the proper sequence of spectral types in 1897. Cannon (1863–1941) joined the project in 1896 and developed an

Table 11.1 *Modern spectral classes in order of decreasing temperature. Note that the temperature scale is for dwarves. The temperature of a more luminous star of the same spectral type is cooler. Temperatures marked with a colon are especially uncertain.*

Type	Temperature range, K For luminosity class V	Main characteristics of absorption line spectra
O	>30 000	Ionized He lines
B	30 000–9800	Neutral He lines, strengthening neutral H
A	9800–7200	Strong neutral H, weak ionized metals
F	7200–6000	H weaker, ionized Ca strong, strong ionized and neutral metals
G	6000–5200	Ionized Ca strong, very strong neutral and ionized metals
K	5200–3900	Very strong neutral metals like Fe, Mn, Si, CH, and CN bands
M	3900–2100:	Strong TiO bands, some neutral Ca, VO strong
L	2100:–1300:	Strong metal hydride molecules, neutral Na, K, Cs
T	1300:–500:	Methane bands, neutral K, weak water
Y	500:	Ammonia bands

uncanny skill at rapid classification. In the four years (1911–14) prior to the completion of the Henry Draper Catalog (HD), Cannon classified an average of 200 stars per day. Her work on the Henry Draper Extension (HDE), from 1923 onward, saw her classify an additional 400 000 stars. She expanded the "Harvard" nomenclature that Fleming (in 1890) had proposed into a system that formed the basis of the modern one. In 1943, William W. Morgan and his collaborators detailed the Morgan–Keenan–Kellman (MKK or, usually, MK) systematization now in general use.

The initial MK system contained only the hottest seven (OBAFGKM) temperature classes and recognized five (I to V) luminosity classes. Many authors have since proposed extensions and refinements of the MK system: For example, luminosity classes now can include VI (subdwarves) and VII (white dwarves) as well as finer subdivisions (e.g. Ia+ = 0 = extremely luminous supergiants, or IVb = faint subgiants). The opening of the IR region has led to the introduction of classes L, T, and Y.

As suggested in Section 1.4, the modern two-dimensional spectral classes can be well-modeled by understanding the physical processes in stellar atmospheres, with the two dimensions corresponding roughly to (a) effective surface temperature (spectral class) and (b) the acceleration due to gravity at that surface (luminosity class, which correlates well with diameter).[1] Table 11.1 gives some

---

[1] The MK types describe most of the variation in the spectra of bright stars. However, very close inspection shows that a third parameter, related to chemical abundance, is needed to account for all variations.

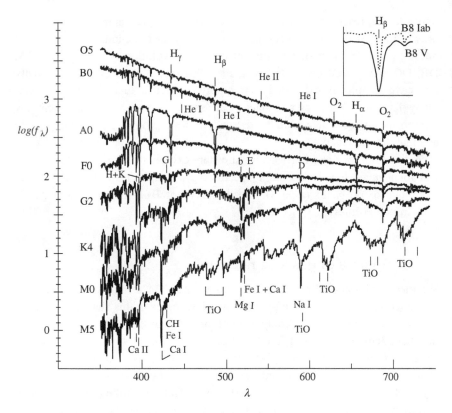

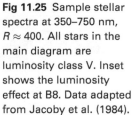

Fig 11.25 Sample stellar spectra at 350–750 nm, $R \approx 400$. All stars in the main diagram are luminosity class V. Inset shows the luminosity effect at B8. Data adapted from Jacoby et al. (1984).

characteristics of the ten "temperature" classes. Appendix K gives colors and absolute magnitudes. Note that within luminosity class V, the actual luminosity changes along **the main sequence** of stars, running from luminous and hot to dim and cool.

At the simplest level, the astrophysical interpretation of the spectral temperature classes rests on the fact that the population of atoms or ions in the energy state that gives rise to a particular absorption line changes drastically with temperature: For example, Hydrogen alpha is strongest at effective temperatures around 10 000K. At lower temperatures, there are fewer electrons in the $n = 2$ state. At higher temperature, there are fewer neutral H atoms, as more and more become ionized. See Figure 11.25.

Astrophysics aside, the classification *process* depends *only* on judging that the lines and bands in a given spectrum are "just like" the lines and bands in the spectrum of a standard. Kaler (1997) discusses the process in detail. In the digital era, this judgment of "just like" can now be made quite quantitative (see below).

Brown dwarves are stars that are massive enough to convert deuterium into helium, but not massive enough to initiate the full-scale thermonuclear reactions that convert hydrogen into helium. Some objects with spectral type M are brown dwarves, as are most of type L, and all of type T. Spectral type Y at present has only a few representatives. This class may very well represent the transition from brown dwarves to large planets (i.e. objects not massive enough for deuterium burning). *White dwarves,* an endpoint of stellar evolution (spectral types DA, DB, DC – or luminosity class VII) are small, dense objects of low luminosity that have completely exhausted their supply of nuclear fuel. We observe relatively few white dwarves and stars in classes L, T, and Y because their low luminosities make them hard to detect. All four groups are probably quite abundant in the Galaxy.

A few luminous and intrinsically rare stars exhibit strikingly unusual chemical abundances, like *carbon stars* (spectral types R and N) and stars with strong zirconium oxide bands (type S), and the very hot *Wolf–Rayet stars* (type W, showing no H lines).

11.9.2 Spectra of gaseous nebulae

We discussed some of the characteristics of the ISM in Section 10.7. In this introduction we can only briefly touch on a few topics in this vast area of study. Optically, the most obvious component of the ISM consists in the diffuse clouds that have temperatures of a few thousand kelvin. The best-studied of these is M42, the Orion nebula, a complex star-forming region that contains both bright H II (ionized hydrogen) clouds and optically invisible molecular clouds. Figure 11.26 is a rough sketch of the brightest features in the optical emission spectrum typical of such an H II region. Some of brightest lines (there are many faint lines not shown) result from the cascade of an electron to the ground state after recombination with an ion (e.g. the Balmer series and the He I lines). Others result from the population and rapid decay of a low-lying energy state as the result of a collision followed by spontaneous photo-emission.

Most of the remaining lines are "forbidden," (O [I], O [II], O [III], N[II], S [II] etc. – the square brackets indicate a forbidden transition) that is, the lines

Fig. 11.26 A sketch of the emission-line spectrum of an H II region like Orion. Note the logarithmic scale. Continuum has been subtracted. Lines intensities are largely based on measurements of M8 by Esteban et al. (1999) and are on a scale in which the intensity of H-β is 100.

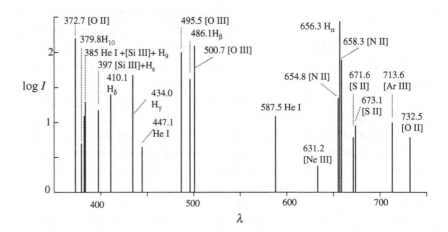

originate from a quantum mechanically metastable state. (The lifetime against spontaneous de-excitation of a metastable excited state is 10^{-3} to 10^3 sec or longer, whereas electrons normally cascade from an unstable state in 10^{-9}–10^{-7} sec.) Nebulae emit these *forbidden lines* because they are such low-density environments. Here is why: a collision bumps an electron into a metastable level. In a normal, dense gas, another collision will exchange energy and quickly bump the electron back to a lower (or higher) energy level – no photon is emitted. But in a nebula, the time between collisions is long compared to the metastable lifetime, so the likelihood of photo-emissive decay becomes large.[2]

Different H II regions will produce different emission spectra, but, like stars, their chemical composition is mostly hydrogen and helium, with the remaining elements present in proportions similar to their terrestrial values. Variations are present. Analysis of the intensities of lines, particularly of forbidden lines, gives insight into the prevailing temperatures, densities, and chemical compositions. Reasonably successful models of particular nebulae have been constructed. The book by Osterbrock and Ferland (2006) gives an extended discussion of the spectra of nebulae and their analysis.

11.9.3 Measuring line strength

> The uniformity of composition of stellar atmospheres appears to be an established fact. The quantitative composition of the atmosphere of a star is derived in the present chapter ... and the inferred composition displays a striking parallel with the composition of the earth.
>
> – Cecelia H. Payne, PhD Thesis, Radcliffe College, 1925

Much of what we now understand about astronomical bodies (like their masses, structures temperatures, environments, motions, chemical composition, and history) stems directly from measurements line strength and position.

The continuum-normalized spectrum as a function of wavelength, $I_N(\lambda)$, when considered only over the region in which a *single* absorption or emission line is present is called the *line profile* of that feature. No line profile is indefinitely narrow, but always extends over a range of wavelengths for reasons we will discuss below. You will realize immediately, however, that every line profile must be at least as wide as the spectrometer's resolution.

We introduce here the idea of the *equivalent width* as a measure of – not the width – but the *strength* of an emission or absorption line. See Figure 11.24d. We define the equivalent width for an absorption line as

---

[2] The same conditions prevail in the upper atmosphere, which explains why our night sky remains so bright at night – atoms and molecules continue to emit forbidden lines long after the sun has set.

$$W_{\text{abs}} = \left| \int_{\text{line}} (1 - I_{\text{N}}(\lambda))\mathrm{d}\lambda \right| \qquad (11.33)$$

The equivalent width has units of wavelength and measures the amount of the star's flux absorbed by the line. The equivalent width should be fairly independent of the spectrograph you use to measure it, and indeed, that is part of its appeal: we have catalogs of the equivalent widths of the lines in the solar spectrum, for example, and everyone can agree on these values (the K line, Ca II, has $W = 2.0$ nm, and the Na I D lines have $W = 0.075$ and 0.056 nm). Great care is needed, however. Spectrometer resolution matters tremendously: at low resolution, weak lines disappear because they become too shallow to recognize, and lines can blend together. Any scattered light in a spectrometer will systematically bias equivalent width measurements to lower values, and uncertainty in the location of the continuum propagates as an uncertainty in equivalent width.

With a spectrum of sufficiently high resolution, the values of absorption line equivalent widths (coupled with a good astrophysical model of the stellar or planetary atmosphere, or of the emission nebula) can produce estimates of chemical composition, temperature, and pressure. ***Classification of stellar spectra*** was traditionally done visually, using criteria like the ratio of critical line strengths, or the presence/absence of certain lines, but always with reference to the same criteria in a set of standard stars. With the onset of digital detectors and the need to deal with the massive flow of data from IFUs and multi-object instruments, astronomers have set completely quantitative definitions for these or related criteria. Computers can now exceed Annie Cannon's classification rate and precision, but the reference to a set of standard stars remains essential.

Measuring the strength of an emission line is much akin to narrow-band photometry – one uses relative or flux-calibrated spectrophotometry to establish the flux within the total line profile above the continuum. Using the notation from Equation (11.32) the relative intensity of an emission line is:

$$I_{\text{em}} = \frac{hc}{K} \int_{\text{line}} (\phi(\lambda, 0) - \phi_C(\lambda, 0)) \frac{d\lambda}{\lambda}. \qquad (11.34)$$

Here, $\phi(\lambda, 0)$ and $\phi_C(\lambda, 0)$ are, respectively, the total photon flux outside the atmosphere and the flux due to the background continuous emission.

11.9.4 Line profiles

The detailed shape of a line profile, $I_{\text{N}}(\lambda)$, depends on the physical and observational factors that produce ***line broadening***. If you make sufficiently precise measurements of the shape, you can often learn about the physical environment in which the line was formed. We examine the most common line broadening mechanisms:

Natural broadening

The emission or absorption lines of isolated atoms (in a very low-pressure gas, for example) are broadened because of the quantum mechanical uncertainty in the energies of the quantum states of the transition. Those uncertainties, and the line's half-width, depend directly on transition probability and, therefore, inversely on the lifetimes of the relevant states. The natural widths of most lines are very narrow; so other broadening mechanisms dominate, and, except for interstellar Ly α, natural broadening is seldom observed in astronomical spectra.

Instrumental broadening

The limited spectral purity of the spectrometer itself will impose an instrumental profile on any infinitely narrow line, and this profile will be convolved with any other broadening profile present. Although some degree of de-convolution is possible in data analysis, it is best to keep the instrumental profile small compared to other broadening mechanisms under investigation.

Rotational broadening

If a source is spinning on an axis making an angle i to the line of sight, you see, relative to the average, some of its material with positive radial velocities, and some with negative velocities. The Doppler effect means that a line will appear over a corresponding range of wavelengths, with a profile (for absorption) given by

$$I_{NR}(\Delta\lambda) = 1 - I_0\left\{1 - \left[\frac{c}{V\sin i}\frac{\Delta\lambda}{\lambda_0}\right]^2\right\}^{\frac{1}{2}}$$ (11.35)

where c is the speed of light and V is the equatorial velocity of rotation of the (spherical) star. You may recognize the term in braces as the expression for an ellipse. Many stars rotate rapidly enough so that this elliptical profile is easy to recognize; see Figure 11.27.

Thermal broadening

On the microscopic level, atoms and molecules will have a spread in velocities that increases with the temperature and decreases with molecular mass. Here again, the Doppler effect means that temperature alters the line profile. The profile for thermal broadening for a molecule of mass m is Gaussian:

$$I_{NT}(\Delta\lambda) = 1 - I_0\exp\left\{-\frac{mc^2}{2kT}\left(\frac{\Delta\lambda}{\lambda_0}\right)^2\right\}$$ (11.36)

Microturbulence

Microturbulence is small-scale fluid motion caused, for example, by convection in a star's atmosphere. Again, relative velocities between different parts of the

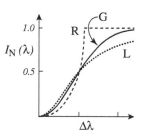

Fig. 11.27 Absorption line profiles produced by different broadening processes. Curves are shown with unit depth and identical FWHM. Profile R is due to solid body rotation. The Gaussian profile, G, is caused by thermal broadening or turbulence. The Lorentzian profile, L, is characteristic of both natural quantum mechanical broadening, as well as pressure broadening. Observed line profiles will be some convolution of the instrumental profile with these shapes at their respective strengths and scales.

source cause line broadening, and microturbulence is expected, like thermal broadening, to produce a Gaussian profile. Since turbulent velocities are independent of molecular mass, the variation in line width as a function of molecular mass in principle allows one to separate the two broadening mechanisms.

Other *organized motions* in a source like expansion, contraction, *macroturbulence* (e.g. rising material might occupy a larger area than falling), and orbital motion can produce asymmetries or secular changes in the line profile.

In emission nebulae, unlike stars, the line broadening is almost always completely dominated by Doppler effects due to large-scale mass motions, temperature, or turbulence.

Pressure broadening

For stellar spectra, pressure effects are most often the dominant cause of broadening away from the central part of the line. Physically, broadening arises because nearby charges, especially free electrons, can perturb the energy of the electron states in an atom. Higher pressures mean more frequent encounters and greater broadening. The pressure-broadened profile is Lorentzian, having the form

$$I_{NL}(\Delta\lambda) = 1 - I_{0L}\left[\frac{A}{\Delta\lambda^2 + B}\right] \qquad (11.37)$$

Natural broadening also has a Lorentzian profile. You can see from Figure 11.27 that compared to the Gaussian, the Lorentzian has more absorption away from the central parts of the line, and in stellar spectra, lines are very frequently Gaussian in their *cores* and pressure broadened in their *wings*. Indeed, the relative strength of the wings of absorption lines permits estimates of the acceleration due to gravity (which determines pressure) in a stellar photosphere. Since gravitational acceleration depends only on mass and radius, and since stellar masses can be determined from the motions of binary stars, spectroscopic observations of the wings of lines permit estimates of stellar radius, and the recognition that stars of the same temperature exist as dwarves, giants, and supergiants. (See the insert in Figure 11.25.)

Abundances

The equivalent width of a line depends on the number of atoms in the initial electron state as well as on the probability of the transition to the final state. You can therefore determine the relative abundance of a chemical element in, say, the atmosphere of a star from its line strengths, provided you can sort out all the other physics relevant to population and transition probability – Cecelia Payne was the first to do this sorting out, and so made the discovery that (except for H and He) the Earth and the stars were all of the same stuff.

Abundance analysis is not trivial. In a stellar atmosphere, for example, once all the light at the core of the line has been absorbed, no more absorption can take place at that wavelength, and the line is saturated, except at very high abundance, where absorption in the wings of the line becomes important. Thus, the **curve of growth** (equivalent width vs. abundance) is non-linear. (See Figure 11.28.)

11.9.5 The redshift parameter

Measuring the shift in the observed wavelength of a line from its at rest value is fundamental to a variety of enterprises, ranging from the discovery exoplanets to determining the structure of the universe on the largest scales.

Because of the relative motion between a source and observer, the wavelength of a photon the source emits, λ_E, will be observed at wavelength λ_O. You will recall from Sections 3.4.3 and 10.8.1 that redshift parameter, z, measures this shift:

$$z \equiv \frac{\lambda_O - \lambda_E}{\lambda_E} \tag{11.38}$$

In the case where the shift is caused by an actual velocity of the source through space (the Doppler effect), the theory of special relativity gives the relationship between z and the radial velocity:

$$z_D = \left(\frac{c+v}{c-v}\right)^{\frac{1}{2}} - 1 \tag{11.39}$$

For small values of v/c the binomial theorem expansion shows:

$$z_D \rightarrow \left(1 + \frac{v}{2c} + ...\right)\left(1 + \frac{v}{2c} + ...\right) - 1 \approx \frac{v}{c} \tag{11.40}$$

This approximation is good to 5% for values of c less than 0.1. Measurements of Doppler velocity usually should be barycentric – corrected for the shifts caused by the orbital and spin motions of the terrestrial observer.

11.9.6 Determination of masses

The only good way to determine the mass of an astronomical object is to observe the gravitational effects it has on another object. For example, we can compute the mass of Jupiter by observing the orbit of one of its satellites. Likewise, almost everything we know about the masses of stars comes from observations of the orbits of binary stars. Kepler's third law reflects the relevant physics for any two bodies in mutual orbit:

$$M_1 + M_2 = \frac{4\pi^2(a_1 + a_2)^3}{GP^2} \tag{11.41}$$

Here M_1 and M_2 are the masses of the two objects, P is the orbit period, and a_1 and a_2 are the mean distances between each object and the center of mass (CM) of the system. The gravitational constant, G, is such that $4\pi^2/G = 1$ if the units are solar masses, years, and astronomical units. For orbit sizes and velocities measured in the CM system:

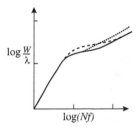

Fig. 11.28 The equivalent width of an absorption line in a stellar atmosphere as a function of abundance, N. The factor f accounts for the probability of the transition. The dashed and dotted curves show, respectively, the effects of increased Doppler and pressure broadening.

Fig. 11.29 (a) Orbits of
two point-masses around
their CM. (b) Velocity
vectors when masses are
located at the open circles
in (a) and thus radial
velocities are at
maximum.

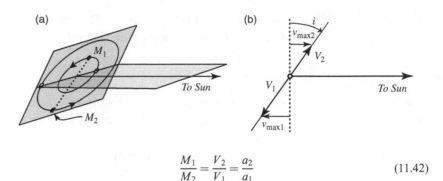

Fig. 11.29 (a) Orbits of two point-masses around their CM. (b) Velocity vectors when masses are located at the open circles in (a) and thus radial velocities are at maximum.

$$\frac{M_1}{M_2} = \frac{V_2}{V_1} = \frac{a_2}{a_1} \tag{11.42}$$

Here V is the magnitude of the instantaneous velocity relative to the center of mass. In the case of a ***visual binary of known distance***,[3] the orbit sizes can be measured directly (with some clever geometric reasoning to remove projection effects) and the masses of both stars follow from Equations (11.41) and (11.42). Of more direct concern here is the case of a ***double-lined spectroscopic binary***. Figure 11.29b, for the case of circular orbits, shows that the maximum radial velocity observed is $v_{\max} = V \sin i$ where i is the inclination of the orbit with respect to the plane of the sky. If we assume circular orbits, then

$$a = \frac{P}{2\pi} V = \frac{P v_{\max}}{2\pi \sin i} \tag{11.43}$$

Substituting into (11.41) gives

$$(M_1 + M_2) \sin^3 i = \frac{P}{2\pi G} (v_{\max,1} + v_{\max,2})^3 \tag{11.44}$$

and making use of (11.42):

$$M_1 \sin^3 i = \frac{P}{2\pi G} (v_{\max,1} + v_{\max,2})^2 v_{\max,2} \tag{11.45}$$

The value of $\sin^3 i$ is difficult to determine except in the rare case of an eclipsing binary ($\sin i \approx 1$). Otherwise, Equation (11.45) only gives a lower limit to the mass. The situation is not completely grim, however, since it is easy to work out the most probable value of $\sin i$ for a large collection of similar stars and to thus obtain a statistical estimate for the mass of, say, an A0V star. An important conclusion from the study of double stars is that, for stars in luminosity class V, there is a correlation between mass and spectral type, and hence, a correlation between mass and luminosity. The mass–luminosity relationship:

$$\frac{L}{L_\odot} = \left(\frac{M}{M_\odot}\right)^\alpha \tag{11.46}$$

[3] The requirement is that the orbit can be determined by astrometry – e.g. AO-assisted measurements in the K band are "visual."

is also supported on theoretical grounds. Empirically, the constant α in Equation (11.46) is 2.3 for masses below 0.43 solar masses, and 4.0 for masses above that limit.

Suppose one star (M_1) is so much brighter than its companion that the companion's spectrum is undetectable. This is a ***single-lined spectroscopic binary***. Such a case, it is possible to estimate only the mass of the unseen companion – combining Equations (11.44) and (11.45):

$$\frac{(M_2 \sin i)^3}{(M_1 + M_2)^2} = \left[\frac{P}{2\pi G}\right] v_{\text{max}, 1}^3 \qquad (11.47)$$

The quantity on the right-hand side is observable, and the quantity on the left is called the ***mass function***. Note that we can determine a lower limit to M_2 if we can guess the mass of the brighter star. (For dwarf stars, the mass can usually be estimated to better than 10% from the spectral type – because we observe others elsewhere in double-lined and in visual systems.) Single-lined binary observations have been important in the detection of compact objects like black hole candidates. However, the most revolutionary application of this method has been in the discovery and investigation of extrasolar planets.

11.9.7 Exoplanets

It plainly follows that there must arise / Distinct and numerous worlds, Earths, men, and skies / In places distant . . .

– T. Lucretius Carus, *De Rerum Natura*, 1:II:1031–33, *c.* 60 BCE

The presence of a Jupiter-mass companion to the star 51 Pegasi is inferred from observations of periodic variations in the star's radial velocity.

– Michel Mayor and Didier Queloz, *Nature*, Vol. 378, 1995

Although thoughtful people since before the time of Lucretius might have assumed the existence of innumerable planets, none were discovered around a Sun-like star outside the Solar System until Mayor and Queloz announced their discovery of 51 Pegasi b at a meeting in the fall of 1995.[4]

Very quickly, two other groups equipped with equally capable spectrographs confirmed the discovery. Mayor and Queloz had applied Equation (11.47) to their very precise measurements from the ELODIE echelle spectrograph ($R = 42\,000$, conventional thorium comparison) at the Observatoire de Haute-Provence. The amplitude of the velocity variations was 59 m s^{-1} with a period of 4.2 days.

Many astronomers were initially skeptical that this discovery was an actual planet. 51 Pegasi b had properties completely inconsistent with the available

---

[4] There was a prior detection of planets around the pulsar PSR 1257+12. HD 114762 b, which now appears to be a planet, was discovered in 1989 but announced then as a brown dwarf.

theories of planet formation: it was a gas giant like Jupiter, but orbited within 8 million km of its star, much closer than Mercury is to our Sun. The skepticism was tenable because of the sin i term in Equation (11.47): Since the orbital inclination was unknown, all that really could be claimed from the observations was that 51 Peg b had a mass larger than 0.46 Jupiter masses. The mass threshold for deuterium-burning brown dwarves lies at around 10–20 Jupiter masses, so 51 Peg b *could* be a brown dwarf in a nearly face-on orbit.

By 1995, several existing spectrographs had reduced radial velocity uncertainties to 5–10 m s^{-1} and the search for more exoplanets expanded. Over the next 10 years, radial velocity monitoring turned up over 100 additional objects. Very few of these had values of $M \sin i$ above the brown dwarf limit, so statistically, most of them *had* to be planets. Many were "hot Jupiters" like 51 Peg b, so theorists had to make (and are at it still) some profound revisions in our understanding of the formation and evolution of planets.

Note that Equation (11.47) applies for only circular orbits (V_1 constant). For the same mass and semi-major axis, the range in orbit velocity will increase with orbit eccentricity (Kepler's second law – V_1 and V_2 maximum at periastron). Including the effect of eccentricity, Equation (11.47) becomes:

$$M_2 \sin i = \left[\frac{P(M_1 + M_2)^2}{2\pi G} \right]^{\frac{1}{3}} \left(1 - e^2 \right)^{\frac{1}{2}} K_1 \qquad (11.48)$$

where K_1 is one-half the total range in radial velocities seen in the spectrum of M_1. The value for the eccentricity, e, can be determined from the shape of the radial velocity curve. See Figure 11.30b. The discovery of the first exoplanets opened a burgeoning field of research where, as in most of astronomy, spectroscopic analysis is essential.

Incidentally, in 2015 the IAU supervised the naming of a selection of 31 exoplanets. 51 Peg b is now officially designated "Dimidium." By 2016, although the world's best dedicated spectrographs approach 1.0 m s^{-1} precision, the exoplanet discovery enterprise had changed, and radial velocity monitoring has been largely supplanted, at least for discoveries, by the ***transit method***: If the inclination is close to 90°, a Jupiter-sized planet passing in front of a solar-type star will cause the star's apparent magnitude to dim by 0.01 mag. Over 2000 exoplanets are now cataloged, most discovered by photometry, many subsequently confirmed by radial velocity measurements. Techniques are tantalizingly close to finding Earth-sized objects. Already the variety of discovered objects: super-Earths, steam worlds, planets in highly eccentric orbits, giant planets of seemingly impossible densities . . . suggests that we still have great deal to learn about planets.

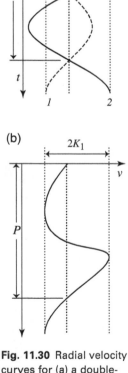

Fig. 11.30 Radial velocity curves for (a) a double-lined spectroscopic binary in a circular orbit and (b) a single-lined spectroscopic binary in an elliptical orbit.

11.9.8 Galaxies and the universe

It would require another book as long as this one to introduce the basic observational approaches to understanding the constituents of the universe on scales larger that our own Galaxy. In this section we confine our discussion to some of the simplest spectroscopic results. We will ignore some rather spectacular achievements like (a) the discovery of "dark matter" from measurements of the masses of individual galaxies and clusters of galaxies (with methods related to those discussed above for binary stars), (b) determination of the present abundances of the isotopes of the light elements (D, He, Li, Be, and B), a measurement which sets strict limits on the possible models for the universe, or (c) the discovery of strong evidence for the presence of black holes at the centers of many galaxies, including our own.

On cursory examination, the spectra of the brighter galaxies look like what you'd expect if you mixed stars of different spectral types together in roughly the proportions we see around the Sun: lots of M stars, hardly any O stars, plus (at least for galaxies where there has been recent star formation) the spectrum of a typical H II region. See Figure 11.31. Galaxies are woefully faint, and accumulating one galaxy spectrum with photographic techniques in the early twentieth century required a multi-night exposure. These early spectra, though, led to a crucial discovery about the nature of the universe.

The Hubble law

"Curiouser and curiouser!" cried Alice (she was so much surprised, that for the moment she quite forgot how to speak good English). "Now I'm opening out like the largest telescope that ever was! Goodbye feet!"

– Lewis Carroll, *Alice's Adventures in Wonderland*, 1865

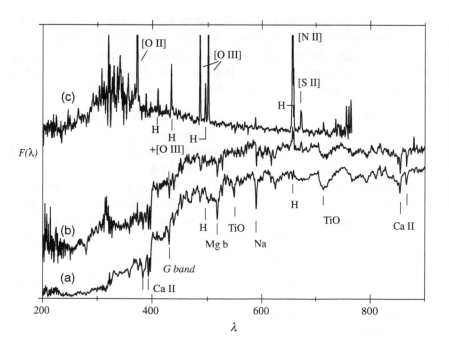

Fig. 11.31 Representative spectra extracted from the templates published by Kinney et al. (1996) for three types of "normal" galaxies: (a) Ellipticals (no recent star formation, no ionized gas). (b) Sb type spirals (weak current star formation), and (c) Sc spirals (strong current star formation, bright H II regions).

In 1914, Vesto Melvin Slipher noticed that the vast majority of the spiral nebulae (galaxies) had redshifted spectra. By 1931, Milton Humason and Edwin Hubble had recorded galaxy radial velocities up to $+20\ 000$ km s^{-1}, and were able to demonstrate that the redshift of a galaxy was directly proportional to its distance, d. Astronomers interpret the ***Hubble law***,

$$v_R = H_0 d \tag{11.49}$$

as indicating that our universe is expanding. "Expanding" in this context means that everywhere in the universe, distances between galaxies are increasing. A simple extrapolation backward in time suggests that at some point in the past, there was a unique event: all galaxies were separated by zero distance. The unique event is now known as the Big Bang. H_0 is called "the present value of the ***Hubble parameter, H***." In Equation (11.49), it is customary to measure v_R in km s^{-1} and d in megaparsecs, so H_0 has units of km s^{-1} Mpc^{-1}. Recent measurements H_0 fall in the range 68–77 in these units.

Most astronomers accept a cosmological theory that assumes (with considerable observational evidence) that the universe on its largest scales is ***homogeneous and isotropic.*** With that assumption, a productive way of thinking about the universe is to imagine that the location of each galaxy is designated by a set of three more or less permanent ***co-moving coordinates*** (r, θ, ϕ) – these coordinates *never change* for a particular object. However, the ***proper distance*** of that object from the origin does change:

$$d = a(t)r \tag{11.50}$$

Where $a(t)$ is called ***the scale factor.*** The relationship in Equation (11.50) is a bit more complicated if the universe has a curved geometry, but we consider only a flat Euclidian space for the present discussion. The scale factor is defined to be unity at the present time $(a(t_0) \equiv 1)$. It is important to recognize two important points about Equations (11.49) and (11.50):

(1) Because the universe is homogeneous and isotropic, *any* point can act as the origin of coordinates, so the Hubble law and the rule about distances, (11.50), would hold no matter where the observer is located. Likewise, there can be no "edge" in a uniform universe and no "center" in an isotropic universe and thus no *place* where the expansion began.

(2) These equations only apply on a large scale. On scales smaller than 100 Mpc or so, ***peculiar motions*** caused by local gravitational forces on the order of several hundred km s^{-1} will seriously modify the "Hubble flow." The Milky Way, for example, is attracted to the center of our local supercluster 18 Mpc away, and recessional velocities in that direction are about 370 km/s lower than predicted by the Hubble law.

So, on large scales, because $a(t)$ is a function of time, the proper distances between clusters of galaxies can change, not because the galaxies move through space and thereby change their co-moving coordinates, but because space itself is expanding or contracting. Thus, at the present time and for the relatively nearby universe, Hubble's law tells us that space is expanding and proper distances are increasing according to:

$$v_R \equiv \dot{d}(t_0) = \dot{a}(t_0)r = H_0 d(t_0) = H_0 a(t_0)r \qquad (11.51)$$

We can generalize the definition of the Hubble parameter from Equation (11.51) to accommodate the idea that the rate of expansion can change over time, so that:

$$H(t) = \frac{\dot{a}(t)}{a(t)} \qquad (11.52)$$

The evolution of the scale factor changes the proper distance between wave crests in a light wave, and neatly accounts for the redshifts of distant galaxies:

$$\frac{\lambda_E}{a(t_E)} = \frac{\lambda_{Obs}}{a(t_0)} = \lambda_{Obs} \qquad (11.53)$$

Thus giving rise to the redshift

$$z = \frac{\lambda_E - \lambda_{Obs}}{\lambda_E} = \frac{1}{a(t_E)} - 1 \qquad (11.54)$$

So $(1 + z)$ is inversely proportional to the scale factor of the universe at the time the light was emitted. Astronomers distinguish this ***cosmological redshift*** from the Doppler effect.

Why should $a(t)$ change over time? Cosmological theory is probably a long way from explaining a cause for the *initial* expansion of the universe. However, if we supply the relevant information about the present state of the universe, modern physics – and especially the theory of general relativity – can describe *how* $a(t)$ should change over time. For example, gravity should work against the expansion, so one might expect $a(t)$ to be increasing at a slower rate now than in the distant past. Cosmology therefore needs observations that measure of the present rate of expansion (i.e. H_0) and other general characteristics of the present universe, like the mean densities of matter (both ordinary and dark) and radiation (photons), as well as its overall pressure (dark energy). Not all of these are easily determined. Moreover, the enterprise of tracing the history of $a(t)$ also depends critically on looking at very distant objects and thereby looking back in time.

Determining the value of H_0

We have space here only to examine only a few observational pieces supporting our current version of the story of the universe. We begin where Hubble did. To measure $H_0 = cz/d_0$, all you need do is (a) measure z from the spectra of a lot of distant-but-not-too-distant galaxies, and (b) determine the distance of each galaxy in your sample. Given a very large telescope and a decent spectrograph, first of these tasks is, if not trivial, at least straightforward. The second, more daunting, task is to measure the distance to each galaxy. The distance in Equation (11.51) is $d_0 = d(t_0)$, the proper distance, or the "distance now." The easier distance to measure in astronomy, however is the luminosity distance, which related to the proper distance by $d_L = d_0(1 + z)$, in a flat geometry. We measure d_L with a **standard candle** – an object whose absolute magnitude is known from criteria other than distance. The distance then follows from the distance modulus equation for large redshifts, Equation (10.56):

$$\mu = 5\log d_L - 5 = m - M + K(z) \tag{11.55}$$

A number of objects have been used as standard candles. Some of the most important are **Cepheid variables**, supergiant stars whose mean luminosities correlate with their pulsation periods and metal content. One of the first tasks of the *HST* was to measure the parallaxes of the ten nearest Cepheids and establish their absolute magnitudes. As of 2015, the absolute magnitude of a Cepheid of known period is uncertain by about 0.09 magnitudes, a value that should shrink as results from *Gaia* and *JWST* become available. If P is in days, and extinction is removed, a **period–luminosity law** holds:

$$M_V = -2.76(\log(P) - 1) - 4.22 \tag{11.56}$$

Cepheid absolute magnitudes are in the interval $-2 < M_V < -6$, so the most distant detectable (with *HST*) are at distances of just over 20 Mpc ($cz \approx 1500$ km s^{-1}). Given galaxy peculiar velocities of several hundred km s^{-1} we need brighter standard candles to pin down H_0.

There are a few brighter candles. We simply list three, should you wish to investigate them: (a) the Tully–Fisher relation for spiral galaxies, (b) the fundamental plane relation for elliptical galaxies, and (c) the surface-brightness fluctuation method. We discuss – briefly – only the most productive standard candle, the Type Ia supernovae (SNe Ia).

SNe Ia are rare, occurring in all types of galaxies at a rate of about 0.3 per century per $10^{10}L_\odot$ galaxy. We distinguish them from other SNe because their spectra exhibit no hydrogen lines, but do show strong absorption due to silicon and other heavier elements. In nearby galaxies of known distance (from Cepheids and other methods) astronomers have accumulated light curves (Figure 11.32) for a sufficient number of SNe Ia that we can be sure their absolute magnitudes at peak lie in the range $M_V = -18.4 \pm 0.6$. Most of the intrinsic variation in luminosity can be calibrated out empirically by examining the shape of the light curve. See Figure 11.32. Although some questions remain in the calibration of SNe Ia peak luminosities, and observations can be difficult, these objects constitute the brightest reliable standard candle, with light curves (4-m telescopes) and spectra (8–10-m telescope) measureable from the ground to redshifts of $z \sim 1.0$. Measurements from HST have collected redshifts of a few more distant SNe Ia (the current record is SN UDS10Wil at $z = 1.91$).

The Hubble Key Project (Freedman et al., 2001) measured the luminosity distances and redshifts of galaxies with $0.02 < z < 0.1$ using five different standard candles, including Cepheids and SNe Ia. They produced a very well-defined linear fit to the z, d_L relationship. Their value, with minor revisions, is the

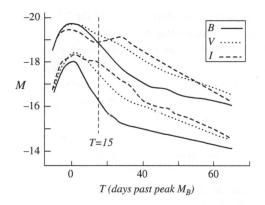

Fig. 11.32 Light curve templates of a SNe Ia. The magnitude change in the first 15 days after the peak B luminosity has been used to calibrate the absolute magnitude at the peak, as have multi-parameter fits to the light curve shapes in different colors. Curves adapted from Jha et al. (2007).

main basis for the most precise optical measurement of the present value of the Hubble constant ($H_0 = 73 \pm 6$). More recent studies of the cosmic microwave background radiation, a completely independent measurement, suggest a slightly lower value near 69 km s^{-1} Mpc.

High z supernovae and dark energy

Two groups at the end of the twentieth century mounted an observing campaign to discover very distant SNe Ia, monitor their light curves, and collect their redshifts. The projects released their initial conclusions, based on 25 SNe, in 1998. The projects, which included many collaborators, won the Nobel prize for the group's leaders in 2011. Their observations measured the evolution of the Hubble parameter over time by constructing a **Hubble diagram**: a plot of distance modulus versus redshift (Figure 11.33). For small z, the Hubble diagram tells the value of H_0. At large z, the diagram shows evidence for changes in the expansion rate. At a given distance, for example, if the universe is decelerating, the expansion rate will have been higher in the past than predicted by $H(t) = H_0$. Therefore, if gravity slows the expansion (as was expected) at large μ (i.e. the universe in the past), the observed redshift should be higher than predicted. This deviation will increase with increasing μ.

The unexpected results of the initial surveys were difficult to accept. Distant SNe had *lower* redshifts than the $H(t) = H_0$ model predicted. The expansion of the universe has *accelerated* over the past 5 Gyrs. The results did solve some long-standing problems (e.g. without acceleration, the age of the oldest stars exceeds the age of the universe). Soon, moreover, microwave observations of the cosmic background at $z \approx 1100$, independently confirmed the acceleration. Observations of hundreds of supernovae since 1998 also strongly confirm the result. The most plausible explanation (there are others) for accelerating expansion is contained in an optional element of Einstein's theory of general relativity: a **cosmological constant** that would manifest as a constant pressure to expand, a pressure exerted by space itself. As space expands further, the effects of

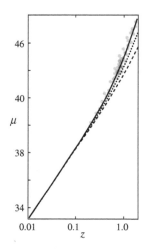

Fig. 11.33 Distance modulus as a function of redshift for high-z SNe Ia. The three curves show the expected relationships in a flat universe for: (dashed line) a universe containing only the observed amount of dark and ordinary matter; (dotted line) a completely empty universe; (solid line) an accelerating universe 70% of whose content is in the form of dark energy.

gravitational attraction become more dilute and the "dark energy" of the cosmological constant becomes more and more dominant.

It is fitting, finally, to reflect on how astronomers built this conception of our universe as a place dominated by utterly fantastic content for which we have no direct experience. The detailed model has strong observational support. But that support rests on a foundation built over hundreds of years by people, many of them nameless, who constructed better and better instruments to measure the sky. Built also by those who dreamed up better ways to use those instruments, and did the very hard work: to measure the first stellar parallaxes, to estimate brightness, to record the first spectra, to probe the composition of the stars, to find planets, to look back in time and see the universe as it was. More work beckons. Observational astronomy can continue to advance, but only if people like you, dear reader, understand and build upon what has already been accomplished.

Summary

- Dispersive spectroscopy relies on optical elements that send light rays in a direction that depends upon wavelength. Concepts:

angular dispersion	*linear dispersion*
spectral purity of optics	*reciprocal linear dispersion, p*
effective spectral purity, $\delta\lambda$	*instrumental profile*
resolving power, R	

- The angular dispersion of a prism varies as λ^{-3}.
- Diffraction gratings depend upon wave interference of diffracted rays. Concepts:

amplitude grating	*transmission grating*	*reflection grating*
grating constant	*groove frequency*	*groove spacing*
order	*free spectral range*	

- The grating equation gives the angle of dispersion as a function of wavelength:
$$\sin\theta + \sin\alpha = \frac{m\lambda}{\sigma}$$
- Phase gratings operate by periodically adjusting the phase of diffracted waves. Concepts:

echellette	*blazed grating*	*blaze angle*
blaze wavelength	*Littrow configuration*	*free spectral range*

- Echelle gratings have steep blaze angles, and usually operate in conjunction with a second (cross-) disperser to separate orders.
- Volumetric phase holographic (VPH) gratings produce phase shifts by periodically a adjusting the index of refraction in a transmitting slab. Concepts:

Bragg diffraction	*Bragg angle*	*superblaze*

- Gratings can be produced by scribing lines on a master blank or by holographic techniques. Concepts:

ruled grating	*holographic grating*	*flat field concave grating*
grating mosaic	*ion etching*	*DCG*

- The objective prism generates a spectrum of every object in the telescopic field of view. The spectra tend to have low resolution, and suffer from high background. Concepts:

  non-objective prism   *grism*   *multiplex advantage*

- Spectrometers with slit or fiber inputs restrict incoming light to increase resolution and suppress background. Concepts:

  slit width   *collimator*   *anamorphic magnification*
  seeing-limited R   *scaling with telescope diameter*   *throughput*

- Resolving power of a conventional slit spectrometer:

$$R = \frac{\lambda}{r_{an}\phi_s} \frac{D_{COL}}{D_{TEL}} \frac{m}{\sigma \cos\theta} = \frac{1}{D_{TEL}\phi_s} \frac{\lambda m}{\sigma} W$$

- Spectrometers for astronomy have special design requirements. Concepts:

  near-Littrow   *off-axis paraboloid*   *quasi-Littrow*
  spectrum widening   *folded Schmidt*   *dekker*
  multiple-slit mask   *imaging spectrometer*   *multi-object spectrometer*
  image slicer   *long-slit spectrometer* *integral field spectrometer*
  flexure   *wavelength calibration* *flat field source*

- Reduction of spectrographic array data varies with one's scientific goals. Concepts:

  flat field   *comparison source*   *spectrum extraction*
  dispersion solution   *spectroscopic standard* *continuum-normalized spectrum*
  linearized spectrum   *photometric standard*   *relative spectrophotometry*

- The classification of stellar spectra is based upon similarities and differences in line strengths. Concepts:

  OBAFGKM(LTY)   *brown dwarf*   *MK system*

- The spectra of gaseous nebulae are dominated by emission lines, some of which are forbidden

- Much of astrophysics concerns measuring line strength. Concepts:

  line profile   *equivalent width*   *line core and wings*
  thermal broadening   *line broadening*   *natural broadening*
  microturbulence   *rotational broadening*   *instrumental broadening*
  pressure broadening   *macroturbulence*   *Lorentzian*
  line core   *abundance analysis*   *curve of growth*

- The measurement of wavelength changes in spectra has produced a number of very important results. Concepts:

  redshift parameter   *spectroscopic binary*   *mass function*
  exoplanet   *transit method*   *galaxy spectrum*
  Hubble law   *co-moving coordinate*   *proper distance*
  luminosity distance *scale factor*   *Hubble flow*
  standard candle   *Cepheid*   *period–luminosity relation*
  SNe Ia   *cosmic acceleration*   *dark energy*

Exercises

1. Derive the grating equation for the transmission grating. Clearly state the rule you adopt for measuring positive and negative angles.

2. Explain quantitatively why the free spectral range for a particular order, m, and maximum wavelength, λ_{max}, is not restricted by overlapping light from order $m - 1$.

3. Compute the free spectral range of grating orders 50, 100, and 101 if $\lambda_{max} = 600$ nm in each case.

4. Compare the angular dispersions of a 600 lines per mm amplitude grating at 400 nm and at 900 nm. Assume you are working in first order and the angle of incidence is $25°$. Do the same for a $60°$ prism with angle of incidence $\alpha = 55°$ and $K_2 = 0.01 \ \mu m^{-2}$.

5. Manufacturers usually describe blazed gratings by specifying the blaze wavelength in the Littrow configuration and the groove density in lines per mm. (a) Compute the blaze angle for a reflection grating of 1000 lines per mm blazed for a wavelength of 400 nm. (b) Compute the blaze wavelength of this grating when it is used at an angle of incidence of $40°$ instead of in Littrow.

6. Explain why the self-shadowing of a grating as a function of wavelength is different if the direction of the ray in Figure 11.5b is reversed: i.e. if instead of $\alpha < \theta$ at the blaze wavelength, we have $\alpha > \theta$.

7. A normal VPH grating has an index modulation frequency of 2000 lines per mm. Sketch a spectrograph design (show the relevant angles) that would permit the most efficient observation of spectra near a wavelength of 400 nm in first order. Now sketch how the spectrograph would have to be adjusted to observe efficiently at a wavelength of 600 nm in first order. What is the minimum number of moving parts required for such an adjustable spectrograph?

8. Show that the anamorphic magnification, $d\theta/d\alpha$, of the simple slit spectrograph in Figure 11.12 is $\cos \alpha/\cos \theta$.

9. An astronomer wishes to build a simple fiber-fed spectrometer, using a reflection grating. She will follow the basic plan illustrated in Figure 11.14. She has a CCD detector measuring 1024×1024 pixels, with each pixel 15 μm on a side. An optical fiber with core diameter 100 μm will sample star images in the telescope focal plane and deliver light to the spectrograph. Tests show that the fiber degrades the telescope focal ratio of $f/7.5$ to $f/$ 7.0. A grating of diameter 50 mm with 600 lines per mm and blaze angle $8.5°$ is available.

   (a) Compute the first-order blaze wavelength of this grating, and its maximum possible resolving power, R_0.

   (b) The astronomer chooses to illuminate the grating at an angle of incidence, α, of $0°$, and to record the first-order spectrum. What is the maximum focal length the collimator mirror must have to avoid light loss at the edge of the grating?

   (c) Compute the value of the anamorphic magnification at the blaze wavelength.

   (d) The astronomer wishes to critically sample the image of the fiber end with her CCD. What is the required focal length for the camera? Compute the resulting plate scale and the wavelength range in nanometers of the complete spectrum the CCD will record.

(e) Compute the actual resolving power, R, of this spectrometer at the blaze wavelength. If R is less than R_0, describe how the astronomer might improve her value for R.

10. A spectrograph has a very narrow slit, and its CCD is oriented so that the image of the slit (for example, when illuminated by the comparison source) is precisely along the detector's y-axis. Explain why the dispersive effect of atmospheric refraction will tilt the trace of the spectrum of an un-trailed exposure so that it is not parallel to the detector x-axis (see Figure 11.18a).

11. You are designing an integral field spectrometer using a fiber bundle, with microlens input. The telescope is 3-m at $f/6$. Your detector is a 4096×4096 CCD with pixels 15 μm on a side. The camera is $f/3$ with a diameter 1.25 times the diameter of the collimator (the anamorphic magnification is 1.25). Median seeing at the telescope is 0.7 arcsec.

   (a) What is the maximum number of fibers you should use for optimum sampling of spectra if there is to be a one-pixel border between spectra? Assume microlenses are available to match the seeing, and fiber cores are 80% of the microlens diameter.

   (b) What are the dimensions (in angular and physical units) of the fiber bundle in the telescope focal plane? (c) what is the diameter of the collimator?

12. Suggest a method for testing for the presence and severity of scattered light in a spectrometer. Consider two absorption lines each with an identical FWHM: one very strong, the other very weak. In equivalent width measurements, which line is more strongly affected by the presence of scattered light?

13. Compute the FWHM of the line profile for a magnesium absorption line of rest wavelength 500 nm on a star spinning with equatorial velocity 100 km s^{-1} and $\sin i = 1$. Compare this with the FWHM due to thermal broadening if the star's temperature is 8000 K.

14. Rapid rotation of a star will distort its shape into an oblate sphere. In a qualitative sketch, explain how the line profile would differ between two model stars with identical values of $V \sin i$. Model A has a perfectly spherical shape, Model B is distorted into an extreme oblate shape because of its rotation.

15. A number of online exercises let you work with spectra of nebulae to determine physical properties like temperature and pressure. Complete one. See especially: https://web.williams.edu/Astronomy/research/PN/nebulae/index.php

16. In Figure 11.30, if $P = 40$ days, $v_{max1} = 27$ km/s and $v_{max2} = 74$km/s, (a) compute the minimum mass for each star. (b) Compute the most probable value (50% chance the actual value is larger) for $\sin^3 i$ and the most probable mass for each star.

17. The range of masses for a brown dwarf is from 0.08 to 0.01 solar masses. Over what range of $\sin i$ would the velocity curve of 51 Peg be consistent with 51 Peg b being a brown dwarf? What is the probability that 51 Peg b is a brown dwarf?

18. Find the second term in Equation (11.40). At what value of z does the second term introduce a 10% error in z?

Appendix A
General reference data

A1 The Greek alphabet

alpha	A	α	iota	I	ι	rho	P	ρ
beta	B	β	kappa	K	κ	sigma	Σ	σ, ς
gamma	Γ	γ	lambda	Λ	λ	tau	T	τ
delta	Δ	δ	mu	M	μ	upsilon	Y	υ
epsilon	E	ε	nu	N	V	phi	Φ	φ
zeta	Z	ζ	xi	Ξ	ξ	chi	X	χ
eta	H	η	omicron	O	o	psi	Ψ	ψ
theta	Θ	θ	pi	Π	π	omega	Ω	ω

A2 Metric system prefixes and symbols

Use with base unit to indicate decimal multiples. A few units not standard in the SI system are in common use in astronomy. See Tables A4 and A5 below.

Factor	Prefix	Symbol	Factor	Prefix	Symbol
10^{24}	yotta	Y	10^{-24}	yocto	y
10^{21}	zetta	Z	10^{-21}	zepto	z
10^{18}	exa	E	10^{-18}	atto	a
10^{15}	peta	P	10^{-15}	femto	f
10^{12}	tera	T	10^{-12}	pico	p
10^{9}	giga	G	10^{-9}	nano	n
10^{6}	mega	M	10^{-6}	micro	μ
10^{3}	kilo	k	10^{-3}	milli	m
10^{2}	hecto	h	10^{-2}	centi	c
10	deca	da	10^{-1}	deci	d

A3 Physical constants

Speed of light in a vacuum	$c = 299\,792\,458$ m s^{-1}
Planck constant	$h = 6.626075 \times 10^{-34}$ J s
Gravitational constant	$G = 6.6726 \times$ m^3 kg^{-1} s^{-1}
Mass of the electron	$m_e = 9.10939 \times 10^{-31}$ kg
Mass of the proton	$m_p = 1.672623 \times 10^{-27}$ kg
Mass of the neutron	$m_n = 1.674929 \times 10^{-27}$ kg
Unit elementary charge	$e = 1.6021773 \times 10^{-19}$ C $= 4.803207 \times 10^{-10}$ esu
Boltzmann constant	$k = 1.380658 \times 10^{-23}$ J K^{-1} $= 8.61733 \times 10^{-5}$ eV K^{-1}
Stefan-Boltzmann constant	$\sigma = 5.6705 \times 10^{-8}$ W m^{-2} K^{-4}
Avogadro number	$N_A = 6.022137 \times 10^{23}$ mol^{-1}

A4 Astronomical constants

Mass of the Sun	1.9891×10^{30} kg
Mass of the Earth	5.975×10^{24} kg
Radius of the solar photosphere	6.9566×10^8 m
Equatorial radius of the Earth	6.378140×10m
Tropical year	365.2421897 days $= 31\,556\,925.19$ sec
TSI (solar flux at top of the atmosphere)	1.361×10^3 W m^{-2}
Solar effective temperature	5772 K
Luminosity of the Sun	3.827×10^{26} W

A5 Conversions

Length

Angstrom	1 Å $= 10^{-10}$ m
Micron	$1\,\mu = 1\,\mu$m $= 10^{-6}$ m
Astronomical unit	1 au $= 1.495978707 \times 10^{11}$ m
Parsec	1 pc $= 3.085678 \times 10^{16}$ m $= 3.2616$ light years
Light year	1 lyr $= 9.46053 \times 10^{15}$ m
Statute mile	1 mi $= 1609.344$ m
Inch	1 in $= 0.0254$ m

Time (See Appendix D)

Mass

Pound (avdp)   1 lb = 0.453592 kg

Pressure

Pascal (SI) $1\ Pa = 1\ N\ m^{-2} = 10^{-5}\ Bar = 9.87 \times 10^{-6}$
                             Atmosphere

Millimeter of mercury   Pressure of 1 mm of Hg = 1 torr = 133.322 Pa

Pound per square inch $1\ lb\ in^{-2} = 6894.7\ Pa$

Energy

Electronvolt $1\ eV = 1.60218 \times 10^{-19}\ J$

Erg (cgs) $1\ erg = 10^{-7}\ J$

Calorie   1 cal = 1.854 J

Kilogram $1\ kg\ c^2 = 8.9876 \times 10^{16}\ J$

Kiloton of TNT $= 4.2 \times 10^{12}\ J$

Monochromatic irradiance
  (flux density)

Jansky $1\ Jy = 10^{-26}\ W\ m^{-2}\ Hz^{-1}$

Velocity

Miles per hour $1\ mph = 0.44704\ m\ sec^{-1}$

au per year $= 4740.6\ m\ sec^{-1} = 4.7406\ km\ sec^{-1}$

Parsec per million years $10^{-6}\ pc\ yr^{-1} = 977.8\ m\ s^{-1} = 0.9778\ km\ s^{-1}$

Appendix B
Light

B1 Photon properties

This table gives the conversion from the photon characteristic in the left-hand column to the corresponding characteristic at the head of each subsequent column. For example, a photon of wavelength of 100 nm has an energy of 1240/100 = 12.4 electronvolts.

To From	λ(nm)	λ(m)	ν(Hz)	E(J)	E(eV)
λ(nm)	1	$10^{-9}\lambda$	$2.99729 \times 10^{17}/\lambda$	$1.98645 \times 10^{-16}/\lambda$	$1239.85/\lambda$
λ(m)	$10^{7}\lambda$	1	$2.99729 \times 10^{8}/\nu$	$1.98645 \times 10^{-25}/\lambda$	$1.2985 \times 10^{-6}/\lambda$
ν(Hz)	$2.99792 \times 10^{17}/\nu$	$2.99792 \times 10^{8}/\nu$	1	$6.62606 \times 10^{-34}\nu$	$4.1357 \times 10^{-15}\nu$
E(J)	$1.98645 \times 10^{-16}/E$	$1.98645 \times 10^{-25}/E$	$1.5092 \times 10^{33}E$	1	$6.2414 \times 10^{28}E$
E(eV)	$1239.85/E$	$1.2985 \times 10^{-6}/E$	$2.4180 \times 10^{20}E$	$1.6022 \times 10^{-19}\,E$	1

B2 The strongest Fraunhofer lines

Telluric lines originate in the Earth's atmosphere, rather than the Sun's. Roman numeral I designates neutral (not ionized) atoms. Wavelengths may vary because of line blending at different spectroscopic resolutions.

Designation	Wavelength (Å)	Identification
A	7593.7	Telluric O_2 Band
a	7160.0	Telluric H_2O Band
B	6867.2	Telluric O_2 Band
C	6562.8	H_α
D1	5895.9	Na I
D2	5890.0	Na I
E	5269.6	Fe I
b1	5183.6	Mg I
b2	5172.7	Mg I
b3	(5169.1 + 5168.9)	Fe I
b4	5167.3	Mg I
F	4861.3	H_β
G	4314.2	CH Band
H	3969.5	Ca I
K	3933.7	Ca I

B3 Sensitivity of human vision

The range of wavelengths detected by human vision is normally 400–760 nm. There are records of individuals detecting light with wavelengths as long as 1050 nm and as short as 310 nm.

Feature	Photopic vision	Scotopic vision
General illumination level for sole operation	Daylight to twilight	Quarter Moon to darkness
Receptor cells	Cones	Rods
Peak sensitivity	555 nm[*]	505 nm
10% of peak, blue cut-on	475 nm	425 nm
10% of peak, red cutoff	650 nm	580 nm
Speed of adaptation	Fast	Slow (up to 30 minutes)
Response time	0.02 seconds	0.1 seconds
Color discrimination	Yes	No
Visual acuity	High	Low
Region of retina	Center (fovea)	Periphery
Threshold of detection	High	Low (10^{-4} photopic)

[*] There are actually three different types of cones, with peak sensitivities at 430, 530, and 560 nm, which permit color discrimination.

B4 The visually brightest stars

Rank	Classical name	Bayer designation	V
1	Sirius	α CMa	−1.46
2	Canopus	α Car	−0.72
3	Rigel Kent	α Cen	−0.27
4	Arcturus	α Boo	−0.04
5	Vega	α Lyr	0.03
6	Capella	α Aur	0.08
7	Rigel	β Ori	0.12
8	Procyon	α CMi	0.38
9	Achernar	α Eri	0.46
10	Betelgeuse	α Ori	0.5 (variable)

Appendix C

C1 The standard normal distribution

The standard normal *probability density function* is defined as

$$G(z) = \frac{1}{\sqrt{2\pi}} e^{-\frac{z^2}{2}}$$

It is related to the Gaussian distribution of mean, μ, and standard deviation, σ, by the transformation

$$z = \frac{x - \mu}{\sigma}$$

Thus, $G(z)$ is simply a Gaussian with a mean of zero and variance of 1. Table C1 gives values for $G(z)$. You can use the table to evaluate any particular Gaussian by applying the desired values of μ, σ, and x to the above transformation.

Table C1 *The standard normal distribution.*

Z	G(z)	P(z)	Q(z)
0.00	0.398942	0.500000	0.000000
0.05	0.398444	0.519939	0.039878
0.10	0.396953	0.539828	0.079656
0.15	0.394479	0.559618	0.119235
0.20	0.391043	0.579260	0.158519
0.25	0.386668	0.598706	0.197413
0.30	0.381388	0.617911	0.235823
0.35	0.375240	0.636831	0.273661
0.40	0.368270	0.655422	0.310843
0.45	0.360527	0.673645	0.347290
0.50	0.352065	0.691462	0.382925
0.55	0.342944	0.708840	0.417681
0.60	0.333225	0.725747	0.451494
0.65	0.322972	0.742154	0.484308
0.70	0.312254	0.758036	0.516073
0.75	0.301137	0.773373	0.546745
0.80	0.289692	0.788145	0.576289
0.85	0.277985	0.802338	0.604675
0.90	0.266085	0.815940	0.631880
0.95	0.254059	0.828944	0.657888
1.00	0.241971	0.841345	0.682689

Table C1 (*cont.*)

Z	G(z)	P(z)	Q(z)
1.10	0.217852	0.864334	0.728668
1.20	0.194186	0.884930	0.769861
1.30	0.171369	0.903199	0.806399
1.40	0.149727	0.919243	0.838487
1.50	0.129518	0.933193	0.866386
1.60	0.110921	0.945201	0.890401
1.70	0.094049	0.955435	0.910869
1.80	0.078950	0.964070	0.928139
1.9	0.065616	0.971284	0.942567
2.00	0.053991	0.977250	0.954500
2.10	0.043984	0.982136	0.964271
2.20	0.035475	0.986097	0.972193
2.30	0.028327	0.989276	0.978552
2.40	0.022395	0.991802	0.983605
2.50	0.017528	0.993790	0.987581
2.60	0.013583	0.995339	0.990678
2.70	0.010421	0.996533	0.993066
2.80	0.007915	0.997445	0.994890
2.90	0.005953	0.998134	0.996268
3.00	0.004432	0.998650	0.997300
3.50	8.7268 E−04	0.999767	0.999535
4.00	1.3383 E−04	0.999968	0.999937
4.50	1.5984 E−05	0.9999966	0.9999932
5.00	1.4867 E−06	0.9999997	0.9999994

A related function is the **cumulative probability**, $P(z)$, also listed in the table:

$$P(z) = \int_{-\infty}^{z} G(t)\,dt,$$

The function $P(z)$ gives the probability that a single sample drawn from a population with a standard normal distribution will be less than or equal to z. A second function of interest is

$$Q(z) = \int_{-z}^{z} G(t)\,dt$$

$Q(z)$ gives the probability that a single sample drawn from a population with a standard normal distribution will be within z of the mean; Q is also tabulated, although it is also easily computed as

$$Q(z) = 2P(z) - 1$$

Note that P is also related to the **error function**:

$$\mathrm{erf}(x) = \frac{2}{\sqrt{\pi}} \int_0^x e^{-t^2} \, dt = 2P(x\sqrt{2}) - 1$$

Appendix D

D1 The nearest stars

Adapted from the RECONS (Georgia State University) 2012 list, which incorporates HIPPARCOS data directly, with recent brown dwarf discoveries added. Parallax in mas, μ in mas yr^{-1}, θ is the position angle of the proper motion, radial velocity is in km s^{-1}, and Spt is the spectral type.

D2 The equation of time

Figure D shows the value of the equation of time,

$$\Delta t_{\mathrm{E}} = \text{local apparent solar time} - \text{local mean solar time}$$
$$= \text{RA of the apparent Sun} - \text{RA of the mean Sun}$$

This is the same as 12^{h} minus the mean solar time of transit for the apparent Sun. More precise values for Δt_{E} in a particular year can be obtained from the Astronomical Almanac. Approximate dates for the extrema of Δt_{E} are Feb 11 (minimum), May 14, July 26, and Nov 3 (maximum).

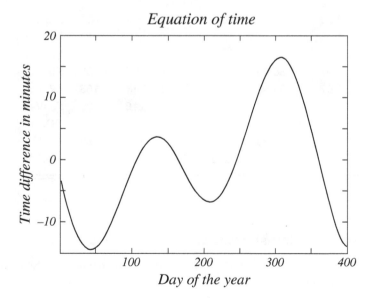

Fig. D2 Equation of time.

Name	Alias	RA 2000	Dec	π	μ	θ	v_R	Spt	V
Sun								G2 V	−26.71
Proxima Cen		14 30	−62 41	772.3	3853	281	−29	M6 V	11.13
α Cen A	Rigel Kent	14 40	−60 50	742.1	3709	277	−32	G2 V	−0.01
α Cen B				742.1	3724	285	−32	K1 V	1.34
Barnard's Star	BD + 4°3561	17 58	4 41	549.0	10358	356	−139	M5 V	9.54
Luhman 16 A		10 49	−53 19	495	2787	277		L8	(K8.8)
Luhman 16 B								T1	
WISE 0855–5319		8 55	−7 14	433	8130	275		Y	
Wolf 359	CN Leo	19 56	7 01	419	4702	235	55	M6.0 V	13.53
BD+36°2147	HD 95735	11 03	35 38	392.4	4802	187	−104	M2 V	7.47
Sirius A	α CMa	6 45	−16 43	374	1339	204	−18	A1 Vm	−1.44
Sirius B								DA2	8.44
GJ-65 A	LB Cet	1 39	−17 56	373	3360	80	52	M5.5 Ve	12.56
GJ-65 B	UV Cet	1 39	−17 56	373	3360	80	53	M5.5 Ve	12.96
Ross 154	V1216 Sag	18 50	−23 50	336.5	6660	107	−10	M3.6 Ve	10.37
Ross 248	HH And	23 42	44 09	316	1588	176	−84	M5.5 Ve	12.27
ε Eri	HD 22049	3 33	−9 27	311	977	271	22	K2 V	3.72
CD-36°15693	HD 217987	23 06	−35 51	303.9	6896	79	108	M2 V	7.35
Ross 128	Fl Vir	11 48	0 48	299.6	1361	154	−26	M4.5 V	11.12
WISE 1506+7027		15 06	70 27	310	1623	310		T6	(J14.3)
GJ 845 A	EZ Aqr	22 39	−15 17	290	3256	47	−80	M5 Ve	12.36
GJ 845 B								M	13.6
GJ 845 C								M	15.1
61 Cyg A	HD201091	21 07	38 45	286.3	5281	52	−108	K5 V	5.02
61 Cyg B	HD201092	21 07	38 45	286.3	5272	53	−108	K7 V	6.05
Procyon A	α CMi	7 39	5 13	285.9	1259	215	−21	F5 IV–V	0.40
Procyon B								DQZ	10.7
BD+59°1915 B	HD173740	18 43	59 38	284.5	2312	323	39	M4 V	9.7
BD+59°1915 A	HD173739			280.3	2238	324	−38	M3.5 V	8.94

D3 Coordinate transformations and relations

To find the angular separation, θ, between two objects having equatorial coordinates (α, δ_1) and $(\alpha + \Delta\alpha, \delta_2)$:

$$\cos\theta = \sin\delta_1 \sin\delta_2 + \cos\delta_1 \cos\delta_2 \cos(\Delta\alpha)$$

With appropriate substitutions, this relation will apply for any similar coordinate system on the surface of a sphere.

To find the altitude, e, and azimuth, a, of an object with equatorial declination, δ, when the object is at hour angle, H, observed from a location with geodetic latitude, β:

$$\sin e = \sin \delta \sin \beta + \cos \delta \cos H \cos \beta$$
$$\sin a = -(\cos \delta \sin H)/\cos e$$

The inverse relationships are

$$\sin \delta = \sin e \sin \beta + \cos e \cos a \cos \beta$$
$$\sin H = -(\cos e \sin a)/\cos \delta$$

D4 Atmospheric refraction

The difference between the true zenith distance, z, and the apparent zenith distance, z', in seconds of arc at visual wavelengths is approximately

$$z - z' = 16.27 \tan (z) \frac{P}{T}$$

where P is the atmospheric pressure in millibars and T is the temperature in kelvin. This formula is reasonably reliable for zenith distances less than 75°. For larger zenith distances, more complex formulas are available – see Problem 3.11. Refraction varies with wavelength (see Table 5.3).

D5 Astrometric catalogs

Four catalogs are recommended by the US Naval Observatory as sources for ICRS coordinates:

1. **The HIPPARCOS Catalog:** See Chapter 3.
2. **Tycho-2 Catalog:** The catalog is based on a mix of 1991 space-based (HIPPARCHOS) data combined with early epoch ground-based astrometry. 2.5 million stars, 99% complete to $V = 11.0$ and 95% complete at $V = 11.5$. Positional accuracies range from about 10 to 100 mas, depending on magnitude. Proper motion accuracies are from 1 to 3 mas.
3. **UCAC2 (USNO CCD Astrograph Catalog, 2nd release + UCAC2 Bright Star Supplement):** 48 million stars primarily in the $R = 8.0$ to 16.0 magnitude range. Positional accuracies are 20 to 70 mas, dependent primarily on magnitude. Proper motion errors are 1 to 7 mas/yr, also magnitude dependent. UCAC2 is about 85% complete in the area covered.
4. **USNO B1.0 Catalog:** Positions of 1 042 618 261 stars and galaxies measured from digitized images of several photographic sky surveys. The estimated positional error is near 200 mas.

D6 Days and years

There are several definitions of the length of time the Earth requires to complete one orbit. In all cases, the following hold:

> 1 day = 24 hours = 86 400 s
> 1 Julian calendar year = 365.25 days
> 1 Gregorian calendar year = 365.2425 days = 3.1556952×10^7 s

The Julian year (introduced by Julius Caesar) and the Gregorian year (introduced by Pope Gregory XIII in 1582) were each meant to approximate the tropical year. The following values are in units of 1 day of 86 400 SI seconds; T is measured in Julian or Gregorian centuries from 2000.0.

1 tropical year (equinox to equinox)	$= 365^d.242193 - 0^d.000\ 0061\ T$
	$= 365^d\ 05^h\ 48^m\ 45^s.5 - 0^s.53\ T$
1 sidereal year (star to star)	$= 365^d.256360 + 0^d.0000001\ T$
	$= 365^d 06^h 09^m 09^s.5 + 0^s.01\ T$
1 anomalistic year (perigee to perigee)	$= 365^d.259635$
	$= 365^d\ 06^h 13^m 52^s.5$
1 eclipse year (lunar nodes to lunar node)	$= 346^d.620076$
	$= 346^d 14^h 52^m 54^s.6$

Julian dates can be computed from calendar dates by the formulas:

$$JD = 2415020.5 + 365(Y - 1900) - L + d + t$$
$$= 2451544.5 + 365(Y - 2000) - L + d + t$$

where Y = current year, L = number of leap years since 1901 or 2001, d = UT day of the year, t = fraction of the UT day.

Appendix E

E1 The constellations

Abbreviation	Nominative	Genitive ending	Meaning	RA	Dec	Area deg^2
And	Andromeda	-dae	Chained princess	1	40 N	722
Ant	Antlia	-liae	Air pump	10	35 S	239
Aps	Apus	-podis	Bird of paradise	16	75 S	206
Aqr	Aquarius	-rii	Water bearer	23	15 S	980
Aql	Aquila	-lae	Eagle	20	5 N	652
Ara	Ara	-rae	Altar	17	55 S	237
Ari	Aries	-ietis	Ram	3	20 N	441
Aur	Auriga	-gae	Charioteer	6	40 N	657
Boo	Böotes	-tis	Herdsman	15	30 N	907
Cae	Caelum	-aeli	Chisel	5	40 S	125
Cam	Camelopardus	-di	Giraffe	6	70 N	757
Cnc	Cancer	-cri	Crab	9	20 N	506
CVn	Canes Venaticium	Canum Venaticorum	Hunting dogs	13	40 N	465
CMa	Canis Major	Canis Majoris	Great dog	7	20 S	380
CMi	Canis Minor	Canis Minoris	Small dog	8	5 N	183
Cap	Capricornus	-ni	Sea goat	21	20 S	414
Car	Carina	-nae	Ship's keel	9	60 S	494
Cas	Cassiopeia	-peiae	Seated queen	1	60 N	598
Cen	Centaurus	-ri	Centaur	13	50 S	1060
Cep	Cepheus	-phei	King	22	70 N	588
Cet	Cetus	-ti	Whale	2	10 S	1231
Cha	Chamaeleon	-ntis	Chameleon	11	80 S	132
Cir	Circinus	-ni	Compasses	15	60 S	93
Col	Columba	-bae	Dove	6	35 S	270
Com	Coma Berenices	Comae Berenicis	Berenice's hair	13	20 N	386
CrA	Corona Australis	-nae lis	Southern crown	19	40 S	128
CrB	Corona Borealis	-nae lis	Northern crown	16	30 N	179
Crv	Corvus	-vi	Crow	12	20 S	184
Crt	Crater	-eris	Cup	11	15 S	282
Cru	Crux	-ucis	Southern cross	12	60 S	68
Cyg	Cygnus	-gni	Swan	21	40 N	804
Del	Delphinus	-ni	Dolphin	21	10 N	189

Abbreviation	Nominative	Genitive ending	Meaning	RA	Dec	Area deg^2
Dor	Dorado	-dus	Swordfish	5	65 S	179
Dra	Draco	-onis	Dragon	17	65 N	1083
Equ	Equuleus	-lei	Small horse	21	10 N	72
Eri	Eridanus	-ni	River	3	20 S	1138
For	Fornax	-acis	Furnace	3	30 S	398
Gem	Gemini	-norum	Twins	7	20 N	514
Gru	Grus	-ruis	Crane	22	45 S	366
Her	Hercules	-lis	Hero	17	30 N	1225
Hor	Horologium	-gii	Clock	3	60 S	249
Hya	Hydra	-drae	Water snake (F)	10	20 S	1303
Hyi	Hydrus	-dri	Water snake (M)	2	75 S	243
Ind	Indus	-di	Indian	21	55 S	294
Lac	Lacerta	-tae	Lizard	22	45 N	201
Leo	Leo	-onis	Lion	11	15 N	947
LMi	Leo Minor	-onis ris	Small lion	10	35 N	232
Lep	Lepus	-poris	Hare	6	20 S	290
Lib	Libra	-rae	Scales	15	15 S	538
Lup	Lupus	-pi	Wolf	15	45 S	334
Lyn	Lynx	-ncis	Lynx	8	45 N	545
Lyr	Lyra	-rae	Lyre	19	40 N	286
Men	Mensa	-sae	Table	5	80 S	153
Mic	Microscopium	-pii	Microscope	21	35 S	210
Mon	Monoceros	-rotis	Unicorn	7	5 S	482
Mus	Musca	-cae	Fly	12	70 S	138
Nor	Norma	-mae	Square	16	50 S	165
Oct	Octans	-ntis	Octant	22	85 S	291
Oph	Ophiuchus	-chi	Serpent-bearer	17	0	948
Ori	Orion	-nis	Hunter	5	5 N	594
Pav	Pavo	-vonis	Peacock	20	65 S	378
Peg	Pegasus	-si	Winged horse	22	20 N	1121
Per	Perseus	-sei	Champion	3	45 N	615
Phe	Phoenix	-nisis	Phoenix	1	50 S	469
Pic	Pictor	-ris	Painter's easel	6	55 S	247
Psc	Pisces	-cium	Fishes	1	15 N	889
PsA	Piscis Austrinus	Piscis Austrini	Southern fish	22	30 S	245
Pup	Puppis	-pis	Ship's stern	8	40 S	673
Pyx	Pyxis	-xidis	Ship's compass	9	30 S	221
Ret	Reticulum	-li	Net	4	60 S	114
Sge	Sagitta	-tae	Arrow	20	10 N	80

Abbreviation	Nominative	Genitive ending	Meaning	RA	Dec	Area deg^2
Sgr	Sagittarius	-rii	Archer	19	25 S	867
Sco	Scorpius	-pii	Scorpion	17	40 S	497
Scl	Sculptor	-ris	Sculptor	0	30 S	475
Sct	Scutum	-ti	Shield	19	10 S	109
Ser	Serpens	-ntis	Serpent	17	10 N	637
Sex	Sextans	-ntis	Sextant	10	0	314
Tau	Taurus	-ri	Bull	4	15 N	797
Tel	Telescopium	-pii	Telescope	19	50 S	252
Tri	Triangulum	-li	Triangle	2	30 N	132
TrA	Triangulum Australe	Trianguli Australis	Southern triangle	16	65 S	110
Tuc	Tucana	-nae	Toucan	0	65 S	295
UMa	Ursa Major	Ursae Majoris	Great Bear	11	50 N	1280
UMi	Ursa Minor	Ursae Minoris	Small Bear	15	70 N	256
Vel	Vela	-lorum	Sails	9	50 S	500
Vir	Virgo	-ginis	Virgin	13	0	1294
Vol	Volans	-ntis	Flying fish	8	70 S	141
Vul	Vulpecula	-lae	Small fox	20	25 N	268

Sources: *Allen's Astrophysical Quantities*, 4th edition, 2000, A. N. Cox, ed., Springer, New York; *The Observer's Handbook*, 2002, Rajive Gupta, ed., The Royal Astronomical Society of Canada, Toronto.

E2 Some named stars

Name	Alternative designation	V	Claim to fame
Albireo	β Cyg	3.08	Telescopic double
Alcor	80 Uma	4.01	Visual double with Mizar
Alcyone	η Tau	2.87	Brightest Pleiad
Aldebaran	α Tau	0.87	Bright red, near ecliptic
Algol	β Per	2.09	Eclipsing variable
Alnilam	ε Ori	1.69	Middle star of Orion's belt
Altair	α Aqi	0.76	
Antares	α Sco	1.06	Very red
Arcturus	α Boo	−0.05	Brightest in northern hemisphere

Name	Alternative designation	V	Claim to fame
Barnard's Star	HIP 87937	9.54	Largest proper motion, 2nd nearest system
Bellatrix	γ Ori	1.64	West shoulder of Orion
Betelgeuse	α Ori	0.45	East shoulder. Very red. Variable
Canopus	α Car	−0.62	Second brightest
Capella	α Aur	0.08	
Castor	α Gem	1.58	
Cor Caroli	α CVn	2.90	Undistinguished white star. Named by Halley to mock Charles I
Deneb	α Cyg	1.25	
Denebola	β Leo	2.14	Tail of the lion
Dubhe	α Uma	1.81	Northern of the two pointer stars
Fomalhaut	α PsA	1.17	
Kapteyn's Star	HD 33793	8.86	Large proper motion (8.8″ yr^{-1})
Luyten's Star	HIP 36208	9.84	Large proper motion
Merak	β UMa	2.34	Southern of the pointers
Mintaka	δ Ori	2.25	Western end of belt
Mizar	ζ UMa	2.23	Visual double with Alcor
Plaskett's Star	HD 47129	6.05	Most massive binary
Polaris	α UMi	1.97	Pole star
Pollux	β Gem	1.16	
Procyon	α CMi	0.40	
Proxima Centauri	α Cen C	11.01	Nearest star. Member of α Cen system
Regulus	α Leo	1.36	
Rigel	β Ori	0.18	West foot of Orion
Rigel Kent	α Cen A + B	−0.01	Nearest system
Saiph	κ Ori	2.07	East foot
Sirius	α CMa A	−1.44	Brightest. Fourth-nearest system
Sirius B	α CMa B	8.4	Nearest white dwarf
Spica	α Vir	0.98	
Thuban	α Dra	3.65	Former pole star
Vega	α Lyr	0.03	Photometric standard
Zubenelgenubi	α Lib	2.75	

E3 Naming small bodies in the Solar System

Minor planets

The *provisional designation* is a four-part name, with all parts related to the date of discovery. Assigned by the Central Bureau for Astronomical Telegrams (CBAT), it combines:

- The year of discovery (all four digits).
- A single uppercase Roman letter, coding the UT half-month of the discovery. Months always divide on the 15th day: e.g. A = Jan 1–15, B = Jan 16–31, D = Feb 16–29. This uses all letters except I and Z.
- A second uppercase letter, indicating the order of discovery within the half-month (A = first, Z = last). The letter I is not used.
- If there are more than 25 discoveries in a half-month (there usually are) append a final number, indicating the number of times the second letter has been recycled. This should be written as a subscript, if practical. So for example, 2002 WZ_5 is the 150th provisional discovery made during the interval November 16–30 in 2002.

The *permanent designation*, assigned after observers establish a definitive orbit, consists of a sequential catalog number followed by a name. Names are proposed by the discoverer and approved by the Committee for Small-Body Nomenclature of the IAU. The temporary designation is retained if a name is not proposed and approved. Example permanent designations are: (1) Ceres, (2) Pallas, (9479) Madres-PlazaMayo, (9548) Fortran, and (134340) Pluto. As of 2015, there were over 450 000 objects with permanent designations, many discovered by automated search programs like **L**incoln **L**aboratory **N**ear **E**arth **A**steroid **R**esearch Project (LINEAR).

Comets

The modern rules for comet designations are similar to those for minor planets. Indeed, the distinction between comets and minor planets is not always clear. Upon discovery, the CBAT assigns a candidate new comet a *provisional designation* based on the date of the discovery. The designation consists of the four-digit year, a single letter designating the half-month, and a final numeral indicating the order within the half-month. It has been traditional since the eighteenth century to name new comets after the observer (or group, program, or satellite) who discovers them, and the provisional designation also contains the name of the discoverer, as determined by the IAU. It may also contain a prefix describing the nature of the orbit, using the codes:

P/, a short-period comet (P < 200 years)

C/, a long-period comet

X/, orbit uncertain

D/, disappeared, destroyed, or lost

A/, an object later determined to be an asteroid

Thus, D/1993 F2 (Shoemaker–Levy 9) was an actual comet discovered by Eugene and Carolyn Shoemaker and David Levy in the second half of March 1993. It was destroyed in a spectacular collision with Jupiter.

Most newly discovered comets are in such large orbits that a reliable ephemeris cannot be computed for the comet's next perihelion passage. The "periodic" (i.e. short period) and destroyed comets are the exception, and these are given a permanent designation that prefixes a catalog number, assigned in order of orbit discovery or comet destruction to the provisional name. For most references, the date segment of the name can be dropped. An example, undoubtedly the most famous, is 1P/1682 Q1 (Halley) = 1P/Halley. As of 2015, there were 330 comets with permanent catalog numbers.

Natural satellites of the major and minor planets

Again, designation of these objects parallels the practice for minor planets. A provisional designation consists of (1) the prefix S/, (2) the year of discovery, (3) a roman letter coding the planet, or a parenthetical numeral of the numbered asteroid, and (4) a numeral giving order of announcement within the year. For example:

S/2000 J 7 is a satellite of Jupiter

S/2002 (3749) 1 is a satellite of the minor planet (3749) Balam

Once an orbit is well defined, the temporary designation is replaced with a sequential roman numeral affixed to the planet name, and a permanent name whose selection is based on mythological or literary themes. For example:

Jupiter I = Io

Jupiter XVI = Metis

Uranus XIII = Belinda

Neptune VI = Galatea

Appendix F

F1 A timeline for optical telescopes

See also: http://amazing-space.stsci.edu/resources/explorations/groundup/

c. 3500 BCE	Invention of glass in Egypt and Mesopotamia.
c. 2000 BCE	Lenses fashioned from rock crystal in Ionia. Use unknown.
424 BCE	Aristophanes (*The Clouds*, Act II, Scene 1) describes the focusing power of a glass globe filled with water.
c. 300 BCE	Euclid gives a rudimentary treatment of the ray theory of light and of refraction at a plane interface. Euclid, following Plato, believed rays moved from the eye of the beholder to the object beheld.
212 BCE	Archimedes is reported to have used curved mirrors to focus sunlight and set fire to the sails of Roman ships during the siege of Syracuse. It is not reported how proponents of Euclid's ray theory accounted for this effect.
c. 300 BCE – 170 CE	Hellenistic astronomers make naked-eye observations with the armillary sphere and the mural quadrant, predecessors of the equatorial mount and the transit telescope. Ptolemy's catalog of 1000 stars, with positions precise to about 15 minutes of arc, becomes a standard for the next millennium.
c. 1000 CE	Ibn al-Haitham (Latin: Alhazen), in Egypt, conducts experiments in optics and writes on spherical mirrors, lenses, and refraction.
c. 1275	Roger Bacon, English philosopher, conducts optics experiments and describes, for example, the magnifying power of a plano-convex lens. Vitello of Silesia publishes a large volume on optics, founded upon and advancing Alhazen.
c. 1285	Spectacles invented (first examples appear in northern Italy, manufactured from high-quality glass produced in Venice).
1565–1601	Tycho designs a spectacular series of instruments that permit naked-eye observations to the unprecedented accuracy of one minute of arc.
1608	Hans Lippershey, a spectacle-maker, petitions the States-General of the Netherlands for a patent on his invention of the spy-glass (Galilean telescope). The patent is denied because "many other persons have a knowledge of the invention."
1609	Galileo, on hearing rumors of Lippershey's device, constructs one for himself. He uses subsequent models to observe the night sky, with momentous consequences.
1611	After acquiring a telescope, Kepler, familiar with the work of Vitello, writes a treatise on optics, *Dioptrice*. This includes the

	first description of spherical aberration and of the Keplerian telescope (objective and ocular both convex lenses).
1612–90	Era of very long focal length refractors, constructed with single-lens objectives and limited by both chromatic aberration and SA. Large focal ratios minimize both aberrations.
1655	Invention of the pendulum clock by Huygens makes positional astronomy with transit instruments a much more precise enterprise. At about the same time, Huygens also introduces a micrometer ocular, and "divided instruments" of the sort used by Tycho begin to appear equipped with telescopic sights.
1663–74	First reflecting telescopes designed by Gregory, Newton, Cassegrain, and Hooke. A number of large (up to 80-cm aperture) speculum-metal reflecting telescopes appear over the next century, but most of the advances in astronomy come from refractors.
1688–1720	Flamsteed observes the positions of 3000 stars with a 7-foot quadrant equipped with telescopic sights. Positional accuracy is about 10 seconds of arc in declination, and about 1 second of time (15 seconds of arc at the equator) in RA.
1721	John Hadley produces a 6-inch $f/10.3$ Newtonian reflector, which rivals the performance of the "long" refractors.
1728	Bradley discovers the velocity aberration of starlight using a zenith sector – a telescope suspended vertically – to measure changes in the apparent declination of stars transiting near the zenith. Bessel (in 1818) shows that Bradley's fundamental catalog, derived from observations with a quadrant and a transit telescope, has a positional accuracy of 4 seconds of arc in declination and 1 second of time in RA.
1729	Chester Moor Hall works out the theory of the achromatic doublet. Doublets will not be used in telescopes until the work of Dolland in the 1750s.
1733–68	James Short in England manufactures a number of excellent speculum-metal reflectors, many in the Gregorian configuration.
c. 1760	The first achromatic objectives begin to appear. These are of small aperture (<10 cm) owing to the difficulty of casting blanks of flint glass that is free from optical flaws.
1761–64	Clairaut gives a rigorous explication of the achromatic doublet, and uses ray tracing to characterize most of the third-order aberrations.
1781	William Herschel discovers the planet Uranus with a 16-cm Newtonian reflector. Herschel's fame and productivity, especially with his 45-cm reflector, led to increasing popularity of speculum-metal reflectors.

1812–26	Fraunhofer combines an ability to produce large flint blanks with practical methods of optical testing, and produces a number of excellent achromatic refractors. The tide at observatories turns away from reflectors. With the "Great Dorpat Refractor" (aperture 24 cm), Fraunhofer introduces the German equatorial mount.
1848	Fighting the tide, William Parsons, Earl of Rosse, builds a Herschel-style reflector with an aperture of 1.8 m. This telescope is heavy, unwieldy, and located at a poor site. Refractors with 0.4-m apertures, installed at Pulkova and at Harvard at about this time, prove more productive. The refractor is supreme.
1852	William Lassell applies the equatorial mount to large reflectors. He moves his 0.6-m telescope from England to the island of Malta in order to obtain better seeing. It will be 50 years before astronomers fully recognize the importance of site selection.
1856	Leon Foucault, von Steinheil, and others introduce methods for depositing silver on glass mirrors.
1892	The refractor reaches its technological limit with the Alvan Clark 1.0-m Yerkes Observatory telescope. At about this time, there is a flurry of construction of astrographs – refractors optimized for photographic work.
1895	Edward Crossley donates his private telescope, a silver-on-glass 0.9-m reflector, to the Lick Observatory, because he recognizes the telescope is wasted at its site at Halifax in northern England. The remounted telescope demonstrates the suitability of large reflectors for photographic work.
1917	With the commission of the 2.5-m (100-inch) reflector on Mt. Wilson, it is clear that telescopes with silvered-glass mirrors have surpassed refractors in light-gathering power and cost effectiveness. From now on, all new major optical telescopes will be reflectors. Because of the world wars, new telescope construction is greatly curtailed in Europe for 60 years.
1948	Reflector technology reaches a plateau with the 5.0-m on Palomar Mountain. Over the next 50 years, gradual technological advances result in 24 large optical–infrared telescopes with apertures between 2.5 and 4.2 m. All are reflectors. Most have equatorial mounts, although altazimuths become common after 1985.
1953	US astronomer Horace Babcock lays out the principle of adaptive optics. No practical system will be developed for another 20 years.

1974	The Anglo-Australian Telescope, a 3.9-m equatorial, is the first large telescope designed with computer-controlled pointing and tracking.
1975	The 6.0-m Bolshoi Teleskop Azimutal'ny is installed in the Caucus Mountains. Generally regarded as only partially successful, it demonstrates the superiority of the altazimuth for large telescopes.
1979	The Multiple-Mirror Telescope (MMT) on Mt. Hopkins, Arizona, uses six 1.8-m mirrors to bring light to a common focus, and demonstrates several concepts employed by future large-aperture systems.
1982	The Advanced Research Projects Agency (ARPA; US Department of Defense) demonstrates the Compensated Imaging System, a practical adaptive optics system for imaging artificial Earth satellites, culminating ten years of development effort.
1984	The 2.6-m Nordic Optical Telescope on La Palma is the first to use effective climate control and active primary mirror support.
1989	The first astronomical adaptive optics images are obtained – diffraction-limited K-band images with the 1.5-m telescope at the Observatoire de Haute-Provence in France. The successor systems, COME-ON/ADONIS, produce a steady stream of practical results from the ESO 3.6-m telescope at la Silla, Chile. The highly developed active optics system on the 3.5-m New Technology Telescope (NTT) achieves seeing disks as small as 0.3 seconds of arc.
1990	The Hubble Space Telescope (HST) is launched. After repair of residual SA in the optics in 1993, this 2.4-m telescope achieves a resolution of 0.1 arcsec in the visible. The HST was followed by three other large space telescopes in the NASA great observatories program: the Compton Gamma-Ray Observatory in 1991, the Chandra X-Ray Observatory in 1999, and the Spitzer Infrared Observatory in 2003.
1993	Keck I 9.8-m aperture telescope with an active optics segmented primary is installed on Mauna Kea.
1997–2008	A half-dozen large optical telescopes with light-gathering power equivalent to apertures in the 6.5–16.4 m range see first light (see Appendix A6).
1999	Adaptive optics system operational on the second Keck telescope (Keck II), achieving resolutions of 0.02 arcsec for bright stars in K band.
2001	Tests successfully combine beams of the Keck I and II telescopes for operation as an interferometer with a resolution of 1 mas.

	Similar tests combine beams of the ESO VLT telescopes in Chile for a similar resolution.
2018	Projected launch of the James Webb Space Telescope (6.6-m aperture).
2018–2025	Projected first lights for the Thirty-Meter Telescope, the Giant Magellan Telescope (24.5-m aperture), and the European Extremely Large Telescope (E-ELT, 42-m aperture).

Appendix G

G1 Websites

AO (ESO): www.eso.org/projects/aot/introduction.html
AO (Keck): www2.keck.hawaii.edu%3A3636/realpublic/inst/ao/ao.html
HET: www.as.utexas.edu/mcdonald/het
HST: www.stsci.edu/hst/
JWST: www.ngst.nasa.gov/
Optical glass (Schott, Inc): www.schott-group.com/english/company/us.html
Spin casting mirrors: http://medusa.as.arizona.edu/mlab
Subaru telescope: http://SubaruTelescope.org/index.html

G2 Largest optical telescopes (2015)*

Name	Organization; location	Aperture, mirror type	Focal ratios	Year	Comments
VLT (Very Large Telescope)	European South Observatory (ESO); Cerro Paranal, Chile	8.2 m × 4, monolithic zerodur meniscus	R:13.4, N:15, C:47.3	1999–2001	Can combine four large beams equivalent to 16.4-m aperture. Four additional 1-m telescopes improve interferometric resolution
Keck I & II	University of California; California Institute of Technology; Mauna Kea, Hawaii	9.82 m × 2, 1.8 m × 36 (each), hexagonal zerodur segments	P:1.75, R:15	1993–1996	Interferometric combination of two beams (2001) gives 13.9-m equivalent aperture
LBT (Large Binocular Telescope)	LBT; Consortium, Mt. Graham, Arizona	8.4 m × 2, spin-cast ribbed borosilicate	P:1.14, 5.4 R:15	1999–2004	Major partners are University of Arizona and the Vatican Observatory. Combined beams give 11.9-m aperture

Name	Organization; location	Aperture, mirror type	Focal ratios	Year	Comments
GTC (Gran Telescopio Canarias)	Spain + others; La Palma, Canary Islands	10.4 m, 1.9 m× 36, hexagonal segments	P:1.75, R:15 N:25	2002	
HET (Hobby– Eberly Telescope)	University of Texas, Pennsylvania State University, Stanford, München, Gottingen; McDonald Observatory	92 m, 1.0 m × 91, hexagonal zerodur segments	P:1.42	1997	Spherical primary is stationary during observations, with tracking done in the focal plane. 9.2-m entrance pupil on 11-m primary
Magellan	Carnegie Consortium Las Campanas, Chile;	2 × 6.5 m, spin-cast honeycomb borosilicate	P:1.25, N:11, C:15	1999 2003	
Subaru	National Observatory of Japan; Mauna Kea, Hawaii	8.3 m, Corning ULE thin meniscus	P:2.0, R:12.2, N:12.6	1998	
Gemini North	USA, UK, Canada, Chile, Australia, Argentina, Brazil,	8.1 m, Corning ULE meniscus	R:16	1998 (N)	Optimized for near infrared
Gemini South	Mauna Kea, Hawaii; Cerro Pachon, Chile			2000 (S)	
SAO (MMT)	Smithsonian Astrophysical Observatory, Mt. Hopkins, Arizona (MMT)	6.5 m, spin-cast ribbed borosilicate	P:3.0	1978	

[*] Focal ratios for prime focus (P), RC (R), Naysmith (N), and coudé (C). Corning ULE is an ultra-low thermal expansion coefficient glass; zerodur and borosilicate are types of glass.

G3 Large Schmidt telescopes

Name	Location	Diameter: corrector / mirror	Focal ratio	Year	Comments
Tautenberg	Tautenberg, Germany	1.34/2.00	f/3.00	1960	Equipped with a Nasmyth and coudé focus, multi-object spectrograph
Oschin	Palomar Mt., California	1.24/1.83	f/2.47	1948	Important because it was the first very large Schmidt. Produced the Palomar Sky Survey (PSS), a basic reference tool
UK Schmidt	Siding Spring Mt., Australia	1.24/1.83	f/2.5	1973	Collaborated on the ESO-Science and Engineering Research Council (UK) survey, extending the PSS project to the southern hemisphere
Kiso	Kiso, Japan	1.05/1.5	f/3.1	1975	
Byurakan	Mt. Aragatz, Armenia	1.0/1.5	f/2.13	1961	Conducted objective prism survey for galaxies with ultraviolet bright nuclei
Uppsala	Kvistaberg, Sweden	1.0/1.35	f/3.00	1963	Surveys for near-Earth objects
ESO	La Silla, Chile	1.0/1.62	f/3.06	1972	Decommissioned in 1998
Venezuela	Merida, Venezuela	1.0/1.52	f/3.0	1978	

Appendix H

H1 The hydrogen atom

The energy levels of the hydrogen atom depend primarily on the principal quantum number, as does the wavelength of a transition between levels:

$$E_n = -\frac{me^4}{8\varepsilon_o^2 h^2}\frac{1}{n^2}$$

$$\frac{1}{\lambda_{ab}} = R\left(\frac{1}{n_a^2} - \frac{1}{n_b^2}\right) = 0.01097\left(\frac{1}{n_a^2} - \frac{1}{n_b^2}\right)nm^{-1}$$

The table gives the wavelengths of several lines in the spectrum of atomic hydrogen:

Line	Transition	Wavelength (nm)	Line	Transition	Wavelength
Lyman Series			Paschen Series		
Ly-α	1–2	121.57	P-α	3–4	1875.1
Ly-β	1–3	102.56	P-β	3–5	1281.8
Limit	1–∞	91.18	Limit	1–∞	820.4
Balmer Series			Brackett Series		
H-α	2–3	656.3	B-α	4–5	4051.2
H-β	2–4	486.1	B-β	4–6	2625.2
H-γ	2–5	434.0	Limit	4–∞	1458.4
H-δ	2–6	410.2	Pfund Series		
H-ε	2–7	397.0	Pf-α	5–6	7460.0
Limit	2–∞	364.3	Limit	5–∞	2279.0

H2 Some common semiconductors

Forbidden band gap energies and cutoff wavelengths are given at room temperature, except where noted. Band gap data from Section 20 of Anderson (1989) or from Kittel (2005).

Material		Band gap (eV)	λ_c (μm)
IV			
Diamond	C	5.48	0.23
Silicon	Si	1.12	1.11
	Si (4 K)	1.17	1.06
	Si (700 K)	0.97	1.28
Germanium	Ge	0.67	1.85

Material		Band gap (eV)	λ_c (μm)
	Ge (1.5 K)	0.744	1.67
Gray tin	αSn	0.0	
Silicon carbide	SiC	2.86	0.43
III–V			
Gallium arsenide	GaAs	1.35	0.92
Gallium antimonide	GaSb	0.68	1.83
Indium phosphide	InP	1.27	0.98
Indium arsenide	InAs	0.36	3.45
Indium antimonide	InSb	0.18	6.89
	InSb(77K)	0.23	5.39
Boron phosphide	BP	2.0	0.62
II–VI			
Cadmium sulfide	CdS	2.4	0.52
Cadmium selenide	CdSe	1.8	0.69
Cadmium telluride	CdTe	1.44	0.86
Mercury cadmium telluride	$Hg_x Cd_{1-x}$ Te	0.1–0.5 ($x = 0.8$–0.5)	12.4–2.5
IV–VI			
Lead sulfide	PbS	0.42	2.95

Appendix I

I1 Characteristics of some commercial CCDs for astronomy

These are advanced devices nevertheless within budgetary reach of a small observatory. See manufacturers websites (I2) for additional examples.

Device	e2v CCD42–90	KAF09000	e2v CCD60
Type	Three-phase BCCD, backthinned, polysilicon gates, three-side buttable	Two-phase, front-illuminated ITO gates, anti-blooming. Dark reference pixels	Backthinned, electron multiplying g = 1000, fram-transfer device
Dimensions, $C \times R$	2048 × 4096	3103 × 3086	128 × 128
Pixel size, μm	13.5 × 13.5	12 × 12	24 × 24
Output amplifiers	Two parallel	1	1
Amplifier responsivity, μV electron^{-1}	4.5	24	1.2
Pixel full well (electrons)	150 000	110 000	530 000
Serial full well (electrons)	600 000	na	800 000
Summing full well or amplifier saturation (electrons)	900 000	110 000	800 000
CTE	0.999 995	0.999 990	na
Read noise, rms	3 (at 20 kHz)	7 (at 1 MHz), 15 (at 8 MHz)	100 (at 11 MHz)
Dark current, electrons pixel^{-1} s^{-1}	0.0003 (at −100 °C)	0.02(at −35 °C)	0.2 (at −40 °C)
AR coating: QE with coating, % at wavelength (nm):	Blue, visual	std	std
350	50, 17	22	20
400	80, 52	42	53
500	85, 92	59	90
650	80, 93	59	90
900	50, 55	24	40

I2 Manufacturers of sensors and cameras for astronomy

Fairchild Imaging: www.fairchildimaging.com

Teledyne Scientific and Imaging (SITe, HAWAII and PICNIC arrays): www.teledyne-si.com/

DALSA Corporation: www.dalsa.com

Hamamatsu Corporation (photo-emissive devices): http://jp.hamamatsu.com/products/sensor-etd/pd007/index_en.html.

Hamamatsu Photonics: www.hamamatsu.com

HORIBA Jobin Yvon Inc.: www.jobinyvon.com

OnSemi (absorbed Kodak CCDs) onsemi.com

Santa Barbara Instrument Group: www.sbig.com/

Apogee: www.ccd.com/

Finger Lakes Instruments: www.flicamera.com/

Pan-STARRS: http://pan-starrs.ifa.hawaii.edu/public/design-features/cameras.*html*

e2v: www.e2v.com

Appendix J

J1 The point-spread function

The point-spread function (PSF) is the two-dimensional brightness distribution produced in the plane of the detector by the image of an unresolved source, such as a distant star. A real detector has finite pixel size, so it records a matrix of pixel-sized samples of the PSF (see, for example, Figures 9.12 and 9.20). Astronomers often face the problem of determining the PSF that best fits the pixilated detector image of one or more point sources. There are two different approaches:

(1) One approach is to assume some analytical function, $P(x-x_0, y-y_0)$, describes the brightness distribution as a function of the displacement from the image centroid at (x_0, y_0). In this case you select a likely function with a sufficient number of free parameters, and select the parameter values that best fit the one or more observed stellar images. For example, images dominated by seeing are often thought to assume the **Moffat** profile:

$$P = B_0 \left[1 + \left(\frac{r}{\alpha} \right)^2 \right]^{-\beta}$$

where

$$r^2 = (x - x_0)^2 + (y - y_0)^2$$

The parameters α and β determine the PSF shape, while B_0 sets the scale, and a least-squares fit can estimate their values. The above profile has circular isophotes. You would require additional parameters to fit the more irregular profiles often seen in practice. For example, you could fit profiles that had elliptical isophotes of ellipticity ε, elongated direction θ_0, by setting

$$\alpha^2 = \alpha_0^2 \left[\cos^2 \left\{ \left(\tan^{-1} \frac{y}{x} \right) - \theta_0 \right\} + \varepsilon \sin^2 \left\{ \left(\tan^{-1} \frac{y}{x} \right) - \theta_0 \right\} \right]$$

in the Moffat formula. **Gaussian** functions are also popular choices for analytical PSFs.

(2) A second approach is to use bilinear interpolation to estimate the brightness values at fractional pixel positions in an observed stellar image (or in several). This produces a completely empirical PSF, and has the advantage of coping nicely with very irregular profiles. It has difficulty near the centroid, where interpolated values are unstable, especially in under-sampled images.

A frequent strategy is to meld the two approaches: First fit the data with an analytical function, then fit the residuals using the empirical method, so the final PSF is the sum of the two fits.

Appendix K

K1 Intrinsic broadband colors for various spectral types

Calibration of MK spectral types with temperatures, absolute magnitudes, bolometric corrections (magnitudes) and photometry. Data for dwarves are directly from Table 5 in Pecaut and Mamajek (2013) as updated by Mamajek at www.pas.rochester.edu/~emamajek/EEM_dwarf_UBVIJHK_colors_Teff.txt.

Data for giants and supergiants are from Table 15.7 in *Allen's Astrophysical Quantities* (Cox, 1999) and Ducati et al. (2001).

Table K1 *Main sequence dwarves (luminosity class V) K band is the "short" K of the 2MASS system.*

Spectral type	T_{eff}	M_v	BC_v	$B - V$	$U - B$	$V - R_c$	$V - I_c$	$V - K_s$	$J - H$	$H - K$
O3	46000	−5.7	−4.05	−0.32	−1.22					
O6	39000	−5.1	−3.57	−0.32	−1.17					
O9	32500	−4.2	−3.09	−0.318	−1.114		−0.369	−1.00	−0.164	-0.071
B0	31500	−4.0	−3.02	−0.307	−1.067	−0.16	−0.355	−0.958	−0.159	-0.067
B1.5	24500	−2.8	−2.43	−0.252	−0.91	−0.114	−0.281	−0.752	−0.132	-0.047
B2	20600	−1.7	−2.06	−0.21	−0.79	−0.094	−0.23	−0.602	−0.113	-0.032
B5	15700	−0.9	−1.35	−0.156	−0.581	−0.07	−0.165	−0.417	−0.089	-0.013
B8	12500	−0.2	−0.81	−0.109	−0.364	−0.048	−0.108	−0.254	−0.067	0.003
B9	10700	0.7	−0.42	−0.07	−0.2	−0.028	−0.061	−0.121	−0.05	0.016
A0	9700	1.11	−0.24	0	−0.005	0.001	0.004	0.041	−0.032	0.028
A2	8840	1.48	−0.1	0.074	0.063	0.042	0.091	0.188	−0.01	0.034
A5	8080	1.84	−0.03	0.16	0.1	0.089	0.186	0.403	0.031	0.038
F0	7200	2.51	−0.01	0.294	0.053	0.166	0.339	0.732	0.098	0.045
F2	6810	2.99	−0.02	0.374	−0.008	0.213	0.432	0.925	0.14	0.05
F5	6510	3.4	−0.04	0.438	−0.029	0.252	0.506	1.079	0.173	0.054
F8	6150	4.01	−0.07	0.53	0.001	0.3	0.599	1.29	0.225	0.061
G0	5920	4.39	−0.09	0.588	0.049	0.331	0.656	1.421	0.258	0.066
G2	5770	4.79	−0.11	0.65	0.133	0.363	0.713	1.564	0.293	0.073
G5	5660	4.98	−0.13	0.68	0.185	0.377	0.738	1.635	0.31	0.076
G8	5490	5.32	−0.17	0.737	0.284	0.404	0.786	1.768	0.342	0.082
K5	5280	5.76	−0.22	0.816	0.436	0.443	0.853	1.953	0.387	0.091
K2	5040	6.19	−0.29	0.893	0.6	0.487	0.929	2.155	0.432	0.099
K5	4450	7.25	−0.67	1.134	1.056	0.671	1.246	2.835	0.56	0.13
K8	3970	8.6	−1.11	1.382	1.213	0.859	1.671	3.554	0.63	0.181
M0	3850	9.16	−1.3	1.431	1.19	0.913	1.848	3.79	0.622	0.203

Table K1 (*cont.*)

Spectral type	T_{eff}	M_v	BC_v	$B - V$	$U - B$	$V - R_c$	$V - I_c$	$V - K_s$	$J - H$	$H - K$
M2	3550	10.3	−1.65	1.5	1.17	1.001	2.173	4.24	0.6	0.234
M3	3400	11.14	−1.97	1.544	1.181	1.079	2.42	4.6	0.579	0.252
M4	3200	12.8	−2.59	1.661	1.215	1.241	2.831	5.25	0.557	0.282
M5	3050	14.3	−3.28	1.874	1.433	1.446	3.277	5.942	0.58	0.311

Table K2 *Lower main sequence dwarves.*

Spectral type	T_{eff}	M_V	M_K	M_{bol}	$V - I_C$	$V\text{-}K$	$J - H$	$H - K_S$
M6	2800	16.62	9.32	12.26	4.10	7.3	0.605	0.352
M8	2570	18.75	10.04	13.09	4.6	8.7	0.675	0.446
L0	2250	19.8	10.38	13.67	4.6	9.3	0.79	0.5
L2	1960	20.8	11.17	14.34	4.9	9.9	0.87	0.57
L5	1590	23.1	11.82	15.34	6	11.4	1.13	0.65
L8	1350		12.7	16.12			1.14	0.63
T0	1260		12.9	16.39			1.02	0.54
T3	1160		13.69	16.67			0.68	0.08
T6	960		15.54	17.54			0.1	−0.03
T9V	530		18.48	20.12			0.1	−0.2
Y0	400:							

Table K3 *Giant stars (luminosity class III) K band is the "long" K of the original Johnson system.*

Spectral type	T_{eff}	M_V	BC	$B - V$	$U - B$	$V - R_C$	$V - I_C$	$V - J$	$V - H$	$V - K$
G5	5050	+0.9	−0.34	+0.86	+0.56	0.47	0.93	0.94	1.44	1.53
G8	4800	+0.8	−0.42	+0.94	+0.70	0.50	0.47	1.11	1.61	1.77
K0	4660	+0.7	−0.50	+1.00	+0.84	0.49	1.00	1.23	1.72	1.94
K2	4390	+0.5	−0.61	+1.16	+1.16	0.84	1.11	1.56	2.08	2.39
K5	4050	−0.2	−1.02	+1.50	+1.81	0.60	1.53	2.25	2.87	3.14
M0	3690	−0.4	−1.25	+1.56	+1.87	0.88	1.78	2.55	3.23	3.46
M2	3540	−0.6	−1.62	+1.60	+1.89	0.92	1.97	2.99	3.76	3.89
M5	3380	−0.3	−2.48	+1.63	+1.58	1.30	3.04	3.95	4.98	4.73
M6	3250						3.80	4.34	5.50	5.04

Table K4 *Supergiant stars (luminosity class I).*

Spectral type	T_{eff}	M_V	BC	$B - V$	$U - B$	$V - R_C$	$V - I_C$	$V - J$	$V - H$	$V - K$
O9	32000	−6.5	−3.18	−0.27	−1.13	−0.08	−0.16	−0.57	−0.75	−0.84
B2	17600	−6.4	−1.58	−0.17	−0.93	−0.06	−0.09	−0.43	−0.56	−0.63
B5	13600	−6.2	−0.95	−0.10	−0.72	−0.01	−0.07	−0.28	−0.34	−0.39
B8	11100	−6.2	−0.66	−0.03	−0.55	0.03	0.07	−0.12	−0.14	−0.15
A0	9980	−6.3	−0.41	−0.01	−0.38	0.04	0.09	0.02	0.06	0.08
A2	9380	−6.5	−0.28	+0.03	−0.25	0.05	0.11	0.11	0.17	0.21
A5	8610	−6.6	−0.13	+0.09	−0.08	0.08	0.20	0.20	0.29	0.35
F0	7460	−6.6	−0.01	+0.17	+0.15	0.13	0.28	0.36	0.51	0.60
F2	7030	−6.6	−0.00	+0.23	+0.18	0.18	0.33	0.44	0.62	0.73
F5	6370	−6.6	−0.03	+0.32	+0.27	0.21	0.42	0.57	0.79	0.91
F8	5750	6.5	0.09	+0.56	+0.41	0.27	0.52	0.87	1.17	1.34
G0	5370	−6.4	−0.15	+0.76	+0.52	0.36	0.66	1.14	1.52	1.71
G2	5190	−6.3	−0.21	+0.87	+0.63	0.40	0.75	1.35	1.80	1.99
G5	4930	−6.2	−0.33	+1.02	+0.83	0.41	0.76	1.61	2.13	2.32
G8	4700	−6.1	−0.42	+1.14	+1.07	0.56	1.03	1.83	2.41	2.59
K0	4550	−6.0	−0.50	+1.25	+1.17	0.61	1.12	2.01	22.64	2.80
K2	4310	−5.9	−0.61	+1.36	+1.32	0.65	1.22	2.20	2.87	3.01
K5	3990	−5.8	−1.01	+1.60	+1.80	0.86	1.72	2.74	3.55	3.59
M0	3620	−5.6	−1.29	+1.67	+1.90	0.97	2.04	3.07	3.97	3.92
M2	3370	−5.6	−1.62	+1.71	+1.95	1.11	2.39	3.45	4.45	4.28
M5	2880	−5.6	−3.47	+1.80	+1.60:	2.7		5.26	6.68	5.73

References

Allen, R. H., *Star Names and their Meanings* (Original edition, New York, Stechert, 1899; reprinted: Washington, DC, Dover, 1963). Online: http://penelope.uchicago.edu/Thayer/E/Gazetteer/Topics/astronomy/home.html

Anderson, H. L., ed., 1989. *A Physicist's Desk Reference*, New York, American Institute of Physics.

Bessell, M. S., 1990. *Publications of the Astronomical Society of the Pacific*, **102**, 1181.

Bessell, M. S., 1991. *Astronomical Journal*, **101**, 662.

Bessell, M. S., 1992. *IAU Colloquium*, **136**, 22.

Bessell, M. S., 2005. *Annual Review of Astronomy & Astrophysics*, **43**, 293.

Bessell, M. S., Caselli, F. and Plez, B., 1998. *Astronomy and Astrophysics*, **333**, 231.

Bevington, P. R., 1969. *Data Reduction and Error Analysis for the Physical Sciences*, New York, McGraw-Hill.

Birney, D. S., Gonzalez, G., and Oesper, D., 2006. *Observational Astronomy*, 2nd edn, Cambridge, Cambridge University Press.

Blundell, S. J., 2009. *Superconductivity: A Very Short Introduction*, New York, Oxford University Press.

Boyle, W. and Smith, G. E., 1971. *IEEE Spectrum*, **8**(7), 18.

Butler, R. P., Marcy, G. W., Williams, E., et al., 1996. *Publications of the Astronomical Society of the Pacific*, **108**, 500.

Colina, L., Bohlin, R., and Castelli, F., 1996. Absolute flux calibration of Vega. Space Telescope Instrument Science Report CAL/SCS-008. See: www.stsci.edu/instruments/observatory/PDF/scs8.rev.pdf.

Cox, A., 1999. *Allen's Astrophysical Quantities*, New York, AIP Press.

Davies, R. and Kasper, M., 2012. *Annual Reviews of Astronomy and Astrophysics*, **50**, 305.

Dekker, H., D'Odorico, S., Kaufer, A., Delabre, B., and Kotzlowski, H., 2000. *Proceedings of the Society of Photo-optical Instrumentation Engineers*, **4008**, 534.

Ducati, J. R., Bevilacqua, C. M., Rembold, S. B., and Ribeiro, D., 2001. *Astrophysics Journal*, **558**, 309.

Eisenhauer, F. and Raab, W., 2015. *Annual Review of Astronomy and Astrophysics*, **53**, 155.

Esteban, C., Peimbert, M., Torres-Peimbert, S., et al., 1999. *Astrophysical Journal Supplement*, **120**, 113.

Freedman, W. L., Madore, B. F., Gibson, B. K., et al., 2001. *Astrophysical Journal*, **553**, 47.

Glass, I. S., 1999. *Handbook of Infrared Astronomy*, Cambridge, Cambridge University Press.

Gordon, K. D., Clayton, G. C., Misselt, K. A., et al., 2003., *Astrophysical Journal*, **594**, 279.

Hardy, J. W., 1998. *Adaptive Optics for Astronomical Telescopes*, New York, Oxford University Press.

Harris, R., 1998. *Nonclassical Physics*, New York, Addison Wesley.

Hearnshaw, J. B., 1986. *The Analysis of Starlight*, Cambridge, Cambridge University Press.

Hearnshaw, J. B., 1996. *The Measurement of Starlight*, Cambridge, Cambridge University Press.

Hearnshaw, J. B., 2009. *Astronomical Spectrographs and their History*, Cambridge, Cambridge University Press.

Hewett, P. C., Warren, S. J., Leggett, S. K., and Hodgkin, S. T., 2006, *Monthly Notices of the Royal Astronomical Society*, **367**, 454.

Hogg, D. W., Baldry, I. K., Blanton, M. R., and Eisenstein, D. J., 2002. eprint arXiv:astro-ph/0210394.

Hogg, V. R., Tanis, E., and Zimmerman, D. L., 2013. *Probability and Inference*, 9th edn, San Francisco: Pearson.

Holmes, R., 2008. *The Age of Wonder: How the Romantic Generation Discovered the Beauty and Terror of Science*, London, Harper Press.

Howell, S. B., 2006. *Handbook of CCD Astronomy*, 2nd edn, Cambridge, Cambridge University Press.

Jacoby, G. H., Hunter, D. A., and Christian, C. A., 1984. *Astrophysical Journal Supplement*, **56**, 257.

Jha, S., Reiss, A. G., and Kirschner, R. P., 2007. *Astrophysical Journal*, **659**, 122.

Johnson, H. L. and Harris, D. L., 1954. *Astrophysical Journal*, **120**, 196.

Kaler, J. B., 1997. *Stars and their Spectra*, Cambridge, Cambridge University Press.

King, Henry C., 1979. *The History of the Telescope*, New York, Dover. (Originally published by Griffin & Co., 1955.)

Kinney, A. L., Calzetti, D., Bohlin, R. C., McQuade, K., Storchi-Bergmann, T. and Schmitt, H. R., 1996. *Astrophysical Journal Letters*, **120**, 196.

Kitchin, C. R., 1995. *Optical Astronomical Spectroscopy*, Bristol, Institute of Physics.

Kitchin, C. R., 2008. *Astrophysical Techniques*, 5th edn, Boca Raton, FL, CRC Press.

Kittel, C., 2005. *Introduction to Solid State Physics*, 8th edn, New York, John Wiley and Sons.

Landolt, A. U., 1983. *Astronomical Journal*, **88**, 439.

Landolt, A. U., 1992. *Astronomical Journal*, **104**, 340.

Landolt, A. U., 2013. *Astronomical Journal*, **146**, 88.

Lasker, B. M. and Lattanzi, Mario G., 2008. *Astronomical Journal*, **236**, 735.

Lyons, L., 1991. *Data Analysis for Physical Science Students*, Cambridge, Cambridge University Press.

Martinez, P. and Klotz, A., 1998. *A Practical Guide to CCD Astronomy*, Cambridge, Cambridge University Press.

Mathis, J. S., 1990. *Annual Reviews of Astronomy and Astrophysics*, **28**, 37.

Mazin, B. A., Bumble, B., Meeker, S. R., et al., 2012. *Optics Express*, **20**, 1503.

McLean, I. S., 2008. *Electronic Imaging in Astronomy*, 2nd edn, Chichester, Springer-Praxis.

Menzies, J. W., Cousins, A. W. J., Banfield, R. M., Laing, J. D., and Coulson, I. M., 1989. *SAAO Circular*, **13**, 1.

Menzies, J. W., Marang, F., Laing, J. D., Coulson, I. M., and Engelbrecht, C. A., 1991. *Monthly Notices of the Royal Astronomical Society*, **248**, 642.

Möller, K. D., 1988. *Optics*, Mill Valley, CA, University Science Books.

Osterbrock, D. E., and Ferland, G. J., 2006. *Astrophysics of Gaseous Nebulae and Active Galactic Nuclei.*, 2nd edn, Sausalito, CA, University Science Books

Palmer, C., 2002. *Diffraction Grating Handbook*, 5th edn, Rochester, NY, Thermo RGL (widely available online).

Pecaut, M. J. and Mamajek, E. E., 2013. *Astrophysical Journal Supplement Series*, **208**, 9.

Pedrotti, F. S., Pedrotti, L. M., and Pedrotti, L. S., 2006. *Introduction to Optics*, 3rd edn, San Francisco, CA, Benjamin Cummings.

Rieke, G., 2003. *Detection of Light: From the Ultraviolet to the Submillimeter*, 2nd edn, Cambridge, Cambridge University Press.

Rieke, G., 2008. *Annual Reviews of Astronomy and Astrophysics*, **45**, 77–115.

Rieke, G. H. and Lebofsky, M. J., 1985. *Astrophysical Journal*, **288**, 618.

Rieke, G. H., Lebofsky, M. J., and Low, F. J., 1985. *Astronomical Journal*, **90**, 900.

Roddier, F., 1999. *Adaptive Optics in Astronomy*, Cambridge, Cambridge University Press.

Rutten, H. G. J. and van Venrooij, M. A. M., 1988. *Telescope Optics: A Comprehensive Manual for Amateur Astronomers*, Richmond, VA, Willmann-Bell, Inc.

Schaefer, B., 2006. *Scientific American*, **295**, 5, 70.

Schroeder, D. J., 1987. *Astronomical Optics*, London, Academic Press.

Smith, J. A., Tucker, D. L., Kent, S., et al., 2002. *Astronomical Journal*, **123**, 2121.

Sterken, C. and Manfroid, J., 1992. *Astronomical Photometry*, Dordrecht, Kluwer.

Thoren, V. E., 1990. *The Lord of Uraniborg: A Biography of Tycho Brahe*, Cambridge, Cambridge University Press, 1990.

Tokunaga, A. T., Simons, D. A., and Vacca, W. D., 2002. *Publications of the Astronomical Society of the Pacific*, **114**, 180.

Van Altena, W. F., 2013. *Astrometry for Astrophysics*, Cambridge, Cambridge University Press.

Verhoeve, P., Martin, D., Brammetz, G., Hijmering, R., and Peacock, A., 2004. *Proceedings of the 5th International Conference on Space Optics, ESA SP-554.*

Wall, J. V. and Jenkins, C. R., 2012. *Practical Statistics for Astronomers*, 2nd edn, Cambridge, Cambridge University Press.

Weaver, H., 1946. *Popular Astronomy*, **24**, 212; 287; 339; 389; 451; 504.

Wilson, R. N., 1996. *Reflecting Telescope Optics*, vol. I, New York, Springer.

Wilson, R. N., 1999. *Reflecting Telescope Optics*, vol. II, New York, Springer.

Young, A., 2006. See http://mintaka.sdsu.edu/GF/bibliog/overview.html.

Index

Boldface entries refer to figures.

Printed in the USA
CPSIA information can be obtained
at www.ICGtesting.com
LVHW081534150324
774517LV00042B/1805

9 781107 572560